AF897925

Gels Horizons: From Science to Smart Materials

Series editor

Vijay Kumar Thakur, School of Aerospace, Transport and Manufacturing, Cranfield University, Cranfield, Bedfordshire, UK

This series aims at providing a comprehensive collection of works on the recent advances and developments in the domain of *Gels*, particularly as applied to the various research fields of sciences and engineering disciplines. It covers a broad range of topics related to *Gels* ranging from *Polymer Gels, Protein Gels, Self-Healing Gels, Colloidal Gels, Composites/Nanocomposites Gels, Organogels, Aerogels, Metallogels* & *Hydrogels* to *Micro/Nano gels*. The series provides timely and detailed information on advanced synthesis methods, characterization and their application in a broad range of interrelated fields such as chemistry, physics, polymer science & engineering, biomedical & biochemical engineering, chemical engineering, molecular biology, mechanical engineering and materials science & engineering.

This Series accepts both edited and authored works, including textbooks, monographs, reference works, and professional books. The books in this series will provide a deep insight into the state-of-art of *Gels* and serve researchers and professionals, practitioners, and students alike.

More information about this series at http://www.springer.com/series/15205

Vijay Kumar Thakur · Manju Kumari Thakur
Stefan Ioan Voicu
Editors

Polymer Gels

Perspectives and Applications

Springer

Editors
Vijay Kumar Thakur
Faculty in Manufacturing, Enhanced Composites and Structures Centre, School of Aerospace, Transport and Manufacturing
Cranfield University
Cranfield
UK

Manju Kumari Thakur
Division of Chemistry
Government Degree College Bhoranj, Himachal Pradesh University
Shimla, Himachal Pradesh
India

Stefan Ioan Voicu
Department of Analytical Chemistry and Environmental Engineering, and Advance Polymer Materials Group, Faculty of Applied Chemistry and Materials Science
Politehnica University of Bucharest
Bucharest
Romania

ISSN 2367-0061 ISSN 2367-007X (electronic)
Gels Horizons: From Science to Smart Materials
ISBN 978-981-10-6079-3 ISBN 978-981-10-6080-9 (eBook)
https://doi.org/10.1007/978-981-10-6080-9

Library of Congress Control Number: 2017962559

© Springer Nature Singapore Pte Ltd. 2018
This work is subject to copyright. All rights are reserved by the Publisher, whether the whole or part of the material is concerned, specifically the rights of translation, reprinting, reuse of illustrations, recitation, broadcasting, reproduction on microfilms or in any other physical way, and transmission or information storage and retrieval, electronic adaptation, computer software, or by similar or dissimilar methodology now known or hereafter developed.
The use of general descriptive names, registered names, trademarks, service marks, etc. in this publication does not imply, even in the absence of a specific statement, that such names are exempt from the relevant protective laws and regulations and therefore free for general use.
The publisher, the authors and the editors are safe to assume that the advice and information in this book are believed to be true and accurate at the date of publication. Neither the publisher nor the authors or the editors give a warranty, express or implied, with respect to the material contained herein or for any errors or omissions that may have been made. The publisher remains neutral with regard to jurisdictional claims in published maps and institutional affiliations.

Printed on acid-free paper

This Springer imprint is published by Springer Nature
The registered company is Springer Nature Singapore Pte Ltd.
The registered company address is: 152 Beach Road, #21-01/04 Gateway East, Singapore 189721, Singapore

Preface

A gel is described as a soft, solid or liquid-like unique condensed material that has a three-dimensional network composed of several components such as long polymers, species of small molecules, and a large amount of solvent. These 3D network condensed materials usually form through chemical, physical, or supramolecular crosslinking. The weight and size of gels are more like a liquid, but they are treated like a solid. Two important characteristics of gels are phase state and their rheological properties. On the other hand, a polymer is defined as a large molecule (macromolecules) composed of repeating structural units that comprise multiple assemblies of simple structural units. In gels, the polymer network can be physically or chemically crosslinked. In case of physical gels, the network formation occurs due to various weak interactions, like the entanglement of the polymer chains, hydrogen bonds, or van der Waals interactions. Such structures are usually not permanent, and they dissolve over the time when immersed in their solvents. However, the polymer chains can also be crosslinked through chemical reactions, leading to strong covalent bonds. The chemically crosslinked network is much more stable and cannot be dissolved without the degradation of the polymer. Therefore, chemical gels are usually preferable in the majority of the application fields. Polymer gels comprise a great variety of different polymeric components that present innumerable industrial applications. Polymers can be naturally produced (sometimes referred as bio-based polymers), in which case the most representative group is polysaccharides. Natural polymers' demand is expected to grow 7.1% every year. Moreover, their low toxicity and excellent biodegradability have also attracted researchers to pay attention toward the widespread application of natural polymers. Polymer obtained from natural sources such as chitosan, alginate, dextran, starch, pectin, cellulose, lignin has shown excellent potential for biomedical and other applications in the form of microsphere, nanoparticles, crosslinked hydrogels, beads, membranes, and granules. On the other hand, a wide variety of synthetic polymers capable of forming gels presents different industrial applications, such as polyacrylamide and polyvinyl alcohol-based gels. Both synthetic and natural polymer-based gels find applications from health sciences such as agents for controlled drug delivery, sustained drug delivery, targeted drug delivery, and various other types of novel drug delivery systems to water purification. Polymer gels due to their several unique characteristics have become an indispensable part of new advanced and smart materials in the twenty-first century for numerous applications including but not limited to biological, biomedical, electronic, and environmental.

Polymer Gels: Perspectives and Applications is a significant collection of findings in a rapidly developing smart materials field. It gives an in-depth look at the current perspectives and applications of research and practice while pointing ahead to new possibilities in potential advanced polymer gel materials. It explores market opportunities due to the versatility and increasing applicability of polymer gels as well as suggestions on the direction of the field from its current point, paving the way for future developments and finding new applications.

Some of the important topics include but not limited to polysaccharide-based gels for drug delivery; stimuli-responsive polymeric nanoparticles for cancer therapy; the application of high polymer materials in the aspects of seed science and technology; nanogels of natural polymers; polymer gel composites for bio-applications; vegetable oils as a chemical platform for advanced applications; imprinted polymeric gels for pharmaceutical and biomedical purposes; applications of biopolymeric gels in agricultural sector; agro-based polymer gel and its application in purification of industrial water wastes; application of polymer gels and their nanocomposite in electrochemistry; smart polymer gels in drug delivery; emerging trends of organogels in drug chemistry; design and application of injectable gels in tissue engineering and drug delivery; thermo-intelligent injectable implants: intricate mechanisms and therapeutic applications, to name a few.

In editing and organizing this volume *Polymer Gels: Perspectives and Applications* of the book series "Gels Horizons: From Science to Smart Materials," we have made our best efforts to cover the growing field of polymer gels and related technologies. It reflects the recent theoretical advances and experimental results and opens new avenues for researchers as well as readers working in the field of polymer and functional materials. In addition, several critical issues and suggestions for future work are comprehensively discussed in this book with the hope that the book will provide a deep insight into the state of the art of "Polymer Gels." We express our sincere thanks to all the authors, who have contributed their extensive experience through their work for the success of this book. We would also like to thank Swati Meherishi and the rest of the team at Springer for their invaluable help in the organization of the editing process.

Cranfield, UK — Vijay Kumar Thakur, Ph.D.
Shimla, India — Manju Kumari Thakur, M.Sc., M.Phil., Ph.D.
Bucharest, Romania — Stefan Ioan Voicu, Ph.D.

Contents

About the Editors

Dr. Vijay Kumar Thakur, Ph.D. prior to commencing in the School of Aerospace, Transport and Manufacturing at Cranfield University, he was working as a Staff Scientist in the School of Mechanical and Materials Engineering at Washington State University, USA (2013–2016). Some of his other prior significant appointments include being a Research Scientist in Temasek Laboratories at Nanyang Technological University, Singapore (2009–2012), and a Visiting Research Fellow in the Department of Chemical and Materials Engineering at LHU, Taiwan. He did his postdoctoral study in Materials Science and Engineering at Iowa State University and received his Ph.D. in Polymer Chemistry (2009).

In his academic career, he has published more than 100 SCI journal research articles in the field of chemical sciences/materials science and holds one US patent. He has also published 33 books and 35 chapters on the advanced state of the art of polymer science/materials science/nanotechnology with numerous publishers. His research interests include the synthesis and processing of bio-based polymers, composites, nanostructured materials, hydrogels, polymer micro-/nanocomposites, nanoelectronic materials, novel high dielectric constant materials, engineering nanomaterials, electrochromic materials, green synthesis of nanomaterials, and surface functionalization of polymers/nanomaterials. Application aspects range from automotive to aerospace, energy storage, water purification, and biomedical fields.

He is an editorial board member of several international journals, as well as a member of scientific bodies around the globe. Some of his significant appointments include Associate Editor for Materials Express (SCI), Advisory Editor for SpringerPlus (SCI), Editor for Energies (SCI), Editor for Cogent Chemistry (SCI), Associate Editor for Current Smart Materials, Associate Editor for Current Applied Polymer Science, Regional Editor for Recent Patents on Materials Science (Scopus), and Regional Editor for Current Biochemical Engineering (CAS). He also serves on the Editorial Advisory Board of Polymers for Advanced Technologies (SCI) and is on the Editorial Board of Journal of Macromolecular Science, Part A: Pure and Applied Chemistry (SCI), International Journal of Industrial Chemistry (SCI), Biointerface Research in Applied Chemistry (SCI), and Advances in Natural Sciences: Nanoscience and Nanotechnology (SCI).

Dr. Manju Kumari Thakur, M.Sc., M.Phil., Ph.D. has been working as an Assistant Professor of Chemistry at the Division of Chemistry, Govt. Degree College Sarkaghat, Himachal Pradesh University, Shimla, India, since June 2010. She received her B.Sc. in Chemistry, Botany, and Zoology; M.Sc. and M. Phil. in Organic Chemistry; and Ph.D. in Polymer Chemistry from the Chemistry Department at Himachal Pradesh University, Shimla, India. She has rich experience in the fields of organic chemistry, biopolymers, composites/nanocomposites, hydrogels, applications of hydrogels in the removal of toxic heavy metal ions, drug delivery, etc. She has published more than 30 research papers in several international journals, coauthored 5 books, and has also published 25 chapters in the field of polymeric materials.

Dr. Stefan Ioan Voicu, Ph.D., (Habilitation) is Associate Professor at Faculty of Applied Chemistry and Materials Science, Politehnica University of Bucharest and working in the Department of Analytical Chemistry and Environmental Engineering in the field of polymeric membrane materials and processes. He received his B.Sc. in Organic Chemistry, M.Sc. in Environmental Engineering, and a Ph.D. in Polymeric Membranes, from 2016 having also Habilitation in Chemical Engineering, all at Politehnica University of Bucharest, Romania. In the fields of polymers, polymer composites and polymeric membranes (for different applications—from water purification to sensors, fuel cells, and biomedical field), he has published over 30 SCI journal articles, 5 chapters, and 3 granted US patents.

Chapter 1
Polymer-Based Responsive Hydrogel for Drug Delivery

Dhanabal Kumarasamy, Miltu Kumar Ghosh and Tapan Kumar Giri

Abstract Responsive hydrogels comprise huge prospective in a variety of applications. The hydrogels change their properties in particular responses. To induce response in hydrogel system, various stimuli (chemical and physical) have been applied. Electric field, temperature, light, solvent composition, pressure, magnetic fields, and sound are used as physical stimuli. The chemical stimuli used are ions and pH. Application of responsive hydrogels has many fields, such as making chemical valves and artificial muscles, immobilization of cells and enzymes, drug delivery, biosensor, and concentrating dilute solutions in bioseparation. Although the concepts are sound, practical utility necessitates major improvements in the properties of hydrogel. The most considerable limitation of these responsive hydrogels is response time and that is too slow. As a result, fast responsive hydrogels are needed. This chapter presents an overview of the advances in the area of responsive hydrogels that have become the interest of most researchers.

1 Introduction

Over the past numerous years, immense advances have been made to deliver the biomolecule. Controlled drug delivery systems are projected to deliver biomacromolecules at determined rates for longer time and surmount the limitations of conventional drug delivery system (Bhoyar et al. 2012; Agarwal et al. 2012). Even though considerable development has been made in this delivery system, further progresses are to come for the treatment of many disorders like heart disorders and diabetes. The drugs have to be administered in the existence of definite

D. Kumarasamy · M. K. Ghosh · T. K. Giri (✉)
NSHM College of Pharmaceutical Technology, NSHM Knowledge Campus, Kolkata Group of Institutions, 124 BL Saha Road, Kolkata 700053, West Bengal, India
e-mail: tapan_ju01@rediffmail.com

© Springer Nature Singapore Pte Ltd. 2018
V. K. Thakur et al. (eds.), *Polymer Gels*, Gels Horizons: From Science to Smart Materials, https://doi.org/10.1007/978-981-10-6080-9_1

biomolecules in the body or in response to metabolic requirements. Biomolecule would be delivered by a system that senses the signal caused by a disease would be extremely beneficial.

Hydrogel has been used widely in the advance of drug delivery systems (Thakur and Thakur 2014a, b, 2015). It is a hydrophilic polymeric network that can swell many times of its original volume and hold a huge amount of water (Behra et al. 2012; Giri et al. 2013a; Badwaik et al. 2013). Hydrogels are prepared from natural or synthetic polymers that are biodegradable (Thakur and Kessler 2015; Thakur et al. 2016). Hydrogels can be obtained by radiation cross-linking, graft polymerization, networks formation of water-soluble polymer, and cross-linking polymerization. Polymer chains are cross-linked to form three-dimensional networks. Cross-linking can be achieved by means of hydrogen bonding, covalent bonding, physical entanglements, or van der Waals interactions (Kamath and Park 1993; Park et al. 1993). Hydrogels can shield the incorporated drug molecules from digestive enzymes and acidity in the stomach. It releases the enclosed drug molecules in a controlled manner in response to environmental stimuli through the changing of gel structure. This property can undergo phase transitions or gel–sol phase transitions even with small environmental condition changes and are called smart hydrogels (Park and Park 1999). Numerous chemical and physical stimuli have been used to persuade a variety of responses in smart hydrogel systems. Physical stimuli consist of electric fields, temperature, light, solvent compositions, pressure, magnetic fields, and sound, whereas the chemical stimuli comprises of pH and ions (Hoffman 1997; Bae 1997).

2 Swelling Properties of Hydrogel

Hydrogels are considered single molecule in spite of its size since polymer chains are cross-linked to each other. Small changes in surrounding environment trigger rapid changes in hydrogel system. These factors are temperature, pH, electric charge, presence of ionic species, and presence of enzyme and ultimately change the size and water content of hydrogel system. Difference in mobile ion concentration between interior of hydrogel and external medium is responsible for pH change, resulting in volume change. Anionic or cationic functional groups of hydrogels are responding to pH of the external medium and ionize. This leads to the swelling and volume change of hydrogel. pH-sensitive hydrogel containing polyacrylic acid swells in basic medium owing to the ionization of the carboxylic acid group in the polymer chain.

3 Mechanical Properties of Hydrogel

Mechanical properties of hydrogels are very vital for their suitable and potential application. Mechanical property evaluation is a vital parameter in a variety of biomedical applications such as wound dressing, ligament and tendon repair, tissue engineering, and drug delivery. Hydrogels should have sufficient mechanical properties to maintain its physical texture during predetermined period of time for delivery of therapeutic molecules. The desired mechanical property of the hydrogel for a particular application can be achieved by altering the extent of cross-linking. Although stronger hydrogels may possibly be achieved by escalating the degree of cross-linking, the resultant hydrogels are more brittle in texture due to decreases in the percentage elongation of hydrogels. Numerous researchers applied copolymerization technique to increase mechanical properties of hydrogel.

4 Biocompatible Properties of Hydrogel

The prerequisite of biomedical application of hydrogels is biocompatibility. Hydrogel containing polymers should pass cytotoxicity and toxicity tests. The capability of a material to do the intended task with a suitable host response in a precise application is known as biocompatibility. Biosafety and bio-functionality are the two elements of biocompatibility. This is mainly appropriate in tissue engineering as the nature of tissue construction is to constantly interact with the body throughout curing and regeneration procedure in addition to scaffold degradation. The hydrogel may lead to damage and scarring of connected tissues if these requirements are not met. Noxious chemicals used in the preparation of synthetic hydrogels pose a problem for in vivo biocompatibility. In addition, organic solvents, initiators, emulsifiers, stabilizers, cross-linkers, and unreacted monomers used in hydrogel synthesis and polymerization are perhaps toxic to host cells. Irgacure 2959 (2-hydroxy-4′-hydroxyethoxy-2-methylpropiophenone), a typical alpha-hydroxy ketone, is used as initiator in photopolymerizations. It has been observed that when used more than 0.1% concentrations diminishes cell viability (Bryant et al. 2000). Diverse refining processes such as dialysis or solvent washing are used to remove the hazardous cross-linkers from preformed hydrogels. Pre-polymers and oligomers used in in situ gelation of scaffolds present an exceptional challenge because reactants used for the synthesis of gel are injected into the body. Employment of this method is perfect for its least invasiveness; however, this technique requires particular consideration to guarantee all components used are nontoxic and safe. Even if natural polymers are normally regarded as better biocompatibility than synthetic one, however the use of initiators and cross-linkers used for the polymerizations of natural pre-polymers and monomers leads to toxicity and is subjected to toxicity tests, similar to that of purely synthetically derived hydrogels.

5 Temperature-Responsive Hydrogels

Temperature-responsive hydrogels are mainly attractive since their gelling and swelling can be activating via change in temperature. This property is utilized, and hydrogels can be directly generated at a desired site in body through change of temperature. These systems allow in situ hydrogel formation, where a biomaterial can be delivered as solution in a minimally invasive manner and gelation and solidification of hydrogel occurs inside the body. When the injected biomaterial attains gelation temperature, hydrogel formation occurs almost instantaneously. Chemical initiator is removed since temperature is the only activator; the net result is a milder, nontoxic process. Various thermoresponsive polymers show a lesser critical solution temperature (Schild and Tirrell 1990) and were transformed into gel upon temperature rise. However, the procedure is reversible as the temperature is diminished below critical solution temperature, and the polymers in gel state return to solution. The thermoresponsive polymer has been chemically cross-linked, and the resulting hydrogel shows augmented swelling beneath the lower critical solution temperature (Ward and Georgiou 2011). There exist some thermosensitive polymers that gel beneath a definite temperature and are soluble above it (Boustta et al. 2014), and these are represented as polymers having higher critical solution temperature performance.

In chitosan-based hydrogel system, addition of β-glycerophosphate to chitosan leads to thermosensitive hydrogel (Chenite et al. 2000; Thakur and Thakur 2014b). Chitosan-based hydrogel was developed for the delivery of doxorubicin in controlled way (Cho et al. 2009). Initially, doxorubicin and acrylated chitosan were chemically conjugated that produced a physical gel when subjected to physiological temperature. Then, the hydrogel was chemically cross-linked and release profiles of doxorubicin were investigated. In this system controlled release of the doxorubicin was observed. Cytotoxicity of the doxorubicin hydrogel to tumor tissue was analogous to that of free doxorubicin in vitro, and improved tumor volume reduction was noted in vivo. Thermogel of chitosan, grafted with poly(L-alanine-co-L-phenylalanine) and polyethylene glycol, demonstrated a complete biodegradation in vivo after 14 days, signifying potential applications in drug delivery systems, where immediate release of therapeutic molecules is desired (Kang et al. 2012). Couto et al. (2009) developed complex hydrogels based on chitosan and bioactive glass nanoparticles. Nanoparticles addition to the system resulted in tougher hydrogels with gelation temperature ranges at physiological conditions. In another study, demineralized bone matrix was delivered through chitosan hydrogel (Tian et al. 2012). It was observed that the osteoinductive activities of the hydrogel were related to the ratio of demineralized bone matrix and chitosan. Particularly, augmented demineralized bone matrix contents led to diminished porosity, higher swelling ratio, and superior elastic modulus. The utmost ratio of chitosan to demineralized bone matrix (1:2) was demonstrated in vivo osteoinductive effect.

Gelling temperature of gelatin is below 25 °C and its molecules undergoing structural changes from coil to helix. Physically formed hydrogels dissolve at physiological temperature. Therefore, gelatin is chemically cross-linked with other polymers to stabilize the resulting hydrogels at 37 °C. The thermal behavior of gelatin at 37 °C was used for introducing porosity (Hwang et al. 2010) or microfluidic networks (Golden and Tien 2007) into hydrogels. The porosity is a vital parameter for a hydrogel scaffold used in tissue engineering since it facilitates transportation and provides gap for cell migration and proliferation. Hwang et al. (2010) reported a method of fabricating controlled pore size alginate hydrogels by incorporating gelatin. The resulting scaffold showed augmented cell proliferation in comparison with nonporous alginate. Leong et al. (2013) prepared alginate hydrogel incorporating gelatin microparticles that act as both porogen and cell carrier. After two-day incubation at 37 °C, the gelatin microparticles were entirely dissolved and lead to generation of pores in alginate hydrogel matrix. Cartilage cells suspended in the alginate matrix were mostly viable, forming islets. Gelatin and alginate are combined to print three-dimensional scaffold that are solidified through chemical cross-linking of alginate and thermal gelation of gelatin (Wuest et al. 2014). Cells included into this structure maintained their viability subsequent to printing and solidification procedure.

The lower critical solution temperature of poly(*N*-isopropylacrylamide) is around 32 °C. The hydrogel was developed by copolymerizing *N*-isopropylacrylamide with propylacrylic acid (Garbern et al. 2010). The resulting system produced a hydrogel in pH below 5.5, and at physiological temperature formation of gel was observed, probably due to ionization of the carboxylate groups of propylacrylic acid. The use of butyl acrylate as a copolymer instead of propylacrylic acid results in formation of hydrogel at temperatures around 21 °C and pH 6. Bioactivity of fibroblast growth factor liberated from the hydrogels based on poly (*N*-isopropylacrylamide) confirmed it's utility. Moreover, vascular endothelial growth factor included in hydrogels displayed a pH-dependent release behavior. This is due to combination of dissolution rate and electrostatic effects of the polymer. The authors applied the hydrogel for drug delivery locally at elevated pH as in wound healing and that hydrogels are subsequently eliminated by dissolution upon sites change to physiological pH. Siegwart et al. (2008) incorporated biodegradable component into the backbone of polymer by copolymerizing *N*-isopropylacrylamide with benzomethylene dioxepane. Hydrogel containing hydrophobic monomer diminished the lower critical solution temperature in comparison with poly(*N*-isopropylacrylamide). However, by changing the monomer composition, the lower critical solution temperature could be adjusted above room temperature. Degradation yield of such hydrogels was nontoxic to cells. Remarkable research based on biodegradable *N*-isopropylacrylamide systems is hydrogels with degradation-dependent lower critical solution temperature. This perception was first implemented by Neradovic et al. (1999). These polymers contain either enzymatically or hydrolytically labile bonds, which facilitates degradation. Before degradation, the polymer's lower critical solution temperature is well below 37 °C, ensuing in gel development inside the body. Following breakdown of these groups, the polymers' overall hydrophilicity increases resulting in an augment in the lower critical solution temperature to above physiological temperature. Then the polymer becomes soluble again and can be eliminated. A series of hydrogels have been described (Li et al. 2011; Guan et al. 2008; Fujimoto et al. 2009) with potential application in drug and cell delivery platforms. Hydrogels based on *N*-isopropylacrylamide, acrylic acid, polylactide–hydroxyethyl methacrylate, and *N*-acryloxysuccinimide were developed. Polylactide was included on the backbone of polymer to facilitate hydrolysis; acrylic acid increases hydrophilicity and lowers the critical solution temperature and *N*-acryloxysuccinimide to give a bioconjugation position for collagen binding. The molar *N*-isopropylacrylamide content used was 75–85%. Hydrogel formulations had lower critical solution temperature in the ambient temperature before degradation and over 40 °C after degradation. Thus, strong hydrogels were made up. Hydrogels without collagen were shown 32–80% water content depending on the composition, however, swelling augmented with protein conjugation. Moreover, mechanical properties are decreased by collagen concentration. On the other hand, it promoted cell adhesion of smooth muscle.

Hydrogels containing polyester and poly(ethylene glycol) have the advantages of degradability and biocompatibility and are useful in biomedical applications. Huynh et al. (2008) have developed a hydrogel using thermosensitive poly(ethylene glycol)-poly(ε-caprolactone). Poly(β-aminoester) was chosen as an added block, giving pH-responsive property to the resulting copolymer [poly(β-aminoester)-poly(ε-caprolactone)-poly(ethylene glycol)-poly(ε-caprolactone)-poly(β-aminoester)]. Poly(β-aminoester) moiety contains positive charge that could be utilized to form complexes with anionic molecules and to assist their controlled release. Poly(β-aminoester) is ionized at low temperature and pH, and poly(ethylene glycol)-poly(ε-caprolactone) is soluble due to more hydrophilicity. By rising temperatures, the poly(ethylene glycol)-poly(ε-caprolactone) copolymer's hydrophilic–hydrophobic balance shifts; at elevated temperature, and it tends to be more hydrophobic. At pH above the pKa of poly(β-aminoester), the polymer is unionized and present in hydrophobic state. Micelle creation occurs in aqueous solutions since poly(β-aminoester)-poly(ε-caprolactone) is a hydrophobic block and poly(ethylene glycol) is a hydrophilic block. Under these circumstance, the micellar structure enables the creation of bridges at adequately high polymer concentrations resulting in formation of gels. Ionic complexation of insulin with poly(β-aminoester) augmented the viscosity of solution and decreased gelation temperatures. These effects were observed through escalating the loading concentration of insulin.

Recently, protein drugs have been delivered in controlled manner using poly(organophosphazene)-based hydrogel by Park et al. (2010a, b, 2013). Burst release of protein occurred from the hydrogels by quick diffusion. Rapid release of drug was independent of the molecular weight of protein and hydrogels viscosity. To circumvent this problem, the group developed poly(organophosphazene) hydrogels by incorporating polyelectrolyte complexes of negatively charged protein with polycationic compounds. Hydrophilic–hydrophobic ratio of the complexes had shown to effect gelation behavior of the hydrogels. However, all gelation occurred at physiologically relevant temperature of 37 °C. Early burst release could be minimized by combined utilization of a second ionic interaction through the polyelectrolyte complex formation and anionic poly(organophosphazene) hydrogel. In vivo studies in rats demonstrated the utility of hydrogel for controlled delivery of human growth hormone. Single injection of the hydrogel in rats showed superior growth rates in comparison with daily injections of solution containing hormone for a seven day period.

6 pH-Responsive Hydrogels

pH is a vital signal in responsive hydrogel system (Giri et al. 2013b, c). Polymers which are ionizable and their pKa value in the range of 3–10 are suitable candidates for preparing pH-responsive hydrogel (Siegel 1993). Ionization state of weak acids like carboxylic acids, phosphoric acid and bases like amines changes with variation in pH. When these groups are connected to the polymer, the swelling behavior of the hydrogel is changed due to a change in conformation of the polymers.

Hydrogels comprised of cross-linked polyelectrolytes exhibit large variation in swelling depending on environment pH. Polyelectrolytes containing suspended basic or acidic groups undergo ionization similar to that of basic or acidic groups of monobases or monoacids. However, ionization of polyelectrolytes is complicated owing to electrostatic effects of adjacent ionizable groups affecting dissociation constant of monobase or monoacid. The presence of ionizable groups on polymer chains results in swelling of the hydrogels much beyond the level, which cannot archived by nonelectrolytic polymer based hydrogels. The swelling of polyelectrolytic hydrogels is mostly owing to electrostatic repulsion between the same charges existing in polymer chain. Extent of swelling is affected by several conditions that reduce electrostatic repulsion, for example ionic strength, pH, type, and extent of counterions present (Firestone and Siegel 1991). The pH-responsiveness and swelling of polyelectrolyte hydrogels can be changed by using neutral co-monomers, such as methyl methacrylate, 2-hydroxyethyl methacrylate, and maleic anhydride. A variety of co-monomers offers different hydrophobicity to the polymer chain resulting in different pH-sensitive performance.

pH-sensitive hydrogels are the most commonly used systems to develop oral controlled release formulations. pH of the stomach is totally different from intestine, and such variation is sufficient to bring out pH-dependent activities of polyelectrolyte hydrogels. The swelling of polycationic hydrogels is lowest at neutral pH resulting in lowering of drug release. It has been utilized to avoid release of foul-tasting drugs in the mouth having neutral pH. Hydrogels loaded with caffeine were prepared using copolymers of *N*,*N*′-dimethylaminoethylmethacrylate and methyl methacrylate. The drug was not released at neutral pH, however released at pH 3–5. The hydrogel was swelled at pH 3–5 due to ionization of *N*,*N*′-dimethylaminoethylmethacrylate (Siegel et al. 1988). Semi-interpenetrating polymer network hydrogel has also been used for the delivery of drug to the stomach. Cross-linked chitosan and polyethylene oxide semi-interpenetrating polymer network hydrogel exhibited more swelling in the stomach. For the treatment of

Helicobacter pylori infection in stomach, this kind of hydrogels is ideal for delivery of amoxicillin and metronidazole (Patel and Amiji 1996).

Nonviral gene therapy is carried out using cationic polymers (Twaites et al. 2005). Through electrostatic interaction the polycations can complex with nucleotides. Responsive behavior of the polymer is essential, when the pH drops during uptake into cell as the polymer becomes progressively charged and triggers endosomolytic, osmotic, or other events consequently. So far, transfection efficiency of cationic polymers loaded with nucleotides is below that of viral vectors. Additionally, currently employed polycations are still having unacceptable toxicity. The search is still on for identifying the exact synthetic vector with tolerable toxicity and high transfection efficiency. So far, the current studies offer a few perceptive on the fundamental mechanism.

Tumor tissue has extracellular pH of 6.5–7.2, lower than the physiological pH (Na and Bae 2005). After cellular uptake, the hydrogel containing drug reaches the lysosomes, where pH of the environment is 4.5–5.0 and triggered release of a drug load due to change in pH. Hydrolytic enzymes cathepsin B is involved in the degradation of hydrogel in the lysosomes, thus facilitating the drug release (Duncan 2003). The nanometer range of the drug conjugates or micellar structures reaches the tumor tissue by passive targeting through enhanced permeability and retention (EPR) effect (Matsumura and Maeda 1986).

Doxorubicin is delivered extracellularly using weakly acidic sulfonamide as trigger (Na and Bae 2005). Micelles with drug load are kept in solution owing to fractional charges at the sulfonamide. If sulfonamide is further protonated through acidification, subsequently the micelle collapses. The collapsed nanoparticles containing doxorubicin build up in the tissue and are consequently taken up by the cells. Poly(L-histidine)-β-poly(ethylene glycol) in amalgamation with polylactic acid-β-poly(ethylene glycol) was used for the extracellular tumor targeting of adriamycin. Micelle system shows an exceedingly pointed transition from stable hydrophobic and unionized (non-protonated) form at pH 7.4 to destabilized and ionized micelle at pH 6.6. At this pH adriamycin is speedily released from the micelles (Lee et al. 2003).

Conjugation of drug molecules and polymer is usually inactive, and these conjugates are called prodrugs. For cytotoxic drugs this is advantageous, e.g., in cancer chemotherapy, since the inclusion of a targeting system avoids the undesirable side effects owing to toxicity. On the other hand, merely a proficient release of the drug at target site gives complete advantage of such prodrugs. Besides from the above-mentioned pH fall in the extracellular tissue, delivery of the polymer prodrug systems in the lysosome following cellular uptake, the active drug is released through hydrolytic degradation by enzymes present in the lysosome (Brocchini and Duncan 1999; Duncan 2003).

Klier and Peppas (1989) developed hydrogels using graft copolymers of ethylene glycol and poly(methacrylic acid) that are susceptible to buffer concentration and pH. The polymethacrylic acid was protonated in acidic media and produced hydrogen-bonded complexes with polyethylene glycol, forming low swelling hydrophobic structures and has little permeability to solutes of hydrophilic nature.

The complexes were broke down, and the resulting gel exhibits a high degree of swelling and consequently solute permeability upon neutralization.

Heller and Trescony (1979) were the first to change polymer erosion rates by using immobilized enzymes to modify local pH. They used enzyme urease to convert urea to NH_4HCO_3 and NH_4OH. This enzymatic reaction raises pH, and consequently erosion of the polymer takes place. Authors recommended copolymer of maleic anhydride and methylvinyl ether that displays pH-dependent release behavior. Polymers dissolve via ionization of the incorporated carboxylic acid group in presence of weak acid salts of ammonia, produced by the action of urease (Heller et al. 1978). Ishihara et al. (1985a, b) developed a non-erodible system using similar idea. They developed pH-sensitive membrane using copolymer of 4-carboxyacrylanilide and methacrylate, placed inside membranes that contain urease. It was observed that permeation of 1,4-bis(2-hydroxyethoxy)benzene changes with the external urea concentration.

Biosensors have also been made by using pH-sensitive hydrogels (Hoffman 1997). The pH-dependent hydrogels are frequently loaded with an enzyme that alters the pH of the hydrogels. Example of one such enzyme is glucose oxidase which converts glucose to glucuronic acid that lowers the pH affecting the swelling of pH-dependent hydrogels.

Nonbiodegradability is the intrinsic limitations of synthetic pH-sensitive polymers. Therefore, hydrogels prepared from nonbiodegradable polymers must be eliminated from the body. Nonbiodegradability and removal of the hydrogel are not a problem in oral drug delivery; however, it becomes a severe problem in implantable biosensors or implantable drug delivery systems. Accordingly, interest has been paying attention on the development of pH-sensitive, biodegradable hydrogels based on polysaccharides, proteins, and polypeptides. Dextran was cross-linked with 1,10-diaminodecane using 4-aminobutyric acid as activator and then grafted with carboxylic acid (Chiu et al. 1999). Resultant hydrogels showed higher swelling at elevated pH conditions. Cyclic deswelling–swelling resulted in changing the pH between 2 and 7.4. Synthetic polypeptides were used in the development of biodegradable hydrogels. Examples of such polypeptide include poly(L-ornithine), poly(hydroxyl-L-glutamate), poly(aspartic acid), poly(L-glutamic acid), and poly(L-lysine) (Markland et al. 1999). Besides usual electrostatic effects connected with the majority of pH-sensitive synthetic hydrogels, the structures of the polypeptide backbone may possibly give pH-sensitive swelling (Markland et al. 1999). In general, the degree of pH-responsive swelling can be modified through changing polypeptide's degree of ionization and hydrophobicity.

7 Glucose-Sensitive Hydrogels

Most difficult problems in controlled drug delivery are the progress of self-regulated insulin delivery systems. Insulin delivery is different from other drug delivery since insulin has to be delivered at the exact time of need and in exact amount. Therefore,

$$\text{Glucose} + O_2 + H_2O \xrightarrow{\text{Glucose oxidase}} \text{Gluconolactone} + H_2O_2$$

$$H_2O_2 \xrightarrow{\text{Catalase}} H_2O + 1/2O_2$$

$$\text{Gluconolactone} \xrightarrow{\text{Gluconolactanase}} \text{Gluconic Acid}$$

Fig. 1 Conversion of glucose to gluconic acid

self-regulated delivery of insulin requires the ability to sense glucose concentration and must have an automatic shutoff system. Numerous hydrogels have been developed for modulated insulin deliveries that have a glucose sensor in the system. The hydrogels are glucose-responsive and show response variability in the presence of glucose.

A series of polybasic hydrogel were prepared through copolymerization of dimethylaminomethacrylate and a neutral monomer that swells with diminishing pH, with supreme sensitivity just lower than physiologic pH (Ishihara et al. 1984; Ishihara and Matsui 1986; Albin et al. 1985, 1987; Cartier et al. 1995; Kost et al. 1985). Turning pH to glucose sensitivity, the enzymes catalase and glucose oxidase were immobilized within the hydrogels. Glucose oxidase as commercially obtained generally contains gluconolactonase. Reactions occur in the presence of these three enzymes that are represented in Fig. 1.

Gluconic acid is entirely dissociated above pH 6 since its pKa is about 3.8; therefore, one proton (H^+) is formed for each glucose molecule. The H^+ ion binds to the tertiary amino group on dimethylaminomethacrylate (pKa $\approx$ 7.9). This leads to swelling of hydrogel and augmented penetration of insulin through the hydrogel. It is to be noted that half of the O_2 molecule is also consumed in this reaction. This is significant, due to the fact, physiologic free O_2 is there at much lesser concentration than glucose. Therefore, glucose oxidase catalyzed reaction might become oxygen limited, thus less responsive to changes in glucose concentration. To a certain extent, the superior diffusion coefficient of oxygen in comparison with glucose balances the relative shortage of free oxygen. Besides reactive oxygen species, catalase guards the enzymes from oxidative damage. This is in combination with the stabilization of enzymes through immobilization (Mateo et al. 2007) offers enhanced life span of the enzymes. Moreover, all the enzymes can be included in excess, so that the activity remains even after a large fraction of the enzymes have been deactivated (Lucisano et al. 1987).

In absence of buffers, low pH can be achieved by exposing the membrane to glucose solution that is unbuffered. However, pH does not go below 6.8 when physiological buffers are used. This was experimentally confirmed. Therefore, prerequisite for a membrane is its sensitivity that lies between pH 6.8 and 7.4 (Klumb and Horbett 1992). In the early inventive studies, optimization of the polymer swelling response has been made using dissimilar neutral and ionizable monomers to enhance pH sensitivity of membranes and swelling differential among the neutral and ionized states (Podual et al. 2000). It has also been shown that by lowering access of glucose to the enzymes compared to oxygen, the deficit of oxygen can be mitigated (Klumb and Horbett 1992).

Hydrogel prepared using poly(methacrylamido phenylboronic acid)-coacylamide releases the therapeutic in a controlled way, when the glucose concentration is high in the neighboring environment causing hydrogel swelling (Eddington and Beebe 2004; Pluta and Karolewicz 2004). Hydrogels have been prepared using cross-linked copolymers of polyethylene glycol and methacrylic acid loaded with insulin by partitioning the concentration of insulin. Hydrogel microparticles showed no leakage in acidic conditions, whereas at pH 7.4 the release was highest (Kumar et al. 2006). Glucose-responsive hydrogel was also prepared by photopolymerization of 3-acrylamido phenylboronic acid and 2-hydroxyethyl methacrylate. It was observed that the release of insulin was glucose concentrations dependent (Lee et al. 2004). The hydrogel showed utmost swelling at pH 7.4 in a 0–300 mg/dl local glucose environment (Kang and Bae 2003) for insulin delivery. It was an enzymatic approach.

Photonic crystal-based hydrogel is developed that acts as a glucose sensor for diabetes mellitus patients (Holtz and Asher 1997). Glucose oxidase is incorporated in polystyrene nanospheres and then polymerized inside a hydrogel matrix. Prepared hydrogel reversibly swells in the existence of glucose, like glucose-sensitive systems. The developed hydrogel can be implanted as ocular inserts or contact lenses to identify changes in blood glucose levels using lachrymal fluid. In a modified system, instead of enzyme boronic acid derivatives are included and polymerized inside a network of polyacrylamide–polyethylene glycol. Glucose binds to the boronic acid derivatives and leads to cross-links so as to shrink the hydrogel resulting in a blueshift. Blood glucose levels are determined through a color chart.

Depending on the glucose concentration in the surroundings, hydrogels undergo sol-gel transformations (Obaidat and Park 1996, 1997). Glucose-responsive cross-linking was required for reversible sol-gel phase transitions. An interaction between concanavalin A and glucose was used to create cross-links between chains of polymer containing glucose. Since concanavalin A is present as a tetramer and every unit has a binding site for glucose, concanavalin A can function as a cross-linking agent for polymer chains containing glucose. The created cross-links are reversible since interaction between concanavalin A and glucose is non-covalent, as represented in Fig. 2. When the exterior glucose enters into the hydrogel the free glucose can compete with the glucose attached to polymer and replace with them, and this process is concentration-dependent. The concentrations of concanavalin A and glucose-containing polymers can be tuned in such a manner that make hydrogels respond (i.e. undergo gel-to-sol transformation) at specific free glucose concentrations. Diffusion of insulin through the sol phase is faster than that of hydrogel in gel state, and release of insulin can be regulated as a function of the environmental glucose concentration.

Glucose-sensitive hydrogels with reversible phase can also be produced without concanavalin A. Polyol polymers and polymers containing phenylboronic groups form a gel by complex formation among the pendant phenyl borate and hydroxyl groups (Kitano et al. 1992; Shiino et al. 1994; Hisamitsu et al. 1997). Pendant

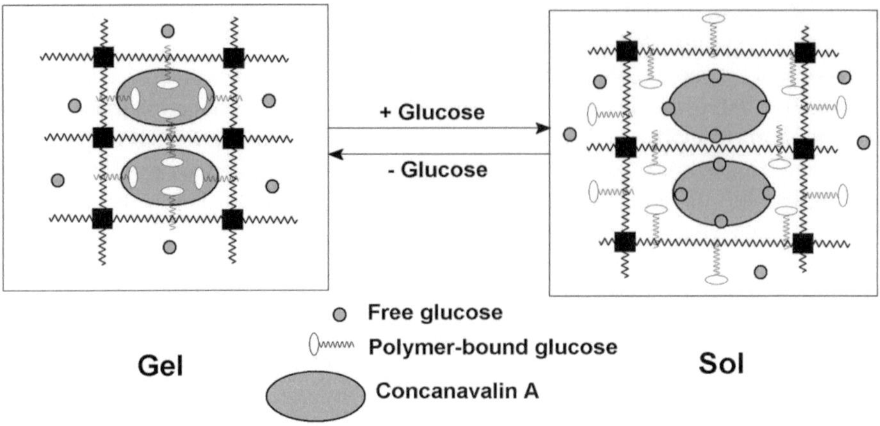

Fig. 2 Sol-gel phase transition of a glucose-sensitive hydrogel

hydroxyl groups containing glucose compete with polyol polymers for the cross-linking of borate. Because glucose is monofunctional, it cannot function as a cross-linking agent, whereas a polyol polymer does. As a result, increasing glucose concentration decreases gels cross-linking density and the gel releases more insulin through swelling. The gel becomes a sol with elevated concentrations of glucose. The exchange of glucose reaction is reversible, and cross-linking of borate–polyol is renewed at lower concentration of glucose. Shorter molecules diglucosylhexanediamine can also be used as a cross-linking agent in spite of polyol polymers containing long chain. Major limitation of phenylboronic acid gel is its sensitivity toward glucose occurs only at pH 9. A variety of copolymers having phenylboronic acid were synthesized to afford sensitivity of glucose at physiological pH. However, this system has low specificity to glucose.

Even though all the investigated glucose susceptible insulin delivery systems are extremely promising, lots of improvements are required to make them useful clinically. At the outset, the responses of these hydrogels are very slow with changes in the surrounding concentration of glucose. In addition, hydrogels do not return to their original states fast enough, after responding to the changing glucose concentration. Reduction of hydrogels' dimensions may be a means of decrease in their response time. Also, the present hydrogel systems need an enhanced reproducibility. The hydrogels required to respond to changing concentrations of glucose at all times necessitate that hydrogels can respond reproducibly along with quick response onset times on a continuing basis.

8 Electric-Responsive Hydrogels

Electric current is also considered as an environmental indicator to stimulate responses of hydrogels. Electric current-sensitive hydrogels are generally prepared from polyelectrolytes and are of pH-sensitive. Electro-responsive hydrogels undergo shrinking or swelling in presence of an electric field. Occasionally, the hydrogels exhibit swelling on one side and shrinking on the other side; such response leads to twisting of the hydrogels. Shapes of hydrogel change depend on numerous conditions. If the surface of hydrogel is in contact with the electrode, the result of applied electric field may be different from systems wherein hydrogel is placed in water (or acetone–water mixture) without direct contact of the electrode. Hydrogel comprised of acrylamide–sodium acrylic acid copolymer is positioned in aqueous solution under electric field devoid of touching the electrodes; the nature of hydrogel bend depends on the electrolyte concentration. Absence of electrolytes or else in the existence of extremely low electrolytes shrinks the hydrogel by the application of electric field. This is owing to the movement of Na^+ to the cathode and polymer carboxyl groups changed from $-COO^-Na^+$ to $-COOH$ (Shiga et al. 1992). However, in existence of high amount of electrolytes in solution, extra Na^+ ion diffuses into hydrogel than the ions migrate out from the hydrogel to cathode (Shiga et al. 1992). Therefore, swelling is more apparent at the hydrogel surface facing the anode, and as a result twisting of the hydrogel occurs.

When microspherical particles of hydrogels are placed in water without any salts, use of an electric field results in hydrogels shrinking owing to movement of water (electroosmosis) and charged ions (electrophoresis) from the hydrogel to cathode (Sawahata et al. 1990). The property has been used for pulsatile drug delivery by switch off and switch on of the electric field. The response of hydrogels depends on the investigational conditions, and accordingly, any generality on the swelling/shrinking activities cannot be made.

Poly(2-acrylamido-2-methylpropanesulfonic acid–co-*n*-butyl methacrylate) hydrogels were proficient to release hydrocortisone and edrophonium chloride (quaternary ammonium compound) using electric current in a pulsatile manner (Kwon et al. 1991). Control of pulsatile drug release was obtained by changing the strength of electric stimulation. The drug release pattern of edrophonium (positively charged drug) was explained as an ion exchange between the hydrogen ion formed by electrolysis of water and positively charged solute.

Living organisms are in motion by the conversion of chemical energy to mechanical work as muscular contraction, flagellar, and ciliary movement. Electrically driven motility has been demonstrated using weakly cross-linked poly (2-acrylamido-2-methylpropanesulfonic acid) based hydrogels. Existence of positively charged surfactant molecules is reducing the overall negative charge of the polyanionic hydrogel facing the cathode. This results in bending of the hydrogel due to local shrinkage. Application of an oscillating electrode polarization could direct the hydrogel to rapidly repeat its oscillatory motion, which leads to worm-like movement (Osada et al. 1992).

Heparin was released from a poly(vinyl alcohol) hydrogel formed on a polypropylene film (Li et al. 2005), while in an another experiment, applying electrical potential between −0.4 and −0.7 V for 90 s heparin was not released into the medium, but was exposed on the surface of the hydrogel (Garner et al. 1999). The anti-inflammatory drug dexamethasone was successfully released from a 50-nm-thin film using cyclic voltammetry (Wadhwa et al. 2006). The release was proportional to the number of stimulation cycles. The expulsion of the molecules generally seems to happen quite rapidly, in just few minutes, which can be considered as an advantage or disadvantage of conductive polymers, depending on the application (Gomez and Schmidt 2007).

There are some factors that restrict the drug release of conductive polymers; for example, the molecules loaded into the polymer have a tendency to leak out through diffusion, to be replaced by new molecules from environment. Passive loss of load is further worsened by the fact that only a moderately small amount of drug can be bound in the polymer in the first place. In addition, both molecular weight and charge dictate in which molecules can be bounded and released (George et al. 2006). The problem can be conquered through the use of streptavidin–biotin pairing. Biotin acts as a dopant, whereas the bioactive molecule bounds to the biotin covalently. The molecule is then released with electrical stimulus. A further bonus is that biotin gives more consistent release kinetics.

Grimshaw et al. (1989) have established four discrete electromechanical and electrochemical mechanisms for discriminating controlled transport of neutral solutes and proteins through hydrogel membranes: (a) chemically and electrically induced membrane swelling to change the effectual permeability and pore size; (b) electrophoretic intensification of solute flux inside a membrane; (c) electroosmotic increase of solute flux inside the membrane; and (d) electrostatic attachment of charged solutes in charged membranes.

Pasechnik et al. (1979) reported an enhancement in the effective pore radius of ultrafiltration membranes owing to electrodynamic effects. The changes in the ionic permeability of polypyrrole redox membranes by an electrochemical reaction controlled by voltage were observed by Burgmcyer and Murray (1982). Bhaskar et al. (1985) studied the liquid-crystalline membrane permeability using transmembrane electric fields that changes arrangement of the polymeric molecules. Ionization states of membrane can be changed through electric field, which indeed affect permeability and hydration of the membranes. Eisenberg and Grodzinsky (1987) reported constrained diffusion of sucrose through collagen membranes by electrodiffusion producing up to 25% flux changes. Nussbaum and Grodzinsky (1981) established reversible changes in the uniaxial swelling of polymethyl methacrylate membranes using electrodiffusion control of intramembrane ionic strength. Osada et al. (1992) induced the release of pilocarpine into the surrounding solution when a direct current electric field was applied to polymethyl methacrylate gels incorporated with pilocarpine.

Electric field applied through a hydrated polyelectrolyte polymeric membrane, such as polymethyl methacrylate, increases the net force in space charge containing excess amount of counterions over co-ions. This force is transferred to the solvent, ensuing in electroosmotic fluid flow in comparison with the solid polymeric membrane. Grodzinsky and Grimshaw (1990) showed a volume flux at the way of the current across polymethyl methacrylate polymeric membranes at pH 7. When the current was stopped the flow ceased, and when the current was inverted the flow reversed. At pH 3 the membrane exhibits a small volume flux. The existence of electroosmotic flow at pH 7 and absence at pH 3 showed the pH dependence of fixed charge of membranes. Therefore, electroosmotic transport of water can increase or resist diffusive transfer of charged or neutral molecule. With a charged polymeric membranes and charged molecules, electric field application can result in control of flux of solute involving a combination of electroosmotic and electrophoretic mechanisms.

The main advantage of electro-responsive hydrogels in drug delivery is controlled release of drug through electric field modulation. Presently, controlled release drug delivery using electro-responsive hydrogels is still in its nascent stage. The most common problem of these hydrogels are slow response and prerequisite of controllable voltage source. In addition, the majority of the electro-responsive hydrogels work in the absence of electrolytes. Development of electro-responsive hydrogels-based drug delivery modules is not easy under physiological conditions, where the presence of electrolytes inevitable.

9 Light-Sensitive Hydrogels

Hydrogels based on light sensitive have prospective applications in the development of display units, optical switches, and ocular drug delivery devices. As the light stimulus can be imposed instantaneously and delivered in exact amounts with high precision, hydrogels based on light sensitivity have some unique advantages over other hydrogels. For example, the sensitivity of temperature-sensitive hydrogels is rate limited by thermal diffusion, while pH sensitive hydrogels can be limited by hydrogen ion diffusion. The ability for instant sol-gel transformation makes the development of light-responsive hydrogels for a variety of applications such as biochemical and engineering fields. Hydrogels based on light sensitivity are visible light-sensitive and UV-sensitive hydrogels. In comparison with UV light, visible light is inexpensive, readily available, clean, safe, and easily modulated.

Absorption of nonionizing radiation energy (light) increases energy levels of reactant molecule's requisite for chemical changes. Energy absorbed by the reactant molecules is adequately sufficient to either propagate/initiate chemical reaction or to perform cross-linking. Consequently, photoirradiation assembly is composed of light source, photoresponsive precursor, and photoinitiator system. Light represents a form of electromagnetic waves that move as photowaves which carries radiant energy. Photoinitiator system acts as a recipient of this photon energy of light and performs a rapid chemical transformation of photoresponsive precursor, principally through electronic excitation.

Photoinitiator system is composed of one or more of photoinitiator (PI) either alone or in amalgamation with catalysts (photoaccelerators and photosensitizers). When the system is exposed to appropriate wavelength of light, photoinitiator initiates the reaction through the creation of free radicals (Baroli 2006; Jeon et al. 2011). These free radicals attack the photosensitive moiety of the photoresponsive precursor (generally the double bonds). Afterward, this reacted precursor successively reacts again with new photoresponsive precursor and moves with chain propagation until end of the photopolymerization/cross-linking process. Table 1 shows some examples of photoresponsive hydrogels using photoinitiator system for the pharmaceutical delivery of biological macromolecules.

Table 1 Photosensitive hydrogel for the delivery of biological macromolecules

Biomolecule	Precursor(s)	Light used	Uses	Reference
Diflunisal	Succinylated/ methacrylated inulin	UV light of short wavelength (254 nm)	Controlled delivery of nonsteroidal anti-inflammatory drugs	Castelli et al. (2008)
Beclomethasone dipropionate	Methacrylated dextran and methacrylated α,β-poly(*N*-2 hydroxyethyl)-DL-aspartamide	Long-wave UV light (313 nm)	Colon-targeted delivery for the treatment of inflammatory bowel diseases	Pitarresi et al. (2007)
Paclitaxel	Azide and lactose functionalized chitosan	UV light (240–380 nm)	Anticancer drug delivery	Obara et al. (2005)
Fluorescein isothiocyanate–dextran	Poly(vinyl alcohol) methacrylate and silk fibroin	UV light of long wavelength (365 nm)	Macromolecular drug delivery	Pang et al. (2011)
Theophylline	Glycidyl methacrylated dextran and glycidyl methacrylated α,β-poly (*N*-2-hydroxyethyl)-DL-aspartamide	UV light of long wavelength (365 nm)	Oral delivery of drug molecules	Pitarresi et al. (2003)
Recombinant human basic fibroblast growth factor	Azide- and lactose-functionalized chitosan	Long-wave UV light (240–380 nm)	Occlusive dressing for wound management	Obara et al. (2003)
Silk fibroin protein	Poly(vinyl alcohol) methacrylate	UV light of long wavelength (365 nm)	Macromolecular hybrid drug delivery	Kundu et al. (2012)
Bovine serum albumin	Sodium alginate, poly (ethylene glycol)-co-poly (epsilon-caprolactone) and *N*-isopropylacrylamide	UV light of long wavelength (365 nm)	Peptide and protein delivery	Zhao et al. (2010)

(continued)

Table 1 (continued)

Biomolecule	Precursor(s)	Light used	Uses	Reference
Immunoglobulin G	Methacrylic anhydride and succinic anhydride functionalized inulin and α,β-poly [*N*-(2-hydroxyethyl)-D,L-aspartamide] methacrylic anhydride	UV light of long wavelength (366 nm)	Immunoglobulin G delivery	Tripodo et al. (2009)
DNA plasmid	Methacrylated hyaluronic acid and diacrylated pluronic F127	UV light of long wavelength (320–500 nm)	Controlled gene delivery	Chun et al. (2005)
siRNA	Methacrylated ethyleneimine and methacrylated dextran	UV light of long wavelength (320–500 nm)	Application in regenerative medicine and gene therapy	Nguyen et al. (2013)
DNA plasmid	Acrylated hyaluronic acid	UV light of long wavelength (365 nm)	Controlled gene delivery	Wieland et al. (2007)
Goat mesenchymal stem cells	Acrylated poly (6-aminohexyl propylene phosphate)	UV light of long wavelength (365 nm)	Application in bone tissue regeneration	Li et al. (2006)
Insulin-secreting pancreatic β-cells (MIN6)	Polyethylene glycol diacrylate	UV light of long wavelength (365 nm)	Pancreatic β-cells restoration for type 1 diabetes	Lin et al. (2011)
Hepatocytes with fibroblast cells	Polyethylene glycol diacrylate	UV light of long wavelength (365 nm)	Hepatic tissue regeneration	Tsang et al. (2007)
Neural progenitor cells	Hyaluronic acid methacrylate	UV light of long wavelength (365 nm)	Treatment of Parkinson's disease	Seidlits et al. (2010)
Fibroblasts	Gelatin methacrylamide and gellan gum methacrylate	UV light of long wavelength (320–500 nm)	Regeneration of load-bearing tissues	Shin et al. (2012)

Guo and Chu (2007) synthesized biodegradable hydrogels loaded with paclitaxel using fumarate-based polyethylene glycol diacrylate and poly(ester amide) via the UV irradiation. Controlled release of drug was achieved for a period of more than two month without initial burst release. Hydrogel nanoparticles containing doxorubicin were prepared using polyethylene diacrylate and acrylated pluronic F127 by laser irradiation, wherein the drug was released for a period of one week in a controlled manner with a minor burst effect (Missirlis et al. 2006).

Chun et al. (2005) effectively developed hydrogels containing DNA plasmid through photoirradiation by UV light using pluronic F127 diacrylate. DNA release from the hydrogel was continued for a period of more than three weeks, and the release depends primarily on cross-linking density. The continual release of DNA and the transfection efficiency of released DNA indicate that photo-cross-linked hydrogels are a potential platform for controlled delivery of therapeutic gene (Chun et al. 2005).

Photoirradiated hydrogels expand the scope of proficient pharmaceutical carriers used in the targeted delivery systems through a variety of strategies. Lee et al. observed the thermogelling character of photoirradiated hydrogels, and it is used as thermosensitive delivery system (Lee et al. 2007). Serra et al. developed acrylic-based hydrogel through irradiation of modified poly(acrylic acid) with polyethylene glycol by UV for site specific mucoadhesive delivery (Serra et al. 2006). In addition, photoirradiation also produced multilayer hydrogels to combine a variety of properties including high water absorption, pH sensitivity, and continuous drug release that may be requisite for a specific application.

UV-responsive hydrogels were prepared by incorporating a leuko derivative molecule into the network of polymer (Mamada et al. 1990). Triphenylmethane leuko derivatives are usually neutral however dissociate into ion pairs under ultraviolet irradiation creating triphenylmethyl cations. At a set temperature, the hydrogels intermittently swelled in presence of UV irradiation but shrank in absence of UV light. The UV induced irregular volume–phase transition is dissimilar from an uninterrupted volume–phase transition without UV irradiation. The UV light induced swelling owed to enhance in osmotic pressure inside the gel due to generation of cyanide ions by UV irradiation.

The visible light induced volume changes are based on the generation of temperature changes through integrated photosensitive molecules. In absence of light absorbing chromophores, infrared light can be used to induce changes in hydrogels. The volume–phase transition was observed along with bending of the gel toward the laser beam when poly(*N*-isopropylacrylamide) hydrogels are irradiated by a CO_2 laser infrared without chromophores (Zhang et al. 1995). The bending was due to temperature gradient, which depended on the CO_2 laser strength, whereas gel relaxation to its original shape followed an exponential pattern.

Light-responsive hydrogels can be utilized for the development of photoresponsive artificial muscles, memory devices, and switches (Suzuki and Tanaka 1990). A possible application of visible light-sensitive hydrogels for drug delivery was using response of cross-linked hyaluronic acid hydrogels that degrade in the existence of methylene blue (Yui et al. 1993).

Although the action of light is instant, the response of hydrogels to such action is still unacceptably slow. In most of the cases, conversion of light energy into thermal energy must precede the restructuring of polymer chains upon temperature change. Additionally, non-covalently linked chromophores may leach out during sol-gel transformation.

10 Conclusion

It is evident that hydrogels have a bright future in the biomedical field and specifically for the drug delivery areas. Many stimuli-sensitive hydrogels have been utilized as a direct means of sensing and responding to a variety of environmental conditions. A wide variety of methods have been demonstrated for the preparation of such responsive hydrogels. We believe that the most promising application of hydrogels in delivery is of biological macromolecules in diverse ways.

References

Agarwal S, Giri TK, Tripathi DK, Ajazuddin, Alexander A (2012) A review on novel therapeutic strategies for the enhancement of solubility for hydrophobic drugs through lipid and surfactant based self-micro emulsifying drug delivery system: a novel approach. Am J Drug Discov Dev 2:143–183

Albin G, Horbett TA, Ratner BD (1985) Glucose sensitive membranes for controlled delivery of insulin: insulin transport studies. J Control Release 2:153–164

Albin G, Horbett TA, Miller SR, Ricker NL (1987) Theoretical and experimental studies of glucose sensitive membranes. J Control Release 7:267–291

Badwaik H, Giri TK, Nakhate KT, Kashyap P, Tripathi DK (2013) Xanthan gum and its derivatives as a potential bio-polymeric carrier for drug delivery system. Curr Drug Deliv 10:587–600

Bae YH (1997) Stimuli-sensitive drug delivery. In: Park K (ed) Controlled drug delivery: challenge and strategies. American Chemical Society, Washington, DC, pp 147–160

Baroli B (2006) Photopolymerization of biomaterials: issues and potentialities in drug delivery, tissue engineering, and cell encapsulation applications. J Chem Technol Biotechnol 81:491–499

Behra A, Giri TK, Tripathi DK, Ajazuddin, Alexander A (2012) An exhaustive review on recent advancement in pharmaceutical bioadhesive used for systemic drug delivery through oral mucosa for achieving maximum pharmacological response and effect. Int J Pharmacol 8:283–305

Bhaskar RK, Sparer RV, Himmelstein KJ (1985) Effect of an applied electric field on liquid crystalline membranes: control of permeability. J Membr Sci 24:83–96

Bhoyar N, Giri TK, Tripathi DK, Alexander A, Ajazuddin (2012) Recent advances in novel drug delivery system through gels: review. J Pharm Allied Health Sci 2:21–39

Boustta M, Colombo PE, Lenglet S, Poujol S, Vert M (2014) Versatile UCST-based thermoresponsive hydrogels for loco-regional sustained drug delivery. J Control Release 174:1–6

Brocchini S, Duncan R (1999) Pendent drugs, release from polymers. In: Mathiowitz E (ed) Encyclopaedia of controlled drug delivery. Wiley, New York, pp 786–816

Bryant SJ, Nuttelman CR, Anseth KS (2000) Cytocompatibility of UV and visible light photoinitiating systems on cultured NIH/3T3 fibroblasts in vitro. J Biomater Sci Polym Ed 11:439–457

Burgmeyer P, Murray RW (1982) An ion gate membrane: electrochemical control of ion permeability through a membrane with an embedded electrode. J Am Chem Soc 104:6139–6140

Cartier S, Horbett TA, Ratner BD (1995) Glucose-sensitive membrane coated porous filters for control of hydraulic permeability and insulin delivery from a pressurized reservoir. J Membr Sci 106:17–24

Castelli F, Sarpietro MG, Micieli D, Ottimo S, Pitarresi G, Tripodo G, Carlisi B, Giammona G (2008) Differential scanning calorimetry study on drug release from an inulin-based hydrogel and its interaction with a biomembrane model: pH and loading effect. Eur J Pharm Sci 35 (2008):76–85

Chenite A, Chaput C, Wang D, Combes C, Buschmann MD, Hoemann CD, Leroux JC, Atkinson BL, Binette F, Selmani A (2000) Novel injectable neutral solutions of chitosan form biodegradable gels in situ. Biomaterials 21:2155–2161

Chiu HC, Hsiue GH, Lee YP, Huang LW (1999) Synthesis and characterization of pH-sensitive dextran hydrogels as a potential colon-specific drug delivery system. J Biomater Sci Polym Ed 10:591–608

Cho YI, Park S, Jeong SY, Yoo HS (2009) In vivo and in vitro anti-cancer activity of thermo-sensitive and photo-crosslinkable doxorubicin hydrogels composed of chitosan-doxorubicin conjugates. Eur J Pharm Biopharm 73:59–65

Chun KW, Lee JB, Kim SH, Park TG (2005) Controlled release of plasmid DNA from photo-cross-linked pluronic hydrogels. Biomaterials 26:3319–3326

Couto DS, Hong Z, Mano JF (2009) Development of bioactive and biodegradable chitosan-based injectable systems containing bioactive glass nanoparticles. Acta Biomater 5:115–123

Duncan R (2003) The dawning era of polymer therapeutics. Nat Rev Drug Discov 2:347–360

Eddington DT, Beebe DJ (2004) Flow control with hydrogels. Adv Drug Deliv Rev 56:199–210

Eisenberg SR, Grodzinsky AJ (1987) The kinetics of chemically induced nonequilibrium swelling of articular cartilage and corneal stroma. J Biomech Eng 109:79–89

Firestone BA, Siegel RA (1991) Kinetics and mechanisms of water sorption in hydrophobic, ionizable copolymer gels. J Appl Polym Sci 43:901–914

Fujimoto KL, Ma Z, Nelson DM, Hashizume R, Guan J, Tobita K, Wagner WR (2009) Synthesis, characterization and therapeutic efficacy of a biodegradable, thermoresponsive hydrogel designed for application in chronic infarcted myocardium. Biomaterials 30:4357–4368

Garbern JC, Hoffman AS, Stayton PS (2010) Injectable pH- and temperature responsive poly (N-isopropylacrylamide-co-propylacrylic acid) copolymers for delivery of angiogenic growth factors. Biomacromol 11:1833–1839

Garner B, Georgevich A, Hodgson AJ, Liu L, Wallace GG (1999) Polypyrrole–heparin composites as stimulus-responsive substrates for endothelial cell growth. Inc J Biomed Mater Res 44: 121–129

George PM, LaVan DA, Burdick JA, Chen CY, Liang E, Langer R (2006) Electrically controlled drug delivery from biotin-doped conductive polypyrrole. Adv Mater 18:577–581

Giri TK, Kumar K, Alexander A, Ajazuddin, Badwaik H, Tripathy M, Tripathi DK (2013a) Novel controlled release solid dispersion for the delivery of diclofenac sodium. Curr Drug Deliv 10:435–443

Giri TK, Thakur D, Alexander A, Ajazuddin, Badwaik H, Tripathy M, Tripathi DK (2013b) Biodegradable IPN hydrogel beads of pectin and grafted alginate for controlled delivery of diclofenac sodium. J Mater Sci Mater Med 24:1179–1190

Giri TK, Verma S, Alexander A, Ajazuddin, Badwaik H, Tripathy M, Tripathi DK (2013c) Crosslinked biodegradable alginate hydrogel floating beads for stomach site specific controlled delivery of metronidazole. Farmacia 61:533–550
Golden A, Tien J (2007) Fabrication of microfluidic hydrogels using molded gelatin as a sacrificial element. Lab Chip 7:720–725
Gomez N, Schmidt CE (2007) Nerve growth factor-immobilized polypyrrole: bioactive electrically conducting polymer for enhanced neurite extension. J Biomed Mater Res A 81:135–149
Grimshaw PE, Grodzinsky AJ, Yarmush ML, Yarmush DM (1989) Dynamic membranes for protein transport: chemical and electrical control. Chem Eng Sci 44:827–840
Grodzinsky AJ, Grimshaw PE (1990) Electrically and chemically controlled hydrogels for drug delivery. In: Kost I (ed) Pulsed and self-regulated drug delivery. CRC Press, Boca Raton, pp 47–64
Guan J, Hong Y, Ma Z, Wagner WR (2008) Protein-reactive, thermoresponsive copolymers with high flexibility and biodegradability. Biomacromol 9:1283–1292
Guo K, Chu CC (2007) Controlled release of paclitaxel from biodegradable unsaturated poly(ester amide)s/poly(ethylene glycol) diacrylate hydrogels. J Biomater Sci Polym Ed 18:489–504
Heller J, Trescony PV (1979) Controlled drug release by polymer dissolution. II. An enzyme mediated delivery system. J Pharm Sci 68:919–921
Heller J, Baker RW, Gale RM, Rodin JO (1978) Controlled drug release by polymer dissolution. I. Partial esters of maleic anhydride copolymers. Properties and theory. J Appl Polym Sci 22:1991–2009
Hisamitsu I, Kataoka K, Okano T, Sakurai Y (1997) Glucose-responsive gel from phenylborate polymer and polyvinyl alcohol: prompt response at physiological pH through the interaction of borate with amino group in the gel. Pharm Res 14:289–293
Hoffman AS (1997) Intelligent polymers. In: Park K (ed) Controlled drug delivery: challenge and strategies. American Chemical Society, Washington, DC, pp 485–497
Holtz JH, Asher SA (1997) Polymerized colloidal crystal hydrogel films as intelligent chemical sensing materials. Nature 389:829–832
Huynh DP, Nguyen MK, Pi BS, Kim MS, Chae SY, Lee KC, Kim BS, Kim SW, Lee DS (2008) Functionalized injectable hydrogels for controlled insulin delivery. Biomaterials 29:2527–2534
Hwang CM, Sant S, Masaeli M, Kachouie NN, Zamanian B, Lee SH, Khademhosseini A (2010) Fabrication of three-dimensional porous cell-laden hydrogel for tissue engineering. Biofabrication 2:035003
Ishihara K, Matsui K (1986) Glucose-responsive insulin release from polymer capsule. J Polym Sci Polym Lett Ed 24:413–417
Ishihara K, Kobayashi M, Ishimaru N, Shinohara I (1984) Glucose induced permeation control of insulin through a complex membrane consisting of immobilized glucose oxidase and a polyamine. Polym J 16:625–631
Ishihara K, Muramoto N, Fuji H, Shinohara I (1985a) pH-induced reversible permeability control of the 4-carboxy acrylanilide-methyl methacrylate copolymer membrane. J Polym Sci Polym Chem Ed 23:2841–2850
Ishihara K, Muramoto N, Fuji H, Shinohara I (1985b) Preparation and permeability of urea-responsive membrane consisting of immobilized urease and poly(aromatic carboxylic acid). J Polym Sci Polym Lett Ed 23:531–535
Jeon O, Powell C, Solorio LD, Krebs MD, Alsberg E (2011) Affinity-based growth factor delivery using biodegradable, photocrosslinked heparin-alginate hydrogels. J Control Release 154: 258–266
Kamath K, Park K (1993) Biodegradable hydrogels in drug delivery. Adv Drug Deliv Rev 11:59–84
Kang SI, Bae YH (2003) A sulfonamide based glucose-responsive hydrogel with covalently immobilized glucose oxidase and catalase. J Control Release 86(1):115–121
Kang EY, Moon HJ, Joo MK, Jeong B (2012) Thermogelling chitosan-g-(PAF-PEG) aqueous solution as an injectable scaffold. Biomacromol 13:1750–1757

Kitano S, Koyama Y, Kataoka K, Okano T, Sakurai Y (1992) A novel drug delivery system utilizing a glucose responsive polymer complex between poly(vinyl alcohol) and poly (N-vinyl-2-pyrrolidone) with a phenyl boronic acid moiety. J Control Release 19:162–170

Klier J, Peppas NA (1989) Complex-forming hydrogels sensitive to physiological conditions. Proc Int Symp Control Release Bioact Mater 16:101–102

Klumb LA, Horbett TA (1992) Design of insulin delivery devices based on glucose sensitive membranes. J Control Release 18:59–80

Kost J, Horbett TA, Ratner BD, Singh M (1985) Glucose-sensitive membranes containing glucose oxidase: swelling, activity, and permeability studies. J Biomed Mater Res 19:1117–1122

Kumar A, Lahiri SS, Singh H (2006) Development of PEGDMA: MAA based hydrogel microparticles for oral insulin delivery. Int J Pharm 323:117–124

Kundu J, Poole-Warren LA, Martens P, Kundu SC (2012) Silk fibroin/poly(vinyl alcohol) photocrosslinked hydrogels for delivery of macromolecular drugs. Acta Biomater 8:1720–1729

Kwon IC, Bae YH, Okano T, Kim SW (1991) Drug release from electric current sensitive polymers. J Control Release 17:149–156

Lee ES, Na K, Bae YH (2003) Polymeric micelle for tumor pH and folate-mediated targeting. J Control Release 91:103–113

Lee YJ, Pruzinsky SA, Braun PV (2004) Glucose sensitive inverse opal hydrogels: analysis of optical diffraction response. Langmuir 20:3096–3106

Lee SY, Tae G, Kim YH (2007) Thermal gellation and photo-polymerization of diacrylated Pluronic F 127. J Biomater Sci Polym Ed 18:1335–1353

Leong W, Lau TT, Wang DA (2013) A temperature-cured dissolvable gelatin microsphere-based cell carrier for chondrocyte delivery in a hydrogel scaffolding system. Acta Biomater 9:6459–6467

Li Y, Neoh KG, Kang ET (2005) Controlled release of heparin from polypyrrole-poly(vinyl alcohol) assembly by electrical stimulation. J Biomed Mater Res 73:171–181

Li Q, Wang J, Shahani S, Sun DD, Sharma B, Elisseeff JH, Leong KW (2006) Biodegradable and photocrosslinkable polyphosphoester hydrogel. Biomaterials 27:1027–1034

Li Z, Guo X, Matsushita S, Guan J (2011) Differentiation of cardiosphere-derived cells into a mature cardiac lineage using biodegradable poly(N-isopropylacrylamide) hydrogels. Biomaterials 32:3220–3232

Lin CC, Raza A, Shih H (2011) PEG hydrogels formed by thiol-ene photo-click chemistry and their effect on the formation and recovery of insulin-secreting cell spheroids. Biomaterials 32:9685–9695

Lucisano JY, Armour JC, Gough DA (1987) In vitro stability of an oxygen sensor. Anal Chem 59:736–739

Mamada A, Tanaka T, Kungwachakun D, Irie M (1990) Photoinduced phase transition of gels. Macromolecules 23:1517–1519

Markland P, Zhang Y, Amidon GL, Yang VC (1999) A pH- and ionic strength-responsive polypeptide hydrogel: synthesis, characterization, and preliminary protein release studies. J Biomed Mater Res 47:595–602

Mateo C, Palomo JM, Fernandez-Lorente G, Guisan JM, Fernandez-Lafuente R (2007) Improvement of enzyme activity, stability and selectivity via immobilization techniques. Enzyme Microb Technol 40:1451–1463

Matsumura Y, Maeda H (1986) A new concept for macromolecular therapeutics in cancer-chemotherapy—mechanism of tumoritropic accumulation of proteins and the antitumor agent SMANCS. Cancer Res 46:6387–6392

Missirlis D, Kawamura R, Tirelli N, Hubbell JA (2006) Doxorubicin encapsulation and diffusional release from stable, polymeric, hydrogel nanoparticles. Eur J Pharm Sci 29:120–129

Na K, Bae YH (2005) pH-sensitive polymers for drug delivery. In: Kwon GS (ed) Polymeric drug delivery systems. Taylor & Francis, Boca Raton, pp 129–194

Neradovic D, Hinrichs WLJ, Kettenes-van den Bosch JJ, Hennink WE (1999) Poly (N-isopropylacrylamide) with hydrolyzable lactic acid ester side groups: a new type of thermosensitive polymer. Macromol Rapid Commun 20:577–581

Nguyen K, Dang PN, Alsberg E (2013) Functionalized, biodegradable hydrogels for control over sustained and localized siRNA delivery to incorporated and surrounding cells. Acta Biomater 9:4487–4495

Nussbaum JH, Grodzinsky AJ (1981) Proton diffusion-reaction in a protein polyelectrolyte membrane and the kinetics of electromechanical forces. J Membr Sci 8:193–219

Obaidat AA, Park K (1996) Characterization of glucose dependent gel–sol phase transition of the polymeric glucose–concanavalin A hydrogel system. Pharm Res 13:989–995

Obaidat AA, Park K (1997) Characterization of protein release through glucose-sensitive hydrogel membranes. Biomaterials 18:801–806

Obara K, Ishihara M, Ishizuka T, Fujita M, Ozeki Y, Maehara T, Saito Y, Yura H, Matsui T, Hattori H, Kikuchi M, Kurita A (2003) Photocrosslinkable chitosan hydrogel containing fibroblast growth factor-2 stimulates wound healing in healing-impaired db/db mice. Biomaterials 24:3437–3444

Obara K, Ishihara M, Ozeki Y, Ishizuka T, Hayashi T, Nakamura S, Saito Y, Yura H, Matsui T, Hattori H, Takase B, Ishihara M, Kikuchi M, Maehara T (2005) Controlled release of paclitaxel from photocrosslinked chitosan hydrogels and its subsequent effect on subcutaneous tumor growth in mice. J Control Release 110:79–89

Osada Y, Okuzaki H, Hori H (1992) A polymer gel with electrically driven motility. Nature 355:242–244

Pang SC, Chin SF, Ay SH, Hong FM (2011) Starch-maleate-polyvinyl alcohol hydrogels with controllable swelling behaviors. Carbohydr Polym 84:424–429

Park K, Park H (1999) Smart hydrogels. In: Salamone JC (ed) Concise polymeric materials encyclopedia. CRC Press, Boca Raton, pp 1476–1478

Park K, Shalaby WSW, Park H (1993) Biodegradable hydrogels for drug delivery. Technomic, Lancaster

Park MR, Chun C, Ahn SW, Ki MH, Cho CS, Song SC (2010a) Sustained delivery of human growth hormone using a polyelectrolyte complex-loaded thermosensitive polyphosphazene hydrogel. J Control Release 147:359–367

Park MR, Chun C, Cho CS, Song SC (2010b) Enhancement of sustained and controlled protein release using polyelectrolyte complex-loaded injectable and thermosensitive hydrogel. Eur J Pharm Biopharm 76:179–188

Park MR, Seo BB, Song SC (2013) Dual ionic interaction system based on polyelectrolyte complex and ionic, injectable, and thermosensitive hydrogel for sustained release of human growth hormone. Biomaterials 34:1327–1336

Pasechnik VA, Cherkosov AN, Zhemko VP, Chechina VV, Alimardinov RS (1979) Effect of direct current on the permeability of ultrafiltration membranes. Kolloid Zh 41:586–590

Patel VR, Amiji MM (1996) Preparation and characterization of freeze-dried chitosan-poly (ethylene oxide) hydrogels for site-specific antibiotic delivery in the stomach. Pharm Res 13:588–593

Pitarresi G, Palumbo FS, Giammona G, Casadei MA, Micheletti MF (2003) Biodegradable hydrogels obtained by photocrosslinking of dextran and polyaspartamide derivatives. Biomaterials 24:4301–4313

Pitarresi G, Casadei MA, Mandracchia D, Paolicelli P, Palumbo FS, Giammona G (2007) Photocrosslinking of dextran and polyaspartamide derivatives: a combination suitable for colon-specific drug delivery. J Control Release 119:328–338

Pluta J, Karolewicz B (2004) Hydrogels: properties and application in the technology of drug form I. The characteristics hydrogels. Polym Med 34:3–19

Podual K, Doyle FJI, Peppas NA (2000) Preparation and dynamic response of cationic copolymer hydorgels containing glucose oxidase. Polymer 41:3975–3983

Sawahata K, Hara M, Yasunaga H, Osada Y (1990) Electrically controlled drug delivery system using polyelectrolyte gels. J Control Release 14:253–262

Schild HG, Tirrell DA (1990) Microcalorimetric detection of lower critical solution temperatures in aqueous polymer solutions. J Phys Chem 94:4352–4356

Seidlits SK, Khaing ZZ, Petersen RR, Nickels JD, Vanscoy JE, Shear JB, Schmidt CE (2010) The effects of hyaluronic acid hydrogels with tunable mechanical properties on neural progenitor cell differentiation. Biomaterials 31:3930–3940

Serra L, Domenech J, Peppas NA (2006) Design of poly (ethylene glycol)-tethered copolymers as novel mucoadhesive drug delivery systems. Eur J Pharm Biopharm 6:11–18

Shiga T, Hirose Y, Okada A, Kurauchi T (1992) Electric field-associated deformation of polyelectrolyte gel near a phase transition point. J Appl Polym Sci 46:635–640

Shiino D, Murata Y, Kataoka K, Koyama Y, Yokoyama M, Okano T, Sakurai Y (1994) Preparation and characterization of a glucose-responsive insulin-releasing polymer device. Biomaterials 15:121–128

Shin H, Olsen BD, Khademhosseini A (2012) The mechanical properties and cytotoxicity of cell-laden double-network hydrogels based on photocrosslinkable gelatin and gellan gum biomacromolecules. Biomaterials 33(11):3143–3152

Siegel RA (1993) Hydrophobic weak polyelectrolyte gels: studies of swelling equilibria and kinetics. Adv Polym Sci 109:233–267

Siegel RA, Falamarzian M, Firestone BA, Moxley BC (1988) pH-controlled release from hydrophobic/polyelectrolyte copolymer hydrogels. J Control Release 8:179–182

Siegwart DJ, Bencherif SA, Srinivasan A, Hollinger JO, Matyjaszewski K (2008) Synthesis, characterization, and in vitro cell culture viability of degradable poly (N-isopropylacrylamide-co-5,6-benzo-2-methylene-1,3-dioxepane)-based polymers and cross-linked gels. J Biomed Mater Res Part A 87:345–358

Suzuki A, Tanaka T (1990) Phase transition in polymer gel induced by visible light. Nature 346 (6282):345–347

Tsang VL, Chen AA, Cho LM, Jadin KD, Sah RL, DeLong S, West JL, Bhatia SN (2007) Fabrication of 3D hepatic tissues by additive photopatterning of cellular hydrogels. FASEB J 21:790–801

Thakur VK, Thakur MK (2014a) Recent trends in hydrogels based on psyllium polysaccharide: a review. J Clean Prod 82:1–15

Thakur VK, Thakur MK (2014b) Recent advances in graft copolymerization and applications of chitosan: a review. ACS Sustainable Chem Eng 2(12):2637–2652

Thakur VK, Kessler MR (2015) Self-healing polymer nanocomposite materials: a review. Polymer 69:369–383

Thakur VK, Thakur MK (2015) Recent advances in green hydrogels from lignin: a review. Int J Biol Macromol 72:834–847

Thakur MK, Thakur VK, Gupta RK, Pappu A (2016) Synthesis and applications of biodegradable soy based graft copolymers: a review. ACS Sustainable Chem Eng 4:1–17

Tian M, Yang Z, Kuwahara K, Nimni ME, Wan C, Han B (2012) Delivery of demineralized bone matrix powder using a thermogelling chitosan carrier. Acta Biomater 8:753–762

Tripodo G, Pitarresi G, Cavallaro G, Palumbo FS, Giammona G (2009) Controlled release of IgG by novel UV induced polysaccharide/poly(amino acid) hydrogels. Macromol Biosci 9(4): 393–401

Twaites B, de las Heras Alarcon C, Alexander C (2005) Synthetic polymers as drugs and therapeutics. J Mater Chem 15:441–455

Wadhwa R, Lagenaur CF, Cui XT (2006) Electrochemically controlled release of dexamethasone from conducting polymer polypyrrole coated electrode. J Control Release 110(3):531–541

Ward MA, Georgiou TK (2011) Thermoresponsive polymers for biomedical applications. Polymers 3:1215–1242

Wieland JA, Houchin-Ray TL, Shea LD (2007) Non-viral vector delivery from PEG-hyaluronic acid hydrogels. J Control Release 120(3):233–241

Wust S, Godla ME, Muller R, Hofmann S (2014) Tunable hydrogel composite with two-step processing in combination with innovative hardware upgrade for cell-based three-dimensional bioprinting. Acta Biomater 10:630–640

Yui N, Okano T, Sakurai Y (1993) Photo-responsive degradation of heterogeneous hydrogels comprising crosslinked hyaluronic acid and lipid microspheres for temporal drug delivery. J Control Release 26:141–145

Zhang X, Li Y, Hu Z, Littler CL (1995) Bending of N-isopropylacrylamide gel under the influence of infrared light. J Chem Phys 102:551–555

Zhao S, Cao M, Li H, Li L, Xu W (2010) Synthesis and characterization of thermo-sensitive semi-IPN hydrogels based on poly(ethylene glycol)-co-poly(epsilon-caprolactone) macromer, N-isopropylacrylamide, and sodium alginate. Carbohydr Res 345(3):425–431

Chapter 2
Stimuli-Responsive Polymeric Nanoparticles for Cancer Therapy

Shantanu V. Lale and Veena Koul

Abstract Anticancer drugs are cytotoxic in nature and hence have severe side effects due to their non-specific action on normal healthy cells. Nanoparticles have been widely studied to minimize these side effects by selective targeting of cancer cells either by passive or active targeting mechanism. Side effects can be further minimized by delivering drugs to the cancer site without leaking it in the blood during nanoparticle circulation. To achieve this, drug delivery systems need to provide secure encapsulation of the drugs without leakage before reaching the site of action and selectively release encapsulated drugs in the cancer tissues. Recent advances and progress in nanotechnology have resulted in development of stimuli-responsive polymeric materials which have demonstrated great potential in cancer therapy. Stimuli-responsive materials significantly change their physical or chemical properties in response to variations in the environment. Stimuli responsiveness allows secure drug encapsulation in absence of stimuli, but allows rapid drug release in presence of the given stimuli. These stimuli can be inherent and intrinsic to the tumor tissue such as acidic condition, higher tissue temperature, higher redox environment or activity of certain enzymes, or they can be externally applied stimuli such as magnetic field, temperature, ultrasonic waves, or other irradiation. In this chapter, recent advances and applications of various stimuli-responsive polymeric nanosystems for cancer therapy are reviewed.

Keywords Stimuli-responsive polymers · Cancer · Nanoparticles
Drug delivery

S. V. Lale (✉)
Department of Chemical Engineering, Texas Tech University,
Lubbock, TX 79409, USA
e-mail: lale.shantanu@gmail.com

V. Koul
Centre for Biomedical Engineering, Indian Institute of Technology Delhi,
New Delhi 110016, India
e-mail: veenak_iitd@yahoo.com

V. Koul
All India Institute of Medical Sciences, New Delhi 110029, India

© Springer Nature Singapore Pte Ltd. 2018

V. K. Thakur et al. (eds.), *Polymer Gels*, Gels Horizons: From Science to Smart Materials, https://doi.org/10.1007/978-981-10-6080-9_2

1 Introduction

Cancer nanotechnology is developing as a promising field of multidisciplinary research in cancer therapeutics with the involvement of medicine, biology, chemistry, and engineering disciplines. Drug delivery nanocarriers have emerged as a new class of therapeutics for cancer therapy with improved therapeutic efficacy, tumor selectivity, and fewer side effects. These nanocarriers have distinguishing properties as compared to conventional cancer chemotherapy which makes them suitable for cancer therapeutics as shown below (Misra et al. 2010; Bertrand et al. 2014; Cheng and Pun 2015)

- Nanocarriers can be customized to selectively target cancer cells by conjugating multiple targeting ligands.
- Nanocarriers can be designed to carry large amount of therapeutic payload.
- Nanocarriers can encapsulate multiple drugs simultaneously enabling combinatorial cancer therapy.
- Nanocarriers can bypass multidrug resistance of cancer cells.
- Nanocarriers protect labile therapeutic drugs during circulation.
- Nanocarriers improve blood circulation and tissue distribution of chemotherapeutic drugs.
- Combination of therapeutic drug and imaging agent can be achieved in nanocarriers for simultaneous theranostic application.
- Nanocarriers can be surface functionalized for various applications.

Various designs of nanoparticles have been explored for their application in cancer therapy because of their limitless versatility in composition and surface chemistry. Nanoparticles can be designed in variety of sizes, shapes, and compositions with multiple functionalities for improved therapeutic efficacy in cancer. Irrespective of drug's inherent properties, pharmacokinetic and distribution patterns of drug candidates are governed by properties of nanocarriers such as particle size, surface charge, surface hydrophilicity. Alteration of these parameters has profound effect on biological fate of nanoparticles and their usefulness as drug delivery carrier (Jabr-Milane et al. 2008; Bertrand et al. 2014). Polymer-based nanoparticles including micelles, polymersomes, nanocapsules, nanospheres are one of the most common nanotechnology platforms for cancer therapy and have garnered much attention because of several advantages. Polymers are versatile macromolecules, and their chemistry can be easily modified to synthesize variety of polymers with different architectures. Size of polymeric nanocarrier can be easily optimized for efficient cellular uptake via endocytosis. Polymer chemistry allows for many variations in polymer structure for desired physical, chemical, and biological properties of the resulting nanocarriers (Crucho 2014).

2 Stimuli-Responsive Polymeric Nanoparticles

Anticancer drugs are cytotoxic in nature and exhibit severe side effects after administration to the patients due to their non-specific action on normal healthy cells. Nanoparticles have been widely studied to minimize these side effects by selective targeting of cancer cells either by passive or active targeting mechanism. Side effects can be further minimized by selective delivering anticancer drugs to the cancer site without leaking it in the blood during nanoparticle circulation. To achieve this, drug delivery systems need to provide secure encapsulation of the drugs without leakage in the bloodstream, before reaching the site of action (tumor) and selectively release encapsulated drugs in the cancer tissues (Mura et al. 2013; Yu et al. 2014).

In order to achieve secure encapsulation of the drugs without leakage in the bloodstream, various materials which are responsive to certain stimuli have been developed. These stimuli-responsive materials significantly change their physical or chemical properties in response to variations in the surrounding environment. Stimuli responsiveness allows secure drug encapsulation in absence of stimuli, but allows rapid drug release in presence of the specified stimuli. As shown in Fig. 1, these stimuli can be intrinsic and inherent to the tumor tissue such as acidic condition, higher tissue temperature, higher redox environment or activity of certain enzymes, or they can be externally applied stimuli such as temperature, magnetic field, and ultrasonic waves (Mura et al. 2013; Cheng et al. 2014; Crucho 2014).

3 pH-Responsive Polymeric Nanoparticles

Tumor tissues are generally characterized by irregular angiogenesis which causes rapid deficit of both nutrients and oxygen. Also, unlike normal cells which produce energy by oxidative phosphorylation in mitochondria, cancer cells predominantly produce energy by rapid glycolysis in cytosol, a phenomenon known as the Warburg effect and is unique property of most of cancer tissues (Warburg 1956). This results in accumulation of lactic acid around the tumor tissues in high concentration which lowers the pH of tumor extracellular environment (Vander Heiden et al. 2009; Choi et al. 2013). Similarly, acidic pH is also found in the endocytic

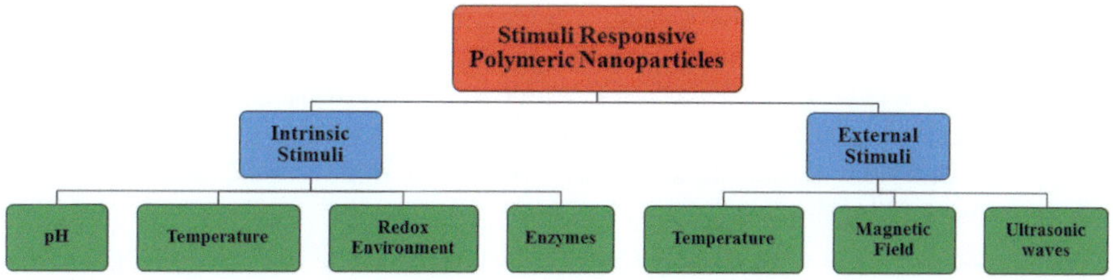

Fig. 1 Classification of stimuli-responsive polymeric nanoparticles

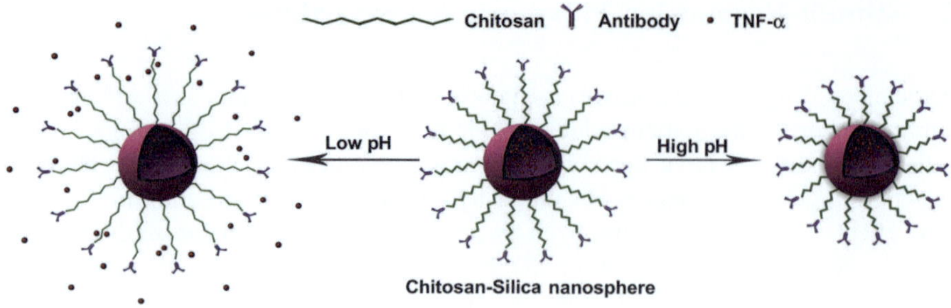

Fig. 2 Schematic diagram of pH responsiveness of chitosan–silica nanospheres reported by Deng et al. Figure adapted from Deng et al. (2011) with modification

compartments (endosomes pH 5–6, lysosomes pH 4–5). Thus, acidic environment of tumors and endocytic compartment forms a potential platform for development of pH-responsive nanocarriers in cancer therapy. Two strategies are widely used—use of polymers with ionizable groups that undergo solubility or conformational changes in response to pH variation in the surrounding environment; and use of acid-sensitive linkages in the polymer backbone, cleavage of which releases the attached drug molecules in the environment (Liu et al. 2014a, b; Kwon et al. 2015).

Pujana et al. prepared folic acid functionalized chitosan nanogels using inverse microemulsion method for targeted delivery of 5-fluorouracil in cancer cells. The hydrogels showed pH-responsive swelling, which can be attributed to the protonation of the free amino groups of chitosan (Pujana et al. 2014). Similarly, Deng et al. showed swelling of hollow chitosan–silica nanospheres due to amino group protonation (pKa $\sim$ 6.3) in chitosan which led to release of encapsulated tumor necrosis factor alpha (TNFα) both in vitro in MCF-7 cells as well as in vivo in athymic nude mice (Deng et al. 2011) (Fig. 2).

Methotrexate-loaded chitosan NPs loaded with surfactant derived from *N*(α),*N*(ε)-dioctanoyl lysine with an inorganic lithium counterion was developed by Nogueira et al. *N*(α),*N*(ε)-dioctanoyl lysine imparts pH sensitivity to the polymer along with membrane-disruptive activity. Triggered release of methotrexate from the nanosystem was observed under acidic conditions. This can be attributed to decreased ionization of *N*(α),*N*(ε)-dioctanoyl lysine in acidic condition which results in reduced amount of available anionic groups that can electrostatically interact with the positive amino groups of chitosan, causing destabilization of the nanoparticle structure (Nogueira et al. 2013).

In another approach, Duong et al. reported pH sensitivity of acetalated dextran-PEG copolymer nanoparticles which can be attributed to hydrolysis of acetalated dextran in the endolysosomal compartment leading to intracellular drug release and showed higher doxorubicin toxicity to tumor cells as compared to normal cells (Duong et al. 2012).

In another strategy, Min et al. achieved pH-responsive delivery of camptothecin with disassembly of PEG-poly(β-amino ester) micelles in acidic conditions.

PEG-poly(β-amino ester) micelles showed pH-dependant micellization/ demicellization transition at pH 6.4. Camptothecin-loaded micelles showed superior therapeutic efficacy in mice model as compared to free camptothecin and camptothecin-encapsulated PEG-PLLA micelles (Min et al. 2010). Quio et al. designed a family of poly(β-amino ester)s copolymers with tertiary amine groups for targeted intracellular drug delivery of doxorubicin in cancer. The pH-responsive behavior of the nanoparticles was studied by dynamic light scattering and Nile red release experiments. Enhanced doxorubicin release was observed in acidic media confirming pH-responsive property of poly(β-amino ester)s copolymers (Qiao et al. 2014). In other studies, Zhang et al. synthesized pH-sensitive poly(β-amino ester)-g-poly(ethylene glycol) methyl ether-cholesterol triblock polymer. pH sensitivity of the copolymer was due to protonation–deprotonation of the amino groups of poly (β-amino ester) which resulted in enhanced doxorubicin release and higher cytotoxicity in HepG2 cells (Zhang et al. 2014a, b). Similarly, Tang et al. constructed pH-sensitive poly (β-amino esters)-based polymeric nanoparticles of poly [(1,4-butanediol)-diacrylate-β-5-hydroxyamylamine] (PDHA) and polyethyleneimine (PEI, MW 1.8 kDa) for co-delivery of Snail siRNA (siSna) and Twist siRNA (siTwi) along with paclitaxel. This resulted in accumulation of both paclitaxel and siRNA in tumor tissue of 4T1 tumor-bearing mice models, leading to inhibition of the tumor growth (Tang et al. 2015).

In another approach, Li et al. developed zeolitic imidazole framework-based nanoparticles for pH-sensitive delivery of 5-fluorouracil. Zeolitic imidazole framework undergoes protonation leading to rapid degradation in acidic environments of tumor and endolysosomal compartments which promotes faster drug release. NPs showed high distribution to the lung with significantly improved the antitumor efficacy in a nude mouse model with tumor lung metastasis (Li et al. 2016a, b).

Ling et al. prepared nano-formulation for delivery of triptolide in hepatocellular carcinoma which comprised of Pluronic F127 and synthetic decamer of aspartic acid with octadecylamine tail along with ionizable imidazole side chains. The imidazole side chains impart pH sensitivity to the nanostructures which resulted in enhanced selective drug release in acidic environment of tumor with reduced toxicity (Ling et al. 2014). Liu et al. showed pH-responsive nature of poly(ethylene glycol)-poly(L-histidine)-poly(L-lactide) (PEG-PH-PLLA) triblock copolymers with higher drug release at acidic pH due to protonation of the imidazole groups in poly(L-histidine). Pegylated poly(L-histidine) micelles show pH responsiveness by protonating histidine residues in the poly(L-histidine) block at endosomal pH resulting in increase in critical micelle concentration (CMC) and micellar destabilization which causes drug release (Liu et al. 2011). Park et al. reported pH-triggered drug-eluting nanocomposite microspheres of poly(lactide-co-glycolide) (PLGA) along with pH-responsive additive consisting of octadecylamine and 1-(3-aminopropyl)imidazole group. This pH-responsive additive is in solid state at a physiological pH (pH $\sim$ 7.4). However, it is destabilized by an acidic environment due to ionization of imidazole functional groups, causing its phase transition into a charged water-soluble form which further triggers drug release from pH-sensitive NPs (Park et al. 2016). Similarly, Gaspar et al.

demonstrated pH sensitivity of chitosan-based nanosystem by incorporating pH-sensitive histidine and cell-penetrating arginine residues in the polymer backbone. This resulted in efficient internalization of polymer NPs loaded with plasmid DNA in the intracellular compartment leading to higher transfection efficiency than unmodified chitosan NPs (Gaspar et al. 2013).

In another report, Sethuraman and coworkers used sulfonamide groups as pH-responsive groups for drug delivery application. Polymeric nanoparticles of pH-sensitive poly(methacryloyl sulfadimethoxine) (PSD)-block-PEG (PSD-b-PEG) diblock copolymer with plasmid DNA complexed with poly(ethylenimine) were prepared. The strategy involved shielding of the DNA–PEI complex during systemic circulation while at the acidic tumor site, sulfonamide groups lose their charge and get detached from the nanocomplex exposing the plasmid DNA complex. The authors observed pH-responsive behavior of the NPs with higher cytotoxicity and transfection due to detachment of PSD-b-PEG with subsequent exposure of PEI to the cell (Sethuraman et al. 2006).

Various acid-sensitive linkages have been explored for chemical conjugation of drugs which results in selective drug release in tumor cells while preventing unwanted drug release from nanoparticles during circulation. Various acid-labile linkages including acetal, β-thiopropionate, hydrazone, imine, benzoic-imine, cis-aconityl have been developed and incorporated in the polymeric nanocarriers for drug delivery applications (Mura et al. 2013; Meng et al. 2014).

Wang et al. exploited acid responsive property of acetal linkage by preparing doxorubicin-loaded nanoparticles of poly(ε-caprolactone)-acetal-poly(ethylene glycol)-acetal-poly(ε-caprolactone) (PCL-a-PEG-a-PCL) containing acid-labile acetal bond. The polymeric NPs released their payload at acidic conditions because of pH-triggered dissolution behavior (Wang et al. 2013). Gu et al. conjugated paclitaxel with poly(ethylene glycol)-b-poly(acrylic acid) (PEG-PAA) block copolymers through pH-sensitive acetal linkage. pH-dependent drug release of the micellar nanosystem was observed along with higher antitumor effect in MTT assay in KB and HeLa cells (Gu et al. 2013).

Alternatively, Qiu et al. conjugated camptothecin with hyperbranched star copolymers through acid-labile β-thiopropionate linkage. The micelles with diameters of about 3–6 nm showed pH-dependent drug release with triggered release at mild acidic pH conditions (Qiu et al. 2016). In study by Zhou et al., paclitaxel was conjugated with poly(ethylene oxide)-block-polyphosphoester via acid-sensitive β-thiopropionate linkage. The nanosystem showed pH-dependent release of paclitaxel in acidic pH and also exhibited five- to eightfold enhanced cytotoxicity against OVCAR-3 and RAW 264.7 cancer cell lines (Zou et al. 2014).

In another approach, Su et al. prepared Pluronic P123-based micelles with docetaxel conjugated via hydrazine linkage as well as physically encapsulated in the micelles (dual drug loading). pH-sensitive hydrazine linkage resulted in higher cellular toxicity in B16F10 cells while in the xenograft B16F10 melanoma model, the micelles suppressed tumor growth more effectively than marketed formulation Duopafei® without causing adverse effects (Su et al. 2016). Lale et al. designed poly(poly(ethylene glycol) methacrylate)-poly(caprolactone)-poly(poly(ethylene glycol) methacrylate) (pPEGMA-PCL-pPEGMA) triblock polymer and conjugated doxorubicin via acid-sensitive hydrazone bond. Due to pH-sensitive hydrazine linkage, higher cumulative doxorubicin release at pH 5.0 as compared to pH 7.4 was observed in in vitro release studies. In vitro and in vivo studies showed higher antitumor efficacy as compared to free doxorubicin (Lale et al. 2014). Similarly, Lane et al. constructed linear copolymers of *N*,*N*-dimethylacrylamide and 2-hydroxyethylacrylamide with hydrazone-linked doxorubicin molecules. Drug release studies of nanoparticles showed enhanced drug release in acidic pH conditions with 90% drug release after 24 h (Lane et al. 2015). Sun et al. developed doxorubicin-conjugated gold nanoparticles with hydrazone linkage for targeting cancer stem cells and showed enhanced drug release and enhanced antitumor efficacy of the nanosystems in MCF-7 cancer cells as well as in female SCID mice (Sun et al. 2014).

Alternatively, in a study by Du et al., degradable pH-responsive polyspermine imine was evaluated as a polycationic siRNA carrier. It is formed by condensing spermine and glyoxal, through Schiff base linkage, which is stable under neutral condition but rapidly dissociates under the acidic endosomal pH. It was observed that the polyplex readily released the loaded siRNA in target cells and effectively silenced target genes in various cell lines as well as in xenograft human cervical carcinoma mouse model, as compared with non-degradable PEI (Du et al. 2014). Sagnella et al. developed pH-responsive dextran nanoparticles by conjugating doxorubicin to the aldehyde groups through acid-labile imine bonds. The nanoparticles were successfully shown to penetrate 3D tumor spheroids. The interactions between dextran NPs and highly glycosylated tumor cell surface may be responsible for superior penetration of the NPs in tumor thus indicating potential of dextran-based nanocarriers pH-triggered drug release in solid tumors (Sagnella et al. 2013). Similarly, Liu et al. also constructed imine-based pH-responsive nanosystem by conjugating doxorubicin to the star-shaped polymer synthesized from poly(oligoethylene glycol) methyl ether acrylate in presence of both cross-linker and vinyl benzaldehyde. Particles of 20 nm obtained after self-assembly showed pH-responsive drug release in in vitro release experiments and showed cytotoxicity similar to free doxorubicin (Liu et al. 2012).

Similarly, Li et al. developed poly(ethylene glycol)-(lysine)-based micelles with doxorubicin conjugated via benzoic-imine bond linkage. The micelles quickly released doxorubicin in acidic conditions in in vitro release experiments and also exerted higher anticancer activity in MCF-7 cells (Li et al. 2016a, b). Wang et al. constructed poly[oligo(ethylene glycol) malicate]-based polymer and conjugated doxorubicin via pH-sensitive benzoic-imine linkage and achieved high doxorubicin loading content of about 38.2%. In vitro release studies showed rapid release of doxorubicin from the micelles at pH 5.8 as compared to that at pH 7.4 (Wang et al. 2014).

In another approach, Wang et al. designed pH-sensitive doxorubicin prodrug nanoparticles based on dextran-poly(ethylene imine) copolymers. Doxorubicin was conjugated to the polymer through acid cleavable cis-aconityl bonds. Cis-aconityl bonds confer pH sensitivity to the NPs which resulted in pH-dependent drug release in acidic condition and also showed higher anticancer efficacy against MCF-7 cells (Wang et al. 2015). Xu et al. prepared polyethylenimine (PEI)-based nanoparticles for simultaneous delivery of doxorubicin and Bcl2 siRNA to the lungs in lung cancer. Doxorubicin was conjugated to PEI through pH-sensitive cis-aconityl linkage. Both in vitro studies and in vivo studies showed enhanced antitumor efficacy as compared to single delivery of doxorubicin or Bcl2 siRNA (Xu et al. 2015). Kaneo et al. synthesized antitumor anthracycline conjugates with cis-aconityl-aconityl-daunorubicin and aconityl-doxorubicin both of which exhibited higher cytotoxic effect on the HL60 cell line as compared to acid-insensitive analogues and also inhibited growth of Sarcoma 180 cells inoculated subcutaneously in mice (Kaneo et al. 2013).

4 Temperature-Responsive Polymeric Nanoparticles

Temperature-responsive polymers have garnered great attention in pharmaceutical and biomedical engineering and are one of the most investigated stimuli-responsive delivery systems in cancer therapy. Temperature-responsive NPs have shown potential as smart materials for drug delivery applications owing to their phase-transition behavior in response to temperature change. In theory, thermosensitive nanocarriers should retain their drug payload at body temperature (~37 °C), while rapidly delivering the drug within a locally heated tumor (~40–42 °C). Diseased and healthy tissues show difference in their temperature. Also, local hyperthermia can be achieved by applying external stimulus such as using radiofrequency oscillators, annular-phased array microwave applicators or temperature-controlled water sacks (Shao et al. 2011; Mura et al. 2013).

Temperature-responsive polymers are characterized by critical solution temperature (CST) at which variation in the hydrophilic and hydrophobic interactions between the aqueous media and polymeric chains suddenly change within a short range of temperatures. This results in disruption of intramolecular and intermolecular hydrophobic and electrostatic interactions and causes polymer chain to collapse or expand to release the encapsulated drug as shown in Fig. 3. Typically, thermo-responsive polymer solutions possess lower critical solution temperature (LCST) below which polymer solutions appear as monophasic and above which phase separation occurs (biphasic) (Bae et al. 1990; Al-Ahmady and Kostarelos 2016).

Mao and other researchers developed temperature-responsive trimethyl chitosan-g-poly (*N*-isopropylacrylamide) (TMC-g-PNIPAAm) copolymer with LCST of 32 °C for gene delivery applications. By using a temperature variation protocol by incubating the cultured cells at 25 °C for a certain period followed by 37 °C incubation resulted in significant gene transfection (Mao et al. 2007). Choi et al. reported temperature-responsive Pluronic/PEI nanoparticles which exhibited swelling/de-swelling behavior over temperature range of 24–37 °C. As the temperature increased from 20 to 37 °C, nanocapsules size decreased from 330 to

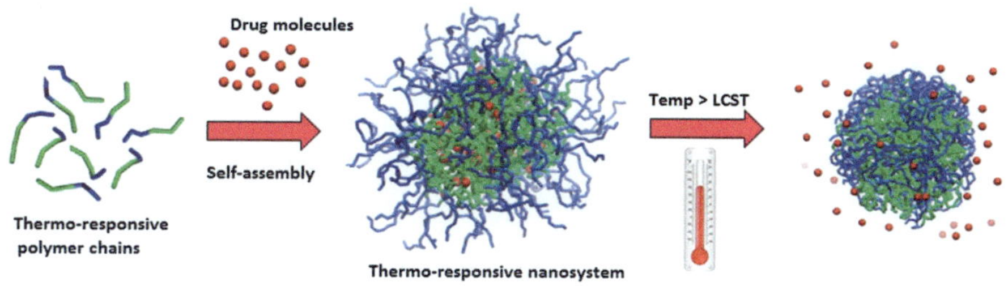

Fig. 3 Schematic representation of thermo-responsive polymeric nanoparticles. Figure adapted from Cheng et al. (2012) with modification

100 nm leading to free diffusion of the encapsulated drugs due to increased porosity (Choi et al. 2006). Similarly, Zhuo's research group synthesized thermosensitive polymeric NPs of poly(*N*,*N*-diethylacrylamide-co-acrylamide)-block-poly(g-benzyl L-glutamate) with LCST in the range of 37 and 42 °C for paclitaxel delivery. After inducing hyperthermia, the nanosystem exhibited similar anticancer activity as that of Taxol with low cytotoxicity (Yuanpei et al. 2009).

In case of polymeric nanocarriers, poly(*N*-isopropyl acrylamide) (PNIPAM) has been widely studied polymer for thermos-responsive drug delivery in cancer. Goodall et al. prepared *N*-isopropylacrylamide (NIPAM)-based thermo-responsive polymeric nanoparticles conjugated with scFv antibody targeting epidermal growth factor receptor (EGFR) commonly expressed in range of cancer types. They reported novel method to prepare nanoparticles of temperature-induced self-assembly of immunoparticles of antibody-conjugated polymer (Goodall et al. 2015). Similarly, Zeighamian et al. reported use of poly(*N*-isopropylacrylamide-co-methacrylic acid) (PNIPAAm–MAA) nanoparticles for curcumin delivery in MCF-7 breast cancer cells. Curcumin-loaded NIPAAm–MAA nanoparticles exhibited higher cytotoxicity on the MCF-7 cell line as compared to free curcumin (Zeighamian et al. 2015). Zhang et al. developed mesoporous silica-coated gold nanoparticles coated with thermos-responsive polymer shell made of poly(*N*-isopropylacrylamide-co-acrylic acid) for NIR-mediated theranostic nanosystem for simultaneous drug delivery, imaging, and hyperthermia therapy in cancer cell lines. NIR laser caused phase transformation of the polymer causing shrinking of the nanoparticles leading to localized drug release in the tumor in Balb/c nude mice (Zhang et al. 2014a, b).

Other polymeric materials such as poly(L-lactic acid)-poly(ethylene glycol)-poly (L-lactic acid) (PLLA-PEG-PLLA) triblock copolymers (Na et al. 2006), poly (ethylene oxide)-poly(propylene oxide)-poly (ethylene oxide) (PEO-PPO-PEO) copolymers (Niu et al. 2011), poly (*N*-vinylalkylamides)-poly(*N*-vinylcaprolactam) (PNVC) (Liu et al. 2014a, b), and poly(γ-2-(2-(2-methoxyethoxy)-ethoxy) ethoxy-ε-caprolactone)-b-poly(γ-octyloxy-ε-caprolactone) have also been studied for their thermo-responsiveness (Cheng et al. 2012).

Poly(trimethylene carbonate)-b-poly(L-glutamic acid) (PTMC-b-PGA) diblock copolymer is another thermo-responsive polymer. Temperature-dependent reversible melting of PTMC-b-PGA vesicles occurs with increase in temperature above the melting point of the PTMC block (34–35 °C), leading to the disruption of vesicular structure occurred (Sanson et al. 2010).

In general, the challenge in the design of thermo-responsive nanosystems lies in the sensitivity of the thermal response of the materials. For clinical applications, they should be sensitive to slight temperature changes within tolerable temperature limits above the physiological temperature of 37 °C.

5 Redox-Responsive Polymeric Nanoparticles

The difference in redox potential of intracellular and extracellular environment and also between normal healthy cells and tumor cells can be exploited for drug delivery in cancer. This significant difference in redox environment has led to development of redox-responsive nanosystems which have recently gained fame in cancer therapy. It uses redox potential as a stimulus for intracellular drug delivery in cancer cells (McCarley 2012; Crucho 2014).

Glutathione (GSH) and glutathione disulfide (GSSG) forms one of the major redox couple in the cells and plays critical role in intracellular redox homeostasis. GSH is a tripeptide which is responsible for reduction of disulfide linkages. GSH is a major antioxidant in mammalian cells which provides protection against oxidative stress. It is also involved in phase II detoxification mechanism of the cells by conjugating with endogenous as well as exogenous toxins in presence of glutathione transferases. GSH conjugates can then be eliminated via an ATP-dependent glutathione-X (GS-X) pump (Balendiran et al. 2004). Cancer cells express about three- to fourfold higher GSH levels with respect to normal healthy tissues as a defense mechanism in order to detoxify the free radicals generated during cellular metabolism as well as to eliminate chemotherapeutic drugs. This results in resistance of cancer cells to the radiotherapy and chemotherapy (Balendiran et al. 2004; McCarley 2012).

Endolysosomal compartment is also redox active with high concentration of reducing enzymes such as enzyme gamma interferon inducible lysosomal thiol reductase (GILT). Intracellular compartment of cell is much more reductive than the extracellular matrix with presence of 100–1000 times higher GSH concentration in cell cytoplasm than that in the extracellular matrix. Physiological body fluids–blood and extracellular matrices have low GSH concentration (2–20 μM), whereas cell cytosol exhibits high GSH concentration (2–10 mM). This significant difference in GSH level in intracellular and extracellular compartment makes redox-responsive nanosystem promising drug delivery platform in cancer.

Redox-responsive nanosystems offer several advantages such as stability in the extracellular environment and high sensitivity to intracellular GSH levels leading to faster drug release in cancer cells. Nanosystem can be tailor-made to be redox responsive by incorporating disulfide linkages in the hydrophobic backbone of amphiphilic copolymers, since they are cleavable in reductive environment. Redox-responsive systems can also use dendrimer-drug conjugates containing thiol-cleavable bonds, capped mesoporous materials, or disulfide cross-linked nanogels. The disulfide linkage has been widely used for rendering redox sensitivity to polymeric nanosystem in cancer therapy. Disulfide linkages can be incorporated in the drug delivery systems using oxidation of thiol groups, use of disulfide-containing cross-linkers, and by exchange reaction of disulfide and thiols (Cheng et al. 2014; Torchilin 2014).

Researchers have shown improved anticancer efficacy of chemotherapeutic drugs with the help of redox-responsive nanocarriers. Han et al. prepared nanoparticles of doxorubicin-loaded amphiphilic hyaluronic acid-polycaprolactone copolymer with redox-responsive disulfide linkage and showed enhanced drug release in reductive environment of 10 mM GSH while achieving improved anti-tumor activity in vitro in SCC7 cells and in vivo in SCC7 tumor-bearing mice (Han et al. 2015). Similarly, Zhong et al. designed galactose-targeted redox-responsive micelles from poly(ethylene glycol)-s-s-poly(ε-caprolactone) (PEG-s-s-PCL) copolymer for targeting hepatocellular carcinoma cells. They observed enhanced antitumor activity of doxorubicin in HeLa and HepG2 cells in MTT assay with redox-responsive micelles as compared to redox-insensitive and non-targeted micelles indicating superior therapeutic efficacy of redox sensitive micelles (Zhong et al. 2013).

Our group at IIT Delhi has worked extensively with disulfide (-s-s-) linkages and demonstrated higher drug release as well as tumor regression in Swiss albino EAT model. As shown in Fig. 4, Lale et al. developed redox-responsive triblock copolymeric polymersomes consisting of poly(ethylene glycol)-polylactic acid-poly (ethylene glycol) [PEG-s-s-PLA-s-s-PLA-s-s-PEG] with multiple disulfide linkages. Both in vitro studies in breast cancer cell lines—BT-474 and MCF-7 cells and in vivo studies in Ehrlich ascites tumor (EAT)-bearing Swiss albino mice showed superior antitumor efficacy and lower cardiotoxicity of the redox-responsive polymersomes as compared to free doxorubicin (Lale et al. 2015). Similarly, Kumar et al. reported nanosystem consisting of poly(polyethylene glycol methacrylate)-poly(caprolactone)-s-s-poly(caprolactone)-poly(polyethylene glycol methacrylate) (pPEGMA-PCL-ss-PCL-pPEGMA). It was observed that higher drug release was observed in presence of redox conditions of 10 mM GSH with better tumor regression in Swiss albino model as compared to control (Kumar et al. 2015).

In another approach, Zhuang et al. designed aptamer-functionalized redox-responsive polymer—poly(2-((2-(acryloyloxy)ethyl)disulfanyl)ethyl 4-cyano-4-(((propylthio)carbonothioyl)-thio)-pentanoate-co-poly(ethylene glycol) methacrylate) (HPAEG). Doxorubicin-loaded hyperbranched polymeric nanoparticles exhibited higher cytotoxicity in cancer cells while exhibiting low cytotoxicity to normal cells due to higher glutathione present in the cancer cells (Zhuang et al. 2016). Similarly, Chen et al. demonstrated redox responsivity of PEGylated hyperbranched polyphosphoester (PPE) nanoparticles encapsulated with doxorubicin. NPs showed redox-responsive drug release behavior with enhanced intracellular drug release in human breast cancer cell line MDA-MB-231 (Chen et al. 2015).

Redox polymers are also very useful for gene delivery since complexation of nucleic acids with a redox polymers such as poly(disulfide amine) or

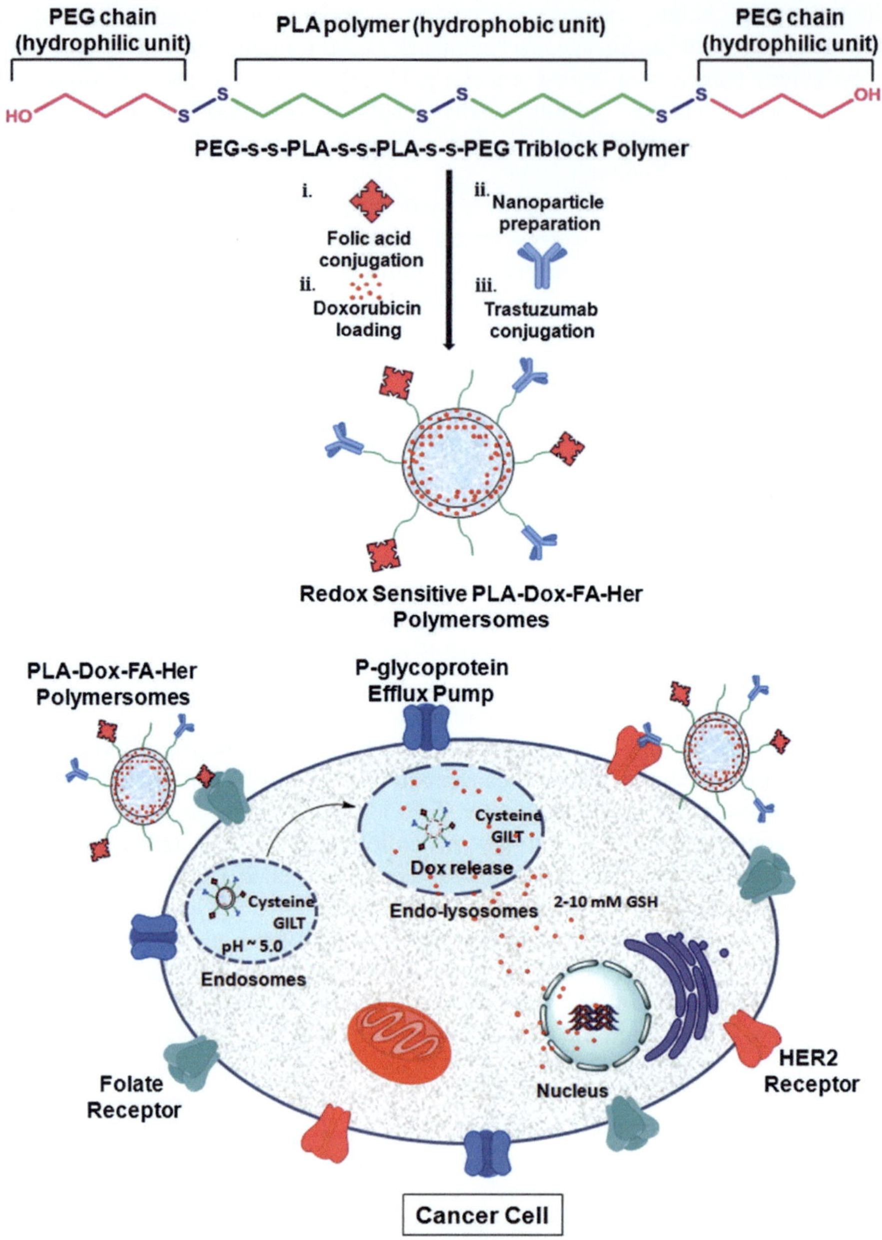

Fig. 4 Schematic illustration of redox-responsive PEG-s-s-PLA-s-s-PLA-s-s-PEG nanocarriers for doxorubicin delivery in breast cancer. Figure adapted from Lale et al. (2015)

disulfide-containing poly(amido amine) can improve transfection by quick disassembly of the complex under reductive intracellular conditions. Feng et al. designed redox-responsive polymeric vesicle from polyethyleneimine-g-poly(disulfide amine)-g-polyethyleneimine (PEIg-PSSA-g-PEI) for gene delivery. They observed higher transfection efficacy and lower cytotoxicity on COS-7 cells as compared to polyethyleneimine and Lipofectamine (Feng et al. 2015). In another approach, Manickam et al. designed redox sensitive polypeptides containing disulfide bonds by an oxidative copolymerization of histidine-rich peptide and nuclear localization sequence peptide. It was observed that the polypeptides showed higher transfection activity while achieving low cytotoxicity as compared to control PEI polyplexes (Manickam and Oupicky 2006). Li et al. demonstrated efficacy of redox-responsive cationic brush block copolymers—based gene delivery nanocarriers. Under reductive conditions, pDNA was able to be released due to breaking of disulfide linkages present in the block copolymer. Also, the nanosystem exhibited low cytotoxicity while achieving high gene transfection efficiency (Li et al. 2014a, b). Alternatively, Suma et al. developed smart multi-layered assembly consisting of siRNA-loaded core, an endosomolytic polycation, core stabilizing silica and detachable PEG shell. The detachable PEG gets shedded in response to a reducing environment which resulted in enhanced siRNA delivery, leading to significant in vivo gene silencing (Suma et al. 2012).

6 Magnetic-Responsive Polymeric Nanoparticles

Magnetic-responsive polymeric nanoparticles are generally composed of nanocarriers with magnetite (Fe_3O_4) or maghemite (γ-Fe_2O_3) core. They are called as superparamagnetic iron oxide nanoparticles (SPIONs) and are widely used in variety of biomedical applications such as drug delivery, magnetic resonance imaging (MRI), and magnetic hyperthermia. Magnetic field can be applied at intended therapeutic site driving magnetic nanocarriers to accumulate at the site resulting in higher local concentration of drug-loaded magnetic nanocarriers at site of action. Improved drug accumulation in solid tumors can be obtained by use of magnetic guidance by focusing magnetic field on the tumor after injection of magnetically responsive nanocarrier. Magnetically, responsive systems allow many variations in the drug delivery mechanisms since they can be magnetically guided under a permanent magnetic field or can cause increase in temperature when an alternating magnetic field is applied. Also, due to magnetic resonance imaging property, it is characterized by theranostics, i.e., therapy and diagnostics within a single system (Hervault and Thanh 2014; Yildiz and Yildiz 2015).

Functionalization of magnetic nanoparticles with biocompatible polymers is necessary for improving their stability and biocompatibility. Surface polymer coating not only improves stability of the magnetic nanoparticles, but also enables their functionalization for various applications. Basuki et al. developed α-D-mannose-functionalized diblock PEG-glycopolymer-coated magnetic nanoparticles for lung cancer targeting. They successfully demonstrated diagnostic application of the magnetic nanosystem by quantifying amount of lectin binding on the nanoparticle surfaces using MRI and also achieved improved cellular uptake in the lung cancer cells (Basuki et al. 2014). Arias et al. prepared chemotherapeutic drug ftorafur-loaded magnetic nanoparticles consisting of iron oxide/chitosan through a coacervation method and showed improved therapeutic activity of the formulation (Arias et al. 2011). Similarly, Licciardi et al. reported doxorubicin-loaded PEGylated squalene grafted inulin copolymer-coated SPIONs for targeted cancer therapy. They demonstrated improved therapeutic efficacy of nano-formulation due to the magnetic targeting ability. They also showed application of external magnetic field as uptake promoter in in vitro drug uptake studies as well as cytotoxicity studies in HCT116 cancer cell line (Scialabba et al. 2014).

Drugs can be covalently linked with magnetic nanocarriers to improve low drug loading and to avoid uncontrolled burst release. Mu et al. developed anticancer drug gemcitabine- and glioblastoma-targeting peptide chlorotoxin-conjugated iron oxide nanoparticles for glioblastoma therapy. The nanoparticles were observed to enter cancer cells without losing potency as compared to free drug and showed crossing of the blood–brain barrier in wild-type mice (Mu et al. 2016). Daglioglu et al. reported 5-aminoimidazole-4-carboxamide-1-β-D-ribofuranoside (AICAR) and doxorubicin conjugated iron oxide multifunctional nanoparticles to neutralize survivin (BIRC5) in order to overcome doxorubicin chemoresistance. In vitro studies in five different tumor-derived cell lines (A549, HCT-116, HeLa, Jurkat, and MIA PaCa-2) showed efficient reduction of cancer cell viability (Daglioglu and Okutucu 2016). Jiang et al. prepared methotrexate-conjugated Fe_3O_4/polypyrrole (PPy) nanospheres and showed controlled drug release and enhanced intracellular drug retention in HeLa cells (Jiang et al. 2015).

Magnetically guided nanocarriers are also helpful for delivery of nucleic acids such as siRNA and genes. Under a permanent magnetic field, nano-assemblies containing cationic coatings to condense nucleic acids and iron oxide nanoparticles results in higher transfection efficiencies. Lee et al. prepared cationic PDMA-block-poly(ε-caprolactone) (PDMA-b-PCL) micelles consisting of SN-38 (7-ethyl-10-hydroxycamptothecin), and human vascular endothelial growth factor (VEGF) targeting small interfering RNA (siRNA) along with superparamagnetic iron oxide nanoparticles. The micelleplexes efficiently suppressed tumor growth and also acted as magnetic resonance imaging (MRI) contrast agent, thus helping in diagnosis (Lee et al. 2016). Li et al. reported polyetherimide-grafted Fe_3O_4–SiO_2 nanoparticles for VEGF siRNA delivery along with MRI cell imaging. The nano-assemblies exhibited higher inhibition of VEGF gene expression in MCF-7 cell line. Also, superparamagnetic iron oxide core of the nanocomposites acted as a contrast agent for cancer magnetic resonance imaging (Li et al. 2015a, b).

In another approach, cell-based gene therapies such as human monocytes (which migrate from bloodstream into tumors naturally) loaded with magnetic nanoparticles and transfected with therapeutic genes under the influence of magnetic field resulted in significant increase in the monocytes delivery in the tumor (Muthana et al. 2008). Similarly, Li et al. reported human-induced pluripotent stem (iPS) cells with fluorescent-labeled magnetic nanoparticles for simultaneous imaging and hyperthermia in gastric cancer therapy. It was observed in in vivo studies that fluorescent magnetic nanoparticles labeled human iPS cells targeted gastric cancer cells and inhibited cancer growth in vivo through the external magnetic field (Li et al. 2015a, b).

Magnetic nanoparticles generate heat in presence of alternating magnetic field (AMF). Hence, they have been widely used for selective heating of tumors (hyperthermia). Thermosensitive polymer-coated magnetic nanoparticles can trigger drug release in response to a magnetically induced heating. Active targeting by means of nanocarrier functionalization along with hyperthermia can achieve synergistic cytotoxic effect in tumor cells (Estelrich et al. 2015). Sadhukha et al. developed EGFR targeted, inhalable SPIONs for targeted hyperthermia in lung cancer. It was observed that magnetic hyperthermia treatment using targeted SPIONs resulted in significant inhibition of lung tumor growth in vivo (Sadhukha et al. 2013). Hu et al. reported core/shell nanocapsules for dual drug delivery with hydrophilic doxorubicin in the core and hydrophobic paclitaxel in the shell in polyvinyl alcohol polymeric shell stabilized by iron oxide nanoparticles. In alternating magnetic field, the generated heat caused triggered release of both drugs. Thus, the combined magneto-hyperthermia and magneto-chemotherapy showed improved therapeutic efficacy in in vitro as well as in vivo experiments (Hu et al. 2012a, b). Similarly, Kuo et al. reported magnetic polymeric nanocarriers for targeted delivery of curcumin and hyperthermia treatments in cancer cells. Amine group-immobilized iron oxides, Fe_3O_4–NH_2, were attached on poly[(acrylic acid)-block-(*N*-isopropylacrylamide)-block-(acrylic acid)] triblock copolymeric self-assembly containing encapsulated curcumin. In vitro uptake experiments of the nanosystem in MCF-7 cell line showed higher cytotoxicity to MCF-7 cells (Kuo et al. 2016).

The heat generated by an AMF can also be used to accomplish pulsatile drug release. Hoare et al. synthesized magnetically triggered composite membranes containing thermo-responsive PNIPAM-based nanogels. The magnetic nanoparticles enabled pulsatile on–off drug delivery due to de-swelling or swelling of the polymer (Hoare et al. 2011). As shown in Fig. 5, Campbell et al. demonstrated enhanced pulsatile drug release from hydrogels containing SPIONs together with thermo-responsive microgels. Pulsatile release of 4 kDa FITC-dextran was achieved after application of AMF with enhanced release immediately after application of AMF, followed by a rapid return to a baseline release rate when AMF was shut off (Campbell et al. 2015).

Similarly, in work carried by Hyun, pulsatile drug release of doxorubicin was achieved from magnetic poly(ε-caprolactone) (PCL) nanoparticles containing hollow interiors with the help of externally applied magnetic field. Efficient intranuclear release of doxorubicin was achieved from the nanoparticles resulting in significantly improved anticancer activity along with the hyperthermia effect (Hyun 2015).

The use of magnetic nanoparticles for magnetic guidance or local hyperthermia is normally limited to localized tumors but is ineffective against metastatic tumors.

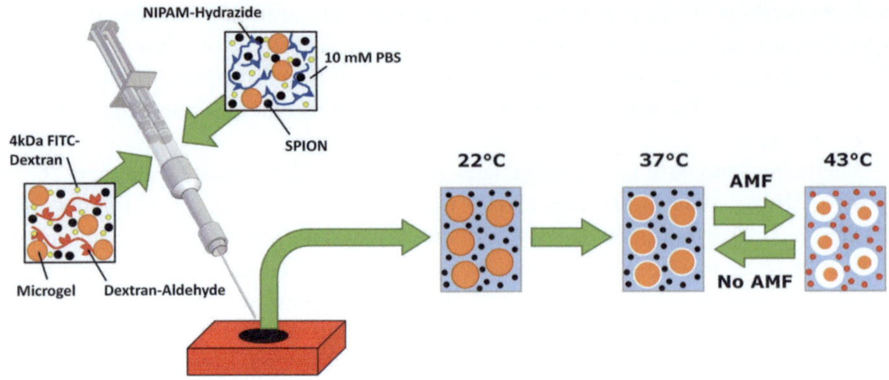

Fig. 5 Schematic representation of AMF controlled drug release from magnetic hydrogels containing thermosensitive microgel reported by Campbell et al. Figure adapted from Campbell et al. (2015)

In cancer cases where surgery cannot be performed because they are localized in tissues with high risk of healthy tissue injury (e.g., brain cancers), magnetic nanoparticles are promising therapeutic option. However, major limitation of magnetic nanosystem is the complexity involved in the setup of magnetic fields and need for deep penetration into the tumor tissues.

7 Light-Responsive Polymeric Nanoparticles

Light-responsive delivery system allows precise and targeted drug release at intended site of action. Photo-responsive chemical groups in the polymeric nanocarriers are activated by the light source which causes conformational changes in the polymeric structure causing triggered drug release at the site of action. Most photo-responsive polymers contain light-sensitive chromophores such as azobenzene groups, nitrobenzyl groups, spiropyran groups. Ultraviolet–visible reversible photoisomerization of azobenzenes from trans to cis on UV irradiation and from cis to trans by visible light enables photoregulated control of drug release. Light-responsive nanocarriers can respond to ultraviolet (UV), visible and infrared radiations depending on the photosensitivity of the polymeric material (Klinger 2013; Claudia et al. 2016).

The major drawback of light-responsive drug delivery is their low penetration depth ($\sim$10 mm) due to strong scattering properties of soft tissues in the UV region. Near-infrared (NIR) radiations penetrate more deeply in the tissues with less harmful effects on the skin and hence is widely used for tumor targeting. NIR laser (700–1000 nm range) has been extremely promising with deeper tissue penetration and minimal harm to tissues, with the help of photosensitive groups that respond to higher wavelengths. Also, NIR-absorbing plasmonic materials can convert adsorbed photon energy into heat to trigger drug release from NIR-responsive nanocarriers (Cho et al. 2015; Liu et al. 2016; Wu and Butt 2016).

Drug release from NIR-responsive nanocarriers can be accomplished through disassembly and disruption of polymeric nanocarriers as shown in Fig. 6.

You et al. demonstrated NIR-mediated enhanced drug release from doxorubicin-loaded hollow gold nanospheres and achieved reduced systemic toxicity as compared to the free doxorubicin (You et al. 2014). Noh et al. developed doxorubicin-loaded Au/Ag hollow nanoshells for photothermal drug delivery under near-infrared (NIR) light for lung cancer treatment. Following irradiation with NIR laser, EGFR-targeted Au/Ag hollow nanoshells caused photothermal ablation of lung cancer cells and also resulted in enhanced doxorubicin release from the nanoshells (Noh et al. 2015). In another approach, Yang et al. designed multifunctional anticancer nanoplatforms based on mesoporous silica-coated core–shell NPs loaded with doxorubicin and conjugated with copper sulfide nanoparticles as a photothermal agent. Synergistic effect of both photothermal therapy and chemotherapy triggered simultaneously by near-infrared (NIR) light source was observed with significant improvement in the therapeutic efficacy as evidenced in both in vitro and in vivo studies (Yang et al. 2015). As shown in Fig. 7, Fomina et al. developed a light-responsive polymeric nanosystem which exhibited photo-responsive release behavior after NIR irradiation due to the presence of photo-cleavable 4-bromo-7-hydroxycoumarin. Photolysis of multiple 4-bromo-7-hydroxycoumarin groups after NIR radiation triggers cascade of rearrangement reactions leading to the degradation of the polymer backbone. This results in achieving drug release after irradiation of biologically small levels of NIR light irradiation (Fomina et al. 2011).

Some light-responsive drug delivery systems are irreversible where the conformational change in the polymer structure is permanent causing release of the entire dose of the payload drug at site of action. Some light-responsive drug delivery systems undergo reversible conformational change in polymer structure by changing light/dark cycles and thus behave as multi-switchable carriers resulting in controlled pulsatile drug release (Alvarez-Lorenzo et al. 2009). Kim et al. synthesized photo-responsive nanocarriers based on o-nitrobenzyl PEI and

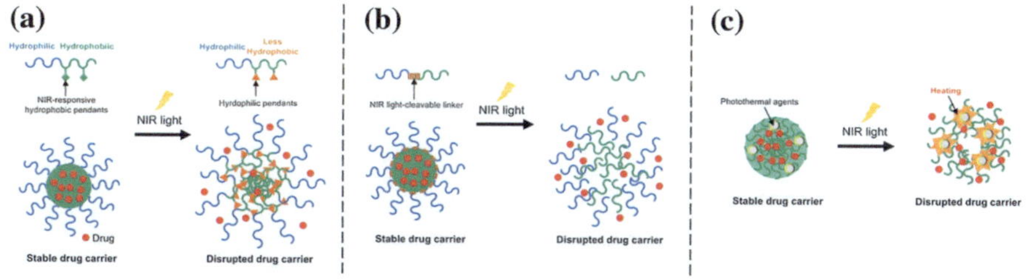

Fig. 6 Schematic representation mechanisms for NIR-responsive drug release. **a** Photoinduced hydrophobic-hydrophilic transition, **b** cleavage of photosensitive linker, and **c** photoinduced heating. Figure adapted from Cho et al. (2015)

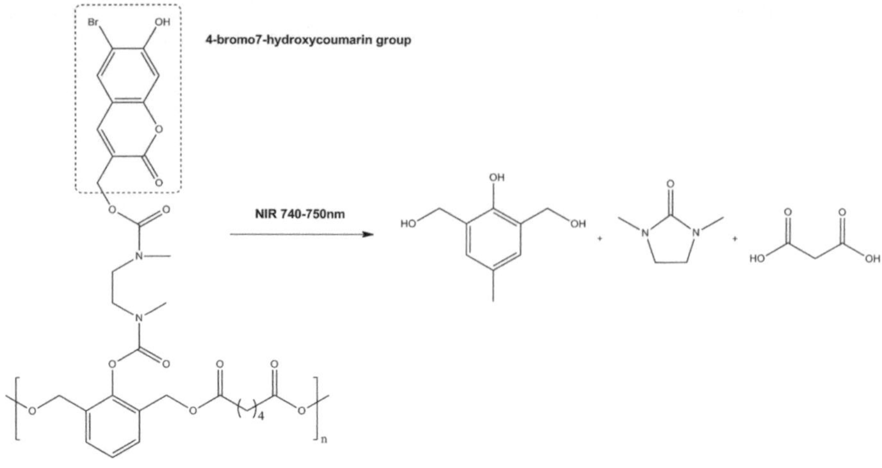

Fig. 7 Degradation mechanism of NIR-responsive polymeric nanosystem containing photo-cleavable group-4-bromo-7-hydroxycoumarin. Image adapted from Fomina et al. (2011) with modification

methacrylate-functionalized PEI cross-linked together for delivery of condensed plasmid DNA. Following exposure to photo-irradiation, cross-linked nanosystem showed threefold increase in gene expression in cells as compared to uncross-linked nanosystem (Kim et al. 2010). Li et al. developed light responsive nanoparticles using a photolabile linker for siRNA delivery. After UV exposure, the photolabile linkage cleavage occurred which led to rapid shell detachment of the nanoparticles, facilitating siRNA release and hence improved gene-silencing efficiency (Li et al. 2014a, b). Cui et al. prepared nanocarriers formed by ionic self-assembly of chitosan and photosensitive 4-oxo-4-(pyren-4-ylmethoxy)butanoic acid which showed photo-responsive drug release under UV or NIR radiations (Wei et al. 2011). Lv et al. designed tagalsin G, a marine drug candidate, encapsulated photolabile nanocarriers based on photodegradable polyurethanes and successfully achieved burst release of drug in the RAW 264.7 cells after photo-activation (Lv et al. 2012).

Light-triggered nanosystems show promising results, but the safety and biocompatibility of light-responsive nanoparticles such as azobenzene, o-nitro benzyl derivatives, spiropyrans is uncertain and questionable. Searching for new non-toxic and biocompatible photosensitive materials is vital for clinical translation of these nanosystems.

8 Ultrasound-Responsive Polymeric Nanoparticles

Ultrasound-responsive polymeric nanoparticles use noninvasive ultrasonic or acoustic waves which can act as a drug delivery trigger by two primary mechanisms—heat and pressure. Ultrasound waves can cause drug release from nanocarriers by cavitation phenomenon through thermal and mechanical effects. Cavitation can destabilize nanocarriers by increasing its permeability leading to drug release. Application of long-pulsed ultrasonic waves to the temperature-responsive nanosystems can be used to trigger drug release by heat. However, application of short pulse signals results in pressure-mediated drug release without generation of heat. Ultrasound-triggered drug delivery is promising because of the absence of ionizing radiations, its noninvasiveness, and easy regulation of tissue penetration depth by changing tuning frequency (Sirsi and Borden 2014; Zhou et al. 2014).

Celsion Corporation has combined high-intensity-focused ultrasound with ThermoDox (doxorubicin-loaded thermosensitive liposomes) for ultrasound-responsive nanosystem and is under clinical trials III for hepatocellular carcinoma and phase II for breast cancer and liver metastases (Celsion 2016a, b). Most of the developed ultrasound-responsive-based systems are lipid based. There are very few polymeric ultrasound-responsive nanoparticles that are reported.

Paris et al. developed polymer grafted mesoporous silica nanoparticles as ultrasound-based drug delivery system. The nanoparticles can be loaded with a cargo at low temperature (4 °C), at which polymer is present in open conformation. At 37 °C, the polymer collapses closing the pore entrances and allowing the nanoparticles to carry the drugs at physiological temperature without premature release. Upon ultrasound irradiation, the sensitive polymer changes its hydrophobicity resulting in conformation change causing release of the drug payload. Authors showed selective doxorubicin release when incubated with LNCaP cells after exposure to ultrasound (Paris et al. 2015). In another study, Rapoport et al. reported development of micelles of poly(ethylene oxide)-co-poly(L-lactide) (PEG-PLLA) block copolymer with echogenic perfluoropentane (PFP) nanoemulsions for ovarian, breast, and pancreatic cancers. In tumor-directed ultrasound, paclitaxel-loaded said nanoemulsions were evaluated in three human tumor xenografts in nu/nu mice (MDA MB231 breast cancer, A2780 ovarian carcinoma, and MiaPaCa-2 pancreatic cancer) (Rapoport et al. 2009).

In another study, Husseini et al. showed that lower ultrasound frequencies ($\sim$20 kHz) resulted in higher doxorubicin release and with increase in frequency up to 90 kHz resulted in slower drug release (Husseini et al. 2005). Same research group also showed that drug release after ultrasound exposure is a reversible process after observing rapid doxorubicin reuptake into the polymer micelles after ultrasound exposure (Husseini et al. 2013) (Fig. 8).

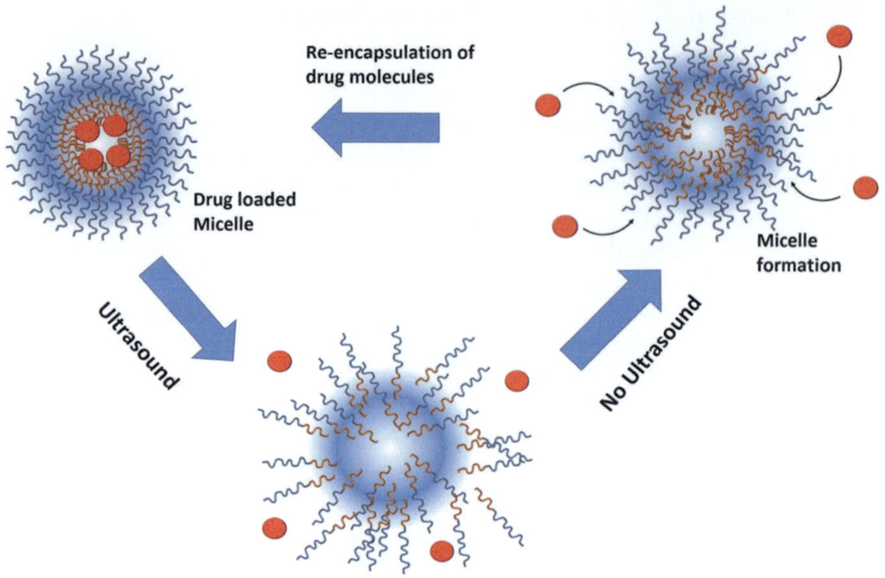

Fig. 8 Schematic illustration of reversible drug release from micelles after exposure to ultrasound. Figure adapted from Sirsi and Borden (2014) with modification

9 Enzyme-Responsive Polymeric Nanoparticles

Enzymes associated with diseased condition or occurring at higher concentration at a particular pathological tissue can be utilized as a trigger for drug release using enzyme-responsive nanocarriers. Enzyme-responsive systems take advantage of the fact that certain enzymes such as proteases, matrix metalloproteinases, phospholipases, glycosidases are up-regulated in cancer, which can be exploited for enzyme-mediated drug release at the desired site (de la Rica et al. 2012; Hu et al. 2012a, b; Kuang et al. 2016) (Fig. 9).

In a study carried by Zhu et al., they studied effect of short peptide sequences cleavable by matrix metalloproteinases (MMPs), as linkers between TAT-functionalized liposomes and PEG chains. After cleavage of the peptide by MMPs, exposed surface bioactive ligands resulted in higher intracellular penetration as compared to nanocarriers without cleavable linkers. This resulted in improved gene-silencing activity of NPs in tumor-bearing mice (Zhu et al. 2012). Similarly, Callmann et al. developed matrix metalloproteinases (MMPs) responsive, paclitaxel-loaded nanoparticles and showed its antitumor efficacy in in vivo in a human HT-1080 fibrosarcoma murine xenograft model (Callmann et al. 2015). Wen et al. developed plasmin (serine proteases) and MMP-responsive nanocarriers for triggered controlled release of bovine serum albumin (BSA) and VEGF from the nanosystem using proteases (Wen et al. 2011).

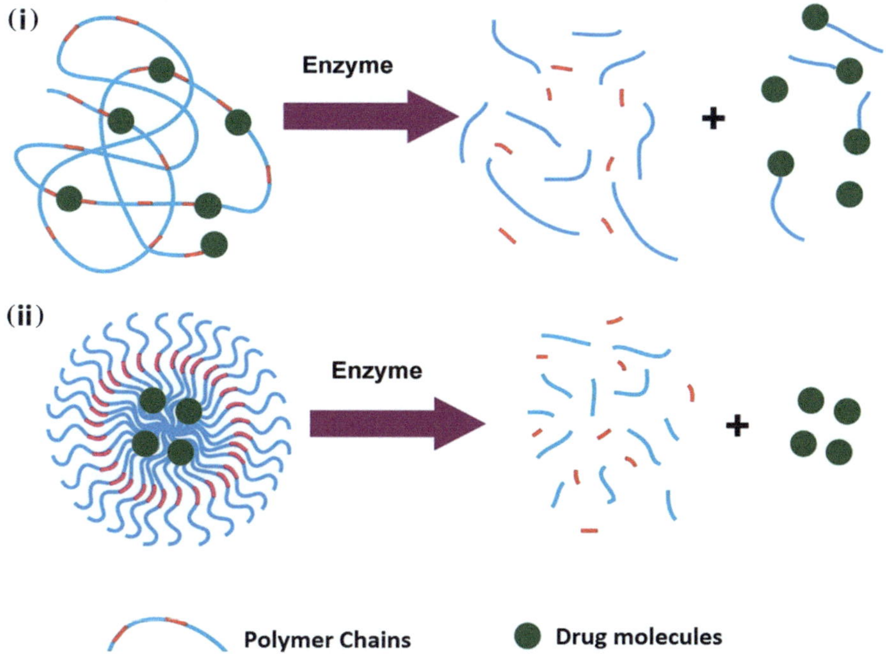

Fig. 9 Schematic representation of enzyme-responsive polymeric nanoparticles. (i) Polymer-drug conjugate system using enzyme-cleavable linker, (ii) drug-encapsulated polymeric system. Figure adapted from de la Rica et al. (2012)

In another approach, Lee et al. showed delivery of drugs to intracellular compartments using enzymes. Lee et al. developed lysosomally cleavable peptide containing polymersomes-methoxy poly(ethylene glycol) (mPEG) and poly(D,L-lactide) (PDLLA) block copolymer with Gly-Phe-Leu-Gly-Phe peptide as linker between two blocks. They showed that lysosomal enzyme cathepsin B, which is often overexpressed in several malignant tumors, helped in payload release due to fast enzymatic degradation of peptide linker of the polymersomes (Lee et al. 2011). In similar study by another group, Lee et al. synthesized methoxy poly(ethylene glycol) (MPEG)-doxorubicin conjugates using a cathepsin B-cleavable peptide (Gly-Phe-Leu-Gly peptide). The conjugate showed improved antitumor activity in vivo in CT26 tumor xenograft model as compared to the control without Gly-Phe-Leu-Gly peptide (Lee et al. 2015).

These representative examples show potential of enzyme-responsive drug delivery systems in cancer therapy. However, it is necessary to determine concentration of target enzymes at the tumor site to control in vivo drug release.

10 Conclusion and Future Perspective

Polymeric nanosystems will continue to play an important role in targeted cancer chemotherapy. Combination of polymeric nanoparticles and a stimulus selective to tumor tissue exhibits synergistic therapeutic effects. The nanocarrier design responsive to exogenous/endogenous stimuli has attractive potential for drug delivery in cancer. Multiple stimuli and the diversity of responsive materials allow greater flexibility in the designing of stimuli-responsive nanosystems. Tumor targeting using stimuli-responsive polymeric nanoparticles increases the therapeutic efficacy of the drug and decreases its toxic side effects by selective targeting of cancer cells and releasing the drug locally in the cancer tissue.

Although many novel stimuli-responsive polymer nanoparticles are being developed, most are just proof-of-concept studies which are not suitable for further clinical development. In case of most of these systems, the complexity of their design, insufficient biocompatibility, and their scale-up difficulties are the current problems which slow down their translation from the bench to the bedside. This could be overcome by further optimization and simplification of the techniques suitable for biological application. Also, cost-effectiveness and regulatory approval are a must for the formulation to be effective as a nanomedicine. Intrinsic stimuli-responsive nanoparticles (pH, temperature, redox, and enzyme stimuli) need to exhibit sharp property changes with respect to small changes between normal and pathological tissues. External stimuli-responsive NPs (magnetic, light, and ultrasound) need to be more responsive toward the external stimuli for better therapeutic effectiveness.

Simple and biocompatible stimuli-responsive polymeric nanoparticles have a broad potential in the cancer therapy and in the near future, they will definitely replace conventional chemotherapy to become primary therapeutic regimen in cancer therapeutics.

References

Al-Ahmady Z, Kostarelos K (2016) Chemical components for the design of temperature-responsive vesicles as cancer therapeutics. Chem Rev 116:3883–3918

Alvarez-Lorenzo C, Bromberg L, Concheiro A (2009) Light-sensitive intelligent drug delivery systems. Photochem Photobiol 85:848–860

Arias J, Lopez-Viota M, Saez-Fernandez E, Ruiz MA, Delgado AV (2011) Engineering of an antitumor (core/shell) magnetic nanoformulation based on the chemotherapy agent ftorafur. Colloids Surf A 384:157–163

Bae YH, Okano T, Kim SW (1990) Temperature dependence of swelling of crosslinked poly(*N,N′*-alkyl substituted acrylamides) in water. J Polym Sci Part B Polym Phys 28:923–936

Balendiran GK, Dabur R, Fraser D (2004) The role of glutathione in cancer. Cell Biochem Funct 22:343–352

Basuki JS, Esser L, Duong HTT, Zhang Q, Wilson P, Whittaker MR, Haddleton DM, Boyer C, Davis TP (2014) Magnetic nanoparticles with diblock glycopolymer shells give lectin concentration-dependent MRI signals and selective cell uptake. Chem Sci 5:715–726

Bertrand N, Wu J, Xu X, Kamaly N, Farokhzad OC (2014) Cancer nanotechnology: the impact of passive and active targeting in the era of modern cancer biology. Adv Drug Deliv Rev 66:2–25

Callmann CE, Barback CV, Thompson MP, Hall DJ, Mattrey RF, Gianneschi NC (2015) Therapeutic enzyme-responsive nanoparticles for targeted delivery and accumulation in tumors. Adv Mater 27:4611–4615

Campbell S, Maitland D, Hoare T (2015) Enhanced pulsatile drug release from injectable magnetic hydrogels with embedded thermosensitive microgels. ACS Macro Lett 4:312–316

Celsion (2016a) Doxorubicin liposomal. Celsion Corporation. From http://adisinsight.springer.com/drugs/800018968

Celsion (2016b) Phase 3 study of ThermoDox with radiofrequency ablation (RFA) in treatment of hepatocellular carcinoma (HCC). From https://clinicaltrials.gov/ct2/show/NCT00617981

Chen C, Zheng P, Cao Z, Ma Y, Li J, Qian H, Tao W, Yang X (2015) PEGylated hyperbranched polyphosphoester based nanocarriers for redox-responsive delivery of doxorubicin. Biomater Sci 4:412–417

Cheng J, Pun SH (2015) Polymeric biomaterials for cancer nanotechnology. Biomater Sci 3:891–893

Cheng Y, Hao J, Lee LA, Biewer MC, Wang Q, Stefan MC (2012) Thermally controlled release of anticancer drug from self-assembled gamma-substituted amphiphilic poly(epsilon-caprolactone) micellar nanoparticles. Biomacromolecules 13:2163–2173

Cheng W, Gu L, Ren W, Liu Y (2014) Stimuli-responsive polymers for anti-cancer drug delivery. Mater Sci Eng C 45:600–608

Cho HJ, Chung M, Shim MS (2015) Engineered photo-responsive materials for near-infrared-triggered drug delivery. J Ind Eng Chem 31:15–25

Choi SH, Lee SH, Park TG (2006) Temperature-sensitive pluronic/poly(ethylenimine) nanocapsules for thermally triggered disruption of intracellular endosomal compartment. Biomacromolecules 7:1864–1870

Choi SYC, Collins CC, Gout PW, Wang Y (2013) Cancer-generated lactic acid: a regulatory, immunosuppressive metabolite? J Pathol 230:350–355

Claudia C, Sara M, Diogo Silva P, Agnese M, Francesca U, Fabiana Q (2016) Polymeric nanoparticles for cancer photodynamic therapy. In: Sortino S (ed) Light-responsive nanostructured systems for applications in nanomedicine, vol 370. Springer International Publishing, pp 61–112

Crucho CIC (2014) Stimuli-responsive polymeric nanoparticles for nanomedicine. ChemMedChem 10:24–38

Daglioglu C, Okutucu B (2016) Synthesis and characterization of AICAR and DOX conjugated multifunctional nanoparticles as a platform for synergistic inhibition of cancer cell growth. Bioconjug Chem 27:1098–1111

de la Rica R, Aili D, Stevens MM (2012) Enzyme-responsive nanoparticles for drug release and diagnostics. Adv Drug Deliv Rev 64:967–978

Deng Z, Zhen Z, Hu X, Wu S, Xu Z, Chu PK (2011) Hollow chitosan-silica nanospheres as pH-sensitive targeted delivery carriers in breast cancer therapy. Biomaterials 32:4976–4986

Du Z, Xiang S, Zang Y, Zhou Y, Wang C, Tang H, Jin T, Zhang X (2014) Polyspermine imine, a pH responsive polycationic siRNA carrier degradable to endogenous metabolites. Mol Pharm 11:3300–3306

Duong HTT, Hughes F, Sagnella S, Kavallaris M, Macmillan A, Whan R, Hook J, Davis TP, Boyer C (2012) Functionalizing biodegradable dextran scaffolds using living radical polymerization: new versatile nanoparticles for the delivery of therapeutic molecules. Mol Pharm 9:3046–3061

Estelrich J, Escribano E, Queralt J, Busquets MA (2015) Iron oxide nanoparticles for magnetically-guided and magnetically-responsive drug delivery. Int J Mol Sci 16:8070–8101

Feng L, Hu X, Xie A, Yu H, Liu Y, Zhang J, Dong W (2015) Cationic charged polymer vesicles from amphiphilic PEI-g-PSSA-g-PEI as potential gene delivery vehicles. Aust J Chem 68: 806–813

Fomina N, McFearin CL, Sermsakdi M, Morachis JM, Almutairi A (2011) Low power, biologically benign NIR light triggers polymer disassembly. Macromolecules 44:8590–8597

Gaspar VM, Marques JG, Sousa F, Louro RO, Queiroz JA, Correia IJ (2013) Biofunctionalized nanoparticles with pH-responsive and cell penetrating blocks for gene delivery. Nanotechnology 24:275101

Goodall S, Howard CB, Jones ML, Munro T, Jia Z, Monteiro MJ, Mahler S (2015) An EGFR targeting nanoparticle self assembled from a thermoresponsive polymer. J Chem Technol Biotechnol 90:1222–1229

Gu Y, Zhong Y, Meng F, Cheng R, Deng C, Zhong Z (2013) Acetal-linked paclitaxel prodrug micellar nanoparticles as a versatile and potent platform for cancer therapy. Biomacromolecules 14:2772–2780

Han HS, Thambi T, Choi KY, Son S, Ko H, Lee MC, Jo DG, Chae YS, Kang YM, Lee JY, Park JH (2015) Bioreducible shell-cross-linked hyaluronic acid nanoparticles for tumor-targeted drug delivery. Biomacromolecules 16:447–456

Hervault A, Thanh NTK (2014) Magnetic nanoparticle-based therapeutic agents for thermo-chemotherapy treatment of cancer. Nanoscale 6:11553–11573

Hoare T, Timko BP, Santamaria J, Goya GF, Irusta S, Lau S, Stefanescu CF, Lin D, Langer R, Kohane DS (2011) Magnetically triggered nanocomposite membranes: a versatile platform for triggered drug release. Nano Lett 11:1395–1400

Hu J, Zhang G, Liu S (2012a) Enzyme-responsive polymeric assemblies, nanoparticles and hydrogels. Chem Soc Rev 41:5933–5949

Hu S-H, Liao B-J, Chiang C-S, Chen P-J, Chen IW, Chen S-Y (2012b) Core-shell nanocapsules stabilized by single-component polymer and nanoparticles for magneto-chemotherapy/ hyperthermia with multiple drugs. Adv Mater 24:3627–3632

Husseini GA, Diaz de la Rosa MA, Richardson ES, Christensen DA, Pitt WG (2005) The role of cavitation in acoustically activated drug delivery. J Control Release 107:253–261

Husseini GA, Velluto D, Kherbeck L, Pitt WG, Hubbell JA, Christensen DA (2013) Investigating the acoustic release of doxorubicin from targeted micelles. Colloids Surf B 101:153–155

Hyun DC (2015) Magnetically-controlled, pulsatile drug release from poly(e-caprolactone) (PCL) particles with hollow interiors. Polymer 74:159–165

Jabr-Milane L, van Vlerken L, Devalapally H, Shenoy D, Komareddy S, Bhavsar M, Amiji M (2008) Multi-functional nanocarriers for targeted delivery of drugs and genes. J Control Release 130:121–128

Jiang H, Zhao L, Gai L, Wang Y, Hou Y, Liu H (2015) Conjugation of methotrexate onto dedoped Fe_3O_4/PPy nanospheres to produce magnetic targeting drug with controlled drug release and targeting specificity for HeLa cells. Synth Met 207:18–25

Kaneo Y, Tanaka T, Yamamoto S, Kikkawa C (2013) Preparation and properties of acid-cleavable poly(vinyl alcohol)cis-aconityl-antitumor anthracycline conjugates. J Drug Deliv Sci Technol 23:143–149

Kim MS, Gruneich J, Jing H, Diamond SL (2010) Photo-induced release of active plasmid from crosslinked nanoparticles: o-nitrobenzyl/methacrylate functionalized polyethyleneimine. J Mater Chem 20:3396–3403

Klinger D (2013) Light-sensitive polymeric nanoparticles based on photo-cleavable chromophores. Springer International Publishing

Kuang T, Liu Y, Gong T, Peng X, Hu X, Yu Z (2016) Enzyme-responsive nanoparticles for anticancer drug delivery. Curr Nanosci 12:38–46

Kumar A, Lale SV, Mahajan S, Choudhary V, Koul V (2015) ROP and ATRP fabricated dual targeted redox sensitive polymersomes based on pPEGMA-PCL-ss-PCL-pPEGMA triblock copolymers for breast cancer therapeutics. ACS Appl Mater Interfaces 7:9211–9227

Kuo C-Y, Liu T-Y, Hardiansyah A, Chiu W-Y (2016) Magnetically polymeric nanocarriers for targeting delivery of curcumin and hyperthermia treatments toward cancer cells. J Polym Sci Part A Polym Chem

Kwon EJ, Lo JH, Bhatia SN (2015) Smart nanosystems: bio-inspired technologies that interact with the host environment. Proc Natl Acad Sci USA 112:14460–14466

Lale SV, Aswathy RG, Aravind A, Kumar DS, Koul V (2014) AS1411 aptamer and folic acid functionalized pH-responsive ATRP fabricated pPEGMA-PCL-pPEGMA polymeric nanoparticles for targeted drug delivery in cancer therapy. Biomacromolecules 15:1737–1752

Lale SV, Kumar A, Prasad S, Bharti AC, Koul V (2015) Folic acid and trastuzumab functionalized redox responsive polymersomes for intracellular doxorubicin delivery in breast cancer. Biomacromolecules 16:1736–1752

Lane DD, Chiu DY, Su FY, Srinivasan S, Kern HB, Press OW, Stayton PS, Convertine AJ (2015) Well-defined single polymer nanoparticles for the antibody-targeted delivery of chemotherapeutic agents. Polym Chem 6:1286–1299

Lee JS, Groothuis T, Cusan C, Mink D, Feijen J (2011) Lysosomally cleavable peptide-containing polymersomes modified with anti-EGFR antibody for systemic cancer chemotherapy. Biomaterials 32:9144–9153

Lee SJ, Jeong Y-I, Park H-K, Kang DH, Oh J-S, Lee S-G, Lee HC (2015) Enzyme-responsive doxorubicin release from dendrimer nanoparticles for anticancer drug delivery. Int J Nanomed 10:5489–5503

Lee S-Y, Yang C-Y, Peng C-L, Wei M-F, Chen K-C, Yao C-J, Shieh M-J (2016) A theranostic micelleplex co-delivering SN-38 and VEGF siRNA for colorectal cancer therapy. Biomaterials 86:92–105

Li H-J, Wang H-X, Sun C-Y, Du J-Z, Wang J (2014a) Shell-detachable nanoparticles based on a light-responsive amphiphile for enhanced siRNA delivery. RSC Adv 4:1961–1964

Li Y, Liu T, Zhang G, Ge Z, Liu S (2014b) Tumor-targeted redox-responsive nonviral gene delivery nanocarriers based on neutral-cationic brush block copolymers. Macromol Rapid Commun 35:466–473

Li C, Ruan J, Yang M, Pan F, Gao G, Qu S, Shen Y-L, Dang Y-J, Wang K, Jin W-L, Cui D-X (2015a) Human induced pluripotent stem cells labeled with fluorescent magnetic nanoparticles for targeted imaging and hyperthermia therapy for gastric cancer. Cancer Biol Med 12:163–174

Li T, Shen X, Chen Y, Zhang C, Yan J, Yang H, Wu C, Zeng H, Liu Y (2015b) Polyetherimide-grafted $Fe_3O_4@SiO_2$ nanoparticles as theranostic agents for simultaneous VEGF siRNA delivery and magnetic resonance cell imaging. Int J Nanomed 10:4279–4291

Li J, Zhang L, Lin Y, Xiao H, Zuo M, Cheng D, Shuai X (2016a) A pH-sensitive prodrug micelle self-assembled from multi-doxorubicin-tailed polyethylene glycol for cancer therapy. RSC Adv 6:9160–9163

Li S, Wang K, Shi Y, Cui Y, Chen B, He B, Dai W, Zhang H, Wang X, Zhong C, Wu H, Yang Q, Zhang Q (2016b) Novel biological functions of ZIF-NP as a delivery vehicle: high pulmonary accumulation, favorable biocompatibility, and improved therapeutic outcome. Adv Func Mater 26:2715–2727

Ling D, Xia H, Park W, Hackett MJ, Song C, Na K, Hui KM, Hyeon T (2014) pH-sensitive nanoformulated triptolide as a targeted therapeutic strategy for hepatocellular carcinoma. ACS Nano 8:8027–8039

Liu R, Li D, He B, Xu X, Sheng M, Lai Y, Wang G, Gu Z (2011) Anti-tumor drug delivery of pH-sensitive poly(ethylene glycol)-poly(L-histidine-)-poly(L-lactide) nanoparticles. J Control Release 152:49–56

Liu J, Duong H, Whittaker MR, Davis TP, Boyer C (2012) Synthesis of functional core, star polymers via RAFT polymerization for drug delivery applications. Macromol Rapid Commun 33:760–766

Liu J, Debuigne A, Detrembleur C, Jérôme C (2014a) Poly(*N*-vinylcaprolactam): a thermoresponsive macromolecule with promising future in biomedical field. Adv Healthc Mater 3:1941–1968

Liu J, Huang Y, Kumar A, Tan A, Jin S, Mozhi A, Liang X-J (2014b) pH-sensitive nano-systems for drug delivery in cancer therapy. Biotechnol Adv 32:693–710
Liu B, Li C, Cheng Z, Hou Z, Huang S, Lin J (2016) Functional nanomaterials for near-infrared-triggered cancer therapy. Biomater Sci 4:890–909
Lv C, Wang Z, Wang P, Tang X (2012) Photodegradable polyurethane self-assembled nanoparticles for photocontrollable release. Langmuir 28:9387–9394
Manickam DS, Oupicky D (2006) Multiblock reducible copolypeptides containing histidine-rich and nuclear localization sequences for gene delivery. Bioconjug Chem 17:1395–1403
Mao Z, Ma L, Yan J, Yan M, Gao C, Shen J (2007) The gene transfection efficiency of thermoresponsive *N,N,N*-trimethyl chitosan chloride-g-poly(*N*-isopropylacrylamide) copolymer. Biomaterials 28:4488–4500
McCarley RL (2012) Redox-responsive delivery systems. Annu Rev Anal Chem 5:391–411
Meng F, Zhong Y, Cheng R, Deng C, Zhong Z (2014) pH-sensitive polymeric nanoparticles for tumor-targeting doxorubicin delivery: concept and recent advances. Nanomedicine 9:487–499
Min KH, Kim JH, Bae SM, Shin H, Kim MS, Park S, Lee H, Park RW, Kim IS, Kim K, Kwon IC, Jeong SY, Lee DS (2010) Tumoral acidic pH-responsive mPEG-poly(beta-amino ester) polymeric micelles for cancer targeting therapy. J Control Release 144:259–266
Misra R, Acharya S, Sahoo SK (2010) Cancer nanotechnology: application of nanotechnology in cancer therapy. Drug Discov Today 15:842–850
Mu Q, Lin G, Patton VK, Wang K, Press OW, Zhang M (2016) Gemcitabine and chlorotoxin conjugated iron oxide nanoparticles for glioblastoma therapy. J Mater Chem B 4:32–36
Mura S, Nicolas J, Couvreur P (2013) Stimuli-responsive nanocarriers for drug delivery. Nat Mater 12:991–1003
Muthana M, Scott SD, Farrow N, Morrow F, Murdoch C, Grubb S, Brown N, Dobson J, Lewis CE (2008) A novel magnetic approach to enhance the efficacy of cell-based gene therapies. Gene Ther 15:902–910
Na K, Lee KH, Lee DH, Bae YH (2006) Biodegradable thermo-sensitive nanoparticles from poly (L-lactic acid)/poly(ethylene glycol) alternating multi-block copolymer for potential anti-cancer drug carrier. Eur J Pharm Sci 27:115–122
Niu G, Djaoui AB, Cohn D (2011) Crosslinkable PEO-PPO-PEO triblocks as building blocks of thermo-responsive nanoshells. Polymer 52:2524–2530
Nogueira DR, Tavano L, Mitjans M, Perez L, Infante MR, Vinardell MP (2013) In vitro antitumor activity of methotrexate via pH-sensitive chitosan nanoparticles. Biomaterials 34:2758–2772
Noh MS, Lee S, Kang H, Yang J-K, Lee H, Hwang D, Lee JW, Jeong S, Jang Y, Jun B-H, Jeong DH, Kim SK, Lee Y-S, Cho M-H (2015) Target-specific near-IR induced drug release and photothermal therapy with accumulated Au/Ag hollow nanoshells on pulmonary cancer cell membranes. Biomaterials 45:81–92
Paris JL, Cabanas MV, Manzano M, Vallet-Regi M (2015) Polymer-grafted mesoporous silica nanoparticles as ultrasound-responsive drug carriers. ACS Nano 9:11023–11033
Park W, Chen J, Cho S, Park S-j, Larson AC, Na K, Kim D-H (2016) Acidic pH-triggered drug-eluting nanocomposites for magnetic resonance imaging-monitored intra-arterial drug delivery to Hepatocellular Carcinoma. ACS Appl Mater Interfaces 8:12711–12719
Pujana MA, Perez Alvarez L, Cesteros Iturbe LC, Katime I (2014) pH-sensitive chitosan-folate nanogels crosslinked with biocompatible dicarboxylic acids. Eur Polym J 61:215–225
Qiao ZY, Qiao SL, Fan G, Fan YS, Chen Y, Wang H (2014) One-pot synthesis of pH-sensitive poly(RGD-co-b-amino ester)s for targeted intracellular drug delivery. Polym Chem 5:844–853
Qiu L, Liu Q, Hong C-Y, Pan C-Y (2016) Unimolecular micelles of camptothecin-bonded hyperbranched star copolymers via [small beta]-thiopropionate linkage: synthesis and drug delivery. J Mater Chem B 4:141–151
Rapoport NY, Kennedy AM, Shea JE, Scaife CL, Nam K-H (2009) Controlled and targeted tumor chemotherapy by ultrasound-activated nanoemulsions/microbubbles. J Control Release 138:268–276
Sadhukha T, Wiedmann TS, Panyam J (2013) Inhalable magnetic nanoparticles for targeted hyperthermia in lung cancer therapy. Biomaterials 34:5163–5171

Sagnella SM, Duong H, MacMillan A, Boyer C, Whan R, McCarroll JA, Davis TP, Kavallaris M (2013) Dextran-based doxorubicin nanocarriers with improved tumor penetration. Biomacromolecules 15:262–275

Sanson C, Le Meins JF, Schatz C, Soum A, Lecommandoux S (2010) Temperature responsive poly(trimethylene carbonate)-block-poly(L-glutamic acid) copolymer: polymersomes fusion and fission. Soft Matter 6:1722–1730

Scialabba C, Licciardi M, Mauro N, Rocco F, Ceruti M, Giammona G (2014) Inulin-based polymer coated SPIONs as potential drug delivery systems for targeted cancer therapy. Eur J Pharm Biopharm 88:695–705

Sethuraman VA, Na K, Bae YH (2006) pH-responsive sulfonamide/PEI system for tumor specific gene delivery: an in vitro study. Biomacromolecules 7:64–70

Shao P, Wang B, Wang Y, Li J, Zhang Y (2011) The application of thermosensitive nanocarriers in controlled drug delivery. J Nanomater. doi:10.1155/2011/389640

Sirsi SR, Borden MA (2014) State-of-the-art materials for ultrasound-triggered drug delivery. Adv Drug Deliv Rev 72:3–14

Su Z, Liang Y, Yao Y, Wang T, Zhang N (2016) Polymeric complex micelles based on the double-hydrazone linkage and dual drug-loading strategy for pH-sensitive docetaxel delivery. J Mater Chem B 4:1122–1133

Suma T, Miyata K, Anraku Y, Watanabe S, Christie RJ, Takemoto H, Shioyama M, Gouda N, Ishii T, Nishiyama N, Kataoka K (2012) Smart multilayered assembly for biocompatible siRNA delivery featuring dissolvable silica, endosome-disrupting polycation, and detachable PEG. ACS Nano 6:6693–6705

Sun TM, Wang YC, Wang F, Du JZ, Mao CQ, Sun CY, Tang RZ, Liu Y, Zhu J, Zhu YH, Yang XZ, Wang J (2014) Cancer stem cell therapy using doxorubicin conjugated to gold nanoparticles via hydrazone bonds. Biomaterials 35:836–845

Tang S, Yin Q, Su J, Sun H, Meng Q, Chen Y, Chen L, Huang Y, Gu W, Xu M, Yu H, Zhang Z, Li Y (2015) Inhibition of metastasis and growth of breast cancer by pH-sensitive poly (b-amino ester) nanoparticles co-delivering two siRNA and paclitaxel. Biomaterials 48:1–15

Torchilin VP (2014) Multifunctional, stimuli-sensitive nanoparticulate systems for drug delivery. Nat Rev Drug Discov 13:813–827

Vander Heiden MG, Cantley LC, Thompson CB (2009) Understanding the Warburg effect: the metabolic requirements of cell proliferation. Science 324:1029–1033

Wang H, He J, Zhang M, Tao Y, Li F, Tam KC, Ni P (2013) Biocompatible and acid-cleavable poly(e-caprolactone)-acetal-poly(ethylene glycol)-acetal-poly(e-caprolactone) triblock copolymers: synthesis, characterization and pH-triggered doxorubicin delivery. J Mater Chem B 1:6596–6607

Wang Y, Luo Q, Sun R, Zha G, Li X, Shen Z, Zhu W (2014) Acid-triggered drug release from micelles based on amphiphilic oligo(ethylene glycol)-doxorubicin alternative copolymers. J Mater Chem B 2:7612–7619

Wang B, Liu P, Shi B, Gao J, Gong P (2015) Preparation of pH-sensitive dextran nanoparticle for doxorubicin delivery. J Nanosci Nanotechnol 15:2613–2618

Warburg O (1956) On the origin of cancer cells. Science 123:309–314

Wei C, Xuemin L, Kun C, Jun W, Yen W, Qinghua L (2011) Photosensitive nanoparticles of chitosan complex for controlled release of dye molecules. Nanotechnology 22:065702

Wen J, Anderson SM, Du J, Yan M, Wang J, Shen M, Lu Y, Segura T (2011) Controlled protein delivery based on enzyme-responsive nanocapsules. Adv Mater 23:4549–4553

Wu S, Butt HJ (2016) Near-infrared-sensitive materials based on upconverting nanoparticles. Adv Mater 28:1208–1226

Xu C, Wang P, Zhang J, Tian H, Park K, Chen X (2015) Pulmonary codelivery of doxorubicin and siRNA by pH-sensitive nanoparticles for therapy of metastatic lung cancer. Small 11:4321–4333

Yang G, Lv R, He F, Qu F, Gai S, Du S, Wei Z, Yang P (2015) A core/shell/satellite anticancer platform for 808 NIR light-driven multimodal imaging and combined chemo-/photothermal therapy. Nanoscale 7:13747–13758

Yildiz I, Yildiz SB (2015) Applications of thermoresponsive magnetic nanoparticles. J Nanomater 2015:12
You J, Zhang R, Xiong C, Zhong M, Melancon M, Gupta S, Nick AM, Sood AK, Li C (2014) Effective photothermal chemotherapy using doxorubicin-loaded gold nanospheres that target EphB4 receptors in tumors. Cancer Res 72:4777–4786
Yu J, Chu X, Hou Y (2014) Stimuli-responsive cancer therapy based on nanoparticles. Chem Commun 50:11614–11630
Yuanpei L, Shirong P, Wei Z, Zhuo D (2009) Novel thermo-sensitive core-shell nanoparticles for targeted paclitaxel delivery. Nanotechnology 20:065104
Zeighamian V, Darabi M, Akbarzadeh A, Rahmati-Yamchi M, Zarghami N, Badrzadeh F, Salehi R, Tabatabaei Mirakabad FS, Taheri-Anganeh M (2015) PNIPAAm-MAA nanoparticles as delivery vehicles for curcumin against MCF-7 breast cancer cells. Artif Cells Nanomed Biotechnol 44:735–742
Zhang CY, Xiong D, Sun Y, Zhao B, Lin WJ, Zhang LJ (2014a) Self-assembled micelles based on pH-sensitive PAE-g-MPEG-cholesterol block copolymer for anticancer drug delivery. Int J Nanomed 9:4923–4933
Zhang Z, Wang J, Nie X, Wen T, Ji Y, Wu X, Zhao Y, Chen C (2014b) Near infrared laser-induced targeted cancer therapy using thermoresponsive polymer encapsulated gold nanorods. J Am Chem Soc 136:7317–7326
Zhong Y, Yang W, Sun H, Cheng R, Meng F, Deng C, Zhong Z (2013) Ligand-directed reduction-sensitive shell-sheddable biodegradable micelles actively deliver doxorubicin into the nuclei of target cancer cells. Biomacromolecules 14:3723–3730
Zhou Q-L, Chen Z-Y, Wang Y-X, Yang F, Lin Y, Liao Y-Y (2014) Ultrasound-mediated local drug and gene delivery using nanocarriers. Biomed Res Int 2014:13
Zhu L, Kate P, Torchilin VP (2012) Matrix metalloprotease 2-responsive multifunctional liposomal nanocarrier for enhanced tumor targeting. ACS Nano 6:3491–3498
Zhuang Y, Deng H, Su Y, He L, Wang R, Tong G, He D, Zhu X (2016) Aptamer-functionalized and backbone redox-responsive hyperbranched polymer for targeted drug delivery in cancer therapy. Biomacromolecules. https://doi.org/10.1021/acs.biomac.6b00262
Zou J, Zhang F, Zhang S, Pollack SF, Elsabahy M, Fan J, Wooley KL (2014) Poly(ethylene oxide)-block-Polyphosphoester-graft-Paclitaxel conjugates with acid-labile linkages as a pH-sensitive and functional nanoscopic platform for paclitaxel delivery. Adv Healthc Mater 3:441–448

Chapter 3
The Application of High Polymer Materials in the Aspects of Seed Technology

Jin Hu and Yajing Guan

Abstract Adverse environmental stresses such as drought and low or high temperature will inhibit seed emergence and delay seedling establishment, resulting in significant yield reduction and poor seed quality. Beneficial chemicals for plant growth, such as fungicides, insecticides, fertilizers, and growth promoters, are usually delivered through seed-coating technology to protect seeds from diseases, pests, environmental stresses, and to enhance field seedling establishment. However, traditional seed-coating agents might not be able to meet the requirements due to their low efficiency. Now high polymer materials have been got lots of attention in many research fields, and here in present chapter we want to give a new angle of view about how to apply the high polymer materials in agriculture research to develop "intelligent" coating agents to make the traditional seed-coating agents more efficient.

Keywords Seed coating · Seed pelleting · Seed germination · Chilling stress
Thermoresponsive hydrogel · Release

1 Seed Coating and Pelleting

Seeds are used as delivery systems for agricultural biotechnology. In general, higher quality seed leads to better seedling performance in the field. Seed quality is a complex trait which is determined by interactions between multiple genetic factors and environmental conditions. Therefore, modern approaches to improve seed

J. Hu (✉) · Y. Guan
Seed Science Center, College of Agriculture and Biotechnology, Zhejiang University, Hangzhou 310058, People's Republic of China
e-mail: jhu@zju.edu.cn

Y. Guan
e-mail: vcguan@zju.edu.cn

© Springer Nature Singapore Pte Ltd. 2018

V. K. Thakur et al. (eds.), *Polymer Gels*, Gels Horizons: From Science to Smart Materials, https://doi.org/10.1007/978-981-10-6080-9_3

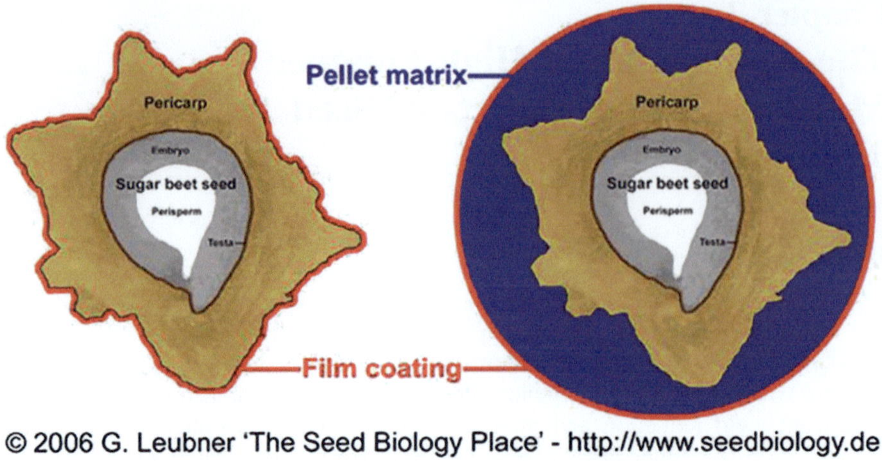

Fig. 1 A film-coated sugar beet fruit (left) and a pelleted plus film-coated sugar beet fruit (right)

quality often need to combine classical genetics, plant molecular biology, and a variety of seed technologies.

Seed technologies known as seed enhancements include seed cleaning, drying, soaking, steeping, tagging, priming, coating, pelleting, encrusting, and other novel seed treatment methods. Important methods of seed enhancements used globally in the seed industry to improve seed and seedling performance in field are seed coating and seed pelleting. These enhancements make seeds rounder or larger and can be used as a carrier for various seed additives such as fungicides, insecticides, fertilizers, growth promoters, and crop protection materials. In recent years, the planting proportion of coated seeds to uncoated seeds increased constantly in the whole world. Figure 1 shows a film-coated sugar beet fruit and a pelleted plus film-coated sugar beet fruit ("The Seed Biology Place" http://www.seedbiology.de/seedtechnology.asp).

1.1 Film Coating

As compared to pelleting, film coating applies less material on seed surface so that the original seed shape is still apparent. Active materials are dispensed or dissolved in a liquid adhesive and then are sprayed onto the seed surface either with a fluidized bed treater or a pharmaceutical coating drum to provide a solid, thin coat covering seeds. The liquid adhesive is often a kind of polymer, such as polyvinyl alcohol (PVA), carboxymethylcellulose (CMC), which can adhere tightly to the seed and prevent loss of active materials like fungicides, nutrients, colorants, or plant hormones. Some novel applications of film coating are used to modify seed imbibition and germination. They can confer temperature-sensitive water permeability to seeds or affect gaseous exchange. By this they control the timing of seed germination and seedling emergence. Certain temperature-dependent water-resistant polymers could prevent seed imbibition at lower temperatures but allow it at higher temperatures. Accordingly, the film-coated seeds can be planted early in a planting season but undergo imbibition until the climatic conditions become suitable for continued seedling growth.

1.2 Seed Pelleting

The term “seed pelleting” refers to the addition of inert fillers to increase the apparent seed size and weight. Seeds for pelleting are first placed in revolving pans or drums of various design and sizes, then blends of powdered materials, active chemicals, and binders are progressively added in a layering process until the desired seed pellet weight or size increase is achieved. Seed pelleting increases seed weight by about twofold up to 50-fold or more. This technology usually is used in small or medium seeds such as vegetable seeds. The pellet matrix consisting of filling materials and loam, starch, cellulose derivative, or polyacrylate/polyacrylamide polymers are commercially used. Some novel synthetic polymers such as superabsorbent polymer (SAP) are used in seed pelleting for seed drought resistance due to the strong water absorption performance of SAP. A film coat can be added onto the pelleting layer as shown in the Fig. 1. In addition, it is very common and useful for greatly increasing the size of very small or fine seeds through pelleting technology to produce bigger, rounder, smoother, and more uniformly sized coated products. This technology improves the planting features of seeds and makes them more suitable for the use of precise planting machines (Grellier et al. 1999). Figure 2 shows the example of pelleted tobacco seeds (provided by the Yuxi Zhongyan Tobacco Seed Co., Ltd, Yunnan, China).

2 Thermoresponsive Hydrogel: High Polymer Materials

Thermoresponsive hydrogels belonging to stimuli-sensitive polymers could change their structure and physical properties in response to external temperatures (Fundueanu et al. 2009). The most popular thermoresponsive polymers, poly(*N*-isopropylacrylamide) (PNIPAm) and its copolymers with other monomers, have been extensively studied for application as a thermal on–off switch for a pulsatile drug release system. PNIPAm exhibits a sharp phase transition at 32 °C in aqueous media (Li and D’Emanuele 2003; Schild 1992). The temperature that phase transition occurs is called the lower critical solution temperature (LCST). Below LCST, the polymer chain is hydrated and adopts an extended coil conformation, while above it the polymer is dehydrated and adopts a globular conformation. Correspondingly, thermoresponsive hydrogels exhibit a reversible phase transition

Fig. 2 Pelleted tobacco seeds

(swelling–shrinking) in response to external temperature changes to realize "on–off" regulation of entrapped drug release. At temperature below LCST, PNIPAm is on the state "on", the entrapped drugs were released by diffusion mechanism. When the temperature rises above LCST, the polymer is on the state "off", the entrapped drug release ceased quickly due to the formation of a dense skin layer on the polymer surface under the strong shrinking force caused by shrinking hydrogel (Yoshida et al. 1992). In addition, PNIPAm polymer has also been widely studied in applications of concentrating and refining dilute solutions (Gehrke et al. 1986), immobilization of enzymes (Dong and Hoffman 1986), and solute separation (Feil et al. 1991) through utilizing its thermosensitivity trait. In recent years, the use of PNIPAm polymers as film coating or pelleting materials on seeds is gradually increasing.

3 Application of Thermoresponsive Hydrogel in Seed Technology for Chilling Resistance

Chilling stress is an important constraint for seed establishment of many crops in field. For example, tobacco (*Nicotiana tabacum* L.) often suffered from chilling injury especially in the South and/or Southeast Asia (Cui et al. 2012). Maize (*Zea mays* L.), one of the most important thermophilic crops, is vulnerable to low temperature. In many places of the world, due to the unstable temperature in the early spring, crop is often subjected to sudden low temperature after sowing, which seriously inhibited seed emergence and delayed seedling establishment, resulting in significant yield reduction and poor seed quality (Guan et al. 2009). Therefore, seeds are generally coated or pelleted with chilling resistance agent before sale to protect seeds from chilling injury to a certain degree. Among several reported cold-resistant agents, salicylic acid (SA) is considered effective. Salicylic acid is an endogenous growth regulator which could induce a protective effect on plants under abiotic stress. And salicylic acid-induced increase in the resistance to low temperature had been reported in many plants such as maize, wheat, banana, cucumber, rice (Janda et al. 1999; Kang and Saltveit 2002; Kang et al. 2003; Tasgin et al. 2003). However, for traditional seed-coating agent, the effective chilling-resistant ingredients such as salicylic acid usually release immediately once coated seeds are sowed into the field regardless of the environmental temperature. Due to the large quantity release and losses of the ingredients before low temperature occurs, the chilling resistance effect of traditional coated seeds is thus unavoidably influenced and becomes unobvious (Zhang et al. 2007). Therefore, controlling exactly the release time and rate of chilling-resistant agent is important for improving seed chilling tolerance.

Although cross-linked poly(*N*-isopropylacrylamide) (PNIPAm) shows a remarkable change in aqueous swelling following a temperature change and is equipped with a thermo "on–off" property for drug release, its poor mechanical property and the high LCST limit its application in seed technology. Fortunately, the mechanical property and LCST of hydrogel could be changed by copolymerizing NIPAm with hydrophilic or hydrophobic monomer (Zhang 2010). And the copolymer hydrogel could raise or decrease its LCST by adjusting the quantity of the hydrophilic or hydrophobic monomer according to practical requirements, providing the copolymer desired mechanical properties and thermo-sensitive swelling behavior for its possibility of the application on seeds (Yoshida et al. 1992). Figure 3 shows that the change in LCST is proportional to the comonomer content. The increase in LCST is the largest for the anionic comonomer (AAc), followed by the cationic and the hydrophilic comonomers (DEAEMA, AAm),

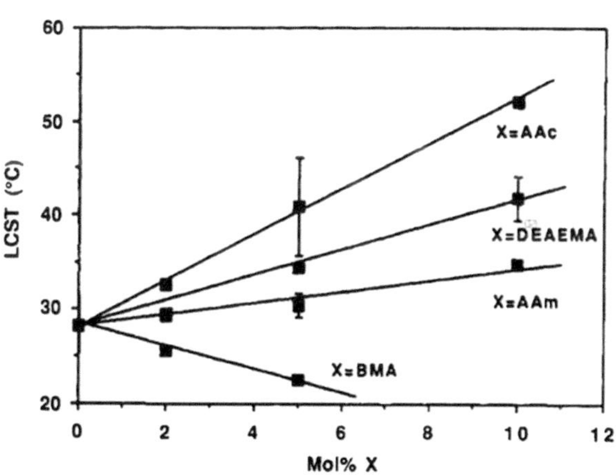

Fig. 3 LCST of P(NIPAm-*co*-BMA-*co*-X) solutions in PBS, pH = 7.4, as a function of comonomer X content. (LCST is the lower critical solution temperature. LCSTs obtained from DSC experiments, $n = 3$. Feil et al. 1993)

while the hydrophobic comonomer butyl methacrylate (BMA) causes a decrease in LCST (Feil et al. 1993). Meanwhile, the hydrophobic shrinking force of the gel is enhanced by introducing hydrophobic BMA into the polymer networks. This leads to formation of more dense compact surface skin layers. As a result, restrictions on water permeation from the gel interior become strong (Kaneko et al. 1995).

For example, when the environmental temperature is lower than 10 °C, the maize seed germination and seedling growth will be inhibited strongly (Hope et al. 1992). With the purpose of effectively improving maize chilling resistance, LCST of hydrogel should be 1–2 °C higher than 10 °C, in that case the hydrogel has a time to release chilling-resistant agent for maize seed protection. Therefore, a copolymer hydrogel consisting of NIPAm and BMA monomer is synthesized by free-radical polymerization, and LCST of P(NIPAm-*co*-BMA) is decreased to 12 °C by modifying the proportion of NIPAm and BMA. Then, the P(NIPAm-*co*-BMA) hydrogel with LCST of 12 °C could be used as a releasing controller in maize seed film-coating agents for effective chilling resistance (Guan et al. 2015).

3.1 *Synthesis of the P(NIPAm-*co*-BMA) Hydrogel*

The N-isopropyl acrylamide (NIPAm) and butyl methacrylate (BMA) are used as the mainly polymeric monomers. Cross-linker and initiator can be selected according to the practical requirements, such as the cross-linkers ethylene dimeth-acrylate (EDMA) and *N,N′*-Methylenebisacrylamide (BIS), initiators *tert*-butyl peroxyoctanoate (BPO), and 2,2′-azobisisobutyronitrile (AIBN) (Bae et al. 1987; Guan et al. 2015). A certain amount of NIPAm, BMA, cross-linker, and initiator are dissolved in 1,4-dioxane solvent. Bubbling nitrogen to remove dissolved oxygen of mixed solution, and then the P(NIPAm-*co*-BMA) hydrogel is polymerized by cross-linking of NIPAm and BMA at 70–80 °C for 12–24 h. Finally, the formed hydrogel is soaked in methanol/distilled deionized water solutions to remove all unreacted compounds and then dried under vacuum at room temperature (Bae et al. 1991).

3.2 *The Characteristics of P(NIPAm-*co*-BMA) Hydrogel*

3.2.1 Lower Critical Solution Temperature Determination

Lower critical solution temperature of hydrogel can be determined by differential scanning calorimetry (Feil et al. 1993) or by investigating the swelling ratio of hydrogel in water of different temperatures (Guan et al. 2015). LCST profile of P (NIPAm-*co*-BMA) hydrogel (86/14 mol%) was shown in Fig. 4. The hydrogel

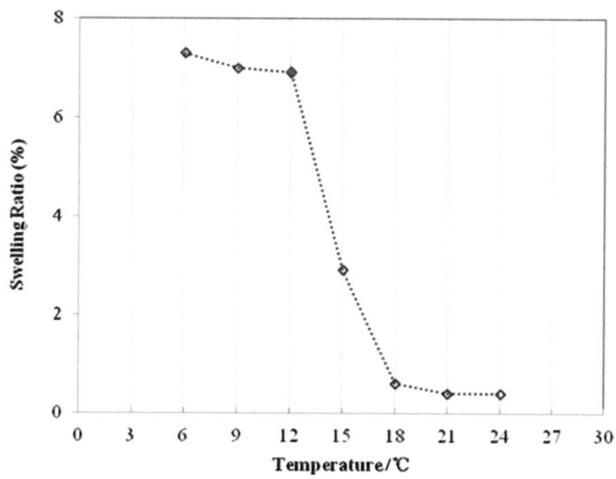

Fig. 4 LCST profile of P (NIPAm-*co*-BMA) hydrogel (86/14 mol%). (Each point is the fully swelling ratio of hydrogel under a certain temperature, and the swelling ratio is defined as the weight of the fully swollen hydrogel in relation to its dried weight. Guan et al. 2015)

changes sharply its swelling ratio at 12 °C, which is considered as LCST of this cross-linked P(NIPAm-co-BMA) (86/14 mol%). When the temperature is below 12 °C, the hydrogel has obviously higher swelling ratio than above 12 °C (Guan et al. 2015).

3.2.2 Swelling Measurement

The swelling characteristic of P(NIPAm-*co*-BMA) hydrogel is usually measured as a function of temperature in phosphate-buffered saline (PBS pH 7.4) or distilled water for various times. The weight changes of the hydrogel (water uptake) are recorded at each time point until equilibrium weight is reached. Figure 5 shows the swelling–deswelling kinetics of cross-linked P(NIPAm-co-BMA) (95/5 mol%) between 20 and 30 °C in PBS (pH 7.4) (Bae et al. 1987). The hydrogel at 20 °C is transparent, and it turns opaque when the temperature is raised to 30 °C. Also, upon raising the temperature to 30 °C, there is an initial rapid shrinking, followed by a slow deswelling. This may be explained as the immediate shrinking of the outer layer of the hydrogel in contact with the solution which may restrict the bulk water outflow from the interior of the gel. Figure 6 shows the swelling and deswelling characteristics of P(NIPAm-*co*-BMA) hydrogel (86/14 mol%) under 5 and 30 °C, respectively, in distilled water (Guan et al. 2015). After swelling in water at 5 °C for 5 h, the swelling ratio is over 300% and even reached to 700% at 30 h. The fully swollen hydrogel loses water quickly at 30 °C and retains only a quarter of the maximum water content after deswelling for 30 min.

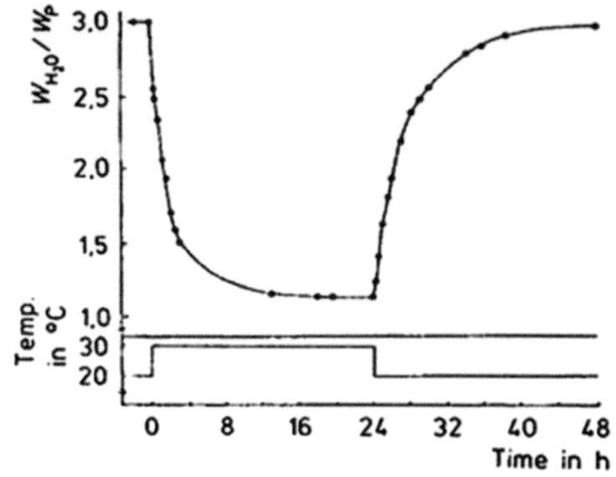

Fig. 5 Swelling–deswelling kinetics of cross-linked P(NIPAm-*co*-BMA) (95/5 mol%) between 20 and 30 °C for 24 h in phosphate-buffered saline. (PBS pH 7.4, W_{H2O}/W_P is the swelling ratio which is defined as the weight of water uptake per unit weight of dried polymer. Bae et al. 1987)

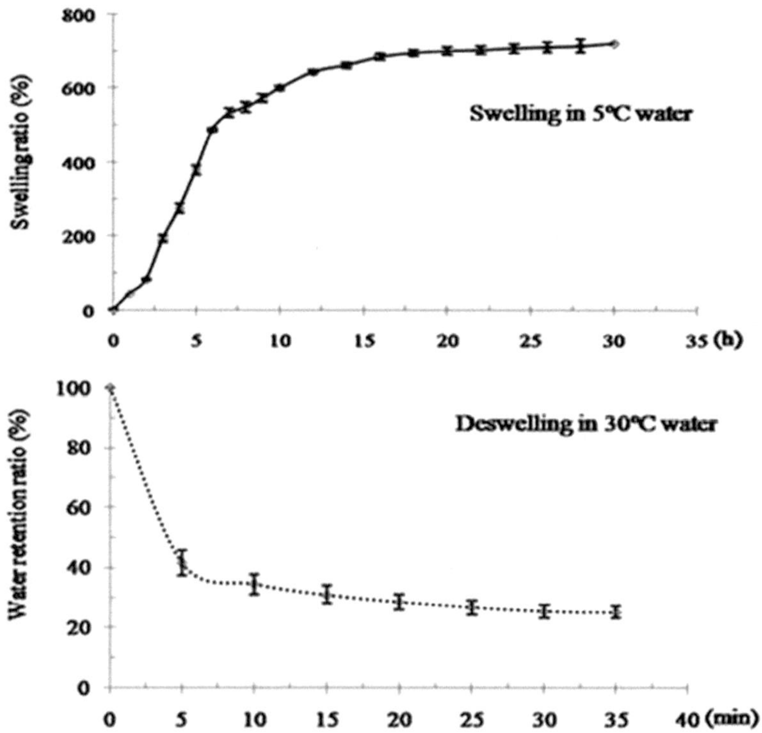

Fig. 6 Swelling and deswelling characteristics of P(NIPAm-*co*-BMA) hydrogel (86/14 mol%) in 5 and 30 °C distilled water (Guan et al. 2015)

3.2.3 Drug Loading and Releasing

Drug is loaded into P(NIPAm-co-BMA) hydrogel usually through a solvent sorption method. Bae et al. (1987) equilibrated dried hydrogel in a saturated indomethacin solution (80/20 v/v% ethanol/water) and then followed by drying it under vacuum. Rao et al. (2006) dissolved dried hydrogel in 10 wt% salicylic acid solution (80/20 v/v% ethanol/water) and in 10 wt% sodium salicylate solution, respectively. The adsorption of sodium salicylate mainly depends on the swelling ratio of the hydrogel; in addition, that of salicylic acid relates to the content of hydrophobic groups in the hydrogel, which is due to that salicylic acid is hydrophobic and sodium salicylate is hydrophilic. Therefore, salicylic acid has higher loading ratio than sodium salicylate in the P(NIPAm-co-BMA) hydrogel.

Drug release from P(NIPAm-*co*-BMA) hydrogel is dependent on surrounding temperature. At lower temperature, drug releases fast as the dried P(NIPAm-*co*-BMA) hydrogel is swelling, while its release slows down and even ceases at higher temperature as the hydrogel is deswelling. Therefore, the P(NIPAm-*co*-BMA) hydrogel provides an "on–off" model as a drug delivery for the drug release due to its temperature-responsive pulsatile performance. The cumulative indomethacin release at 20 and 30 °C is shown in Fig. 7. An abnormal release pattern of indomethacin occurs at 20 °C, however a negligible release rate is observed at 30 °C. It is considered being predominantly governed by the hydration of the hydrogel (free volume mechanism), rather than a partition mechanism or thermal energy effect (Bae et al. 1987). Figure 8 shows a stepwise state of SA releases in response to the circulative pulse temperatures. The slope of SA release curve was obviously higher under 9 °C than 15 °C (Fig. 8, Guan et al. 2015). In addition, the drug release can be influenced by different values of pH (Fig. 9), which is mainly induced by the changed swelling properties of hydrogel. However, SA release quantities and

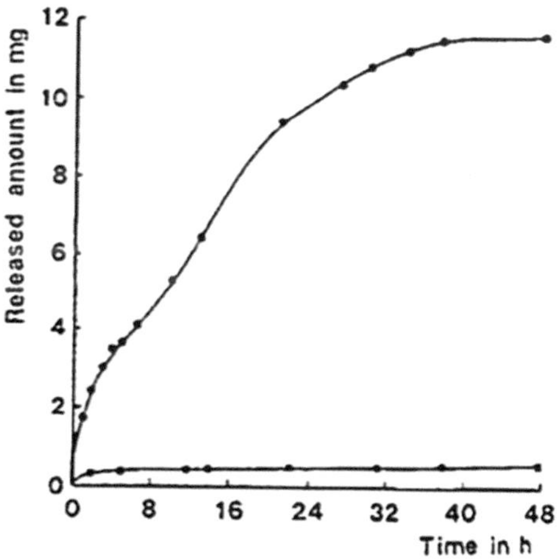

Fig. 7 Release profiles of indomethacin from cross-linked P(NIPAm-*co*-BMA) (95/5 mol%) at 20 °C (above) and 30 °C (below) in PBS (pH 7.4. Bae et al. 1987)

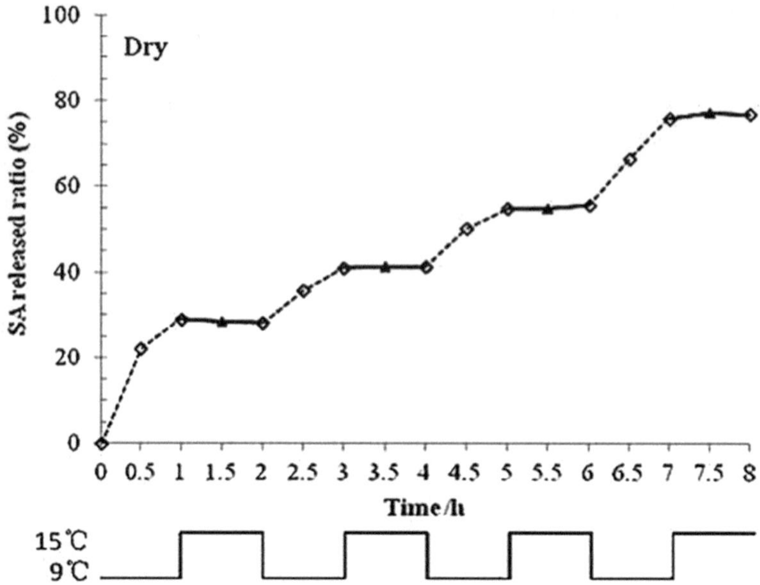

Fig. 8 Effects of circulative pulse temperatures on salicylic acid release from P(NIPAm-*co*-BMA) hydrogel (86/14 mol%). (9 °C (◇) and 15 °C (▲). Guan et al. 2015)

Fig. 9 Salicylic acid releasing curves for dry PAAc/P(NIPAm-*co*-BMA) hydrogel at different pH solutions at 25 °C (Zhang 2010)

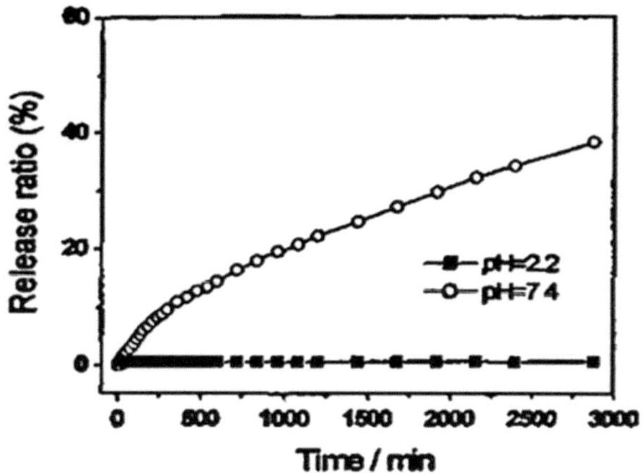

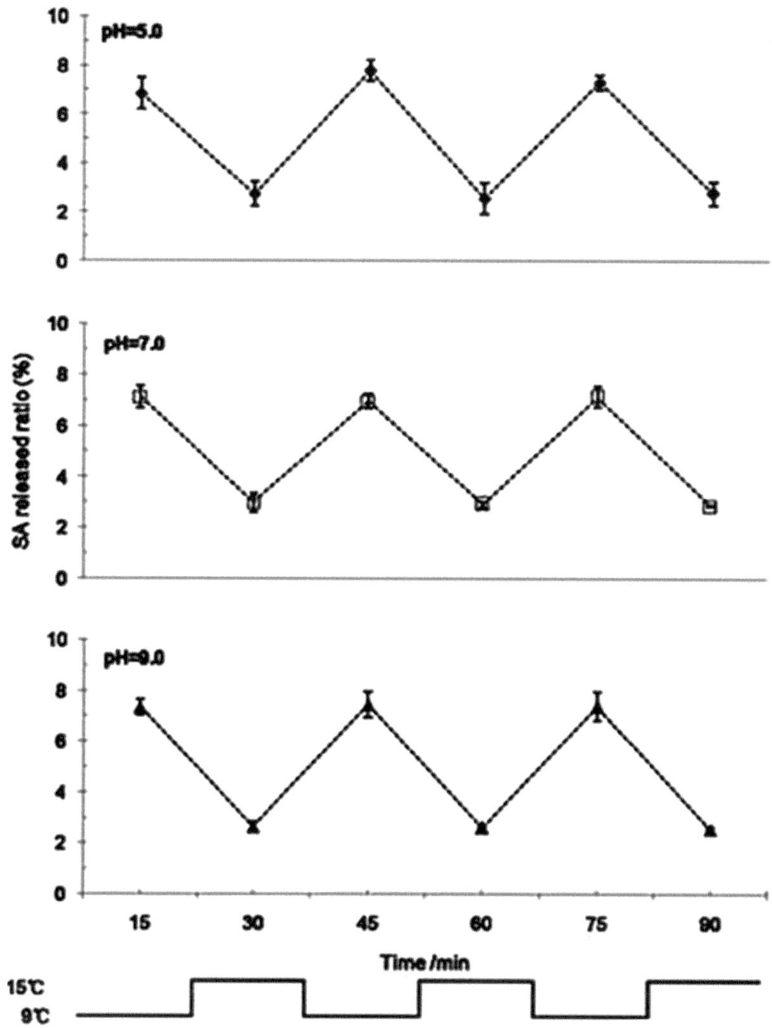

Fig. 10 Effects of pH values on salicylic acid release from P(NIPAm-co-BMA) hydrogel under circulative pulse temperatures 9 and 15 °C. (Error bars represented ± S.E. Guan et al. 2015)

change modes are basically the same in pH values of aqueous solutions at 5.0, 7.0, and 9.0 (Fig. 10). It suggests that the practical application scope of SA-loaded P (NIPAm-*co*-BMA) hydrogel could be greatly exaggerated to soils with different pH values from 5.0 to 9.0.

3.3 The Application of P(NIPAm-Co-BMA) Hydrogel on Seeds

The coated or pelleted seeds are usually prepared according to the given formulas in coating machine. A small amount of drug-loaded P(NIPAm-*co*-BMA) hydrogel powders can be added into the general coating agents to produce a kind of "intelligent" coating agent, which releases effectively the loaded drugs just under certain temperature to protect seeds from chilling injury to a maximum extent. The better "intelligent" chilling-resistant coating treatments should firstly have no negative influence on seed germination and seedling growth. Secondly, it could play the greatest efficiency of chilling-resistant agent to protect the plant from low-temperature stress. Then it would be considered as better "intelligent" coating agent. For example, the coating agents containing 0.15–1.5 g% of SA-loaded P(NIPAm-*co*-BMA) hydrogel (86/14 mol%) powder are used in maize seeds for chilling resistance (Guan et al. 2015). And those containing 0.15–3.0 g% of SA-loaded P(NIPAm-*co*-BMA) hydrogel (86/14 g%) powder are used in tobacco seed pelleting (Cui et al. 2012). The SA-loaded P(NIPAm-*co*-BMA) hydrogel has been proved to be no damage on maize or tobacco seed establishment, and its appropriate dosage could improve obviously seed germination and seedling growth under low temperature. The effect of SA-loaded P(NIPAm-*co*-BMA) hydrogel coating agent on seedling growth of maize is shown in Fig. 11. Also, Table 1 shows the effects of seed pelleting on tobacco seed germination under chilling stress (Cui et al. 2012). The P(NIPAm-*co*-BMA) itself is nontoxic to seed germination (compared data in Fig. 11 and Table 1). SA plays an important role in the activation of plant resistant responses toward most kinds of abiotic stresses to alleviate the injury to plants (Lukatkin 2003; Kunkel and Brooks 2002). However, the treated concentration of SA is very crucial. Lower SA concentration improves seed germination, while SA with higher concentration inhibits seedling growth. The P(NIPAm-*co*-BMA) hydrogel could intelligently control SA release according to the external temperature fluctuations, thereby preventing excessive SA release under non-chilling stress and avoiding any resultant plant damage. Therefore, a remarkably positive effect on seeds and seedlings can be found in coating treatments with a higher concentration of SA-loaded P(NIPAm-*co*-BMA) hydrogel.

4 Conclusion and Future Perspective

In conclusion, high polymer materials such as P(NIPAm-*co*-BMA) hydrogel can be used as a temperature-responsive carrier to realize "on–off" regulation of entrapped drug release in response to external temperature changes. It is considered as a novel and intelligent method to enhance the chilling tolerance of coated or pelleted seeds. The efficacy of the "intelligent" thermoresponsive coating agent should be verified in the practice. Moreover, the release performance of other hormones or compounds

Fig. 11 Effects of "intelligent" coating agents on seedling growth of maize varieties Huang C and Mo17 after chilling stress (Guan et al. 2015). **a** and **b** are Huang C; **c** and **d** are Mo17. The seedlings grew for 11 days including five-day germination and growth at 25 °C, three-day chilling stress at 5 °C, and three-day recover growth at 25 °C. Ck. Control: maize seeds coated with mixture of 70 g talc and 30 g bentonite; J1. maize seeds coated with mixture of 0.1 g P(NIPAm-*co*-BMA), 69.9 g talc and 30 g bentonite; J2. maize seeds coated with mixture of 1.0 g P(NIPAm-*co*-BMA), 69 g talc and 30 g bentonite; S1. maize seeds coated with mixture of 0.1 g salicylic acid, 69.9 g talc and 30 g bentonite; S2. maize seeds coated with mixture of 1.0 g salicylic acid, 69 g talc and 30 g bentonite; J1S1. maize seeds coated with mixture of 0.1 g salicylic acid-loaded P(NIPAm-*co*-BMA), 69.9 g talc and 30 g bentonite; J2S2. maize seeds coated with mixture of 1.0 g salicylic acid-loaded P(NIPAm-*co*-BMA), 69 g talc and 30 g bentonite. The coating agents were applied at a rate of 1 g per 5 g naked seeds for all of the treatments

loaded in P(NIPAm-*co*-BMA) hydrogel needs further study to enhance seed chilling tolerance. And the formula and polymerization process of thermoresponsive hydrogel should be improved to prepare different "on–off" regulated temperatures according to different temperature of chilling stress in different crops.

Table 1 Effects of seed pelleting with "intelligent" coating agents on seed germination in tobacco varieties HHDJY and MSK326 under chilling stress (Cui et al. 2012)

Treatment	HHDJY			MSK326		
	GP** (%)	GI	MGT (d)	GP (%)	GI	MGT (d)
Control*	72.0 d***	11.93 de	7.12 ab	72.5 ef	10.94 de	7.57 b
P1	73.2 cd	12.15 d	6.61 bc	73.2 ef	10.85 e	7.51 b
P2	74.0 cd	11.49 e	7.56 a	71.1 f	11.65 cd	7.46 b
P3	74.8 cd	11.43 e	6.89 bc	72.0 ef	10.92 de	7.74 b
S1	75.5 cd	12.34 cd	6.70 bc	79.2 cd	12.02 bc	7.77 b
S2	88.4 b	14.45 b	6.73 bc	82.2 bc	12.27 bc	7.96 b
S3	46.5 e	7.66 f	7.69 a	36.8 g	4.72 f	9.21 a
P1S1	77.4 c	12.83 c	6.78 bc	75.5 de	11.81 c	7.90 b
P2S2	89.2 ab	15.08 a	6.55 bc	83.6 b	12.7 b	7.39 bc
P3S3	93.4 a	14.90 ab	6.45 c	91.9 a	14.12 a	6.80 c

*Control represented as 2 g seeds pelleted only with basic pelleting agents which consisted of 70 g talc powder and 30 g bentonite powder; P1: pelleted seeds with 99.9 g basic pelleting agents and 0.1 g P(NIPAm-*co*-BMA); P2: pelleted with 99.5 g basic pelleting agents and 0.5 g P(NIPAm-*co*-BMA); P3: pelleted with 98 g basic pelleting agents and 2 g P(NIPAm-*co*-BMA); S1: pelleted with 99.95 g basic pelleting agents and 0.05 g salicylic acid; S2: pelleted with 99.75 g basic pelleting agents and 0.25 g salicylic acid; S3: pelleted with 99 g basic pelleting agents and 1 g salicylic acid; P1S1: pelleted with 99.85 g basic pelleting agents and 0.1 g P(NIPAm-*co*-BMA) and 0.05 g salicylic acid; P2S2: pelleted with 99.25 g basic pelleting agents and 0.5 g P(NIPAm-*co*-BMA) and 0.25 g salicylic acid; P3S3: pelleted with 97 g basic pelleting agents and 2 g P (NIPAm-*co*-BMA) and 1 g salicylic acid

**GP represented as germination percentage, GI represented as germination index, and MGT represented as mean germination time

***Different letters after the data indicate significant differences between treatments of the same variety in the same column, $\alpha = 0.05$, LSD

References

Bae YH, Okano L, Hsu R (1987) Thermo-sensitive polymers as on–off switches for drug release. Makromol Chem, Rapid Commun 8:481–486

Bae YH, Okano T, Kim SW (1991) "On–off" thermocontrol of solute transport. I. Temperature dependence of swelling of *N*-isopropylacrylamide networks modified with hydrophobic components in water. Pharmaceut Res 8:531–537

Cui HW, Ma WG, Guan YJ, Li YP, Zheng YY, Hu J (2012) "Intelligent" seed pellets may improve chilling tolerance in tobacco. Front Life Sci 6(3–4):87–95

Dong LC, Hoffman AS (1986) Thermally reversible hydrogels: III. Immobilization of enzymes for feedback reaction control. J Control Release 4:223–227

Feil H, Bae YH, Feijen J, Kim SW (1991) Molecular separation by thermosensitive hydrogel membranes. J Membr Sci 64:283–294

Feil H, Bae YH, Feijen J, Kim SW (1993) Effect of comonomer hydrophilicity and ionization on the lower critical solution temperature of *N*-isopropylacrylamide copolymers. Macromolecules 26:2496–2500

Fundueanu G, Constantin M, Ascenzi P (2009) Poly (*N*-isopropylacrylamide-*co*-acrylamide) cross-linked thermoresponsive microspheres obtained from preformed polymers: influence of

the physico-chemical characteristics of drugs on their release profiles. Acta Biomater 5:363–373
Gehrke SH, Andrews GP, Cussler EL (1986) Chemical aspects of gel extraction. Chem Eng Sci 41:2153–2160
Grellier P, Riviere LM, Renault P (1999) Transfer and water-retention properties of seed-pelleting materials. Eur J Agron 10:57–65
Guan YJ, Hu J, Wang XJ, Shao CX (2009) Seed priming with chitosan improves maize germination and seedling growth in relation to physiological changes under low temperature stress. J Zhejiang Univ -SCI B 10:427–433
Guan YJ, Li Z, He F, Huang YT, Song WJ, Hu J (2015) "On–off" thermoresponsive coating agent containing salicylic acid applied to maize seeds for chilling tolerance. PLoS ONE 10(3): e0120695
Hope HJ, White RP, Dwyer LM, Maamari R, Seguin S, Hamiltonr RI (1992) Low temperature emergence potential of short season corn hybrids grown under controlled environment and plot conditions. Can J Plant Sci 72:83–91
Janda T, Szalai G, Tari I, Paldi E (1999) Hydroponic treatment with salicylic acid decreases the effects of chilling injury in maize (*Zea mays* L.) plants. Planta 208:175–180
Kaneko Y, Yoshida R, Sakai K, Sakurai Y, Okano T (1995) Temperature-responsive shrinking kinetics of poly (*N*-isopropylacrylamide) copolymer gels with hydrophilic and hydrophobic comonomers. J Membr Sci 101:13–22
Kang GZ, Wang CH, Sun GC, Wang Z (2003) Salicylic acid changes activities of H_2O_2-metabolizing enzymes and increases the chilling tolerance of banana seedlings. Environ Exp Bot 50:9–15
Kang HM, Saltveit ME (2002) Chilling tolerance of maize, cucumber and rice seedling leaves and roots are differentially affected by salicylic acid. Physiol Plant 115:571–576
Kunkel BN, Brooks DM (2002) Cross talk between signaling pathways in pathogen defense. Curr Opin Plant Biol 5:325–331
Li SK, D'Emanuele A (2003) Effect of thermal cycling on the properties of thermoresponsive poly (*N*-isopropylacrylamide) hydrogels. Int J Pharm 267:27–34
Lukatkin AS (2003) Contribution of oxidative stress to the development of cold-induced damage to leaves of chilling-sensitive plants: 3. injury of cell membranes by chilling temperatures. Russ J Plant Physiol 50:243–246
Rao YP, Pan CY, Lai JH (2006) Swelling and drug releasing properties of P(NIPAAm-*co*-AAm) hydrogels and PAAc/P(NIPAAm-*co*-AAm) IPN hydrogels. Chinese J Appl Chem 23:1129–1134
Schild HG (1992) Poly (*N*-isopropylacrylamide): experiment, theory and application. Prog Polym Sci 17:163–249
Tasgin E, Attici O, Nalbantoglu B (2003) Effect of salicylic acid and cold on freezing tolerance in winter wheat leaves. J Plant Growth Regul 41:231–236
Yoshida R, Sakai K, Okano T, Sakurai Y (1992) Drug release profiles in the shrinking process of thermorsponsive poly (*N*-isopropylacrylamide-*co*-alkyl methacrylate) gels. Ind Eng Chem Res 31:2339–2345
Zhang HQ, Zou YB, Xiao GC, Xiong YF (2007) Effect and mechanism of cold tolerant seed-coating agents on the cold tolerance of early Indica rice seedlings. Agr Sci China 7:792–801
Zhang Q (2010) Preparation and analysis of thermo sensitive gels. M.Sc. Thesis, Harbin University of Science and Technology

Chapter 4
Nanogels of Natural Polymers

Daniela Pamfil and Cornelia Vasile

Abstract Over the past few decades, a considerable interest was shown in developing nanogels (NGs) based on natural polymers or their derivatives to serve as generally regarded as safe (GRAS) biomedical devices that fuse the properties of both hydrogels and nanoparticles. This chapter presents different strategies to synthesize NGs via physical self-assembly or covalent crosslinking and their applications in therapeutic/biomedical fields, such as drug delivery systems, tissue engineering, artificial chaperone, imaging probes.

Keywords Nanogels · Biopolymers · Polysaccharides · Protein
Drug delivery · Artificial chaperone

1 Definition, Classification, and General Characteristics

Nanogels (NG), three-dimensional chemically or physically crosslinked nanoscale polymer networks with high capacity to hold water, without actually dissolving into the aqueous medium, are of great interest in various biomedical fields. Nanogels are nanosized counterparts of hydrogels which combine the properties of gels with those of colloids such as high surface–volume ratio, microheterogeneous structure and small sizes. Called as "nanoscalar polymer networks," "gel nanoparticles," "nanoscale hydrogels," etc., nanogels (NGs) are stable, soft, and swollen in good solvents. Like hydrogels, they exhibit tunable chemical and physical structures, good mechanical properties, and biocompatibility. Their large surface area is very important for bioconjugation, and their time of circulation in blood is long, being possible to be actively or passively distributed to target site, a special locations of the body such as a tumor site (Gao et al. 2014b). Because of their nanodimensions, nanogels appear to dissolve in solvents, like linear macromolecular chain polymers

---

D. Pamfil · C. Vasile (✉)
Department of Physical Chemistry of Polymers, "Petru Poni" Institute of Macromolecular Chemistry of Romanian Academy, 41A Grigore Ghica Voda Alley, 700487 Iasi, Romania
e-mail: cvasile@icmpp.ro

© Springer Nature Singapore Pte Ltd. 2018

V. K. Thakur et al. (eds.), *Polymer Gels*, Gels Horizons: From Science to Smart Materials, https://doi.org/10.1007/978-981-10-6080-9_4

in a good solvent, but preserve an almost fixed molecular conformation (Dispenza et al. 2015). Depending on the polymer nature and preparation method, NGs may possess tunable size from nanometers (1–250 nm) to micrometers. They are mostly spherical particles but some synthetic strategies allow for the fabrication of nanogels of different shapes. Actually, when the self-assembling approach is used, the nanostructured gels are sometimes defined also as micelles, due to a certain similarity with surfactant micelles bearing a hydrophobic core and a hydrophilic shell (Zhang et al. 2013).

Nanogels may be classified on the basis of their synthesis method, structure, and functions. In respect with method of preparation, two important classes of nanogels, namely chemically and physically crosslinked nanogels, are known. Usually, the synthesis techniques are established according to the application and type of study in which the designed nanogels are intended to be involved.

Chemical crosslinked nanogels are formed through inter- or/and intramolecular covalent bonds.

There are simple nanogels (artificial chaperons), hollow nanogels, crosslinked core–shell nanogels, hairy crosslinked nanogels, multilayer nanogels, and functionalized nanogels.

One of the most important classes of nanogels according to their functions is based on response to an external stimulus. Smart nanogels exhibit the ability to respond to physiological stimuli like temperature, pH, ionic strength, redox environment, enzymes, by changing their volume (swelling degree), refractive index, and hydrophilicity/hydrophobicity balance.

Multifunctional hybrid nanogels which mix together different polymers and/or nanoparticles such as plasmonic, magnetic, and carbonaceous nanoparticles are also an indispensable class of nanomaterials, especially in nanomedicine, being used as drug carriers and also as imaging and theranostic agents. Molina et al. (2015) classified hybrid nanogels into two classes: nanomaterial–nanogel and polymer–nanogel composites.

In the nanomaterial–nanogel composites, a new component is hybridized with the nanogel; therefore, in their synthesis, nanosized materials such as plasmonic, magnetic, and carbonaceous nanoparticles (e.g., carbon nanotubes, graphene, and fullerenes) are incorporated. In nanomedicine, they are useful in imaging, guided therapy, triggered and controlled drug release, hyperthermia, etc.

The polymer–nanogel composites as interpenetrated networks (IPNs) and core–shell particles are especially advantageous for the formation of multiresponsive nanogels and provide a controlled release profiles of different active principles and for targeting therapy, respectively.

As it concerns their function, the NGs are very important in nanomedicine and they are known as plasmonic hybrid nanogels as theranostic agents, magnetic nanogels as magnetic resonance imaging contrast agents, gold hybrid nanogels as relevant bioimaging devices, silver containing nanogels well known for their antimicrobial properties and optical properties as potential biodiagnostic imaging, superparamagnetic iron nanoparticles (SPION) as magneto-transducers (MTs) and Magnetic Resonance Imaging (MRI), quantum dots (QDs) nanogels as drug delivery systems with simultaneous imaging, and biosensing and possibilities are practically unlimited.

The advantageous characteristic properties of nanogels based on natural polymers are:

- **Higher drug loading capacity**. Nanogels are extremely valuable as drug delivery systems in view of the fact that they display both incorporation and release behavior of a large variety of active compounds (cargo), such as bioactive molecules (drugs, peptides, proteins, antigens, oligonucleotides, genes, carbohydrates, DNA) and inorganic molecules (quantum dots, silver and gold nanoparticle, magnetic nanoparticles) as compared to bulk hydrogels that have some limitations. Moreover, both hydrophilic and hydrophobic drugs can be loaded in nanogels which are able to solubilize hydrophobic drugs. The capacity of nanogels to incorporate an impressive amount of drugs is influenced by the type of functional groups present in the polymeric chains that have a remarkable impact on drug delivery and drug releasing properties. A part of the functional groups are able to conjugate with some therapeutic agents through physical interactions of hydrogen bonding or van der Waals forces that facilitate the drug-carrying efficiency having a good potential for targeting applications, while the functional groups at drug/protein molecules interface assure a higher loading.
- **Nanoparticle size**. The particle size is very important for nanogel use in a particular biomedical application. For example, the nanoparticles with smaller sizes ranging between 10 and 100 nm are more advantageous and more suitable for crossing biological barriers through tissue diffusion, extravasation, improving the blood circulation time, and escape from hepatic filtration (Alexis et al. 2008). Usually, an enhancement in the extravasation rate (good permeation capabilities) along the tissues is required, especially in the tumors therapy and also they are able to overpass the blood-brain barrier (BBB).

Other properties are:

- swelling property in aqueous media in a controlled manner
- electromobility
- superior colloidal stability, while those based natural polymers show good biocompatibility and degradability.

However, nanogels show same disadvantages such as expensive procedures to completely clean the final product of by the traces of residual solvent, surfactants, unreacted monomers at the end of synthesis process; these residues can impart toxicity (Rossetti et al. 2002; Xu et al. 2007).

2 Preparation Methodologies

Table 1 presented the possible ways of preparation for natural polymer-based nanogels. A valuable route that may provide opportunities to control the spatial distribution of polymeric chains at the nanoscale dimensions is the using of micelles controlled self-assembly and simultaneous crosslinking procedures. Hybrid polymerizable nanogels that are obtained by both physical and chemical crosslinks have been synthesized.

2.1 Physical Methods Involving Self-assembly of Interactive Polymers

Actually, macromolecular self-assembly leading to nanostructures is being investigated since many years (Whitesides et al. 1991) because such processes are rather common in nature and also because of the wide variety of practical applications that such systems may have (Kabanov and Vinogradov 2009), in particular in the field of pharmaceutics.

Self-assembly is the spontaneous disorder-to-order transition of amphiphilic copolymers driven by non-covalent interactions, such as the van der Waals and hydrogen bonding forces, hydrophobic effect, stereocomplexation, charge transfer, and polyelectrolyte complexation.

Throughout the duration of self-assembly process, drugs of both small and large dimensions, genes and proteins, can be entrapped into the polymeric nanogels.

2.1.1 Hydrogen Bonding

The crosslinked NGs' networks can be obtained through the formation of supramolecular hydrogen interactions produced between the functional groups of the polymers, including –OH, $-NH_2$, etc. However, these kinds of NG networks built via hydrogen bonding have poor stability, and the hydrogen bonding depends on the type of solvents and pH of the solution (Zhang et al. 2015). Bae et al. (2008) fabricated heparin-based nanogels through the nanocomplexation of thiolated heparin with polyethylene glycol (PEG) via hydrogen bonding, followed by the intermolecular crosslinking of disulfide bonds between thiolated heparin molecules by ultrasonication. The resulted nanogels were able to enhance the internalization and the release of heparin in the reductive cytosol environment causing the inhibition of cell proliferation and caspase-mediated apoptotic cell death.

Table 1 Preparation methods of natural polymeric nanogels and examples

Preparation methods			Examples	Characteristics	Applications	Reference
Physical self-assembly methods	Electrostatic interaction	Ionotropic gelation	Cationic chitosan crosslinked with anionic tripolyphosphate (TPP)	– Ability to change their properties under (volume) electron bombardment – Particle size = 118.54 ± 15.93 nm	– Specific drug release kinetic mechanisms to enhance and predict the drug delivery efficacy	Ashrafi and Azadi (2016)
			Curcumin-loaded alginate-carrageenan nanoparticles	– Particle size = 480 nm	– Improve the curcumin dissolution under gastrointestinal conditions	Guzman et al. (2013)
		Polyelectrolyte complexation	Egg yolk lipoprotein/pectin nanogels obtained by reaction heat-induced complexation	– Particle size = 58 nm, zeta potential = −41 mV, spherical shape, smooth surface, homogeneous size distribution	– Promising features for curcumin encapsulation and controlled release	Zhou et al. (2016)
	Hydrogen bonding		Multifunctional chitosan-poly (methacrylic acid)-QDs hybrid nanogels	– Formed by non-covalent associations such as secondary forces (hydrogen bonding or hydrophobic association) and physical entanglements – Particle size = 74.7–174.5 nm	– Tumor cell imaging and controlled drug delivery	Wu et al. (2010b)
	Hydrophobic effect		Nanogels based on deoxycholic acid-modified glycol chitosan with palmityl acylated exendin-4	– Particle size ~220 nm	– Offer a long-acting inhalation delivery system for treating type 2 diabetes	Lee et al. (2012)

Table 1 (continued)

Preparation methods			Examples	Characteristics	Applications	Reference
Chemical methods	Emulsion/ solvent evaporation	Direct emulsion (o/w)	Poly(L-amino acid) nanogels prepared by crosslinking of poly[L-aspartic acid-g-(3-diethylaminopropyl)]-b-poly(ethylene glycol)-maleimide and poly(L-aspartic acid-g-ethyl thiol)-b-PEG	– Controllable pH-responsive properties exhibiting an intelligent volume change corresponding to a change in the lysosomal pH (pH 5.0) – Particle size = 125 nm	– Tumor therapy	Oh et al. (2013)
		Inverse (mini) emulsion (w/o)	Alginate aldehydes–gelatin nanogels	– Particle size = 320–390 nm – Enhance the water solubility and bioavailability of curcumin	– Drug delivery system for the treatment of breast cancer	Sarika et al. (2016)
		Reverse micellar emulsion or reverse (micro) emulsion	k-carrageenan-based nanogels	– Particle size <100 nm – Obtained by sonication of a system comprising n-heptane as the organic phase, cetyltrimethyl ammonium bromide as surfactant and 1-butanol as co-surfactant – Thermo-sensitive in a temperature range acceptable for living cells (37–45 °C)	– Smart therapeutics such as controlled thermo-sensitive drug carriers	Daniel-da-Silva et al. (2011)
			Nanogels based on silica, polyethylenimine (PEI), and chitosan	– Rigid and stable gene carrier material in the reverse microemulsion system – Particle size = 70 nm	– Potential applications in non-viral gene delivery	Tian et al. (2016)
	Carbodiimide coupling		Glycol chitosan/sodium alginate-poly (L-glutamate-co-*N*-3-L-glutamyl phenyl boronic acid) graft polymer double-layered nanogel	– Particle size = 767.9 nm – Exhibits a strong glucose-responsive characteristic at diabetic glucose levels	– Potential platform for advanced insulin delivery systems	Lee et al. (2015)

Table 1 (continued)

Preparation methods			Examples	Characteristics	Applications	Reference
	Michael addition reaction		Pluronic F127—hyaluronic acid nanogels	– Proceeded by reaction of the thiol-modified HA to Pluronic diacrylate – Particle size = 150 nm	– Delivering anticancer drug without hampering their toxicity value over longer durations and reducing the total dose amount in anticancer therapy	Gurav et al. (2016)
	Free radical polymerization		Nanohydrogels based on poly(*N*-isopropylacrylamide)—chitosan incorporated with Fe_3O_4 magnetic nanoparticles	– Dual stimuli pH and temperature-responsive – Particle size = 250 nm – In the presence of a magnetic field, a heat induces chain collapse takes place, with the expulsion of water facilitating the release of entrapped therapeutics in acidic medium.	– Promising candidate for a combination therapy (cancer hyperthermia)	Jaiswal et al. (2014)
	Precipitation polymerization (direct emulsion)		Nanogels composed of natural gelatin, biological protein macromolecules, and poly (acrylamidoglycolic acid)	– Interpenetrating polymeric network – pH-sensitive properties – Particle size = 100 nm – Release curcumin in a controlled manner	– Colorectal cancer drug delivery applications	Rao et al. (2015)
	Inverse emulsion polymerization (w/o)	Inverse (mini) emulsion polymerization	Nanogels based on gelatine and poly(acrylic acid)	– Particle size = 255 nm – Interpenetrating polymer network (IPN) – Stable droplets of acrylic acid (AA), gelatine and NaOH mixture were obtained by sonication using APS as initiator for radical generation. Then, accelerator (TEMED) and crosslinker (BIS) have been employed to polymerize and crosslink AA. The formed sIPN are further used to make full IPN by the addition of glutaraldehyde as crosslinker for gelatine	– Drug delivery vehicles for cancer targeting – Treatment of solid tumors	Koul et al. (2011)
		Inverse (micro) emulsion polymerization	Nanogels of acrylic acid and *N*-isopropyl acrylamide onto carboxymethyl cellulose	– Synthesized by graft co-polymerization – Ethylene glycol dimethacrylate as crosslinker – Particle size = 50–90 nm	– Remove heavy metal ions from various environments	(Farag et al. 2016)

Table 1 (continued)

Preparation methods			Examples	Characteristics	Applications	Reference
	Heterogeneous controlled/ living radical polymerization	Atom-transfer radical polymerization (ATRP)	Nanostructured HA hydrogels through a thiol–Michael reaction of thiolated HA with acrylate-functionalized nanogels	– These nanogels showed controlled release of biomolecules like rhodamine B and dextran – Particle size = 203 nm	– Injectable biocompatible matrix for cell and protein encapsulation in both tissue engineering and drug delivery applications	Bencherif et al. (2009)
		Reversible addition-fragmentation chain transfer (RAFT)	Nanogel based on PNIPAM and pullulan	– Nanogel building blocks were obtained by grafting short PNIPAM chains onto pullulan by RAFT polymerization of NIPAM in water controlled by sulfanylthiocarbonylsulfanyl groups covalently attached to the pullulan framework acting as chain transfer agents – Particle size = 50–80 nm	– Stimuli-responsive nanobiomaterials	Morimoto et al. (2008)
	Radiation crosslinking		Carboxymethyl starch nanogel	– Particle size <50 nm – Radiation crosslinked process on the electron beam (EB) linear accelerator	– Carrier to bioactive substances (probiotics and/ or microbial biopesticides)	Binh et al. (2012)

Table 1 (continued)

Preparation methods		Examples	Characteristics	Applications	Reference
	Microfluidic system	Alginate-based nanogels	– Microfabricated with a crossjunction poly(dimethylsiloxane) microfluidic device with three inlets (for $CaCl_2$ and particles) and one outlet (for biopolymer solution) – Controlling the flow ratio, the mixing time of the ionic crosslinking can be adjusted and the size of the formed nanogels can be modulated. Considering alginate chains as anionic polyelectrolyte, nanogels are formed through diffusion-mediated mass transfer of Ca^{2+} ions into the focused polymer solution stream in a controlled manner – Particle size = 68–138 nm	– Efficient polypeptides encapsulation and their sustained release	Bazban-Shotorbani et al. (2016)
	Micromolding	Poly(ethylene glycol)-based hydrogel microparticles containing an amino polysaccharide chitosan	– Highly uniform	– Various biomedical applications	Jung and Yi (2015)

2.1.2 Electrostatic Interaction (Ionotropic Gelation, Polyelectrolyte Complexation)

Ionotropic Gelation

Ionotropic gelation technique consists in the ionic crosslinking of a polyelectrolyte using opposite multivalent counter ions (Patil et al. 2010) to form a homogeneous matrix with reversible structure depending on the environmental conditions. This strategy is simple and occurs under mild conditions in the aqueous environment and preserves the native structure of the biopolymers keeping safe their biological activity. This technique implies the pouring of polymer solution to the aqueous solution containing counter ions and then the formation of electrostatic interaction between electrolyte groups of natural polymer and charge groups of counter ions, forming a three-dimensional lattice of ionically crosslinked moieties (Ahmad et al. 2016). The size and surface charge of nanogels can be changed via variation of the polymeric ratio and of possible stabilizing agents.

Chitosan–sodium tripolyphosphate–glycolipide nanogels, prepared by ionic gelation technique with smooth and spherical aspects and sizes ranging in the interval 200–550 nm, have been obtained by magnetic stirring (600 rpm) at a temperature of 25 °C (Ravi and Baskaran 2015).

Polyelectrolyte Complexation (PEC)

Polyelectrolyte complexation or self-assembly polyelectrolytes is a technique that involves polyelectrolytes of opposite charges mixed in an aqueous medium. The primarily driven forces between the oppositely charged polymers are an increase in entropy caused by the elimination of the counter ions in the surrounding reaction medium (Rolland et al. 2012), hydrogen bonding or hydrophobic interactions, all participate to the PEC formation.

The synthesis of polyelectrolyte complexes is influenced by pH and temperature of the reaction, the ratio between polymers and charges, the molecular weight and degree of ionization of the involved polymers, the ionic strength and polymer charge density (Sarika and James 2016a).

Colloidal PECs are formed under non-stoichiometric conditions, with charge mixing ratios $n^+/n^- \sim 1$, by a one-shot addition (Polexe and Delair 2013) of the appropriate volume of the oppositely charged polymer solutions in order to obtain the required charge mixing ratio under magnetic stirring (between 750 and 800 rpm) at room temperature. It was demonstrated that the formation of these complexes is controlled only by the mixing ratios of the polymers and not on their concentrations (Boddohi et al. 2009).

The "one-shot" addition can prevent the potential flocculation of polyelectrolyte complex nanoparticles (PCN) when the charge mixing ratio is approximately 1. The size of the resulting PCN does not depend on the order of mixing (Schatz et al. 2004).

2.1.3 Hydrophobic Effect

Cholesterol is often considered as a hydrophobic molecule leading to the self-assembly of amphiphilic copolymers and to nanogels formation characterized by a good balance between hydrophilicity and hydrophobicity. Hyaluronic acid and pullulan are two biopolymers that were modified with cholesterol in order to provide a high degree of physical crosslinking to form stable nanogels. Ultrasonication is required for nanogel preparation through hydrophobic interactions. The synthesized cholesteryl–hyaluronic acid–drug conjugates easily formed compact nanogels with diameter of 20–40 nm after ultrasonication in aqueous media. By controlling bath sonicator conditions (30–60 min, 70 W, 42 kHz ±6%), aggregate formations can be avoided (Wei et al. 2013).

Cholesterol-modified pullulan (CHP) is also a usual amphiphilic product used to form nanogels by physically crosslinking strategy (Fig. 1).

An example of CHP-based nanogel formed by self-assembly via hydrophobic interactions was reported in a recent work of Hashimoto et al. (2016). This group has proposed amphiphilic pullulan backbone as natural component which was hydrophobized with cholesterol branches that are capable to provide physical crosslinking points by hydrophobic interactions in aqueous media.

2.2 *Chemical Methods*

Chemically crosslinked nanogels by covalent bonds show colloidal stability under in vivo conditions, preventing the fast release of drugs caused by unwanted dissociation of the gel network, and usually require the use of toxics surfactants and organic solvents, initiators and catalysts that may have a negative impact on the biocompatibility and of the nanogels.

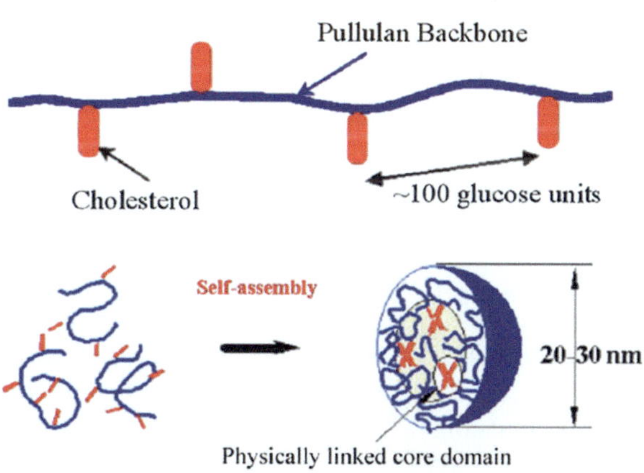

Fig. 1 Schematic representation of cholesterol-modified pullulan-based nanogel preparation by physical self-assembly (hydrophobic effect) (with permission from Elsevier for Lee and Akiyoshi 2004)

The chemical crosslinking is generated between both polymer precursors and monomers.

2.2.1 Emulsion Methods (Solvent Evaporation)

Direct Emulsion Method (o/w)

The emulsion method involves in the first step: the preparation of an emulsion with droplets of uniform size. Nanogels formed by emulsion method imply polymer precursors (Zhang et al. 2016) which are lipophilic (water insoluble) and present some reactive sites which are directly engaged in the chemical crosslinking. In the oil-in-water emulsion (o/w), the formed particles are more readily wetted by the water molecules and they would reduce their contact with the oily phase, inducing the interface to curve and form spherical oil droplets. Contrary, in the water-in-oil emulsion (w/o), more lipophilic particles will occur.

The emulsion stability against phase separation and oil droplet coalescence depends on many factors including surface charge density, emulsifier surface coverage, emulsifier layer thickness as well as bulk physicochemical conditions (Ray and Rousseau 2013).

To produce a stronger complex film in oil–water interface, heat treatment may be an efficient method. The consequence of heating on the physicochemical properties of emulsions synthetized with acid-soluble soy protein (ASSP), and soy-soluble polysaccharide (SSPS) was investigated (Liu et al. 2016). Droplet nanogels were formed under stirring for 3.5 h, and then the mixture was pre-emulsified using a homogenizer at 10,000 rpm for 1 min and was immediately emulsified employing a high-pressure homogenizer at 500 bar for two cycles, followed by heating treatment in water bath with stirring at 80 °C for 60 min or not. Their method of adopting a heating process (80 °C, 1 h) for weight ratio 1:4 ASSP/SSPS complex emulsion at pH 3.5 produced fairly stable emulsion without any flocculation. After 90-day storage at 4 °C, the mean diameter of emulsions after heating at 80 °C for 60 min was 314 ± 1 nm compared with 341 ± 3 nm of emulsions unheated.

Pickering emulsion is a particular class of emulsions, introduced in early twentieth centuries (Pickering 1907; Tang et al. 2015), and ensures an enhanced stability against coalescence compared to surfactant stabilized emulsions.

Inverse Emulsion Method (w/o)

Water-in-oil (w/o) emulsion involves the droplets emulsification of water-soluble biopolymers in continuous oil phase in the presence of oil-soluble surfactant and the biopolymers crosslinking with water-soluble crosslinkers (Oha et al. 2009). There are two types of inverse emulsion, namely inverse (mini) emulsion and inverse (micro) emulsion.

Inverse (Mini) Emulsion Method

This method implies the adding of aqueous biopolymer droplets to a continuous lipophilic solvent using surfactant to produce a homogenous emulsion. A high-speed mechanical stirrer is useful for good homogenization. The aqueous droplets of biopolymer are then crosslinked with the corresponding crosslinkers. The next step is the formation of crosslinked nanogels as dispersion in organic solvents, followed by a purification by precipitation, centrifugation, washing (usually with organic solvents, such as isopropanol), and lastly freeze-drying. The nanogel diameter dimensions depend on the concentration of surfactants and crosslinking agents and on the stirring speed necessary for the inverse emulsion formation.

Hyaluronic-acid-based nanogels with a 200–500 nm size distribution were obtained in an inverse emulsion solution that was subjected to 15 s of sonication and 2 h of stirring with the formation of stable disulfide bonds within the nanosized droplets (Lee et al. 2007).

Alginic aldehyde–gelatine (AA–gel) nanogels (Sarika et al. 2015), with 103–256 nm diameter sizes, were obtained when the mixture was added to the surfactant (Span 20) dissolved in cyclohexane following a sonication process (5 min) to obtain inverse (mini) emulsions. Ultrasonication was chosen in order to obtain a good nanogel dispersion and also for the reduction of droplets sizes and polydispersity. The surfactant stabilizes the resultant nanogels from growth by collision, coalescence, and Ostwald ripening. The crosslinked AA nanogels were produced inside the surfactant micelle.

According to the surface charge impaired by the used emulsifiers and/or lipids, anionic and cationic emulsions are known (de la Fuente et al. 2010). Negatively charged emulsions are obtained by adding anionic lipids and surfactants in the aqueous reaction medium, while positively charged emulsions are synthetized with the aid of cationic lipids. Moreover, cationic polysaccharides such as chitosan contribute to the formation of a layer all around the oily droplets impairing a positive charge to the emulsion.

This technique is not proper for loading of hydrophobic drugs because the continuous phase is a lipophilic solvent (Asadi and Khoee 2016).

Reverse Micellar Method or Inverse (Micro) Emulsion

Like inverse (mini) emulsion technique, the reverse micellar emulsion consists in a water-in-oil dispersion, but a high concentration of oil-soluble surfactants is used to obtain a thermodynamically stable micellar solution which is represented by aqueous droplets dispersed in the continuous oily phase (Maitra 1984). The concentration of amphiphilic surfactants must be above the critical micelle concentration in order to increase the number of reverse micelles (Uthaman et al. 2015). The diameter of the final micellar droplets ranged between tens and hundreds of nanometers.

Reverse (micro) emulsion is an efficient modality to obtain nanosized gel particles with uniform particle size distribution.

Daniel-da-Silva et al. (2011) synthesized nanogels based on k-carrageenan using reverse (micro) emulsions combined with heat-induced gelation. Their method consisted in the dissolving of surfactant in heptane/1-butanol mixture by sonication at a temperature of 75 °C. The aqueous phase consists in a solution of 0.1 M k-carrageenan dissolved in KCl. A quantity of the aqueous phase containing surfactant was mixed with the organic phase, 2 min and sonication a constant temperature of 75 °C. Then the obtained mixture was cooled to the room temperature in order to induce the heat gelation process of the nanospheres. Finally, the gel particles were collected by centrifugation.

2.2.2 Michael Addition Reaction

The Michael addition represents a facile and rapid procedure that takes place at room temperature by the reaction of nucleophiles with the activated olefins and alkynes, where the nucleophile attaches over the multiple C–C bonds (Michael 1887). This reaction is suitable for a broad range of polymers, it offers low cure times and involves less toxic precursors. Michael addition networks based on thiol and acrylate precursors are ideal because of the possibility to control the gelation process using a basic catalyst (Mather et al. 2006).

Michael-type addition methodology was used for the preparation of Pluronic F 127—hyaluronic acid (HA) nanogels and was proceeded by reaction of the thiol-modified HA with Pluronic diacrylate (Gurav et al. 2016). The obtained nanogels have diameters around 150 nm and narrow size distributions.

Chitosan is another biopolymer that was involved in the formation of nanogels using the Michael addition reaction. Chitosan thiomer was first crosslinked via ionic gelation with chondroitin sulfate or dermatan sulfate, and then *N,N′*-bis (acryloyl) cystamine, as disulfide bond-containing crosslinker, was added and a Michael addition reaction between the reactive nucleophilic groups of chitosan thiomer (including SH, NH_2 or OH) and the acrylic groups of *N,N′*-bis (acryloyl) cystamine took place (Curcioa et al. 2015).

2.2.3 Free Radical Polymerization

When the free radical polymerization is used for the nanogel preparation, by a two steps process consisting of the monomer polymerization in a homogeneous phase or in a microscale or nanoscale heterogeneous medium and the nanogels formation. Usually, these steps are performed simultaneously. The nanogels designed by monomer polymerization show a greater efficiency in comparison with the nanogels obtained by means of preformed polymers (polymer precursors) (Zhang et al. 2016). The polymerization reaction in nanogels can be afforded through direct or reverse emulsion polymerization techniques. Radical polymerization may be also performed by three of the most successful methodologies, namely atom-transfer radical polymerization (ATRP), stable free radical polymerization (SFRP), and reversible addition-fragmentation chain transfer (RAFT) polymerization.

Direct Emulsion Polymerization (o/w) or Precipitation Polymerization

The synthesis of the nanogel by *surfactant-free emulsion polymerization* involves water-soluble monomers that are subjected to a homogeneous nucleation (initiation) followed by polymerization with the formation of a colloidal suspension. The charge imparted by the initiator will stabilize the gels (Tan et al. 2010). The reaction requires the utilization of crosslinkers because of the predisposition of the resulted polymers to solubilize in the reaction medium. On the other hand, crosslinking agents may cause some unwanted effects such as the formation of nanogels with irregular shapes and high polydispersity even though they play an essential role on the isolation process of the obtained particles. It can be obtained a higher control over particle dimensions and particle size distribution by changing the monomer concentration in water.

To obtain nanogels with sizes as small as possible, during reaction the precursor particles must be stabilized earlier, and also, to confer colloidal stability, an appropriate amount of ionic surfactants is introduced. It was confirmed that the reduction of the surfactant concentration of the reaction leads to the formation of nanogels with higher dimensions (Nayak and Lyon 2005).

The reactions of (micro) emulsion not only behave as microreactors for the nanogel particles synthesis but also prevent the excessive aggregation of the gel particles (Liu et al. 2007a). The nanogels obtained via emulsion polymerization reaction may contain some unreacted monomers and oligomers that cause toxic effects with the manifestation of serious health problems (Tang et al. 2016).

Precipitation polymerization represents a convenient method in case of the hydrophobic materials because they can easily hang to the collapsed precursor particle. Moreover, this kind of polymerization is valid only for materials that manifest good stability at high temperatures, and thus, there is no possibility to entrap biological macromolecules. When a number of hydrophilic comonomers undergo a co-polymerization reaction, only a few weight percentages can be incorporated (Debele et al. 2016). To overcome these disadvantages, the utilization of inverse emulsion polymerization may be a good option.

Inverse (Mini) Emulsion Polymerization

Inverse (mini) emulsion polymerization represents a water-in-oil polymerization technique that is based on the formation of aqueous droplets containing water-soluble monomers uniformly dispersed and the aid of oil-soluble surfactants in the continuous organic phase. After radical initiation, the polymerization reaction inside the monomer droplets takes place resulting in the nanogels formation.

To obtain kinetically stable dispersions, the mechanical stirring is used in case of the inverse emulsion process and sonification in case of the inverse (mini) emulsions polymerization.

Inverse (Micro) Emulsion Polymerization

If inverse (mini) emulsion polymerization creates kinetically stable emulsions below or close to the critical micellar concentration, inverse (micro) emulsion polymerization forms thermodynamically stable (micro) emulsions upon further addition of emulsifier above the critical threshold (Oh et al. 2008). The inverse (micro) emulsion polymerization allowed the synthesis of nanogels with high molecular weights of *cca.* 107 g/mol and diameters smaller than 50 nm (Puig et al. 2012). This technique was also explored for the formation of Ca-alginate-based nanogels (Zhu et al. 2015) that possessed charged negatively surface with a zeta potential of −44.7 mV and diameter sizes of 42.54 nm. The very small nanoparticle sizes were obtained by stirring the w/o emulsion at a speed of 1000 rpm at a temperature of 40 °C.

Semicontinuous o/w and w/o (micro) emulsion polymerization is another process that, compared to inverse (micro) emulsion, allows the increasing of polymer content without adding more surfactant to the original reacting formulation (Puig and Rabelero 2016).

2.2.4 Radiation Crosslinking

Nanogels can be obtained by *high-energy radiation crosslinking* as well. This technique has been intensely used for the generation of hydrogels with uniform particle size at the micro- and nanoscale that depends on the polymer concentration and on the dose per pulse. If more than one radical is formed on the same chain in favorable position and the polymer concentration is low, intramolecular combination is the prevailing termination reaction and nanogels form as a result.

More recent studies suggest that several experimental parameters can contribute to the final structure and properties of radiation-induced crosslinked polymer nanogels, in particular: the polymer concentration in solution; the dynamics and diffusion kinetics of polymeric segments in solution, i.e., the temperature, nature of the solvent, polymer molecular weight distribution; the nature and half-life of formed macroradicals; the duration of each single pulse, i.e., pulse width; the pulse repetition rate (or frequency); the total dose absorbed.

For example, the electron beam irradiation in the range of absorbed doses of 10–16 kGy conducted to the decrease of some parameters, including intrinsic viscosity of carboxymethyl starch (CMS) solution, particle size, and hydrodynamic radius of irradiated CMS nanogels, and to the increase of average molecular weight of irradiated CMS macromolecules (Binh et al. 2012). The hydrodynamic radius of

CMS nanogel was approximately 13 nm for a CMS solution irradiated at 16 kGy, which was 10 times less than that of unirradiated CMS.

2.3 Modern Techniques

The modern strategies used for nanogels synthesis belong to the microfabrication methodologies such as photolithography, microfluidic system, and micromolding implying different crosslinking chemistries, as physical crosslinking, covalent crosslinking, and photo-crosslinking. When nanogels are obtained by microfabrication, their sizes present a narrow distribution, are mono dispersed, have the ability to load a significant quantity of drug, and provide a better batch-to-batch uniformity in comparison with the conventional methodologies (Jahn et al. 2008).

2.3.1 Photolithographic Technique

Photolithography is a top-down method and utilizes photochemical reactions (photo-polymerization) for activation, and subsequent reaction has been explored to fabricate nanogels. By offering the possibility to have a high level of control and reproducibility, lithographic method can be an attractive option in comparison with the older and well-established methods. Three important lithographic techniques were identified: photolithography, imprint lithography and flow lithography (Helgeson et al. 2011). All these photolithographic methods require the realization of special techniques for surface treatment of stamps or new materials for replica molds to allow the molded gels separation from stamps or replica molds.

This technique involves several stages. At the beginning, the UV-crosslinkable polymer, which has a decreased surface energy as a substrate, is released on the pre-heated photoresist-coated water. The following action is the polymer molding into silicon patterns, pressing of the quartz template onto the polymer and its exposure to the intense UV light. In the next step, the particles with a thin residual interconnecting film layer are uncovered by removing the quartz template. Further, the formed residual coating is detached by oxidation with a plasma containing oxygen. In the last stage, the resulted particles are directly harvested by dissolving the substrate in water buffer.

2.3.2 Micromolding Method

This procedure is similar to photolithographic technique but is more advantageous minimizing the need to use costly lithographic equipment and clean room facilities.

In a method described by Yeh et al. (2006), the live cells were suspended in a hydrogel precursor solution containing methacrylated hyaluronic acid or PEGDA (poly(ethylene glycol) diacrylate) and a photo-initiator. The resulting mixture was molded using a poly(dimethyl siloxane) (PDMS) patterns and then photo-crosslinked by exposure to ultraviolet radiations to form a gel. The mold was then removed to generate an array of micromolding hydrogels. The formed microgels were hydrated and collected using the simple wash. They were also molded into various shapes like square prisms, disks, or strings.

In a study of Jung and Yi (2015), the UV light exposure for 5 s of (ethylene glycol) based hydrogel microparticles containing an amino polysaccharide (chitosan) conducted to the formation of droplets with two different phases (core–shell design) and then the droplets were polymerized preserving the core–shell structure. Minimal differences were observed in the morphology of crosslinked droplets after 3 min of UV exposure with an 8 W handheld UV lamp.

2.3.3 Microfluidic Systems

Various microfluidic systems provide rapid and homogeneous mixing without any stimulator (Capretto et al. 2011). The flow squeezes the solvent stream between two anti-solvent streams and a rapid solvent exchange by diffusion takes place.

Generally, either emulsion-based methods or flow lithography-based methods are employed. In a microfluidic system, multiple phases, including aqueous phase, oil phase, crosslinker, are obtained which meet in a junction geometry (like T-junction) leading to the droplet formations (Liu and Garcia 2016). After droplet formations, the crosslinking takes place usually by chemically or ionically induced gelation. In a flow lithography-based microfluidic system, the crosslinking process occurs by photo-induced polymerization (Chung et al. 2012).

3 Stimuli-Responsive Nanogels

Some nanogels show responsiveness to environmental stimuli such as pH, temperature, enzymes, light, ion strength, magnetic field (Peng et al. 2012). Such nanogels can be obtained by incorporating stimuli-responsive polymers or stimuli-sensitive units into their composition/structures. The affected tissues or organs are different from healthy ones, as it concerns low pH, redox potential, or enzymes; therefore, the features of the stimuli-responsive nanogels should be exploited depending on these modifications (Wu and Wang 2016).

A pH-sensitive nanogel showing swelling–shrinking changed dependent on pH was obtained from natural gelatine and acrylamidoglycolic acid via free radical emulsion polymerization. This swelling–shrinking nature was resulted in the ionization of carboxylic groups as the pH increasing from 5.0 to 8.5. The repulsion interaction increased with the ionization and triggered the volume expansion of the nanogel. This nanogel can be loaded with 42–48% of curcumin (Rao et al. 2015).

Thermo-responsive nanogels from poly(L-lactide)-g-pullulan copolymers with phase transition temperature of 35 °C were investigated as DOX carrier (Seo et al. 2012). Between 37 and 42 °C, they show a high release.

Hyaluronic acid (HA) is an enzyme sensitivity polymer. The enzyme-sensitive groups in its structure can be disassembled by action of lipase and hyaluronidase. Targeting biocompatible and biodegradable HA nanogels containing enzyme-sensitive groups have been prepared by a methacrylation strategy (Yang et al. 2015). These nanogels with negative surface potential had a spherical morphology with an average diameter of about 70 nm. The diameter decreased up to 50 nm when 16% doxorubicin (DOX) was loaded into the nanogels, and encapsulation efficiency was of 62%. The DOX-loaded nanogels showed superior antitumor efficacy in H22 tumor-bearing mice having a great potential in drug delivery for cancer therapy.

Redox-responsive nanogels disassemble and release drugs in the biological medium. HA-based redox-responsive nanogels (200–500 nm) physically encapsulating small interfering RNA (siRNA) have been obtained by Lee et al. (2007) by an inverse water-in-oil emulsion method where thiol-conjugated HA in aqueous emulsion droplets was ultrasonically crosslinked via the formation of disulfide linkages. Release rates of siRNA from the HA nanogels were regulated by the glutathione (GSH) concentration. GHS is an intracellular reducing agent in the cytoplasm and therefore is an ideal internal stimulus for rapid destabilization of nanocarriers inside cells to accomplish efficient intracellular drug release. Upon exposure to GSH of the disulfide crosslinked HA nanogels, their degradation and disintegration occur.

Li and co-workers (2009) prepared a biocompatible and degradable nanovehicle with potential for tumor-targeted chemotherapy composed from dextran-lipoic acid derivatives (Dex-LAs). Lipoic acid is naturally produced in the human body and used as an antioxidant drug for treating diabetes and HIV (Biewenga et al. 1997). The obtained multifunctional crosslinked dextran nanoparticles are stable under extracellular conditions but are rapidly de-crosslinked under reductive environments (cancer cells). A high concentration of GSH tripeptides affects the rapid release of the drugs.

4 Polysaccharide-Based Nanogels

Polysaccharides that feature cationic and anionic groups, like chitosan, alginate, heparin, hyaluronic acid, dextran sulfate, chondroitin sulfate and cyclodextrins, have been used to form nanogels with oppositely charged polymers by intermolecular electrostatic interaction resulting ionic complexes (Fredheim and Christensen 2003; Berger et al. 2004). Such nanogels are often pH responsive and can respond to physiological pH differences.

While polysaccharides are very customizable through grafting and modification, their natural variability in composition and molecular weight can be a disadvantage when compared with well-defined synthetic polymers prepared with techniques such as controlled radical polymerization (Rocha et al. 2013), or other synthetic nanostructures such as dendrimers (Zhu and Shi 2013) and peptide-based nanoparticles (Ischakov et al. 2013).

Modification of hydrophilic polysaccharides with hydrophobic groups results in amphiphilic polymers that, when exposed to particular conditions in an aqueous environment, can self-assemble to form nanogels (Myrick et al. 2014). These derivatizations of polysaccharides with hydrophobic moieties have been recognized as a promising drug carrier, since their core parts, surrounded by a hydrophilic outer shell, can serve to encapsulate hydrophobic drugs (Mendes et al. 2013).

4.1 Chitosan-Based Nanogels

Active research has concentrated on the preparation of nanogels with natural hydrophilic polymers, such as chitosan (CS) (Tiyaboonchai 2013). CS is a positively charged polymer interacts with negatively charged polymers (polyanions) and even certain inorganic polyanions (Hamidi et al. 2012). CS can be grafted with various hydrophobic groups to form polymeric amphiphiles, which are able to self-assemble into NGs (Lee et al. 2012). By the free amino groups, it carries a positive charge and reacts with many negatively charged surfaces such as the cell membranes (Onsoyen and Skaugrud 1990; Kunjachan et al. 2010).

Chitosan-based nanogels may be obtained by oil-in-water lipid emulsion technique since chitosan is known to promote the stability of an emulsion and also prevents coalescence by steric and electrostatic hindrance, without the help of any additional surfactant or protein (Payet and Terentjev 2008).

Chitosan nanogels can be also prepared by chemical crosslinking leading to quite stable matrices. Calvo et al. (1997) prepared CS nanoparticles by an ion gelation technique between positive charges chitosan and negative charged tripolyphosphate (TPP). Some CS nanogels were prepared by reaction with bifunctional agents such as diisocyanate, diepoxy compounds, dialdehyde (Wenling et al. 2005; Rinaudo 2010) and glutaraldehyde. Because of the toxicity of such chemical agents, biocompatible crosslinking agents should be preferred for biomedical applications. An example of alternative natural crosslinker is genipin which reacts under mild conditions with primary amine groups from chitosan and other biopolymers such as collagen and gelatine. The use of genipin as a crosslinking agent to develop chitosan nanogels for biomedical applications was reported by Pujana et al. (2013) which prepared highly monodisperse chitosan nanogels by reverse microemulsion.

4.2 Pullulan-Based Nanogels

Only few polysaccharides have been exploited for nanohydrogel formation using different strategies, and pullulan is one of them being extensively studied in the topical biomedicine (Alhaique et al. 2015).

Because pullulan is a highly hydrophilic polysaccharide, it was hydrophobically modified with cholesterol (Lee and Akiyoshi 2004) or spiropyran (Hirakura et al. 2004) in order to obtain self-assembled nanogels from the resultant amphiphilic polysaccharide which are capable to form complexes with various globular and soluble proteins such as hemoglobin, peroxidase, myoglobin, and cytochrome.

Cholesterol-modified pullulan-based nanogels with functional groups may provide a better template for various purposes. A series of functional groups like tricarboxylate groups (Sasaki et al. 2012), acryloyl groups (Hashimoto et al. 2016), or urocanic acid (a pH-sensitive grafted moiety) (Wang et al. 2014) were added to the pullulan backbone.

The size and density of the pullulan-based hydrogel nanoparticles were related to the degree of substitution of the hydrophobic moiety (i.e., cholesterol).

Different proteins may be loaded in hydrogel nanoparticles under mild conditions just by mixing them. Akiyoshi et al. (1995, 1998) prepared insulin hydrogels using cholesterol-modified pullulan, where they obtained particle size of 20–30 nm. The nanogels were prepared by self-associating hydrophilic polymers, and it was reported that they are very stable and reduce the side effects of protein by significantly protection from enzymatic degradation and suppression the aggregation. Furthermore, in vivo studies demonstrated that insulin physiological activity was preserved after in vivo injection.

In order to induce a pH response to self-assembled cholesteryl pullulan nanogel, an acid-labile cholesterol derivative was bound to pullulan by vinyl ether groups and the obtained product was then hydrolyzed in acidic medium (Morimoto et al. 2013). When complexed with a model protein (BSA) in a neutral environment, the bioactive compound was released at a rate which was controlled by the degradation of the nanogel.

5 Hyaluronic Acid (HA)-Based Nanogels

HA, a polysaccharide of animal origin, is composed by units of glucuronic acid and *N*-acetylglucosamine and exhibits unique features, such as biocompatibility, mucoadhesion, non-immunogenicity, which make it suitable for the formation of nanogels (Alhaique et al. 2016). Besides, the HA shows a high bioactivity, in particular a selective interaction with several cell receptors, the most important being the CD44 (Clusters of Differentiation, CD). CD44 mediates the internalization of HA into the cell via endocytosis and the subsequent degradation within the lysosomes by means of hyaluronidases; an important feature of CD44 is to be a receptor over-expressed in various isoforms of several tumor cells, particularly in human solid tumors (Platt and Szoka 2008).

Cholesterol, as hydrophobic component, can form self-assemblies with amphiphilic copolymers where the hydrophilic/hydrophobic balance favors the synthesis of nanogels (Zhou et al. 2009). The cholesteryl-hyaluronic acid nanogel was found efficient for treatment of cancer cells (Wei et al. 2013). The nanogel particles of 20–40 nm diameter, having a hydrophobic core loaded up to 20% drug (such as etoposide, salinomycin, and curcumin), were formed after ultrasonication and showed a prolonged drug release following the hydrolysis of biodegradable ester linkage.

Taking into account that the most antimicrobials are not capable of diffusing through cellular membranes while bacteria are often able to penetrate inside host cells, innovative self-assembled NGs have been studied as drug carriers for intracellular compartment targeting (Alhaique et al. 2016). The system consisted in HA-cholesterol nanogels loaded with levofloxacin, is one of the most widely used fluoroquinolone antibiotics. These new nanogels were specifically designed to increase the cellular uptake of the drug, thus improving its efficacy in the treatment of intracellular infections caused, for instance, by *Staphylococcus aureus* and *Pseudomonas aeruginosa* (Montanari et al. 2014). Levofloxacin loading within NGs was carried out by means of the coprecipitation method, and the obtained networks were so stable that autoclaving sterilization and freeze-drying, in the presence of dextrose, could be carried out without significant variations of the original structure.

5.1 Alginate-Based Nanogels

Alginate composed of alternating blocks of 1–4 linked α-L-guluronic and β-D-mannuronic acid residues undergoes gelation and crosslinking by the exchange of sodium ions from the guluronic acid with divalent cations such as Ca^{2+}. Alginate also possesses highly reactive multifunctional groups like OH and COOH groups, which can be chemically modified by oxidation, amidation, esterification, sulfation, or grafting (Debele et al. 2016). By these chemical modifications, it is possible to control the biological and physico-chemical properties such as solubility and hydrophobicity giving a wide range of potential applications, including NGs.

The use of the pH-sensitive alginate nanogels is an effective way for successful oral delivery of peptides and proteins which are commonly instable in the harsh acidic environment of stomach (Lu et al. 2014).

5.2 Cyclodextrin-Based Nanogels

Cyclodextrins (CDs), biocompatible cyclic oligosaccharides consisting of α-1,4-glycosidic bonded D-glucopyranose units (van de Manakker et al. 2009), can be modified via esterification and etherification reactions of their primary and secondary hydroxyl groups resulting in many CD derivatives.

Thus, Liu et al. (2007b) synthesized linear methyl methacrylate (MMA)/β-CD copolymers by radical co-polymerization of hydrophobic MMA with mono-vinyl β-CD. Ultrafine β-CD nanogels (50–150 nm) were prepared by Kettel et al. (2012) by reaction between six-arm isocyanate (NCO)-terminated poly(ethylene oxide-stat-propyleneoxide) as functional hydrophilic prepolymers and β-CD in aqueous medium. The nanogels showed a high uptake capacity of hydrophobic phenolphthalein. Later, the same research team has obtained nanogels based on poly(methyl methacrylate) and α-, β-, and γ-cyclodextrins functionalized with methacrylic groups (CD-MAs) by radical precipitation polymerization technique in organic solvent under stirring at 200 rpm and 65 °C for 15 min (Kettel et al. 2016). Their diameter sizes varied between 36 and 363 nm depending on the β-CD-MA content and the number of vinyl groups per β-CD molecule in the nanogels.

Self-assembled CD-based nanogels were obtained by the association of a hydrophobically modified dextran (MD) and a cyclodextrin polymer (pβCD) (Daoud-Mahammed et al. 2009). These nanogels are able to include into the CD cavities of the pβCD both hydrophobic molecules (benzophenone and tamoxifen) and also were solubilized by MD into their hydrophobic microdomains.

5.3 Gum Acacia-Based Nanogels

Gum acacia (GA), also known as arabic gum, is a natural polysaccharide obtained from *Acacia nilotica* tree. It was used for the preparation of microparticles and nanoparticles. High water solubility, biocompatibility, and low cost are the main attractions for selecting this polysaccharide for this purpose. In order to obtain nanogels, gum acacia was used in combination with proteins, like gelatine (Sarika and James 2015), or with polysaccharide, like alginate (Chopra et al. 2015) and chitosan (Avadi et al. 2010). Compared with the chemical agents, arabic gum, besides its biocompatibility and biodegradability, provides more interaction sites and negative charge for interaction with polycationic polymers such as chitosan. The release properties of gum arabic nanogels were investigated with various drugs, especially curcumin, and intensely used for cancer therapy.

6 Protein-Based Nanogels

6.1 Gelatine

Gelatine, a hydrolyzed form of collagen, has been used for nanogels preparation for different applications. There are more reasons using gelatine in nanogel synthesis. First, such biopolymer has demonstrated excellent biocompatibility. Second, its multiple functionality (with COOH and NH_2 groups) is easily accessible for modification. Third, the NGs can penetrate through the pores of the endothelial junctions found in tumor cells. Gelatine-based nanogels have been obtained via several methods as precipitation polymerization (Wu and Wan 2008) and inverse miniemulsions polymerization (Sarika et al. 2015). Generally, the crosslinking of natural nanogels is required to give biopolymer stability, shape, and an enhanced circulation time in vivo as compared to uncrosslinked particles (Elzoghby et al. 2012). A number of crosslinkers have been used in previous works for this purpose such as aldehydes, genipin, carbodiimide/N-hydroxysuccinimide, $CaCl_2$ (Narayanan et al. 2013), or enzymatic crosslinker as transglutaminase (Fuchs et al. 2010). Gelatine-based nanogels prepared without crosslinking were found to be unstable and tended to aggregate upon aging.

Gum arabic (GA) was oxidized to gum arabic aldehyde and then was crosslinked with gelatine to obtain nanogels (Sarika and James 2016b).

Gelatine-based nanogels with thermo-responsiveness may be obtained if the chemical crosslinking density is low allowing the formation of triple helical structures at low temperatures (Bode et al. 2011). Gandhi et al. (2014) demonstrated that tunable gelatine nanogels prepared via a one-step dissolution approach are thermo-responsive showing a volume transition at 32 °C.

6.2 Soy Protein-Based Nanogels

Soy protein, an abundant by-product of the soybean oil industry, shows good functional properties for food processing due to its high nutritional value and emulsifying and gelling properties (Garcia et al. 1997). Its isoelectric point is about pH 4.8 (Hofland et al. 2000).

Many synthetic polymers used for fabricating nanoparticles show poor protein encapsulation, suboptimal dosing characteristics, and limited stability (Gao et al. 2014a) and can form nanoparticles only with a subset of proteins. A versatile protein antigen delivery system is necessary which should be able to form nanoparticles with any protein antigen and to allow a precise control over particle size within nanometer range. Singh et al. (2014) developed self-assembling nanogels of poly(hydroxyethyl methacrylate) (pHEMA) backbone functionalized with a hydrophobic side chains of pyridine. The pyridine is known to be toxic and its traces are relatively difficult to remove; therefore, an advanced purification of the final product is necessary. The particle sizes of the nanogels are dependent on the protein–polymer ratio and can be varied from 200 to 900 nm. The nanogel–protein retention time is longer than 14 days in the rat stifle joints through single intra-articular injections. The immune-stimulatory self-assembly nanogel vaccines have been developed using a modified catalyst for pHEMA-pyridine reaction that provided a fine control of nanogel sizes within a narrow 145–160 nm range (Purwada et al. 2016). These nanogels delivered protein antigen to dendritic cells being efficient priming of ovalbumin-specific CD8+ T cells.

Soy protein and soy polysaccharide can form stable oil-in-water emulsions (Yin et al. 2012). Nanogels prepared using such emulsions were loaded with folic acid (FA) (Ding and Yao 2013) by applying a high-pressure homogenization to break down the original aggregates of soy protein, which benefits the binding of the soy protein with soy polysaccharide and FA at pH 4. A thermal treatment was then applied that induced the soy protein gelation which assures the formation of stable FA-loaded nanogels with a soy polysaccharide surface.

Some soy proteins can be involved in nanogel synthesis after the Maillard reaction, which is a non-enzymatic browning reaction between the ε-amino groups (in proteins, peptides, or amino acids) and the reactive carbonyl group (in reducing sugars) (Kato et al. 1993). The Maillard reaction requires no chemical reagents and is safe and economical. Maillard reaction products (glycated proteins) are amphiphilic copolymers, which possess inspiring potential to construct ordered micro- or nanoscale particles by self-assembly under appropriate conditions. Feng and co-workers (2015) have chosen soy β-conglycinin protein, which has various advantages such as lowering serum triacylgycerol, cholesterol, and lipid levels (de Souza Ferreira et al. 2010), and dextran as the sources to fabricate nanogels. The obtained nanogels exhibited spherical shape with core–shell structures and diameters of approximately 90 nm being fairly stable against long-term storage, pH change, lyophilization, or dilution.

7 Properties

As it was already mentioned, the NGs containing natural polymers show improved bioavailability, large surface area for multivalent bioconjugation, high drug loading capacity, targeted and very efficient delivery of drugs which are combined with hydrophilicity, biocompatibility and biodegradability of hydrogels. The stimuli-responsive nanogels offer besides these characteristics more effective disease treatments because of their efficient ability to act as intelligent drug delivery systems, which can sense and respond directly to pathophysiological conditions (Soni et al. 2015). The rapid swelling/de-swelling of the nanogels is reflected in changes in their dimensions and properties which also offers many advantages in drug delivery. The swelling of nanogels in water is controlled by crosslinker concentration, polymer nature, ionic strength, temperature, polyelectrolyte charge (related to osmotic pressure), biological parameters, etc. Hydrophobic and

Fig. 2 Hydrodynamic diameter ($Z_{average}$) of 1 wt% (open squares) and 4 wt% (full squares) carrageenan nanoparticles in PBS during alternating increasing and decreasing temperature cycles, as measured by DLS. Dashed lines are shown for visual guidance only. (permission from Elsevier for Daniel-da-Silva et al. 2011)

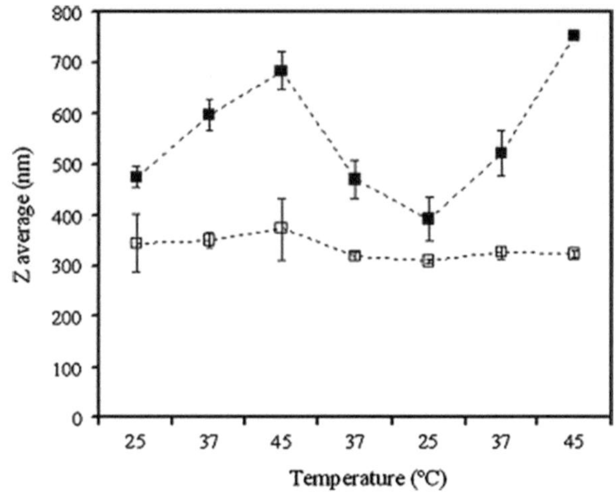

hydrogen bonding interactions between polymeric chains and solutes also contribute to swelling behavior (Kabanov and Vinogradov 2008).

Because of the dependence on mechanism of diffusion, the swelling depends on timescale. While at macroscale level, the swelling occurs during days; in nanoscale, the nanogel response is very fast in a time period of milliseconds.

The temperature-responsive swollen k-carrageenan nanogels undergo a gel-to-sol transition in the 25–45 °C temperature range which involves the breaking up of physical junctions in the gel (Fig. 2). The transition is reversible.

As the polysaccharide k-carrageenan is a polyelectrolyte, therefore the zeta potential of the nanospheres at physiological pH was found to be invariable with k-carrageenan content.

The flexibility (evaluated by Young's modulus) and polymer density of the nanogel particles influence the association/dissociation rates constant and equilibrium constants of multipoint protein recognition process which increased with decreasing crosslinking degree for swollen phase NG, but the crosslinking degree did not affect association/dissociation rate constants of the collapsed phase of NGs (Nakamoto et al. 2014; Manirujjaman et al. 2013).

Electromobility is related with **zeta potential** of some nanogels. It is known that the electric potential at the boundary of the double layer is known as the zeta potential. The values of the zeta potential of nanoparticles greater than +25 mV or less than −25 mV predict a good colloidal stability, at lower values the aggregation occurs. The size (200–1000 nm) and zeta potential (+20 and +60 mV) of the NPs produced by ionic gelation of chitosan with charged amino groups and diblock copolymer in presence of a polyanion as sodium tripolyphosphate were modulated by varying the composition. The obtained NPs shown good association with proteins, such as bovine serum albumin, tetanus toxoid and diphtheria toxoid, insulin as well as oligonucleotide (Soppimath et al. 2001). The cyclodextrin-coated NPs of some NGs drastically reduced the surface negativity helping to avoid the mononuclear phagocytic system uptake. NPs with particle size of 150 nm and negative charges more efficiently accumulate in tumors (He et al. 2010).

Colloidal stability. Nanogels show better colloidal stability in comparison with the surfactant micelles and exhibit lower critical micelle concentrations, slower rates of dissociation, and longer retention of loaded drugs.

Non-immunologic response. Nanogel-based drug delivery systems did not produce any immunological response.

Morphology. Morphology of the nanogels depends on their composition and preparation method. TQ-loaded myristic acid-chitosan (MA-chitosan) nanogels prepared by the technique of self-assembly (Dehghania et al. 2015) show round shapes and smooth surfaces with particle size ranged between 150 and 200 nm.

A multi-(amino acid) copolymer (MAC) based on ω-aminocaproic acid, γ-aminobutyric acid, L-alanine, L-lysine, L-glutamate, and hydroxyproline and MAC microspheres encapsulating bovine serum albumin (BSA) prepared by a double-emulsion solvent extraction method gave microspheres with dimensions ranged from 60.2 to 79.7 μm and different porous structures (Chen et al. 2014). The size of microspheres decreased as the surfactant ratio of Tween 80 to Span 80 decreased (Fig. 3).

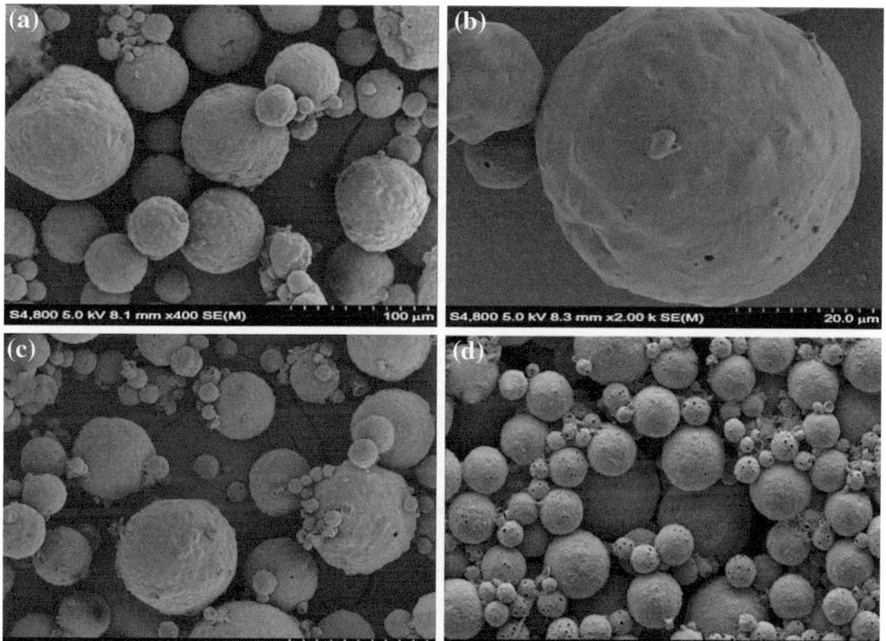

Fig. 3 SEM images of morphology of BSA-loaded microspheres prepared with different surfactant ratios of Tween 80 to Span 80 of 3:2 (**a** and **b**), 7:3 (**c**), and 4:1 (**d**) (with permission from Dove Medical Press for Chen et al. 2014)

The BSA release profile from microspheres shows an initial burst effect in the first week followed by constant slow release for the next 7 weeks. BSA-loaded microspheres could degrade gradually in phosphate-buffered saline with pH 7.1, but the degradation products did not change significantly the pH of the medium. Therefore, the MAC microspheres might have great potential as carriers for protein delivery.

8 Applications of Natural-Based Nanogels in Therapeutic/Biomedical Fields

It is known that natural polymers offer certain specific advantageous and exceptional characteristics over synthetic polymers like good biocompatibility, non-toxicity, degradability, non-antigenicity, high nutritional value, are obtained from abundant renewable sources, and show an extraordinary binding capacity of various drugs and possibility of less opsonization by the reticuloendothelial system (RES) and also good stability during storage and in vivo conditions (Elzoghby 2013). At their turn, nanogels exhibit hydrophilicity, flexibility, versatility, high water absorption, and biocompatibility, as well as the small size and are capable of actively or passively targeting active compounds to the desired biophase. The size of nanoparticles affects biodegradability, cellular uptake, and drug release. The nanogels preserve a high stable structure when are dispersed in water.

However, sometimes a strong interaction between drug and polymer decreases the hydrophilicity of the nanogels and causes the structure to collapse, hence irreversibly entrapping the drug molecules and enhancing the hydrophilicity of the nanogel matrix (Kabanov and Alakhov 2002; Vinogradov et al. 2002) and also surfactant or monomers present in nanogel may cause adverse effects in the formulations (Garg and Goyal 2014).

Some nanogels show limited drug loading efficiency and suboptimal regulation of drug release (Wang and von Recum 2011).

8.1 Drug Delivery Systems

Conventional drug administration forms are pills, ointments, eye drops, and intravenous solutions. Various new drug delivery systems have been developed. These systems include chemically modified drugs, drug entrapment in polymeric materials, drug entrapment in small pores that are injected later into the bloodstream (Sali 2016).

Nanohydrogels demonstrate high therapeutic activity at the targeted site with sustained effects on the target area over a period of days or even weeks (Thampanchira et al. 2016).

Nanogels ensure internal three-dimensional environment for (bio) molecular drugs loading and, compared to conventional hydrogels and other nanoparticles, they are expected to have a more drug loading capacity and provide one or many functions as: drug protection by minimizing drug degradation and inactivation upon administration; site recognition by increasing the fraction of drug delivered in the pathological area; and actuation by drug release activation through an external stimuli. Because size distribution plays an important role in drug delivery application, it is necessary to prepare uniformly sized nanoparticles (Pujana et al. 2013). Nanoscale drug delivery systems can be produced to mix and match different biological and synthetic modules, for various applications, comprising of injectable, oral, topical, implantable, inhalable, and transdermal drug delivery.

Sivaram et al. (2015) showed that some nanogels interact stronger with proteins by formation of complexes, than liposomes or amphiphilic peptide carriers. These complexes are able to uptake intracellular proteins. Shimoda et al. (2011) developed a nanogel based on ethylenediamine and cholesteryl group-bearing pullulan useful for the intracellular delivery of proteins. An hybrid nanogel composed of cadmium chloride quantum dot, super paramagnetic Fe_3O_4, and chitosan loaded with antidiabetic drug insulin was synthetized (Shen et al. 2012) by a two-step procedure. By hydrothermal route, thioglycolic acid-capped cadmium chloride quantum dots were prepared and then Fe_3O_4 nanoparticles incorporation assured superparamagnetic character. By crosslinking with chitosan, hybrid nanogel was fabricated. Insulin loading hybrid nanogels exhibited optimum insulin release profile at pH 7.4.

Lim et al. (2010) used a new injectable bone morphogenic protein-loaded alginate nanohydrogel to enhance osteoblasts growth and differentiation of human bone marrow stromal cells.

Nanogels composed of hyaluronic acid are able to increase water solubility of hydrophobic drugs, controlling drug release at the target site in response to external stimuli, improving biocompatibility and bioavailability, and reducing non-specific side effects (Zhang et al. 2014).

Loading of metallic nanoparticles in polymeric materials has a wide range of applications in the medicine and pharmacy fields (Jovanovic et al. 2011; Juby et al. 2012). Zinc oxide (ZnO) has been popular among nanosized metal oxides because of its high antimicrobial, antitumor activity (Petchthanasombat et al. 2012), and low cost of production. ZnO at the micron size range is non-toxic but when is reduced to the nanoscale dimensions, it may develop new kind of toxicity. Incorporation of the nanosized metal-oxide nanoparticles in polymeric nanogels facilitates the gradual and sustained release instead of burst release. It was reported the preparation of a covalently crosslinked sodium alginate/gum acacia nanogel loaded with ZnO nanoparticles capable for a sustained release behavior (Chopra et al. 2015).

One of the major challenges in cancer chemotherapy is non-specific drug accumulation and its related side effects. Nanotechnology helps in developing drug delivery agents that offer the possibility of destroying cancer cells with minimal damage to healthy normal cells and also the elimination of cancer cells before they can proliferate into a tumor (Uthaman et al. 2014).

Gellan-based nanohydrogels were studied for the treatment of cancer cells by carrying an anticancer drug (paclitaxel, PCT) together with the steroid prednisolone for a combined therapy (D'Arrigo et al. 2014). For this purpose, prednisolone was chemically bound to the polymer, while the anticancer drug was loaded within the network by means of the solvent casting technique. It has been suggested that PCT was loaded within the hydrophobic domains formed by the interaction with prednisolone.

8.2 *Artificial Chaperone*

Chaperones proteins assist the non-covalent folding/unfolding and the assembly/disassembly of other macromolecular structures. These phenomena do not occur in these structures when the structures are performing their normal biological functions.

Nanogel ensures a better carrier as artificial chaperones as they effectively prevent protein aggregation by forming nanogel–protein complexes during protein refolding (Singh et al. 2013). Chaperone nanogels interact much strongly and selectively with denatured proteins than with native ones. The simple amphiphilic nanogels are stimulating the same function as complex proteins are released in refolded state upon dissociation. This is a two-step mechanism by which molecular chaperones capture the denatured protein and release of refolded protein.

A novel hybrid hydrogel was designed by encapsulating cholesteryl group-bearing pullulan (CHP) nanogels through physical interactions in a three-dimensional network of chemically crosslinked hyaluronan (HA) gel via Michael addition reaction for sustained delivery of therapeutic peptides and proteins such as glucagon-like peptide-1, insulin, and erythropoietin, which were trapped in the CHP nanogels (Tai et al. 2010). Both in vitro and in vivo tests demonstrated that the sustained and controlled release occurs from the hybrid hydrogels depending on the crosslinking density and the degradability of the HA gel, modulated by the initial gelation condition without protein denaturation. This biodegradable controlled release matrix with molecular chaperone-like activity could have potential application for therapeutic proteins or drugs for tissue engineering.

8.3 *Tissue Engineering*

Nanogels are also useful for tissue engineering including bone tissue engineering, cartilage tissue engineering, vascular tissue engineering, and wound dressing. The tissue engineering techniques for osteogenesis involve the combinations of biocompatible carriers and specific substances. Fujioka-Kobayashi et al. (2012)

prepared hybrid hydrogels by crosslinking protein-containing nanogels. By the crosslinking of acryloyl group-bearing CHP (CHPOA) nanogels with pentaerythritol tetra (mercaptoethyl) polyoxyethylene (thiol group-modified polyethylene glycol, PEGSH), a biodegradable hydrogel was obtained. Good results were obtained when two distinct growth factors, recombinant human bone morphogenetic protein 2 (BMP2) and recombinant human fibroblast growth factor 18 (FGF18), were applied for bone repair.

A series of artificial extracellular matrices (aECMs) constructed from naturally derived nanogels have been used to provide cellular microenvironments (Rice et al. 2013). Such systems act as a reservoir for growth factors and allow controlled diffusion of the molecules trapped within their network (Discher et al. 2009). Hashimoto et al. (2016) constructed a novel thin (16–24 μm), transparent and flexible film (NanoClik) used as an aECM based on amphiphilic nanogel tectonics (41 nm). They are prepared through self-assembly of hydrophobized polysaccharides. More specifically, the aECM was obtained by combining a Michael addition reaction between the acryloyl groups of modified cholesterol-bearing pullulan and the thiol group of the pentaerythritol tetra (mercaptoethyl) polyoxyethylene crosslinker with solvent evaporation. The films with amphiphilic nanogels are able to induce the absorption with proteins loading having applications as scaffolds for tissue engineering, carriers for protein delivery, and surface coating materials for implants.

8.4 Imaging Diagnostics

Many imaging diagnostic techniques are available (Maya et al. 2013; Baker 2010). When a fluorescent probe is immobilized or conjugated with polymer nanogel networks, it can serve as imaging probes for bioimaging applications, such as early cancer detection (Popovic et al. 2010; Estévez et al. 2010). The polymer coating protects the loaded probes from the environment and also improves the pharmaco-kinetics, biodistribution, and significantly amplifying the diagnostic imaging signals (Tang et al. 2016).

Knowing that human mesenchymal stem cells (hMSCs) have been found to improve cancer therapy and diagnosis due to their tumor-homing properties (Cheng et al. 2013), early tumors (~ 3 mm^3) were detected, using magnetic resonance and optical imaging both in vitro and in vivo, by means of pullulan-deoxycholic acid-diethylene triamine pentaacetic gadolinium-chelate nanoparticles photochemically internalized in hMSCs (Kim et al. 2015).

Nanogels intended for integration of optical sensing, cell imaging, and therapy combine the properties from both optical moieties and polymer gels (Wu and Zhou 2010). Numerous nanogels were developed for intracellular optical probing (Wu et al. 2010a; Gota et al. 2009). Stimuli-responsive nanogels allow a real-time monitoring of the change of pH, temperature, glucose concentration, and common ions, closely associated with cancer, diabetes, and other diseases. The volume phase transition and change of the environment of the optical moieties convert the received signal into optical signal. The optical moieties are organic dye, conjugated polymer or fluorophore-contained polymer, semiconductor quantum dots, or noble metal nanoparticles.

Chitosan, hydroxypropyl cellulose and cholesterol-bearing pullulan modified with amino groups are the most used natural polymers for imaging probes based on nanogel preparation. In contrast with the stable chemically crosslinked nanogels, the physically associated nanogels are not ideal candidates for optical sensing. For example, chitosan was involved in physical interactions with poly(acrylic acid) to form nanogels (Wu et al. 2010b); however, at a pH variation, these weak polyelectrolyte nanogels may involve a pH-induced accumulation of excess charge within the nanogels (Sui and Schlenoff 2004; Kharlampieva et al. 2008) with the chitosan chain expulsion. When a crosslinking agent is added, the nanogel networks can exhibit a reversible structural change in response to a pH change, leading to a reversible optical response of the hybrid nanogels.

9 Conclusion

The synergy between nanoscience and biomedical applications has lead to great developments in nanomedicine namely in biomedical research as well as clinical practices, including drug delivery systems, artificial chaperone, imaging probes, tissue engineering. Natural polymer-based nanogels, the next generation on polymeric nanoparticulate systems, have received and will have a great renown as leading candidates for biomaterials due to their unique compositional and structural similarities to the natural extracellular matrix, in addition to their desirable framework for cellular proliferation and survival. According to this, an extensive overview on the published studies on nanogels based on natural polymers for various applications was presented in this chapter.

By applying nanotechnology to medicine, the superior therapeutics and diagnostic ways of different diseases are developed. The research in this area occurs very fast, almost on a daily basis. The multifunctional hybrid nanogel synthesis is focused on the drug delivery systems because of their ability to efficiently encapsulate therapeutics and environmental stimuli controlled release.

Acknowledgements The authors thank to Romanian UEFISCDI for the financial support given through the research project 164/2012 BIO-NANOMED.

References

Ahmad M, Rai SM, Mahmood A (2016) Hydrogel microparticles as an emerging tool in pharmaceutical field: a review. Adv Polym Tech 35(2):121–128

Akiyoshi K, Nishikawa T, Shichibe S, Sunamoto J (1995) Stabilization of insulin upon supramolecular complexation with hydrophobized polysaccharide nanoparticle. Chem Lett 24:707–708

Akiyoshi K, Kobayashi S, Shichibe S, Mix D, Baudys M, Kim SW, Sunamoto J (1998) Self-assembled hydrogel nanoparticle of cholesterol-bearing pullulan as a carrier of protein drugs: complexation and stabilization of insulin. J Control Release 54:313–320

Alexis F, Pridgen E, Molnar LK, Farokhzad OC (2008) Factors affecting the clearance and biodistribution of polymeric nanoparticles. Mol Pharmacol 5:505–515

Alhaique F, Matricardi P, Di Meo C, Coviello T, Montanari E (2015) Polysaccharide-based self-assembling nanohydrogels: an overview on 25-years research on pullulan. J Drug Deliv Sci Technol 30:300–309

Alhaique F, Casadei MA, Cencetti C, Coviello T, Di Meo C, Matricardi P, Montanari E, Pacelli S, Paolicelli P (2016) From macro to nano polysaccharide hydrogels: an opportunity for the delivery of drugs. J Drug Deliv Sci Technol 32:88–99

Asadi H, Khoee S (2016) Dual responsive nanogels for intracellular doxorubicin delivery. Int J Pharm 511:424–435

Ashrafi H, Azadi A (2016) Chitosan-based hydrogel nanoparticle amazing behaviors during transmission electron microscopy. Int J Biol Macromol 84:31–34

Avadi MR, Sadeghi AMM, Mohammadpour N, Abedin S, Atyab F, Dinarvand R, Rafiee-Tehrani M (2010) Preparation and characterization of insulin nanoparticles using chitosan and Arabic gum with ionic gelation method. Nanomedicine 6:58–63

Bae KH, Mok H, Park TG (2008) Synthesis, characterization, and intracellular delivery of reducible heparin nanogels for apoptotic cell death. Biomaterials 29(23):3376–3383

Baker M (2010) Whole-animal imaging: the whole picture. Nature 463:977–980

Bazban-Shotorbani S, Dashtimoghadam E, Karkhaneh A, Hasani-Sadrabadi MM, Jacob KI (2016) Microfluidic directed synthesis of alginate nanogels with tunable pore size for efficient protein delivery. Langmuir 32(19):4996–5003

Bencherif SA, Washburn NR, Matyjaszewski K (2009) Synthesis by AGET ATRP of degradable nanogel precursors for in situ formation of nanostructured hyaluronic acid hydrogel. Biomacromolecules 10:2499–2507

Berger J, Reist M, Mayer JM, Felt O, Peppas NA, Gurny R (2004) Structure and interactions in covalently and ionically crosslinked chitosan hydrogels for biomedical applications. Eur J Pharm Biopharm 57:19–34

Biewenga GP, Haenen G, Bast A (1997) The pharmacology of the antioxidant lipoic acid. Gen Pharmacol 29(3):315–331

Binh D, Hong PTT, Duy NN, Duoc NT, Dieu NN (2012) A study on size effect of carboxymethyl starch nanogel crosslinked by electron beam radiation. Radiat Phys Chem 81:906–912

Boddohi S, Moore N, Johnson PA, Kipper MJ (2009) Polysaccharide-based polyelectrolyte complex nanoparticles from chitosan, heparin, and hyaluronan. Biomacromolecules 10:1402–1409

Bode F, da Silva MA, Drake AF, Ross-Murphy SB, Dreiss CA (2011) Enzymatically cross-linked tilapia gelatin hydrogels: physical, chemical, and hybrid networks. Biomacromolecules 12:3741–3752

Calvo P, Remunan-Lopez C, Vila-Jato JL, Alonso MJ (1997) Novel hydrophilic chitosan-polyethylene oxide nanoparticles as protein carriers. J Appl Polym Sci 63:125–132

Capretto L, Cheng W, Hill M, Zhang X (2011) Micromixing within microfluidic devices. Top Curr Chem 304:27–68

Chen X, Lv G, Zhang J, Tang S, Yan Y, Wu Z, Su J, Wei J (2014) Preparation and properties of BSA-loaded microspheres based on multi-(amino acid) copolymer for protein delivery. Int J Nanomed 9:1957–1965

Cheng L, Wang C, Ma X, Wang Q, Cheng Y, Wang H, Li Y, Liu Z (2013) Multifunctional upconversion nanoparticles for dual-modal imaging-guided stem cell therapy under remote magnetic control. Adv Funct Mater 23:272–280

Chopra M, Bernela M, Kaur P, Manuja A, Kumar B, Thakur R (2015) Alginate/gum acacia bipolymeric nanohydrogels-promising carrier for Zinc oxide nanoparticles. Int J Biol Macromol 72:827–833

Chung BG, Lee KH, Khademhosseini A, Lee SH (2012) Microfluidic fabrication of microengineered hydrogels and their application in tissue engineering. Lab Chip 12:45–59

Curcioa M, Blanco-Fernández B, Costoya A, Concheiro A, Puoci F, Alvarez-Lorenzo C (2015) Glucose cryoprotectant affects glutathione-responsive antitumor drug release from polysaccharide nanoparticles. Eur J Pharm Biopharm 93:281–292

Daniel-da-Silva AL, Ferreira L, Gil AM, Trindade T (2011) Synthesis and swelling behavior of temperature responsive j-carrageenan nanogels. J Colloid Interface Sci 355:512–517

Daoud-Mahammed S, Couvreur P, Bouchemal K, Chéron M, Lebas G, Amiel C, Gref R (2009) Cyclodextrin and polysaccharide-based nanogels: entrapment of two hydrophobic molecules, benzophenone and tamoxifen. Biomacromolecules 10:547–554

D'Arrigo G, Navarro G, Di Meo C, Matricardi P, Torchilin V (2014) Gellan Gum nanohydrogel containing anti-inflammatory and anti-cancer drugs: a multidrug delivery system for a combination therapy in cancer treatment. Eur J Pharm Biopharm 87:208–216

de la Fuente M, Raviña M, Paolicelli P, Sanchez A, Seijo B, Alonso MJ (2010) Chitosan-based nanostructures: a delivery platform for ocular therapeutics. Adv Drug Deliv Rev 62:100–117

de Souza Ferreira E, Silva MA, Demonte A, Neves VA (2010) β-Conglycinin (7S) and glycinin (11S) exert a hypocholesterolemic effect comparable to that of fenofibrate in rats fed a high-cholesterol diet. J Funct Foods 2:275–283

Debele TA, Mekuria SL, Tsai H-C (2016) Polysaccharide based nanogels in the drug delivery system: application as the carrier of pharmaceutical agents. Mater Sci Eng C Mater Biol Appl. https://doi.org/10.1016/j.msec.2016.05.121

Dehghania H, Hashemi M, Entezari M, Mohsenifar A (2015) The comparison of anticancer activity of thymoquinone and nanothymoquinone on human breast adenocarcinoma. Iran J Pharm Res 14(2):539–546

Ding X, Yao P (2013) Soy protein/soy polysaccharide complex nanogels: folic acid loading, protection, and controlled delivery. Langmuir 29:8636–8644

Discher DE, Mooney DJ, Zandstra PW (2009) Growth factors, matrices, and forces combine and control stem cells. Science 324:1673–1677

Dispenza C, Grimaldi N, Sabatino MA, Soroka IL, Jonsson M (2015) Radiation-engineered functional nanoparticles in aqueous systems. J Nanosci Nanotechnol 15:3445–3467

Elzoghby AO (2013) Gelatin-based nanoparticles as drug and gene delivery systems: reviewing three decades of research. J Control Release 172:1075–1091

Elzoghby AO, Samy WM, Elgindy NA (2012) Protein-based nanocarriers as promising drug and gene delivery systems. J Control Release 161:38–49

Estévez MC, Huang YF, Kang H, O'Donoghue MB, Bamrungsap S, Yan J, Chen X, Tan W (2010) Nanoparticle-aptamer conjugates for cancer cell targeting and detection. Methods Mol Biol 624:235–248

Farag RK, EL-Saeed SM, Abdel-Raouf ME (2016) Synthesis and investigation of hydrogel nanoparticles based on natural polymer for removal of lead and copper(II) ions. Desalination Water Treat 57(34):16150–16160

Feng J-L, Qi J-R, Yin S-W, Wang J-M, Guo J, Weng J-Y, Liu Q-R, Yang X-Q (2015) Fabrication and characterization of stable soy β-conglycinin–dextran core–shell nanogels prepared via a self-assembly approach at the isoelectric point. J Agric Food Chem 63:6075–6083

Fredheim GE, Christensen BE (2003) Polyelectrolyte complexes: interactions between lignosulfonate and chitosan. Biomacromolecules 4:232–239

Fuchs S, Kutscher M, Hertel T, Winter G, Pietzsch M, Coester C (2010) Transglutaminase: new insights into gelatin nanoparticle crosslinking. J Microencapsul 27:747–754

Fujioka-Kobayashi M, Ota MS, Shimoda A, Nakahama K, Akiyoshi K, Miyamoto Y, Iseki S (2012) Cholesteryl group- and acryloyl group-bearing pullulan nanogel to deliver BMP2 and FGF18 for bone tissue engineering. Biomaterials 33:7613–7620

Gandhi SS, Yan H, Kim C (2014) Thermoresponsive gelatin nanogels. ACS Macro Lett 3:1210–1214
Gao W, Vecchio D, Li J, Zhu J, Zhang Q, Fu V, Thamphiwatana S, Lu D, Zhang L (2014a) Hydrogel containing nanoparticle-stabilized liposomes for topical antimicrobial delivery. ACS Nano 8:2900–2907
Gao Y, Xie J, Chen H, Gu S, Zhao R, Shao J, Jia L (2014b) Nanotechnology-based intelligent drug design for cancer metastasis treatment. Biotechnol Adv 32:761–777
Garcia M, Torre M, Marina M, Laborda F, Rodriquez AR (1997) Composition and characterization of soyabean and related products. Crit Rev Food Sci 37:361–391
Garg T, Goyal AK (2014) Biomaterial-based scaffolds—current status and future directions. Expert Opin Drug Deliv 11:767–789
Gota C, Okabe K, Funatsu T, Harada Y, Uchiyama S (2009) Hydrophilic fluorescent nanogel thermometer for intracellular thermometry. J Am Chem Soc 131:2766–2767
Gurav DD, Kulkarni AS, Khan A, Shinde VS (2016) pH-responsive targeted and controlled doxorubicin delivery using hyaluronic acid nanocarriers. Colloids Surf B Biointerfaces 143:352–358
Guzman V, El-Sherbiny IM, Herrera D, Smyth HD (2013) Design and in vitro evaluation of a new nano-microparticulate system for enhanced aqueous-phase solubility of curcumin. Biomed Res Int 2013:724763
Hamidi M, Ashrafi H, Azadi A (2012) Surface functionalized hydrogel nanoparticles. In: Tiwari A, Ramalingam M, Kobayashi H, Turner APF (eds) Biomedical materials and diagnostic devices. Wiley, Hoboken, NJ, USA, pp 191–213
Hashimoto Y, Mukai S, Sawada S, Sasaki Y, Akiyoshi K (2016) Advanced artificial extracellular matrices using amphiphilic nanogel-crosslinked thin films to anchor adhesion proteins and cytokines. ACS Biomater Sci Eng 2(3):375–384
He C, Hu Y, Yin L, Tang C, Yin C (2010) Effects of particle size and surface charge on cellular uptake and biodistribution of polymeric nanoparticles. Biomaterials 31:3657–3666
Helgeson ME, Chapin SC, Doyle PS (2011) Hydrogel microparticles from lithographic processes: novel materials for fundamental and applied colloid science. Curr Opin Colloid Interface Sci 16:106–117
Hirakura T, Nomura Y, Aoyama Y, Akiyoshi K (2004) Photoresponsive nanogels formed by the self-assembly of spiropyrane-bearing pullulan that act as artificial molecular chaperones. Biomacromolecules 5:1804–1809
Hofland GW, de Rijke A, Thiering R, van der Wielen LA, Witkamp G-J (2000) Isoelectric precipitation of soybean protein using carbon dioxide as a volatile acid. J Chromatogr B 743:357–368
Ischakov R, Adler-Abramovich L, Buzhansky L, Shekhter T, Gazit E (2013) Peptide-based hydrogel nanoparticles as effective drug delivery agents. Bioorg Med Chem 21:3517–3522
Jahn A, Reiner JE, Vreeland WN, DeVoe DL, Locascio LE, Gaitan M (2008) Preparation of nanoparticles by continuous-flow microfluidics. J Nanopart Res 10(6):925–934
Jaiswal MK, Pradhan A, Banerjee R, Bahadur D (2014) Dual pH and temperature stimuli-responsive magnetic nanohydrogels for thermo-chemotherapy. J Nanosci Nanotechnol 14:4082–4089
Jovanovic Z, Krkljes A, Stojkovska J, Tomic S, Obradovic B, Iskovic- Stankovic V (2011) Synthesis and characterization of silver/poly(N-vinyl-2-pyrrolidone) hydrogel nano composite obtained by in situ radiolytic method. Radiat Phys Chem 80:1208–1211
Juby KA, Dwivedia C, Kumara M, Kotab S, Misrab HS, Bajaj PN (2012) Silver nanoparticle-loaded PVA/gum acacia hydrogel: synthesis, characterization and antibacterial study. Carbohyd Polym 89:906–913
Jung S, Yi H (2015) Facile micromolding-based fabrication of biopolymeric–synthetic hydrogel microspheres with controlled structures for improved protein conjugation. Chem Mater 27:3988–3998
Kabanov AV, Alakhov VY (2002) Pluronic block copolymers in drug delivery: from micellar nanocontainers to biological response modifiers. Crit Rev Ther Drug Carrier Syst 19:1–72

Kabanov AV, Vinogradov SV (2008) Nanogels as pharmaceutical carriers, chapter 3. In: Torchilin V (ed) Multifunctional pharmaceutical nanocarriers. Springer, pp 67–81
Kabanov AV, Vinogradov SV (2009) Nanogels as pharmaceutical carriers: finite networks of infinite capabilities. Angew Chem Int Ed 48:5418–5429
Kato A, Minaki K, Kobayashi K (1993) Improvement of emulsifying properties of egg white proteins by the attachment of polysaccharide through Maillard reaction in a dry state. J Agric Food Chem 41:540–543
Kettel MJ, Hildebrandt H, Schaefer K, Moeller M, Groll J (2012) Tenside-free preparation of nanogels with high functional β-cyclodextrin content. ACS Nano 6:8087–8093
Kettel MJ, Schaefer K, Pich A, Moeller M (2016) Functional PMMA nanogels by cross-linking with cyclodextrin methacrylate. Polymer 86:176–188
Kharlampieva E, Ankner JF, Rubinstein M, Sukhishvili SA (2008) pH-Induced release of polyanions from multilayer films. Phys Rev Lett 100:128303
Kim KS, Park W, Na K (2015) Gadolinium-chelate nanoparticle entrapped human mesenchymal stem cell via photochemical internalization for cancer diagnosis. Biomaterials 36:90–97
Koul V, Mohamed R, Kuckling D, Adler H-JP, Choudhary V (2011) Interpenetrating polymer network (IPN) nanogels based on gelatin and poly(acrylic acid) by inverse miniemulsion technique: synthesis and characterization. Colloids Surf B Biointerfaces 83:204–213
Kunjachan S, Jose S, Lammers T (2010) Understanding the mechanism of ionic gelation for synthesis of chitosan nanoparticles using qualitative techniques. Asian J Pharm 4:148–153
Lee IS, Akiyoshi K (2004) Single molecular mechanics of a cholesterol-bearing pullulan nanogel at the hydrophobic interfaces. Biomaterials 25:2911–2918
Lee H, Mok H, Lee S, Oh Y-K, Park TG (2007) Target-specific intracellular delivery of siRNA using degradable hyaluronic acid nanogels. J Control Release 119:245–252
Lee J, Lee C, Kim TH, Lee ES, Shin BS, Chi SC, Park ES, Lee KC, Youn YS (2012) Self-assembled glycol chitosan nanogels containing palmityl-acylated exendin-4 peptide as a long-acting anti-diabetic inhalation system. J Control Release 161:728–734
Lee DY, Choe K, Jeong YJ, Yoo J, Lee SM, Park J-H, Kim P, Kim Y-C (2015) Establishment of a controlled insulin delivery system using a glucose-responsive double-layered nanogel. RSC Adv 5:14482–14491
Li Y-L, Zhu L, Liu Z, Cheng R, Meng F, Cui J-H, Ji S-J, Zhong Z (2009) Reversibly stabilized multifunctional dextran nanoparticles efficiently deliver doxorubicin into the nuclei of cancer cells. Angew Chem Int Ed 48:9914–9918
Lim HJ, Ghim HD, Choi JH, Chung HY, Lim JO (2010) Controlled release of BMP-2 from alginate nanohydrogels enhanced osteogenic differentiation of human bone marrow stromal cells. Macromol Res 18:787–792
Liu AL, Garcia AJ (2016) Methods for generating hydrogel particles for protein delivery. Ann Biomed Eng 44(6):1946–1958
Liu G, Shao L, Gea F, Chena J (2007a) Preparation of ultrafine chitosan particles by reverse microemulsion. China Particuol 5:384–390
Liu Y-Y, Yu Y, Zhang G-B, Tang M-F (2007b) Preparation, characterization, and controlled release of novel nanoparticles based on MMA/β-CD copolymers. Macromol Biosci 7:1250–1257
Liu QR, Qi J-R, Yin S-W, Wang J-M, Guo J, Feng J-L, Cheng M, Cao J, Weng J-Y, Yang X-Q (2016) The influence of heat treatment on acid-tolerant emulsions prepared from acid soluble soy protein and soy soluble polysaccharide complexes. Food Res Int. https://doi.org/10.1016/j.foodres.2016.07.001
Lu Y, Sun W, Gu Z (2014) Stimuli-responsive nanomaterials for therapeutic protein delivery. J Controll Release 194:1–19
Maitra A (1984) Determination of size parameters of water-Aerosol OT-oil reverse micelles from their nuclear magnetic resonance data. J Phys Chem 88:5122–5125
Manirujjaman FS, Imran-Ul-Haque Md, Arafat M, Sharmin S (2013) An overview of nanogel drug delivery system. JAPS 3:S95–S105

Mather BD, Viswanathan K, Miller KM, Long TE (2006) Michael addition reactions in macromolecular design for emerging technologies. Prog Polym Sci 31:487–531
Maya S, Sarmento B, Nair A, Rejinold NS, Nair SV, Jayakumar R (2013) Smart stimuli-sensitive nanogels in cancer drug delivery and imaging: a review. Curr Pharm Des 19:7203–7218
Mendes AC, Baran ET, Reis RL, Azevedo HS (2013) Self-assembly in nature: using the principles of nature to create complex nanobiomaterials. Wiley Interdiscip Rev Nanomed Nanobiotechnol 5:582–612
Michael A (1887) On the addition of sodium acetacetic ether and analogous sodium compounds to unsaturated organic ethers. Am Chem J 9:115
Molina M, Asadian-Birjand M, Balach J, Bergueiro J, Miceli E, Calderón M (2015) Stimuli-responsive nanogel composites and their application in nanomedicine. Chem Soc Rev 44:6161–6186
Montanari E, D'Arrigo G, Di Meo C, Virga A, Coviello T, Passariello C, Matricardi P (2014) Chasing bacteria within the cells using levofloxacin-loaded hyaluronic acid nanohydrogels. Eur J Pharm Biopharm 87:518–523
Morimoto N, Qui XP, Winnik FM, Akiyoshi K (2008) Dual stimuli-responsive nanogels by self-assembly of polysaccharides lightly grafted with thiol-terminated poly (N-isopropylacrylamide) chains. Macromolecules 41:5985–5987
Morimoto N, Hirano S, Takahashi H, Loethen S, Thompson DH, Akiyoshi K (2013) Self-assembled pH-sensitive cholesteryl pullulan nanogel as a protein delivery vehicle. Biomacromolecules 14:56–63
Myrick JM, Vendra VK, Krishnan S (2014) Self-assembled polysaccharide nanostructures for controlled-release applications. Nanotechnol Rev 3(4):319–346
Nakamoto M, Hoshino Y, Miura Y (2014) Effect of physical properties of nanogel particles on the kinetic constants of multipoint protein recognition process. Biomacromolecules 15(2):541–547
Narayanan D, Geena MG, Lakshmi H, Koyakutty M, Nair S, Menon D (2013) Poly-(ethylene glycol) modified gelatin nanoparticles for sustained delivery of the anti-inflammatory drug ibuprofen-sodium: an in vitro and in vivo analysis. Nanomedicine 9:818–828
Nayak S, Lyon LA (2005) Soft nanotechnology with soft nanoparticles. Angew Chem Int Ed Engl 44:7686–7708
Oh JK, Drumright R, Siegwart DJ, Matyjaszewski K (2008) The development of microgels/ nanogels for drug delivery applications. Prog Polym Sci 33:448–477
Oh NM, Oh KT, Youn YS, Lee D-K, Cha K-H, Lee DH, Lee ES (2013) Poly(l-aspartic acid) nanogels for lysosome-selective antitumor drug delivery. Colloids Surf B Biointerfaces 101:298–306
Oha JK, Leea DI, Park JM (2009) Biopolymer-based microgels/nanogels for drug delivery applications. Prog Polym Sci 34:1261–1282
Onsoyen E, Skaugrud O (1990) Metal recovery using chitosan. J Chem Technol Biotechnol 49:395–404
Patil JS, Kamalapur MV, Marapur SC, Kadam DV (2010) Ionotropic gelation and polyelectrolyte complexation: the novel techniques to design hydrogel particulate sustained, modulated drug delivery system: a review. Dig J Nanomater Bios 5(1):241–248
Payet L, Terentjev EM (2008) Emulsification and stabilization mechanisms of o/w emulsions in the presence of chitosan. Langmuir 24:12247–12252
Peng J, Qi T, Liao J, Fan M, Luo F, Li H, Qian Z (2012) Synthesis and characterization of novel dual-responsive nanogels and their application as drug delivery systems. Nanoscale 4:2694
Petchthanasombat C, Tiensing T, Sunintaboon P (2012) Synthesis of zinc oxide encapsulated poly (methyl methacrylate)–chitosan core–shell hybrid particles and their electrochemical property. J Colloid Interface Sci 369:52–57
Pickering SU (1907) Emulsions. J Chem Soc 91:2001–2021
Platt VM, Szoka FC (2008) Anticancer therapeutics: targeting macromolecules and nanocarriers to hyaluronan or CD44, a hyaluronan receptor. Mol Pharmacol 5:474

Polexe RC, Delair T (2013) Elaboration of stable and antibody functionalized positively charged colloids by polyelectrolyte complexation between chitosan and hyaluronic acid. Molecules 18 (7):8563–8578

Popović Z, Liu W, Chauhan VP, Lee J, Wong C, Greytak AB, Insin N, Nocera DG, Fukumura D, Jain RK, Bawendi MG (2010) A nanoparticle size series for in vivo fluorescence imaging. Angew Chem Int Ed Engl 49:8649–8652

Puig JE, Rabelero M (2016) Semicontinuous microemulsion polymerization. Curr Opin Colloid Interface Sci 25:83–88

Puig JE, Mendizábal E, López-Serrano F, López RG (2012) Surfactant-assisted polymerization processes. In: Somasundaram P (ed) Encyclopedia of surface and colloid science, 2nd edn. Taylor and Francis. https://doi.org/10.1081/eescs-120047407

Pujana MA, Perez-Alvarez L, Cesteros Iturbe LC, Katime I (2013) Biodegradable chitosan nanogels crosslinked with genipin. Carbohydr Polym 94(2):836–842

Purwada A, Tian YF, Huang W, Rohrbach KM, Deol S, August A, Singh A (2016) Self-assembly protein nanogels for safer cancer immunotherapy. Adv Healthcare Mater 5:1413–1419

Rao KM, Rao KSVK, Ramanjaneyulu G, Ha C-S (2015) Curcumin encapsulated pH sensitive gelatin based interpenetrating polymeric network nanogels for anti cancer drug delivery. Int J Pharm 478:788–795

Ravi H, Baskaran V (2015) Biodegradable chitosan-glycolipid hybrid nanogels: a novel approach to encapsulate fucoxanthin for improved stability and bioavailability. Food Hydrocol 43:717–725

Ray M, Rousseau R (2013) Stabilization of oil-in-water emulsions using mixtures of denatured soy whey proteins and soluble soybean polysaccharides. Food Res Int 52:298–307

Rice JJ, Martino MM, De Laporte L, Tortelli F, Briquez PS, Hubbell JA (2013) Engineering the regenerative microenvironment with biomaterials. Adv Healthcare Mater 2:57–71

Rinaudo M (2010) New way to crosslink chitosan in aqueous solution. Eur Polym J 46:1537–1544

Rocha N, Mendonça P, Góis JR, Cordeiro R, Fonesca A, Ferreira P, Guliashvili T, Matyjaszewski K, Serra A, Coehlo J (2013) The importance of controlled/living radical polymerization in the design of tailor made nanoparticles for drug delivery systems. In: Coelho J (ed) Drug delivery systems: advanced technologies potentially applicable in personalised treatment. Springer, Netherlands, pp 315–357

Rolland J, Guillet P, Schumers JM, Duhem N, Préat V, Gohy JF (2012) Polyelectrolyte complex nanoparticles from chitosan and poly (acrylic acid) and Polystyrene-block-poly (acrylic acid). J Polym Sci A Polym Chem 50:4484–4493

Rossetti H, Albizzati D, Alfano M (2002) Decomposition of formic acid in a water solution employing the photo-fenton reaction. Ind Eng Chem Res 41:1436–1444

Sali SS (2016) Natural calcium carbonate for biomedical applications, M. Tech (Biotechnology) dissertation thesis project, arXiv:1606.07779

Sarika PR, James NR (2015) Preparation and characterisation of gelatin–gum arabic aldehyde-nanogels via inverse miniemulsion technique. Int J Biol Macromol 76:181–187

Sarika PR, James NR (2016a) Polyelectrolyte complex nanoparticles from cationised gelatin and sodium alginate for curcumin delivery. Carbohyd Polym 148:354–361

Sarika PR, James RN (2016b) Curcumin loaded gum arabic aldehyde-gelatin nanogels for breast cancer therapy. Mater Sci Eng C 65:331–337

Sarika PR, Anil Kumar PR, Raj DK, James NR (2015) Nanogels based on alginic aldehyde and gelatin by inverse mini emulsion technique: synthesis and characterization. Carbohydr Polym 119:118–125

Sarika PR, James NR, Anil Kumar PR, Raj DK (2016) Preparation, characterization and biological evaluation of curcumin loaded alginate aldehyde-gelatin nanogels. Mater Sci Eng C Mater Biol Appl 68:251–257

Sasaki Y, Yamane S, Kurosu K, Sawada S-I, Akiyoshi K (2012) Templated formation of hydroxyapatite nanoparticles from self-assembled nanogels containing tricarboxylate groups. Polymers 4:1056–1064

Schatz C, Domard A, Viton C, Pichot C, Delair T (2004) Versatile and efficient formation of colloids of biopolymer-based polyelectrolyte complexes. Biomacromolecules 5:1882–1892

Seo S, Lee CS, Jung YS, Na K (2012) Thermo-sensitivity and triggered drug release of polysaccharide nanogels derived from pullulan-g-poly (L-lactide) copolymers. Carbohyd Polym 87:1105–1111

Shen JM, Xu L, Lu Y, Cao HM, Xu ZG, Chen T, Zhang HX (2012) Chitosan-based luminescent/magnetic hybrid nanogels for insulin delivery, cell imaging, and anti diabetic research of dietary supplements. Int J Pharm 427:400–409

Shimoda A, Sawada S, Akiyoshi K (2011) Cell specific peptide conjugated polysaccharide nanogels for protein delivery. Macromol Biosci 11:882–888

Singh N, Nisha Gill V, Gill P (2013) Nanogel based artificial chaperone technology: an overview. Am J Adv Drug Deliv 1:271–276

Singh A, Agarwal R, Diaz-Ruiz CA, Willett NJ, Wang P, Lee LA, Wang Q, Guldberg RE, García AJ (2014) Nanoengineered particles for enhanced intra-articular retention and delivery of proteins. Adv Healthcare Mater 3:1562–1567

Sivaram AJ, Rajitha P, Maya S, Jayakumar R, Sabitha M (2015) Nanogels for delivery, imaging and therapy. Wiley Interdiscip Rev Nanomed Nanobiotechnol 7:509–533

Soni KS, Desale SS, Bronich TK (2015) Nanogels: An overview of properties, biomedical applications and obstacles to clinical translation. J Controll Release S0168–3659(15):30233–30239

Soppimath KS, Aminabhavi TM, Kulkarni AR, Rudzinski WE (2001) Biodegradable polymeric nanoparticles as drug delivery devices review. J Controll Release 70:1–20

Sui ZJ, Schlenoff JB (2004) Phase separations in pH-responsive polyelectrolyte multilayers: charge extrusion versus charge expulsion. Langmuir 20:6026–6031

Tai H, Kenji Y, Takayuki N, Mika S, Tsuyoshi S, Yoshinori A, Nobuyuki M, Kazunari A (2010) Hybrid hyaluronan hydrogel encapsulating nanogel as a protein nanocarrier: new system for sustained delivery of protein with a chaperone-like function. J Controll Release 142:483–489

Tan JPK, Tan MBH, Tam MKC (2010) Application of nanogel systems in the administration of local anesthetics. Local Reg Anesth 3:93–100

Tang J, Quinlan PJ, Tam KC (2015) Stimuli-responsive pickering emulsions: recent advances and potential applications. Soft Matter 11:3512–3529

Tang Z, He C, Tian H, Ding J, Hsiao BS, Chu B, Chen X (2016) Polymeric nanostructured materials for biomedical applications. Prog Polym Sci. https://doi.org/10.1016/j.progpolymsci.2016.05.005

Thampanchira D, Krishnakumar K, Dineshkumar B, John A (2016) Hydrogel and nanohydrogel—a review. Int J Pharm 6(2):78–81

Tian R, Xian L, Li Y, Zheng X (2016) Silica modified chitosan/polyethylenimine nanogel for improved stability and gene carrier ability. J Nanosci Nanotechnol 16(5):5426–5431

Tiyaboonchai W (2013) Chitosan nanoparticles: a promising system for drug delivery. Naresuan Univ J Sci Technol 11:51–66

Uthaman S, Maya S, Jayakumar R, Cho C-S, Park I-K (2014) Carbohydrate-based nanogels as drug and gene delivery systems. J Nanosci Nanotechnol 14:694–704

Uthaman S, Lee SJ, Cherukula K, Cho C-S, Park I-K (2015) Polysaccharide-coated magnetic nanoparticles for imaging and gene therapy. Biomed Res Int 2015:959175

van de Manakker F, Vermonden T, van Nostrum CF, Hennink WE (2009) Cyclodextrin-based polymeric materials: synthesis, properties, and pharmaceutical/biomedical applications. Biomacromolecules 10:3157–3175

Vinogradov SV, Bronich TK, Kabanov AV (2002) Nanosized cationic hydrogels for drug delivery: preparation, properties and interactions with cells. Adv Drug Deliv Rev 54:135–147

Wang NX, von Recum HA (2011) Affinity-based drug delivery. Macromol Biosci 11:321–332

Wang Y, Liu Y, Liu Y, Wang Y, Wu J, Li R, Yanga J, Zhang N (2014) pH-sensitive pullulan-based nanoparticles for intracellular drug delivery. Polym Chem 5:423–432

Wei X, Senanayake TH, Warren G, Vinogradov SV (2013) Hyaluronic acid-based nanogeldrug conjugates with enhanced anticancer activity designed for targeting of CD44-positive and drug-resistant tumors. Bioconjugate Chem 24:658–668

Wenling C, Mingyu C, Qiang A, Yandao G, Nanming Z, Xiufang Z (2005) Physical, mechanical and degradation properties, and Schwann cell affinity of cross-linked chitosan films. J Biomater Sci Polym Ed 16:791–807

Whitesides GM, Mathias JP, Seto CT (1991) Molecular self assembly and nanochemistry: a chemical strategy for the synthesis of nanostructures. Science 29:1312–1319

Wu D, Wan M (2008) A novel fluoride anion modified gelatin nanogel system for ultrasound-triggered drug release. J Pharm Pharmaceut Sci 11(4):32–45

Wu H-Q, Wang C (2016) Biodegradable smart nanogels: a new platform for targeting drug delivery and biomedical diagnostics. Langmuir 32:6211–6225

Wu W, Zhou S (2010) Hybrid micro-/nanogels for optical sensing and intracellular imaging. Nano Reviews 1:5730

Wu W, Aiello M, Zhou T, Berliner A, Banerjee P, Zhou SQ (2010a) In situ immobilization of quantum dots in polysaccharide-based nanogels for integration of optical pH-sensing, tumor cell imaging, and drug delivery. Biomaterials 31:3023–3031

Wu W, Shen J, Banerjee P, Zhou S (2010b) Chitosan-based responsive hybrid nanogels for integration of optical pH-sensing, tumor cell imaging and controlled drug delivery. Biomaterials 31(32):8371–8381

Xu D, Hong J, Yao S, Dong L, Sheng K (2007) Preparation of polyethyleneimine nanogels via photo-Fenton reaction. Radiat Phys Chem 76:1606–1611

Yang C, Wang X, Yao X, Zhang Y, Wu W, Jiang X (2015) Hyaluronic acid nanogels with enzyme-sensitive cross-linking group for drug delivery. J Control Release 205:206–217

Yeh J, Ling Y, Karp JM, Gantz J, Chandawarkar A, Eng G, Blumling J, Langer R, Khademhosseini A (2006) Micromolding of shape-controlled, harvestable cell-laden hydrogels. Biomaterials 27:5391–5398

Yin B, Deng W, Xu K, Huang L, Yao P (2012) Stable nano-sized emulsions produced from soy protein and soy polysaccharide complexes. J Colloid Interface Sci 380:51–59

Zhang N, Wardwell PR, Bader RA (2013) Polysaccharide-based micelles for drug delivery. Pharmaceutics 5:329–352

Zhang X, Achazi K, Steinhilber D, Kratz F, Dernedde J, Haag R (2014) A facile approach for dual-responsive prodrug nanogels based on dendritic polyglycerols with minimal leaching. J Controll Release 174:209–216

Zhang X, Malhotra S, Molina M, Haag R (2015) Micro- and nanogels with labile crosslinks—from synthesis to biomedical applications. Chem Soc Rev 44:1948–1973

Zhang H, Zhai Y, Wang J, Zhai G (2016) New progress and prospects: the application of nanogel in drug delivery. Mater Sci Eng C 60:560–568

Zhou Y, Briand V, Sharma N, Ahn S, Kasi R (2009) Polymers comprising cholesterol: synthesis, self-assembly, and applications. Materials 2:636

Zhou M, Wang T, Hu Q, Luo Y (2016) Low density lipoprotein/pectin complex nanogels as potential oral delivery vehicles for curcumin. Food Hydrocoll 57:20–29

Zhu J, Shi X (2013) Dendrimer-based nanodevices for targeted drug delivery applications. J Mater Chem B 1:4199–4211

Zhu X-N, Gao L-N, Cui Y-L (2015) A Novel Ca-Alginate Nanogel mediated by glycyrrhizic acid. In: 5th international conference on information engineering for mechanics and materials (ICIMM 2015)

Chapter 5
Polymer Gel Composites for Bio-Applications

Luminita Ioana Buruiana and Silvia Ioan

Abstract Current trends in bio-applications of polymer gel composites consist in the use of composites made by embedding of different types of micro- and/or nanoparticles into polymer gel matrices. The obtained complex system must provide a very attractive material that, due to its specific property and low toxicity, is recommended in versatile domain uses. On the other hand, cross-linked polymeric gels possess properties that make them suitable for different biomedical applications, due to their similarity to natural living tissue and inherent biocompatibility, attributed in part to their flexible and soft nature, and high water content, as well. Development of gel composites with good response properties and controllability contributes to significant advances in biomedical applications, especially cancer therapy, tissue engineering, food industry, and magnetic hyperthermia.

Keywords Polymer gel · Composite · Biomedical applications
Magnetic gels

Biopolymers have been used in food and packaging industry to ensure desirable texture and structural stability to a variety of products, including emulsions (Pappu et al. 2015, 2016; Trache et al. 2017; Corobea et al. 2016). Biocompatible polymeric gels have been applied in biomedical and pharmaceutical applications, namely delivery drug release, tissue engineering, and regenerative medicine. They also are shown to be very useful for controlled drug delivery, reconstruction of soft tissues, as biosensors or actuators (Moller et al. 2007; Lee et al. 2008; Richter et al. 2008; Andersson et al. 2009; Bassik et al. 2010). Biocompatible polymers could improve quality of life, being tools to treat injury and diseases.

Polymer gels usually represent a binary system (polymer–solvent) with a three-dimensional network composed of macromolecules or their derivatives, capable to retain a considerable quantity of the solvent. This type of system is maintaining its size and shape, being different by this feature from solutions of

L. I. Buruiana (✉) · S. Ioan
"Petru Poni" Institute of Macromolecular Chemistry, Iasi, Romania
e-mail: buruiana.luminita@icmpp.ro; luminitab25@yahoo.com

© Springer Nature Singapore Pte Ltd. 2018

V. K. Thakur et al. (eds.), *Polymer Gels*, Gels Horizons: From Science to Smart Materials, https://doi.org/10.1007/978-981-10-6080-9_5

polymers. Natural polymers used are negatively charged (i.e., alginate, hyaluronic acid, carrageenan), due to the carboxyl or sulfate groups existent along the chain. This type of materials is preferred because it do not induce an important inflammatory response in the body (Remminghorst and Rehm 2006; Goh et al. 2012). Biopolymers can be modified to be photo-cross-linkable ones by functionalization with acrylates groups. The photo-cross-linking process implies the presence of photoinitiator that generates, under irradiation with UV light, free radicals; subsequently, they react with functional groups from the pendant polymer chain and form intermolecular bonds (Ferreira et al. 2009).

Fluid gels contain a concentrated suspension of gelled particles, formed by a suitable flow field applied to a biopolymer solution.

The structures have physical properties and sizes similar to oil droplets in emulsion, which might be applied in offering an enhanced sensorial perception for some newly discovered foods (Norton et al. 2006).

Properties of biocompatible polymers can be changed to meet specific requirements. An important characteristic regarding their application is the specific rheological properties of fluid gels, i.e., flow-like behavior above a critical value of the applied stress. Also, by varying the gel composition, cross-link density, particle size, or interaction degree between the particles, the processing conditions can be controlled (Garrec and Norton 2012).

A classification of biopolymer gel composite is summarized in Scheme 1.

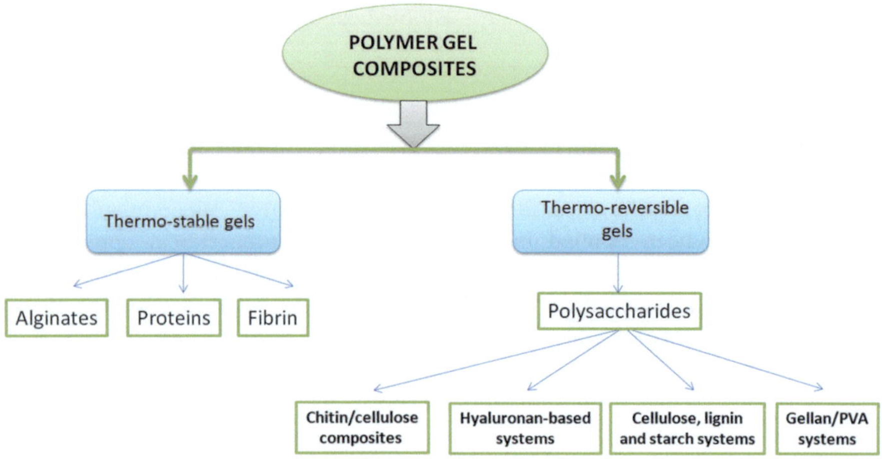

Scheme 1 Main classes of polymer gel composites

1 Biopolymer Fluid Gels

The development of novel methods for obtaining of thermostable and thermo-reversible fluid gels is in the progress in latest years. The obtaining of fluid gels is limited to saccharides, but lately literature has reported also the production of thermally stable fluid gels, which is able to participate at applications in the food and packaging industry, tissue engineering, and cancer therapy. A schematic representation of the most important applications regarding polymer gel composites is depicted in Scheme 2.

1.1 *Thermostable Fluid Gels: Alginate, Protein, and Fibrin Fluid Gels*

Thermostable fluid gels were produced by induced calcium alginate gelation or whey protein materials. The interesting issue in the obtaining process is represented by the relation between processing conditions and the realized microstructures.

(a) ***Alginates*** are a linear polysaccharide extracted from brown algae, consisting of (1-4)-linked blocks of poly-b-D-mannuronic acid and poly-a-L-guluronic acid residues, in different ratios and sequences (Draget et al. 2006). Their ability to form hydrogels in the presence of calcium ions can be fitted in different shapes, the enhanced in vitro and in vivo biocompatibility and reduced toxicity make alginate suitable candidates for drug delivery, tissue engineering, or in food industry (Andersen et al. 2008; Rokstad et al. 2011; Veriter et al. 2010). Alginate is not degradable, but it dissolves in physiological media of mammals and remains in the body, which represents a deficiency in its utilization in biomedical applications. Other limitation in tissue engineering application is the fact that it does not promote cell interactions (Nicodemus and Bryant 2008).

The alginate fluid gel is described in the literature as being obtained by adding of calcium ions into the alginate solution with a pin-stirrer device, under shear (Fernández Farrés et al. 2013). Thus, anisotropic gelled particles with broad size

Scheme 2 Applications of biopolymer gel composites

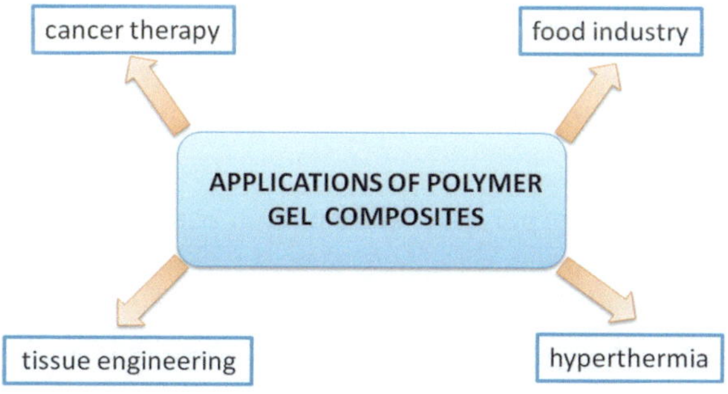

distributions are obtained by cross-linking reaction between the carboxyl groups of the polymer chain and cations. Particles with large dimensions (higher then 10 mm) are formed when concentration is decreasing, and lower shear stresses are applied on the particles obtained in the initial phase of the method. In addition, the enhanced diffusion grade of calcium by the polymer matrix determines the increasing of time necessary to mix the two liquids at molecular level. Thus, by this method, gels with increased particle size are formed. Still, this efficiency of this method is limited because of the slight control of the characteristics parameters, i.e., gelation and mixing and by the location of the injection point that determines the size of the formed particles.

A new technique that uses in situ calcium release is been developed, in which alginate fluid gels are obtained by controlling gelation kinetics (Fernández Farrés and Norton 2014). The method consists in the gradual release of calcium ions from insoluble calcium salts, this introduction into the alginate solutions leading to minimize non-uniform gel networks and preventing large clusters formation (Ström and Williams 2003).

Alginate fluid gels exhibit a specific flow behavior, with G′ greater than G″, more dependent on frequency that in the case of other inactive gels. These rheological measurements show that the gelation time increases once the applied shear rate is enhanced—that means an increased rate of particle growth. Conversely, at the shear removing after complete forming of the fluid gel structure, the percent of reordering due to inter-particle interaction seems less affected by the value of applied shear.

Control of the gelation kinetics and rheological behavior of alginates fluid gels with different molecular weight certify the great impact of the chain length and fraction of polymer loose ends on the cross-linking degree. Thus, alginate fluid gels with high molecular weight display faster formation kinetics, increased viscosities, and greater G′ than those of fluid gels with low molecular weight alginate.

It is proved that inter-particle interactions from alginate gel obtained by this method of calcium-releasing can be overset for giving viscoelastic responses. Conversely, other issue important to mention in this context is the presence of yield stress values, determined by applying a constant stress on the fluid gels and recording the strain as a function of time. At the same time, materials behave differently as a function of the applied stress:

- if it is below the yield stress—with the small value strain—the material behaves as a solid,
- if the value is higher than the yield stress, the strain tends to infinity until constant strain rate is accomplished (corresponding to flow condition of the material).

Increasing polymer concentration determines an enhanced interaction between gel particles and, also, increasing of yield stress. Therefore, the rheological behavior of alginate fluid gels permits overlapping between particles, which impede gels' close packing. From all these reasons, alginate fluid gels could be used as fat replacement in food industry, their advantage being the ability to keep the structure and to not melt at higher temperatures (like pasteurization temperature).

Besides food industry, the literature describes a method to release a drug in gastrointestinal tract of a wax-incorporated alginate gel beads. Thus, methods for prolonging retention of active substances in the stomach are based on different mechanisms, and the floating one being intensely studied because does not interfere in the motility of the stomach. The method employs a modified emulsion–gelation method that includes preparation of wax-incorporated alginate-based emulsion gel. Metformin—a model drug—is extruded into calcium chloride solution and then is studied the impact of different amounts of wax onto drug release behavior of emulsion gel beads. One can see that the increased amount of incorporated wax improved the drug release, while the beads remained floating. The obtained data confirm that sodium alginate gel beads are more suitable in comparison with pectinate gel beads (Ganesh et al. 2014).

In tissue engineering, combination between alginate dialdehyde, gelatin, and bioactive glass is assumed that it will exhibit good cellular adhesion, proliferation, differentiation properties, and also better biocompatibility. In this context, it was evaluated in vitro and in vivo biocompatibility of these materials in order to use them as scaffolds in engineering of hard tissue (Rottensteiner et al. 2014). One can conclude that in vitro evaluation exhibits good cell proliferation from bone marrow that was originated from mesenchymal stem cells spread on alginate dialdehyde–gelatin films, with or without bioglass. Conversely, in vivo subcutaneous implantation in rats does not cause an inflammatory response; underway degradation could be seen in several weeks, as well as ongoing vascularization. Good biocompatibility supports further investigations using dialdehyde–gelatin and nano-bioglass for bone tissue engineering application (Veriter et al. 2010).

One can conclude that the important applications of alginates were based on their gel-forming ability, being used as food additives for improving and stabilizing the structure of food. Also, alginates are used as additives capable of viscosifying, stabilizing, or emulsifying solutions (Rinaudo 2008, 2014).

(b) ***Whey protein material*** is presented in the literature as a precursor in fluid gels preparation due to its good nutritional value and thermal stability (Moakes et al. 2015). Gels are prepared by gelation process induced by heating of a 10 wt% protein solution, under controlled shear and thermal conditions. The size and interactions between gel particles depend on the shear and thermal process conditions. Thus, large clusters (higher than 120 mm) are obtained at low shear, their dimension displaying a decrease at higher shear. Also, one can observe that enhancing of heating rate that is applied to the formed gels determines the initiation of a higher number of small nuclei (Bromley et al. 2006), while larger clusters appear due to increasing of energy applied to the gel system. This energy increasing permits inter-particle bonding appearance. Flow characteristics of the fluid gels are related to the microstructure, shape, and particle interactions. In this context, one can notice that the suspension obtained can be considered a pseudo-plastic material with non-Newtonian behavior and pronounced shear thinning characteristics. The flow behavior is characterized by a reduced phase volume, and the obtained particles tending to occupy different volumes as a function of aggregation degree. Therefore, controlling the microstructure of the formed gel particles and the volume fraction leads to a possible control over the desired rheological properties. Because all of that issues, protein fluid gels could find applications as food supplements with thermostable behavior.

(c) ***Fibrin*** represents one of the most known polymers in nature that reopens constantly novel features of self-assembly and soft elasticity; they exhibit different uses as hemostatic agent in medical applications (prevent bleeding, stimulate wound healing due to trauma, or other lesions of the normal tissue) or scaffold for tissue engineering. Fibrin is designed to assemble under physiological conditions by the activation of the soluble fibrinogen monomer and by the thrombin enzyme, being the most important plasma protein involved in blood clotting (Weisel 2007). Fibrin has a specific polymerization mechanism by which it can be controlled the gelling process and network architecture.

Fibrin gels present interesting rheologic behavior due to their non-Newtonian characteristic, with low strains and significant stiffening to resist at high deformation conditions. At the same time, at interfaces for cells grown and proliferation on 3D matrices, fibrin possesses unique characteristics, uncommon to other materials. These properties are required for their use as hemostatic agents and matrices in wound healing process (Janmey et al. 2009). Future trends in the study of the combined effects of mechanics and biochemical fibrin network consist of developing the already known properties that lead to potential applications in medicine and bioengineering.

1.2 Thermo-Reversible Fluid Gels from Polysaccharides

Fluid gels are obtained by a flow field application of gelling biopolymer solution, determining particle aggregation. The control of the gel preparation process kinetics provides better final particle physical properties, their size and shape being mainly caused by the applied temperature and shear rate. At high cooling rate, large particles with irregular shape are formed, while at lower cooling rate, small and spherical particles are obtained. In this case, the gel systems show a more fluid-like response with reduced storage modulus that in the case of fluid gels obtained at higher cooling rates.

The literature reported obtaining and characterizations of composite gels based on chitin/cellulose (Takegawa et al. 2010). The obtained films were characterized by X-ray diffraction and thermogravimetric analysis, the results showing good miscibility between those two components of the system (polysaccharides and ionic liquids). The mechanical properties of the materials are influenced by the ratios of chitin and cellulose from the materials. Thus, at low chitin/cellulose molar ratio, 1:9, the diffraction peaks of the gel composite were very difficult to observe. Enhancing of chitin/cellulose molar ratio determines the small diffraction peaks due to the crystalline structure of chitin. This indicates a better miscibility of cellulose chains with ionic liquids and causes the break of crystalline structure, while the chitin network is rather formed, increasing its content in the gel films.

Other gel polymer system investigated for special osmotic and scattering properties is composite gels based on hyaluronan compounds (carboxymethylated thiolated hyaluronan—stiff biopolymer chain), cross-linked with polyethylene glycol diacrylate (a polymer with flexible chains). This gel system is specific for many medical uses, including tissue engineering, wound healing, drug delivery, or soft tissue augmentation (Thakur and Kessler 2015; Shu et al. 2004). The viscoelastic behavior manifests a stronger dependence on the pre-gel concentration against cross-link density: osmotic pressure of the mixing depends on the gel weight ratio, being independent by the pre-gel concentration. At the same time, scattering properties highlight the influence of the total polymer concentration, indicating a less significant role of the specific interactions between composite components (Horkay et al. 2010).

Cellulose, lignin, and starch are some of the most commonly used biopolymers in nature, having numerous advantages like availability from renewable resource, ability to replace some synthetic polymers, and low cost (Thakur and Thakur 2015). Besides that, they have also some disadvantages that restrict their larger field of applications. In this category are included high water absorption of starch, cellulose poor solubility and ability to process, or relatively low tensile property. At the same time, they cannot melt, or be processed like thermoplastic polymers, representing a drawback, as well.

Until now, literature presents numerous researches regarding natural polymers, i.e., PCL/starch (Ikeo et al. 2006), PHB/lignin (Kai et al. 2004), PCL/cellulose (Sanchez-Garcia et al. 2008). But it is proved that these polymers are expensive; for improving interfacial interactions of the gel composites and compatibility between hydrophobic and natural polymers, future research must be focused on finding proper reactive coupling agent or compatibilizer. In the past years, a special attention was paid on preparation of totally biodegradable blends containing natural polymers and their derivatives, such as cellulose diacetate/starch (Lee et al. 2006), cationic starch/cellulose (Bratskaya et al. 2006), or chitosan/cellulose acetate (Liu and Bai 2005). The obtained blends although exhibit good compatibility and biodegradability will determine cost and complexity enhancing in comparison with environmental friendly materials based on natural polymers. For improving these aspects, latest research was focused on developing new bio-based composite films obtained by an ionic liquid (1-allyl-3-methylimidazolium chloride) that coagulate in a non-solvent condition (Wu et al. 2009). The advantage of this ionic liquid is that it can be recycled with a high yield and purity immediately after the preparation of the green film. The composite gel formed is analyzed from mechanical point of view and by the interactive effect of each component. Starch polymer could form amorphous flexible films, with good transparence. The measurements illustrate that composite films showed peculiar mechanical properties in dry and wet states due to the reciprocal properties of different components. Composite films have high gas permeability values and a CO_2:O_2 permeability ratio close to 1, so they potentially could be used as fresh food packaging.

A natural polysaccharide that was studied for its intrinsic excellent properties is deacetylated gellan with an ordered, helical chain structure. For its application in tissue engineering, the main property is to be structural resemble native tissue (Boschetti et al. 2004; Liu and Bhatia 2004). For this aim, the researchers are focused on the designing of double hydrogel network with enhanced mechanical strength (Smith et al. 2007). The measurements prove that the complex system formed combine each property of individual polymer, forming a material with suitable characteristics in order to closely mimic the natural complexity of tissue. The inclusion of poly(vinyl alcohol) in gellan systems is shown to increase the gel strength up to a critical concentration. As the poly(vinyl alcohol) introduced in the gellan matrix form a separate phase, the maximum gel strength and stiffness correspond to the phase inversion point, the material properties being caused in that case by the dominant polymer (Norton et al. 2014). Thus, at low concentration, a linear viscoelastic regime is observed while at higher concentration, one can observe a separate linear dependency. This viscoelastic behavior points out the appearance of a chain entanglement at the crossover point, which proves that the decreasing of gel strength is not owing to the addition of PVA, but to the polymer entanglement. So this study demonstrates that the strength of gellan hydrogels can be enhanced by the addition of a second polymer, PVA, in order to mimic the properties of human tissues, for replacing current regenerative procedures that are traumatic and painful.

2 Magnetic Gel Composites

The use of biocompatible thermo-responsive magnetic gels in cancer therapy constitutes an interesting approach in latest decades, due to their good biocompatibility and intrinsic multi-functionality. Recent applications, namely magnetic hyperthermia, are based on the ability of magnetic composites to release thermal energy when are placed in an external oscillating magnetic field. At the international level, such experimental investigations date back from to 1957, when Gilchrist et al. used particles γ-Fe_2O_3 (20–100 nm size) exposed to a 1.2 MHz magnetic field to heat various tissues (Gilchrist 1957). Hyperthermia cancer therapy consists in achieving a uniform, localized temperature inside the tumor, monitoring also the surrounding tissue temperature. Therefore, the incorporation of these magnetic particles into a thermo-responsive hydrogel is a promising approach for magnetic hyperthermia cancer treatment.

Magnetite (Fe_3O_4) represents the most attractive magnetic material used in cancer therapy, due to its strong magnetic property and low toxicity to desired location. Conversely, cross-linked polymeric gels have specific properties that recommend them for various biomedical applications owing to similarity to natural living tissue and intrinsic biocompatibility, attributed partially to their soft and flexible nature and to high content of water (Boucard et al. 2007; Choi et al. 2007; Gao et al. 2008). In this context, the literature reported several types of magnetic fluids used for magnetic hyperthermia, the most known being composed by dextran-coated magnetite injected into the tail of rats vein under a magnetic field of 450 kHz, for 12 min, being observed a tumor shrinkage and tissue necrosis (Tassa et al. 2011).

The most used synthesis method for preparation of magnetic particles of uniform size and shape from solutions is the precipitation method. The particles begin to form when the solution reaches the critical saturation point. In the reaction, there must be used biocompatible polymers to impede cluster formation and, at the same time, to stabilize the formed particles (Roca et al. 2009). The literature reported several types of particle suspensions, stabilized with natural polymers like dextran, starch, alginate, cellulose, chitosan (Zamora-Mora et al. 2015). Thus, multidomain ferrite particle suspensions, coated with dextran (with diameters between 3 and 10 nm and several micrometers), embedded in a 2% agar hydrogel matrix were studied for magnetic hyperthermia (Jordan et al. 2001). An important requirement for better cancer treatment is the accumulation of magnetic fluids inside the tumor. Therefore, it is described in the literature that magnetic fluid based on gelatin could immobilize into tumor around 85% of the administered particle gel (Lao and Ramanujan 2004; Dutz and Hergt 2013).

Other studies display the ability of alginate to obtain macroporous ferrogels to drug delivery under the control of a moderate magnetic field. Magnetite particles with 10 nm diameter are coated with Pluronic F-127 (co-polymer) for impeding cluster formation, subsequently being functionalizate with alginate (Zhao et al. 2011). The porous gel scaffold facilitates in vitro drug releasing of different active substances such as Mitoxantrone (anticancer drug), DNA plasmid, or some types of chemokine. For a better efficiency of this system and control of various required cell types, future studies must be focused on adjustments of different parameters like magnetic particle size and density, the strength of applied magnetic field, or frequency of magnetic stimulation.

Lately, research was focused on the obtaining of magnetic gels based on synthetic polymers used in tumor therapy. In all these cases, the magnetic particle used was magnetite coated with different biocompatible synthetic polymers, the most used being poly(vinyl alcohol) (PVA) and polyethylene glycol (PEG) (Meenach et al. 2013; Ghosh et al. 2008). Combination of magnetite particles embedded in a synthetic polymer provides a composite system suitable for hyperthermia therapy along with drug delivery in the synergistic treatment of solid tumors, allowing heating when a magnetic field is applied (Ang et al. 2007; Ramanujan et al. 2009).

As a conclusion, one can say that the versatility of magnetic gel composites makes them proper candidates for synergistic treatment of cancer, involving drug release and magnetic thermal therapy. Still, there are still several issues that must be better understood in the future research before clinical applications to be extended. Developing new formulations to avoid overheating effects, a significant decrease of the magnetic particles amount used in the in vivo tests, designing of new targeting particles and hydrogels able to improve specific molecular recognition, or foresight testing concerning some physiological parameters that affect the functionality of implanted composites in vascular network are just several of the important features that must be taking into consideration.

3 Concluding Remarks and Future Perspectives

The area of used polymer gel composites attracted the interest of scientists in the past years, as reflected in considerable amount of research focused on the optimization of the gel system and the manufacturing process. Polymer gels are viscoelastic solids that exhibit specific rheological behavior, with parameters which depend on the gel microstructure. Cross-linked polymeric gels possess properties that recommend them as candidates for various biomedical applications, namely cancer therapies or tissue engineering, due to their similarity with natural living tissues. All the experiments performed by now can constitute models to further improvements of the polymer gel systems, thus having a huge impact in the field of biotechnology. However, there are aspects that are still not addressed in the literature and should be considered in future studies, such as the involvement in supramolecular chemical features of the matrix in controlling the molecular recognition and heat conduction abilities to solve problems like targeting and overheating. Such investigations could open new perspectives in perfecting nowadays approaches used in cancer treatment.

References

Andersen T, Strand BL, Formo K, Alsberg E, Christensen BE (2008) Alginates as biomaterials in tissue engineering. Carbohydr Chem 37:227–258

Andersson O, Larsson A, Ekbald T, Liedberg B (2009) Gradient hydrogel matrix for microarray and biosensor applications: an imaging SPR study. Biomacromolecules 10:142–148

Ang KL, Venkatraman S, Ramanujan RV (2007) Magnetic PNIPA hydrogels for hyperthermia applications in cancer therapy. Mater Sci Eng, C 27:347–351

Bassik N, Abebec BT, Laflin KE, Gracias DH (2010) Photolithographically patterned smart hydrogel based bilayer actuators. Polym Commun 51:6093–6098

Boschetti F, Pennati G, Gervaso F, Peretti GM, Dubini G (2004) Biomechanical properties of human articular cartilage under compressive loads. Biorheology 41:159–166

Boucard N, Viton C, Agay D, Mari E, Roger T, Chancerelle Y, Domard A (2007) The use of physical hydrogels of chitosan for skin regeneration following third-degree burns. Biomaterials 28:3478–3488

Bratskaya S, Schwarz S, Petzold G, Liebert T, Heinze T (2006) Cationic starches of high degree of functionalization: 12. Modification of cellulose fibers toward high filler technology in papermaking. Ind Eng Chem Res 45:7374–7379

Bromley E, Krebs M, Donald A (2006) Mechanisms of structure formation in particulate gels of b-lactoglobulin formed near the isoelectric point. Eur Phys J Part E 21:145–152

Choi NW, Cabodi M, Held B, Gleghorn JP, Bonassar LJ, Stroock AD (2007) Microfluidic scaffolds for tissue engineering. Nat Mater 6:908–915

Corobea MC, Muhulet O, Miculescu F, Antoniac IV, Vuluga Z, Florea D et al (2016) Novel nanocomposite membranes from cellulose acetate and clay-silica nanowires. Polym Adv Technol 27(12):1586–1595

Draget KI, Skjåk-Bræk G, Stokke BT (2006) Similarities and differences between alginic acid gels and ionically crosslinked alginate gels. Food Hydrocolloids 20:170–175

Dutz S, Hergt R (2013) Magnetic nanoparticle heating and heat transfer on a microscale: basic principles, realities and physical limitations of hyperthermia for tumour therapy. Int J Hyperth 29:790–800

Fernández Farrés I, Norton IT (2014) Formation kinetics and rheology of alginate fluid gels produced by in-situ calcium release. Food Hydrocolloids 40:76–84

Fernández Farrés I, Douaire M, Norton IT (2013) Rheology and tribological properties of Ca-alginate fluid gels produced by diffusion-controlled method. Food Hydrocolloids 32: 115–122

Ferreira P, Coelho JFJ, Almeida JF, Gil MH (2009) Photocrosslinkable polymers for biomedical applications. In: Fazel R (ed) Biomedical engineering—frontiers and challenges. InTech, Rijeka, Croatia, pp 55–74

Ganesh NS, Bharathi G, Jayanthi C, Hanumanthachar J (2014) Comparative evaluation of wax incorporated alginate and pectinate gel beads of Metformin. Der Pharmacia Lettre 6:10–16

Gao D, Xu H, Philbert MA, Kopelman R (2008) Bio-eliminable nano-hydrogels for drug delivery. Nano Lett 8:3320–3324

Garrec DA, Norton IT (2012) Understanding fluid gel formation and properties. J Food Eng 112:175–182

Ghosh S, Cai T, Mitra SG, Neogi A, Hu Z, Mills N (2008) Multifunctional magnetic hydrogels with polyethylene glycol-derivative colloidal nanospheres for drug delivery and hyperthermia applications. NSTI-Nanotech 2:708–710

Gilchrist RK (1957) Selective inductive heating of lymph. Annal Surgery 146:596–606

Goh CH, Heng PWS, Chan LW (2012) Alginates as a useful natural polymer for microencapsulation and therapeutic applications. Carbohydr Polym 88:1–12

Horkay F, Magda J, Alcoutlabi M, Atzet S, Zarembinski T (2010) Structural, mechanical and osmotic properties of injectable hyaluronan-based composite hydrogels. Polymer 51: 4424–4430

Ikeo Y, Aoki K, Kishi H, Matsuda S, Murakami A (2006) Nano clay reinforced biodegradable plastics of PCL starch blends. Polym Adv Technol 17:940–944

Janmey PA, Winer JP, Weisel JW (2009) Fibrin gels and their clinical and bioengineering applications. J R Soc Interface 6:1–10

Jordan A, Scholz R, Maier-Hauff K, Johannsen M, Wust P, Nadobny J (2001) Presentation of a new magnetic field therapy system for the treatment of human solid tumors with magnetic fluid hyperthermia. J Magn Magn Mater 225:118–126

Kai WH, He Y, Asakawa N, Inoue Y (2004) Effect of lignin particles as a nucleating agent on crystallization of poly(3-hydroxybutyrate). J Appl Polym Sci 94:2466–2474

Lao LL, Ramanujan RV (2004) Magnetic and hydrogel composite materials for hyperthermia applications. J Mater Sci Mater Med 1061–1064

Lee SY, Cho MS, Nam JD, Lee Y (2006) Melting processing of biodegradable cellulose diacetate/ starch composites. Macromol Symp 242:126–130

Lee W, Choi D, Lee Y, Kim DN, Park J, Koh WG (2008) Preparation of micropatterned hydrogel substrate via surface graft polymerization combined with photolithography for biosensor application. Sens Actuators B Chem 129:841–849

Liu CX, Bai RB (2005) Preparation of chitosan/cellulose acetate blend hollow fibers for adsorptive performance. J Membr Sci 267:68–77

Liu TV, Bhatia SN (2004) Three-dimensional tissue fabrication. Adv Drug Deliv Rev 56: 1635–1647

Meenach SA, Shapiro JM, Hilt JZ, Anderson KW (2013) Characterization of PEG-iron oxide hydrogel nanocomposites for dual hyperthermia and paclitaxel delivery. J Biomater Sci Polym Ed 24:1112–1126

Moakes RJA, Sullo A, Norton IT (2015) Preparation and characterisation of whey protein fluid gels: The effects of shear and thermal history. Food Hydrocolloids 45:227–235

Moller S, Weisser J, Bischoff S, Schnabelrauch M (2007) Dextran and hyaluronan methacrylate based hydrogels as matrices for soft tissue reconstruction. Biomol Eng 24:496–504

Nicodemus GD, Bryant SJ (2008) Cell encapsulation in biodegradable hydrogels for tissue engineering applications. Tissue Eng Part B Rev 14:149–165

Norton AB, Hancocks RD, Grover LM (2014) Poly (vinyl alcohol) modification of low acyl gellan hydrogels for applications in tissue regeneration. Food Hydrocolloids 42:373–377

Norton IT, Frith WJ, Ablett S (2006) Fluid gels, mixed fluid gels and satiety. Food Hydrocolloids 20:229–239

Pappu A, Patil V, Jain S, Mahindrakar A, Haque R, Thakur VK (2015) Advances in industrial prospective of cellulosic macromolecules enriched banana biofibre resources: a review. Int J Biol Macromol 79:449–458

Pappu A, Saxena M, Thakur VK, Sharma A, Haque R (2016) Facile extraction, processing and characterization of biorenewable sisal fibers for multifunctional applications. J Macromol Sci, Part A 53(7):424–432

Ramanujan RV, Ang KL, Venkatraman S (2009) Magnet-PNIPA hydrogels for bioengineering applications. J Mater Sci 44:1381–1387

Remminghorst U, Rehm BHA (2006) Bacterial alginates: from biosynthesis to applications. Biotechnol Lett 28:1701–1712

Richter A, Paschew G, Klatt S, Lienig J, Arndt KF, Adler HJP (2008) Review on hydrogel-based pH sensors and microsensors. Sensors 8:561–581

Rinaudo M (2008) Main properties and current applications of some polysaccharides as biomaterials. Polym Int 57:397–430

Rinaudo M (2014) Biomaterials based on a natural polysaccharide: alginate. TIP Rev Esp Cienc Quím Biol 17:92–96

Roca AG, Costo R, Rebolledo AF, Veintemillas-Verdaguer S, Tartaj P, González-Carreño T, Morales MP, Serna CJ (2009) Progress in the preparation of magnetic nanoparticles for applications in biomedicine. J Phys Part D: Appl Phys 42:224002

Rokstad AB, Brekke OL, Steinkjer B, Ryan L, Kollarikova G, Strand BL, Skjak-Braek G, Lacik I, Espevik T, Mollnes TE (2011) Alginate microbeads are complement compatible, in contrast to polycation containing microcapsules, as revealed in a humand whole blood model. Acta Biomater 7:2566–2578

Rottensteiner U, Sarker B, Heusinger D, Dafinova D, Rath SN, Beier JP, Kneser U, Horch RE, Detsch R, Boccaccini AR, Arkudas A (2014) In vitro and in vivo biocompatibility of alginate dialdehyde/gelatin hydrogels with and without nanoscaled bioactive glass for bone tissue engineering applications. Materials 7:1957–1974

Sanchez-Garcia MD, Gimenez E, Lagaron JM (2008) Morphology and barrier properties of solvent cast composites of thermoplastic biopolymers and purified cellulose fibers. Carbohydr Polym 71:235–244

Shu XZ, Liu Y, Palumbo FS, Luo Y, Prestwich GD (2004) In situ crosslinkable hyaluronan hydrogels for tissue engineering. Biomaterials 25:1339–1348

Smith AM, Shelton RM, Perrie Y, Harris JJ (2007) An initial evaluation of gellan gum as a material for tissue engineering applications. J Biomater Appl 22:241–254

Ström A, Williams MA (2003) Controlled calcium release in the absence and presence of an ion-binding polymer. J Phys Chem B 107:10995–10999

Takegawa A, Murakami M, Kaneko Y, Kadokawa J (2010) Preparation of chitin/cellulose composite gels and films with ionic liquids. Carbohydr Polym 79:85–90

Trache D, Hazwan Hussin M, Mohamad Haafiz MK, Kumar Thakur V (2017) Recent progress in cellulose nanocrystals: sources and production. Nanoscale 9(5):1763–1786

Tassa C, Shaw SY, Weissleder R (2011) Dextran-coated iron oxide nanoparticles: a versatile platform for targeted molecular imaging, molecular diagnostics and therapy. Acc Chem Res 44:842–852

Thakur VK, Kessler MR (2015) Self-healing polymer nanocomposite materials: a review. Polymer 69:369–383

Thakur VK, Thakur MK (2015) Recent advances in green hydrogels from lignin: a review. Int J Biol Macromol 72:834–847

Veriter S, Mergen J, Goebbels RM, Aouassar N, Gregoire C, Jordan B, Leveque P, Gallez B, Gianello P, Dufrane D (2010) In vivo selection of biocompatible alginates for islet encapsulation and subcutaneous transplantation. Tissue Eng Part A 16:1503–1513

Weisel JW (2007) Structure of fibrin: impact on clot stability. J Thromb Haemost 5(Suppl. 1): 116–124

Wu RL, Wang XL, Li F, Li HZ, Wang YZ (2009) Green composite films prepared from cellulose, starch and lignin in room-temperature ionic liquid. Biores Technol 100:2569–2574

Zamora-Mora V, Soares PIP, Echeverria C, Hernández R, Mijangos C (2015) Composite chitosan/agarose ferrogels for potential applications in magnetic hyperthermia. Gels 1:69–80

Zhao X, Kim J, Cezar CA, Huebsch N, Lee K, Bouhadir K, Mooney DJ (2011) Active scaffolds for on-demand drug and cell delivery. Proc Natl Acad Sci USA 108:67–72

Chapter 6
Vegetable Oils as a Chemical Platform

Amanda Murawski and Rafael Lopes Quirino

Abstract As dwindling crude oil reserves around the planet bring uncertainty about the supply of energy and raw materials to mankind, many scientists are focusing their efforts in developing inventive solutions to diminish our dependency on petroleum. Besides appeasing concerns related to a very vulnerable economy strongly tied to crude oil markets, alternative technologies and sources of chemicals have the potential to provide more environmentally friendly solutions, including lower CO_2 emissions, improved atom economy, and the use of renewable resources. In that context, vegetable oils can be used as a chemical platform, leading to a variety of products through chemical modifications. This chapter covers the use of vegetable oils for the preparation of various materials, including polymers, composites, coatings, gels, emulsions, binders in ink formulations, and other minor applications.

1 Introduction

Vegetable oils are an invaluable tool to the polymeric industry. As crude oil reserves decline worldwide, it is imperative to seek alternative renewable resources. Many scientists speculate about how long the world's current natural petroleum reserves will last. Regardless of the life span left of those resources, they have created an economic dependency, leaving the global community subjected to fluctuating oil prices. Outside of escaping the financial constraints of a petroleum-based economy, utilizing vegetable oils as an alternative resource provides more environmentally friendly options *in lieu* of global warming and other environmental concerns. Vegetable oils are bio-based, renewable, cost-efficient, easily accessible, environmentally friendly, and less biologically hazardous (Lu and Larock 2009), making them ideal candidates for alternative bio-based material research.

A. Murawski · R. L. Quirino (✉)
Georgia Southern University, Statesboro, GA, USA
e-mail: rquirino@georgiasouthern.edu

© Springer Nature Singapore Pte Ltd. 2018
V. K. Thakur et al. (eds.), *Polymer Gels*, Gels Horizons: From Science to Smart Materials, https://doi.org/10.1007/978-981-10-6080-9_6

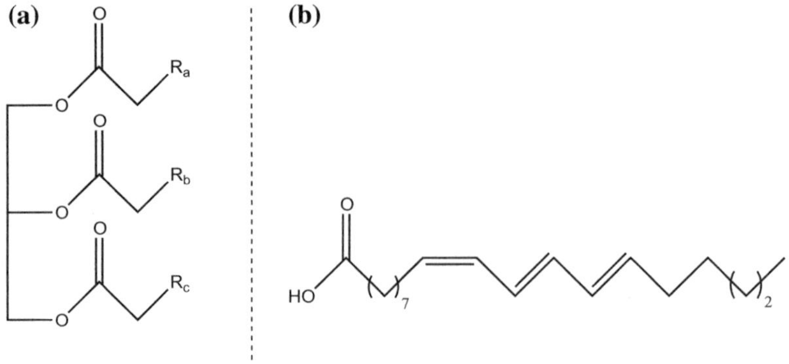

Fig. 1 Chemical structures of **a** triglyceride, where R_a, R_b, and R_c are aliphatic chains, and **b** α-eleostearic acid

Vegetable oils are made up of many triglyceride units. A single triglyceride unit consists of a glycerol unit that is attached to three fatty acid chains. Depending on the origin of the oil, there can be many different compositions of fatty acids (Meier et al. 2007). Even within oils of the same plant, but from different locations of the plant, there still can be variations within the fatty acid compositions (Meier et al. 2007). The applications of vegetable oils are largely based on their fatty acid profile. The distinction between fatty acids occurs in the length of the fatty acid chain, number and position of the carbon–carbon double bonds, and the presence of specific functional groups (Andjelkovic and Larock 2006). Generally, most naturally occurring fatty acids contain an even number of carbon atoms and a *cis* configuration of the carbon–carbon double bonds, which are typically not in conjugation (Andjelkovic and Larock 2006). However, some exceptions to this generalization do exist, such as in the case of tung oil, which is mainly made up of (>84%) of α-eleostearic acid (Fig. 1) and contains *trans* carbon–carbon double bonds on the 11th and 13th carbons, exhibiting a conjugated triene, which makes tung oil an exceptionally attractive oil for polymerization. Reactivity within the fatty acid chains is highly reliant on the conjugation of the carbon–carbon double bonds and the presence of reactive functional groups, such as hydroxyl groups seen in ricinoleic acid (Eren et al. 2003).

Since there is such a variety of fatty acid composition between different vegetable oils, a wide range of applications have been proposed over the years. One such example occurs in the coating industry. Unsaturated vegetable oils have an auto-oxidative property when exposed to air (Tallman et al. 2004), making them the ideal starting material for fast-drying paints. In this chapter, the use and application of vegetable oil-based materials in the polymer, composites, coatings, gels, emulsions, and other relevant industries are reviewed. Along with a focused discussion on the current literature involving thermoset polymers, their potential chemical modifications for improved reactivity and the different methods of polymerization will be covered. Then, this chapter continues with a brief summary of relevant vegetable oil derivatives, such as polyhydroxyurethanes and polycaprolactones. Lastly, this chapter concludes with a critical analysis and evaluation on the future of the discussed systems.

2 Thermosetting Polymeric Systems

Polymers are commonly classified into three distinct categories: elastomers, thermosets, or thermoplastics. We will be mainly focusing on thermoset polymers for the purpose of this section. In order to be classified as a thermosetting polymer, there must be a three-dimensional cross-linked network interconnecting all of the polymer chains of the system. These cross-links are covalently bonded, making thermosetting polymers more resistant to heat than their thermoplastic counterparts (Ratna 2009). Thermoplastics, on the other hand, are assembled by well-compacted, non-covalent bonds that have the ability to reshape and reform in the presence of heat. Thermosets do not possess this quality. They cannot be reshaped or reformed after curing because the polymer chains cannot flow at elevated temperatures and do not exhibit a melting point due to the strongly formed interconnections. They also cannot be dissolved in a solvent (Goodman 1999). The versatility of thermosets lies in the ability of one to adjust the cross-link density of the polymer network without having to change the chemical structure of the starting material. Currently, the most successful commercially available thermosets are phenolic and epoxy resins, which are produced from non-renewable fossil feedstocks (Williams and Hillmyer 2008; Xia and Larock 2010b). Thus, as mentioned earlier in the text, there has been a significant increase in the interest on alternative resources for economic and environmental reasons. This is particularly true in reference to thermosetting polymers.

Over the years, a variety of starting materials, such as sugars, polysaccharides, vegetable oils, lignin, pine resin derivatives, and proteins have been investigated for the preparation of polymeric materials (Gandini 2011, 2008). This section focuses primarily on thermosets based on vegetable oils due to their widespread use. Researchers have taken advantage of the oils' naturally reactive carbon–carbon double bonds and functional groups in polymerization reactions to covalently create cross-linked polymer chains. These triglyceride-based polymers exhibit remarkable thermomechanical properties when compared to other bio-based polymer systems. The denser the cross-links are, the more rigid and stronger the polymers are (Belgacem and Gandini 2008; Van De Mark and Sandefur 2009). Two commonly used polymerization reactions that can take place to cross-link the fatty acid chains are free radical and cationic polymerization (Turley 2008). More recent discoveries have led to new methods of polymerization such as acyclic diene metathesis (ADMET), ring-opening metathesis (ROMP), cycloaddition, epoxy curing, and other miscellaneous reactions (Andjelkovic and Larock 2006). Vegetable oils with multiple reactive hydroxyl function groups (also known as polyols) have also been reacted with anhydrides, diacids, or diisocyanates to yield polyester or polyurethane thermosets (Xia and Larock 2010a). Polyurethane systems will be discussed further when describing vegetable oil-based coatings and emulsions, in Sect. 4.

2.1 Thermosets from Free Radical Polymerization

2.1.1 Unmodified Oils for Polymerization

More often than not, natural vegetable oils must be modified before polymerization due to their intrinsic low reactivity. There are a few exceptions to this generalization, such as highly unsaturated oils, which are often used in the paint and coating industries due to their auto-oxidative behavior when exposed to oxygen in the air (Tallman et al. 2004). During this auto-oxidative process, carbon–carbon double bonds in the oil react with oxygen and polymerize to create a cross-linked network within the polymer. The polymerization process follows a free radical mechanism (Cakmakli et al. 2005). Poly(methyl methacrylate) (PMMA) and poly (*n*-butyl methacrylate) (PBMA) have been grafted onto linseed and soybean oils in order to form macro-initiators (Cakmakli et al. 2004). Other studies have found that in thermosets of linseed oil, styrene (ST), and divinylbenzene (DVB), there is a decrease in mechanical properties when the vegetable oil content increased in the copolymer (Sharma et al. 2008). Another study's results showed that when 70 wt% or more of conjugated low-saturation soybean oil (CLS) was copolymerized with acrylonitrile (AN), DVB, and/or dicyclopentadiene (DCPD) in the presence of azobisisobutyronitrile (AIBN), a significant amount of unreacted oil was recovered from the final thermoset (Valverde et al. 2008). The unreacted oil has the potential to negatively affect the polymer's physical and mechanical properties.

Tung oil (TUN) is another exceptional oil that does not have to be modified before polymerization. This is because of its high percentage (>84%) of a triply conjugated and readily reactive fatty acid (α-eleostearic acid). Studies have shown tung oil to react with comonomers such as DVB, ST, DCPD, limonene, and myrcene through either cationic, thermal, or free radical polymerization. Outside of the realm of multi-component thermosets, tung oil has also been used as a protective coating for ships and other wood surfaces, protein films, or even as an additive to strengthen unsaturated polyester resins (Humar and Lesar 2013). In another study, a series of copolymers containing 30–70 wt% of TUN was prepared by thermal polymerization at temperatures ranging from 85 to 160 °C (Li and Larock 2003). The physical properties ranged from elastomeric to tough and rigid (Li and Larock 2003). The glass transition temperatures ranged from −2 to 116 °C, and cross-link densities from 1×10^3 to 2.5×10^4 mol/m^3 (Li and Larock 2003). The same study saw increased thermal polymerization reaction rate and improved properties with the addition of catalytic amounts of Co, Ca, and Zr salts (Kundu and Larock 2009). Catalysts used in polymeric systems are discussed further when describing conjugated vegetable oils, in Sect. 2.3.

A less reactive oil, conjugated linseed oil (CLO), has been used to make comparable samples in reference to TUN. A commercial CLO, containing approximately 87% of conjugated carbon–carbon double bonds, exhibited enough reactivity to undergo polymerization, leading to similar products obtained with TUN (Kundu and Larock 2005). Unlike most TUN-based systems, however, a micro-phase separation between an oil-rich and a DVB-rich phase was observed (Kundu and Larock 2005). Also, there were traces of unreacted oil evenly distributed throughout the final polymer, which could have negative effects on mechanical properties. As seen with TUN, catalytic amounts of Co, Ca, and Zr salts increased the properties of the CLO-based polymer by increasing its cross-link density (Kundu and Larock 2009).

2.1.2 Oil Modification for Polymerization

The conjugated carbon–carbon double bonds' resonance ability to better distribute the charge stabilizes the radical transition state, lowering the activation energy (Larock et al. 2001). This phenomenon is not possible in unconjugated double bonds, therefore increasing the activation energy in comparison with the conjugated bonds. In order to increase the reactivity of vegetable oils toward free radical reactions, carbon–carbon double bonds in the fatty acid chains can be isomerized and brought into conjugation. Studies have shown positive results in reference to bringing the carbon–carbon double bonds into conjugation in the presence of rhodium catalysts (Andjelkovic et al. 2006; Henna et al. 2007).

Other studies have made use of polymerizable functional groups, such as acrylates, to modify the starting vegetable oil, converting it into a more reactive species (Bonnaillie and Wool 2007; de Espinosa et al. 2009). For example, acrylic acid and epoxidized soybean oil (ESO) react together to form acrylated epoxidized soybean oil (AESO—Fig. 2). Then, AESO can be copolymerized with other comonomers such as ST (La Scala and Wool 2005). The polymer properties can be easily manipulated by adjusting the amount of acrylate and ST, respectively. ST is often used in thermosets for structural applications (Xia et al. 2013). Since these biopolymers have such variable properties, they can easily replace numerous petroleum-based products, such as thermosetting foams. Currently, AESO has been used to manufacture foams with high bio-based content resulting in mechanical properties comparable to the current market's petroleum-based semirigid industrial foams (Bonnaillie and Wool 2007).

AESO contains unreacted epoxy and hydroxyl groups, two highly reactive groups used in oil modification. AESO can react with maleic anhydride (MA) to create maleated epoxidized soybean oil (MAESO), a very reactive species containing many polar hydroxyl and anhydride groups. Studies have shown that the copolymerization of MAESO and ST thermosets produces higher cross-link densities, T_g's, and storage moduli than AESO-based thermosets (Lu et al. 2005).

Other options for the preparation of acrylated fatty acid chains or triglycerides include the oxidation of carbon–carbon double bonds in the oils with singlet oxygen

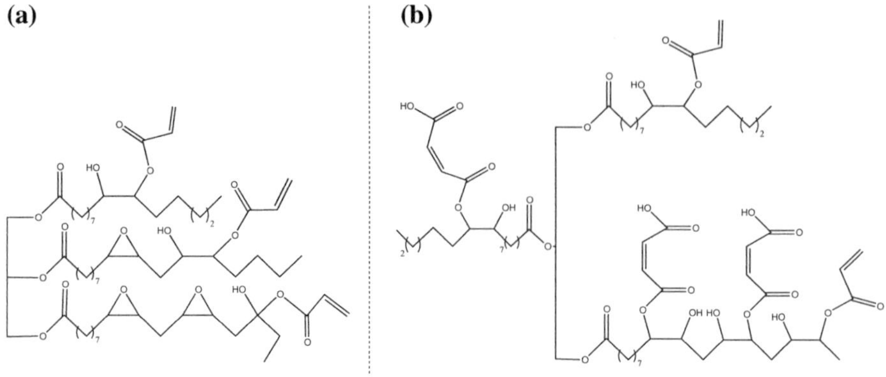

Fig. 2 Chemical structures of **a** acrylated epoxidized soybean oil (AESO) and **b** maleated acrylated epoxidized soybean oil (MAESO)

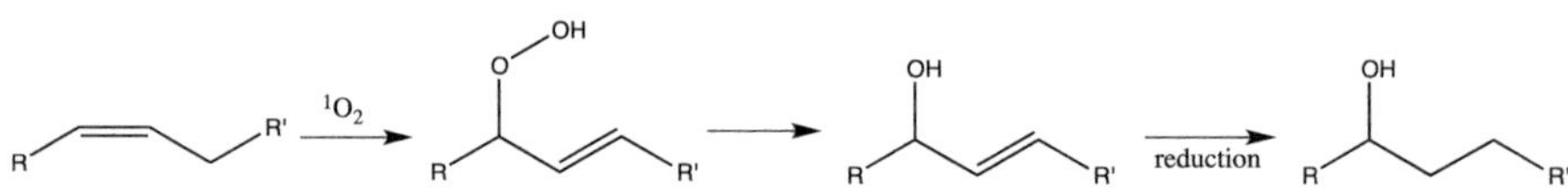

Fig. 3 Steps involved in the conversion of unsaturated fatty acids or triglycerides into saturated fatty alcohols, where R and R′ represent the remainder of the triglyceride or the fatty acid

to produce hydroperoxides, a reaction very similar to the oils' natural auto-oxidative behavior. Then, the newly formed hydroperoxides can be converted into α, β-unsaturated secondary alcohols, which can in return be converted into saturated secondary alcohols by a reduction reaction (Fig. 3) (Samadi et al. 2001). These newly formed fatty alcohols can react with acrylates and be free radically polymerized in the presence of pentaerythritol tetraacrylate. As in previous studies, varying the amounts of pentaerythritol tetraacrylate directly affects the cross-link density, leaving a promising future in route to bio-based polymer networks. These polymers exhibit similar properties to other reported acrylated oil-based materials (de Espinosa et al. 2009).

Aside from modification and functionalization of the carbon–carbon double bonds, structural changes surrounding the fatty acid ester functional group have been investigated as another promising potential approach in the search for more reactive bio-based monomers. A series of rigid thermosets with T_gs of approximately 135 °C and storage moduli at 35 °C of approximately 0.9 GPa have been developed by the copolymerization of soybean oil monoglyceride (SOMG) maleates with 35 wt% ST (Can et al. 2006a, b). In order to enhance the properties of these SOMG maleate-ST polymers, neopentyl glycol (NPG) and bisphenol A (BPA) were combined with SOMG prior to being maleinized (Can et al. 2002). Similar materials have been prepared from linseed oil, leading to linseed oil monoglycerides (LOMGs), which have been then copolymerized with 20–80 wt% of ST (Mosiewicki et al. 2005).

Another system that can be designed based on structural modifications of triglycerides is that of unsaturated polyester resins prepared from ESO and unsaturated diacids. The preparation of copolymers of soybean oil-based vinyl ester resins from the esterification, or the ring opening of ESO with fatty acid dimers, has been recently reported (Yang et al. 2013a). FT-IR results verified the preparation of the vinyl ester components, and copolymers with varying weight ratios were then developed through thermal polymerization (Yang et al. (2013a). The materials obtained exhibit brittle behavior and a maximum flexural strength of 20.6 MPa, along with high thermal stability, as shown through thermogravimetric analysis (Yang et al. (2013a). When the same process was applied to castor oil, there was an overall increase in the modulus and the strength of the resulting polymer. The polymer chains consisted of cross-linked maleates, preventing them from functioning as a plasticizer in the system (Can et al. 2006a, b). The castor oil-based thermosets prepared in this fashion exhibit significantly improved properties compared to the soybean oil-based polymers (Can et al. 2006a).

2.2 *Thermosets from Cationic Polymerization*

2.2.1 Carbon–Carbon Double Bond Polymerization

Apart from the previously described free radical polymerization in vegetable oils, cationic polymerization of triglycerides in the presence of petroleum-based comonomers also leads to thermosetting materials with a range of properties depending on the resin composition. When compared to free radical polymerization with AIBN, the use of strong Lewis acids as initiators, like boron trifluoride dietherate (BFE), in the polymerization process avoids the entrapment of bubbles in the cured system. Cationic polymerization also limits crack formation related to shrinkage of the resin upon cure (Andjelkovic et al. 2005).

During cationic polymerization of triglycerides, each unsaturated fatty acid chain has the ability to participate in the reaction, therefore creating a three-dimensional cross-linked polymer network, similar to the free radical polymerization reaction. It is possible to obtain materials exhibiting room temperature storage modulus (E') of approximately 2 GPa through the cationic copolymerization of 50–55 wt% of tung oil and DVB (Li and Larock 2000). Less unsaturated and less reactive oils, such as soybean oil (SOY), low saturated soybean oil (LSS), or CLS, have been substituted by up to 25 wt% of TUN in order to control the gel time (Li and Larock 2000). Combinations containing 35–55 wt% of regular corn oil (COR) or conjugated corn oil (CCO) with various amounts of DVB and ST result in polymeric materials with mechanical properties that range from soft and rubbery to rigid (Li et al. 2003). The gel time can be affected by the stoichiometry of the comonomers in relation to each other. When oil content is increased above 40 wt%, there is a significant increase in the gel time (Li et al. 2003).

It comes to no surprise that the resins containing conjugated oils have significantly shorter gel times than the oils in which the carbon–carbon double bonds are not in conjugation (Li et al. 2003). To support this claim, experimental data has revealed that the gel times for CCO are shorter than COR-based resins (Li et al. 2003). Curing temperatures also have an effect on the gel time of the cationic resins. In more recent developments, the cationic polymerization of soybean oil, ST, and DVB with either 4-vinylphenyl boronic acid (Sacristan et al. 2010), or 4-trimethylsilyl styrene (Sacristan et al. 2009), initiators have resulted in boron- and silicon-containing thermosets with self-extinguishing properties that have the potential to be used in flame-retardant materials.

Considerable research has been conducted on the cationic polymerization of SOY, LSS, and CLS with various petroleum-based olefin cross-linking agents. In an early study, 50–60 wt% of SOY, LSS, or CLS was copolymerized with DVB and a densely cross-linked network was formed (Li et al. 2001). Despite the high cross-link density, 12–31 wt% of the original oils remained unreacted after cure. The poor miscibility between the oil and the BFE initiator resulted in an initial heterogeneous mixture (Li et al. 2001). Modification of the BFE initiator with a less reactive oil, such as SOY, helped in the formation of a more homogeneous material and the conversion of free unreacted oil and oligomers into cross-linked thermosets. As observed previously, the CLS-based thermosets had the highest storage moduli and thermal stabilities when compared to SOY- and LSS-based polymers (Li et al. 2001).

ST is often used a comonomer in vegetable oil-based resins due to its ability to uniformly cross-link polymers and provide structural support. Substitution of 25–50 wt% of DVB by ST resulted in a significant overall increase in properties (Li and Larock 2001). After a thorough comparison of different cross-linking agents, namely norbornadiene, DCPD, and DVB, it was determined that DVB has the highest reactivity and therefore gives the most promising results with promising potential applications requiring damping and shape memory properties (Li and Larock 2002). Cure studies of DVB-cross-linked vegetable oil-based systems demonstrated that the ideal cure temperature is highly dependent on the resin composition and that the optimum initiator concentration of the Lewis acid is 2 wt% (Badrinarayanan et al. 2009). In the study of the cationic copolymerization of SOY and CSO with DCPD in the presence of modified BFE, it was found that the commonly used fish oil ethyl ester modifier is not needed when the polymerization involves the more reactive CSO (Andjelkovic and Larock 2006).

Other oils such as olive, peanut, sesame, canola, grapeseed, sunflower, safflower, walnut, and linseed oils have been cationically copolymerized with DVB and/or ST to form versatile thermosets with a range of properties that can be modified according to the desired application (Can et al. 2006b). Studies have also shown an increase in mechanical properties of SOY-, LSS-, and CLS-based thermosets with increases in DCPD content in the resin composition (Li and Larock 2001). In general, the properties of thermosets gradually increase with the number of carbon–carbon double bonds present in the oil because of the greater cross-linked network formed. However, it has been found that the gel times of these copolymers are

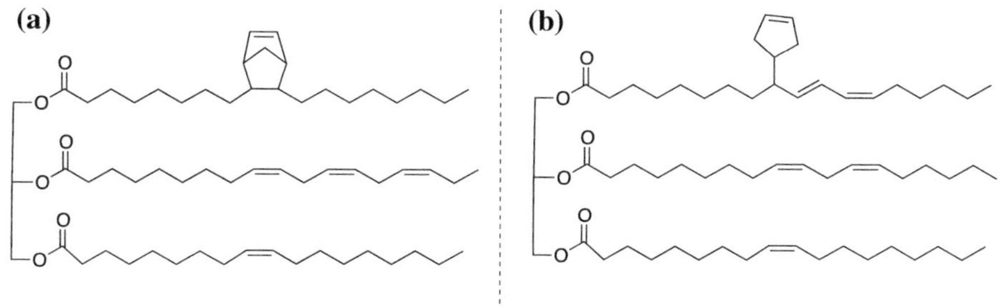

Fig. 4 Chemical structures of **a** Dilulin, and **b** ML189

independent of the reactivity of the vegetable oil used (Andjelkovic and Larock 2006). In 2009, the compounds commercially known as Dilulin and ML189 (Fig. 4) were copolymerized with DCPD under cationic conditions (Xia et al. 2009). These reactions yielded very similar results as previous bio-based systems containing DCPD.

2.2.2 Polymerization of Epoxidized Oils

Latent or delayed initiators, like benzylpyrazinium salts, allow for more controlled polymerizations. Unlike the previous reactions discussed in this chapter thus far, latent cationic initiators are only activated under very specific conditions, such as heating or photo-irradiation. In benzylpyrazinium salts, the activity is controlled by electronic modifications of the benzyl and pyrazine moieties (Kim et al. 2004). In previous studies, ESO and epoxidized castor oil (ECO) were cationically polymerized in the presence of N-benzylpyrazinium hexafluoroantimonate (Park et al. 2004). A reaction scheme for the process is provided in Fig. 5. As a result of increased intermolecular interactions, cationic resins prepared from ECO have been found to have higher T_g's and lower coefficients of thermal expansion than corresponding polymers synthesized from ESO (Park et al. 2004). ESO and ECO have also been copolymerized with the diglycidyl ether of bisphenol A (BPA)—Fig. 6—and resulted in polymers with improved mechanical properties in comparison with the BPA-free system (Jin and Park 2008a, b). Adding large amounts of

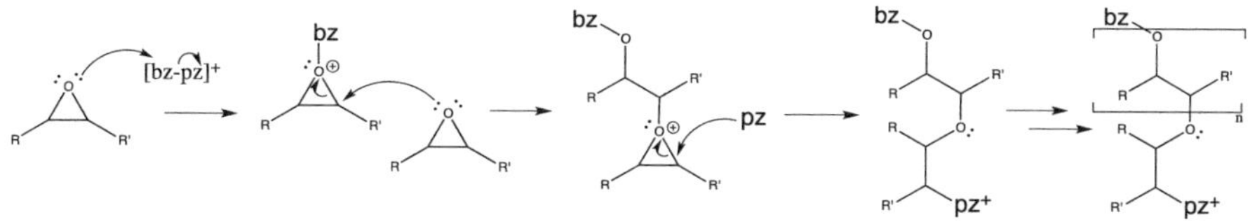

Fig. 5 Reaction scheme for the polymerization of epoxidized oils with benzyl pyrazinium salts, where R and R′ are the remainder of the epoxidized oil, and $[bz\text{-}pz]^+$ is the benzyl pyrazinium cation

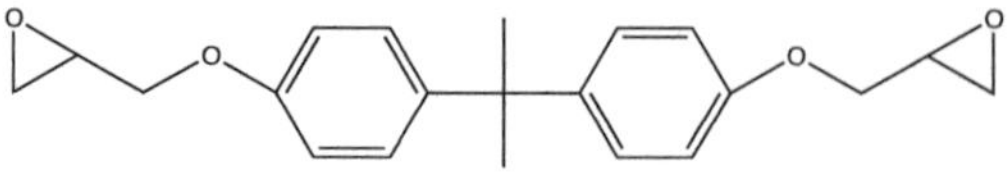

Fig. 6 Chemical structure of diglycidyl ether of bisphenol A

soft segments to the ECO-based polymer resulted in a decrease in the cross-link density and an increase in the toughness of the final material (Park et al. 2004).

2.3 *Thermosets from Olefin Metathesis Polymerization*

Olefin metathesis was first discovered in 1964 and has since grown in popularity in the organic and polymer industries. $WCl_6/(CH_3)_4Sn$ and $WOCl_4/(CH_3)_4Sn$ were the first catalysts developed to perform olefin metathesis on vegetable oil-based derivatives containing functional groups (Mol 2002). Because these catalysts' efficiency was affected by the oil's functional groups, carbene complexes of Ru, Mo, and W were developed (Schrock and Czekelius 2007). Ru carbenes can tolerate most of the influential functional groups in vegetable oils and are also stable when exposed to air and moisture. Vegetable oil-based thermosets have been successfully prepared by ring-opening metathesis polymerization (ROMP) and acyclic diene metathesis (ADMET) polymerization, as discussed in the following sections.

2.3.1 Ring-Opening Metathesis Polymerization (ROMP)

Generally, monomers suitable for ROMP are strained, unsaturated rings, such as norbornene (Fig. 7) groups, and are catalyzed by a ruthenium carbene catalyst. In previous studies, Dilulin and DCPD (Henna et al. 2007; Henna and Larock 2009), as well as a bicyclic norbornene-based cross-linking agent (CL), have been copolymerized independently (Haman et al. 2009). These thermosets have shown phase separation because of the difference in reactivity between Dilulin and DCPD (Cui and Kessler 2012). It was also shown that when norbornene content is increased in the resin composition, the corresponding thermosets' thermal stability and cross-link density also increase (Mauldin et al. 2008). Another study observed that increasing Dilulin content results in lower T_gs, decreased thermal stabilities, and reduced mechanical properties (Haman et al. 2009). A cure kinetics experiment

Fig. 7 Chemical structure of norbornene

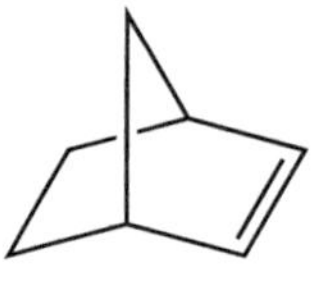

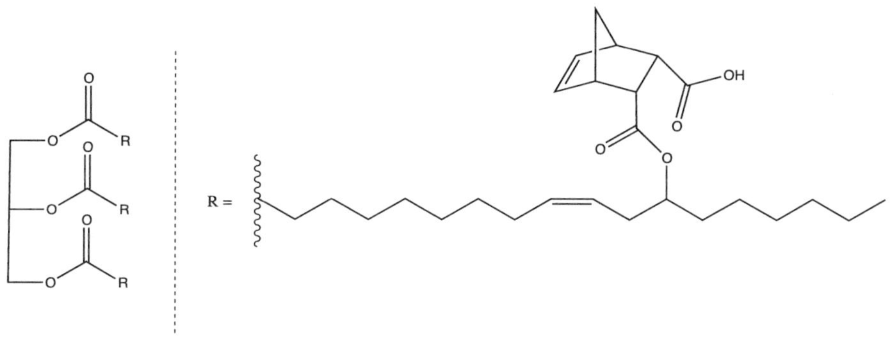

Fig. 8 Chemical structure of bicyclo [2.2.1] hept-5-ene-2,3-dicarboxylic anhydride-modified castor oil (BCO)

was conducted with Dilulin and CL, revealing that the cure rate increases while the gel time decreases whenever CL content increases. However, CL content exhibits little to no effect on the polymer's degradation's activation energy (Haman et al. 2009). Thermosets of Dilulin and 5-norbornene-endo-2,3-dicarboxylate (NBDC), ranging from brittle to flexible, have been prepared. In these systems, it was observed that increases in NBDC content inherently decrease both tensile stress and thermal stability of the final thermosets (Xia et al. 2010).

Aside from Dilulin, numerous other chemical modifications have led to strained ring-functionalized starting materials. Castor oil has been modified with bicyclo [2.2.1] hept-5-ene-2,3-dicarboxylic anhydride to yield a norbornenyl-functionalized bicyclic castor oil derivative (BCO—Fig. 8) (Henna and Larock 2007). BCO (55–85 wt%) has been copolymerized with cyclooctene in the presence of 0.5 wt% of the second-generation Grubbs' catalyst (Henna and Larock 2007). The T_gs of these thermosets ranges from 1 to 14 °C, giving them potential applications as elastomers. An increase in thermal stability was observed upon an increase in BCO content (Henna and Larock 2007). Norbornenyl-functionalized castor oil (NCO—Fig. 9a) and norbornenyl-functionalized castor fatty alcohol (NCA—Fig. 9b) have been synthesized from reduced ricinoleic acid and norbornene carbonyl chloride. Thermosets with varying ratios of NCO/NCA exhibit rubbery to rigid behavior and a large variety of cross-linked densities (Xia and Larock 2010b).

Fatty alcohols from SOY, Dilulin, ML189, and castor oil have been synthesized and used to prepare the respective norbornenyl-functionalized monomers, NMSA, NMDA, NMMA, and NMCA (Fig. 10) (Xia et al. 2010). Thermosetting polymers have been synthesized with these monomers using 0.5 wt% of the second-generation Grubbs' catalyst (G2), leading to homogeneous mixtures, free of phase separation complications. Thermosets made with NMDA and NMMA exhibit very similar mechanical properties, making them potential candidates for high-performance bioplastics (Xia et al. 2010). Different groups can be grafted on the fatty acid chains, which affect the thermomechanical properties of the final polymers.

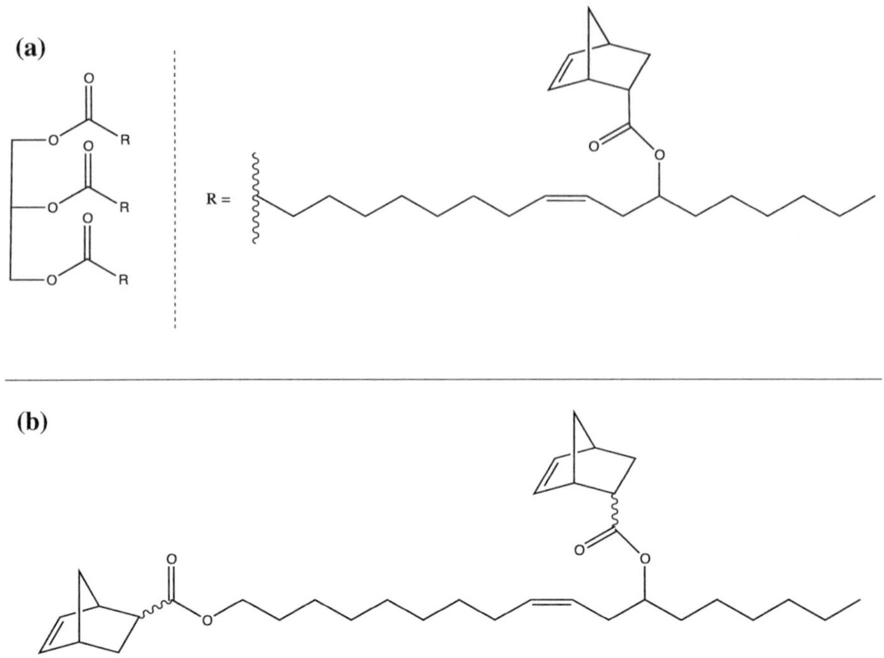

Fig. 9 Chemical structures of **a** norbornenyl-functionalized castor oil (NCO) and **b** norbornenyl-functionalized castor fatty alcohol (NCA)

2.3.2 Acyclic Diene Metathesis (ADMET) Polymerization

Acyclic diene metathesis (ADMET) polymerization is primarily based on the metathesis of acyclic dienes through an organometallic, such as a ruthenium catalyst. Materials ranging from tacky oils to rubbers have been synthesized using the Grubbs' first-generation catalyst and SOY (Tian and Larock 2002). Undec-10-enyl undec-10-enoate, which contains long-chained dienes derived from plant oils, under ADMET polymerization conditions, yielded high-density and block copolymers (Rybak and Meier 2008). ESO and 4-vinylbenzene sulfonic acid can be reacted together to produce a SOY-based monomer (SESO) suitable for ADMET polymerization (Ozturk et al. 2011). ADMET of SESO in the presence of G2 results in rigid thermosets with a high conversion. Although these polymers exhibit non-hygroscopic properties, their hydrolytic stability is low and they quickly degrade at 60 °C in the presence of water, finding potential application as rigid plastics susceptible to heat and moisture degradation (Ozturk et al. 2011).

Castor oil can be polymerized by ATMET (acyclic triene metathesis) with glyceryl triundec-10-enoate, yielding branched polymers (Fokou and Meier 2008). Similar to the previous studies with dienes, the molecular weight can be manipulated by changing the resin composition (Fokou and Meier 2008). Additionally, the ATMET polymerization of glyceryl triundec-10-enoate, and using 10-undecenol as

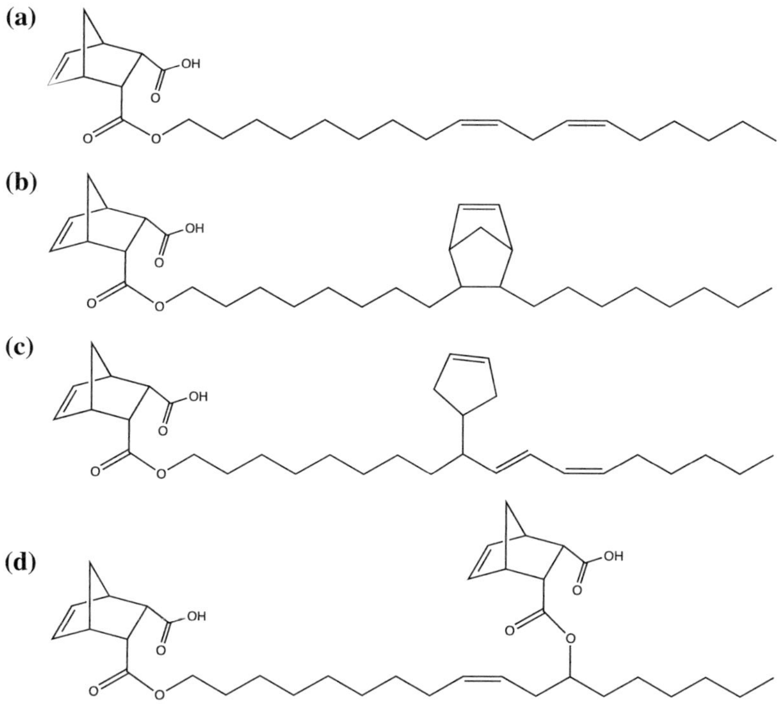

Fig. 10 Chemical structures of **a** norbornenyl-functionalized soybean oil fatty alcohol (NMSA), **b** norbornenyl-functionalized Dilulin fatty alcohol (NMDA), **c** norbornenyl-functionalized ML189 fatty alcohol (NMMA), and **d** norbornenyl-functionalized castor oil fatty alcohol (NMCA)

the chain stopper, results in castor oil-derived branched polyols (del Rio et al. 2011a). These polyols can then be condensed with 4,4'-methylenebis (phenyl isocyanate) (MDI) to prepare polyurethane thermosets with properties suitable for shape memory applications (del Rio et al. 2011a, b). High oleic sunflower oil has also been polymerized by ATMET with G2 to yield highly branched and functionalized polyesters (Biermann et al. 2010).

2.4 *Thermosets from Condensation Polymerization*

Vegetable oils are ideal candidates for condensation polymerization reactions leading to polyesters, polyamides, and polyurethanes because of their ester functional groups (Sharma and Kundu 2008). Additionally, carbon–carbon double bonds in the fatty acid chains can be functionalized into polyols and also be used in condensation reactions. This section focuses on the reactions regarding polyesters and polyamides formation and applications. Polyurethanes will be discussed further in Sect. 4, when describing vegetable oil-based coatings and emulsions.

2.4.1 Polyesters

There are three main approaches for successfully synthesizing polyesters from vegetable oils, namely (1) the condensation of dicarboxylic acids, anhydrides, or diols, (2) the condensation of hydroxyl groups, or (3) the ring opening of lactones. For example, Nahar seed oil can be reacted with a triol, such as glycerol, to produce monoglycerides, which naturally bear two hydroxyl groups (Dutta et al. 2004). Then in return, the newly formed diol can be reacted with acid anhydrides such as phthalic and maleic anhydrides to yield polyesters, as depicted in Fig. 11. These polyesters were found to have a high resistance to dilute strong acids, water, and salt solutions, making them ideal for coating and composite binder applications (Dutta et al. 2004). In a different approach, maleate units can be grafted onto SOY through an ene reaction leading to triglyceride anhydrides (Fig. 12) that can be further reacted with polyols (Eren et al. 2003). These thermosetting polymers have a variety of applications depending on the polyol used (Eren et al. 2003).

Polyesters with appending hydroxyl functional groups can be prepared from castor oil. Such polyesters can then undergo condensation with MDI to yield flexible polyurethane thermosets (Petrovic et al. 2008). These polymers have low T_gs and mechanical properties similar to those of petrochemical polyurethanes, except for lower shore A hardness values. Such properties match applications as sealant due to their elasticity and pliability (Petrovic et al. 2008). Tough and flexible thermosets have been manufactured by reacting epoxy resins with hyper-branched polyester vegetable oil-based derivatives in the presence of poly(amido amine) (Roy and Karak 2012). Because of polyester vegetable oil-based derivatives' low viscosity, they are often used as diluents for epoxy resins, resulting in paints with low volatile organic compounds (VOCs) finding use in the automotive and wood industries (Konwar et al. 2012). The addition of organo-nanoclay has been shown to significantly enhance the thermomechanical, as well as the biodegradability of polyesters derived from vegetable oils (Konwar et al. 2009).

2.4.2 Polyamides

Vegetable oil-derived polyamides have been synthesized by condensation polymerization. Polyesteramides have been created by reacting ethylenediamine

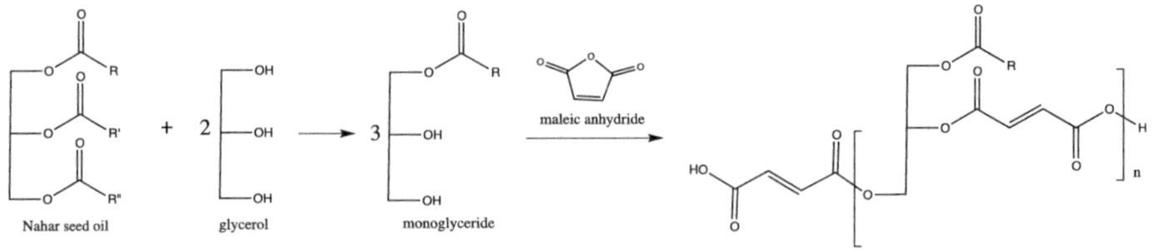

Fig. 11 Steps involved in the synthesis of polyester resin from the reaction of Nahar seed oil monoglycerides with maleic anhydride

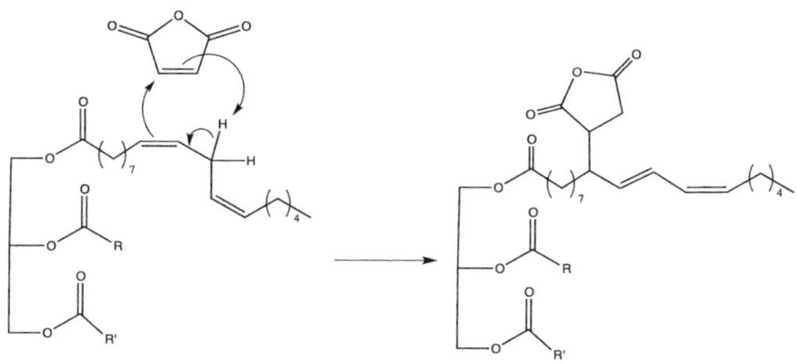

Fig. 12 Reaction scheme for the synthesis of soybean oil triglyceride anhydride, where R and R′ are unsaturated aliphatic chains

tetraacetic acid (EDTA) and linseed oil. These compounds can be cured along with poly(styrene-*co*-maleic anhydride) (SMA) at room temperature (Alam and Alandis 2011). Polyesteramides with 45 phr (parts per hundred) of SMA displayed the most promising decomposition resistance and physico-mechanical characteristics (Alam and Alandis 2011). Block co-polyamides have been created by the condensation of SOY-based dimer acids, diamines, and amino acids (Deng et al. 1999). The amino acids determine the chain structure and crystalline morphology of the final polyamide polymers with no change in biodegradability (Deng et al. 1999). Another polymer, nylon 11, was developed from an amino acid-derived compound and manufactured from the pyrolysis of castor oil (Nayak 2000). This bio-based polyamide is highly important to the automotive industry particularly because of its flexibility, chemical resistance, dimensional stability for usage in fuel lines, brake tubing, and gas and oil pipes, as well as other applications (Nayak 2000).

2.5 *Thermosets from Epoxides*

In order to create rigid thermosetting polymers from epoxides, ethylenediamine, *p*-xylylenediamine, and triethylenetetramine have been independently used in the ring-opening reaction of epoxidized linseed oil (ELO) (Tellez et al. 2008). Varying the amine precursors and ratios has afforded a variety of products (Tellez et al. 2008). Epoxidized oils have also been used as reactive diluents for industrial BPA-based epoxy resins (Das and Karak 2009). Cure kinetic studies were conducted on ESO/diglycidyl ether of BPA with *p*-aminobenzoic acid (Mustata et al. 2011). The results showed that the degradation activation energy increased concurrently with the increase in ESO composition, while T_g was impacted negatively (Mustata et al. 2011). Polyamine derived from grapeseed oil (GSO) synthesized by a thiol-ene and cysteamine chloride has been used as a curing agent for ELO in order to yield bio-based epoxy resins with a T_g of −38 °C (Stemmelen et al. 2011). High molecular weight resins can be created from BPA and epoxidized oils.

These resins are viscous and used mostly in the coating industry (Czub 2009). Currently, pressure-sensitive adhesives (PSAs) can be manufactured from dihydroxyl SOY, ESO, and phosphoric acid under moderate conditions (Ahn et al. 2011). ESO-based PSA tapes were found to air dry within 30 s when cured at 110 °C (Ahn et al. 2011). These bio-based PSAs are comparable to the commercially available products, making them ideal alternatives. Also, due to the polymers' biodegradability, they can potentially find applications in the medical field.

ELO has been reacted with Pripol 1009 (a vegetable oil-based diacid cross-linker) under several amine-catalyzed conditions to produce an almost 100% bio-derived, flexible film with water resistance (Supanchaiyamat et al. 2012). The amine catalysts not only decreased the degradation activation energy but also enhanced the mechanical characteristics of the films (Supanchaiyamat et al. 2012). Cross-linked elastomers have been prepared with ESO, ELO, and a phosphorylated castor oil (PCO) under green conditions (cure temperature of 37 °C, and no initiator or solvent) (Liu et al. 2012). These biodegradable and biocompatible polymers had cytotoxicity screenings in vitro that indicated the polymers were compatible with the L929 cell line. The impregnation of these thermosets in animal tissue studies displayed complete absorption within 3 months (Liu et al. 2012).

2.6 *Miscellaneous Polymerizations*

In other approaches, ESO has been reacted through post-living anionic polymerization with poly(butadienyl) lithium (PBDLi) (Boonkerd et al. 2014). Varying the ratios of monomers yielded a spectrum of molecular weights for PBDLi from 1,000 to 15,000 g/mol with potential application as high-density processing aids for commercial polybutadienyl (PBD) (Boonkerd et al. 2014). Contrary to previous studies, the PBD anion attacks the carbonyl carbon of the ester in ESO, leading to a ketone, instead of attacking the epoxide group (Boonkerd et al. 2014). A proposed reaction mechanism is provided in Fig. 13.

In order to prepare a SOY-based precursor with increased molecular weight, it has been a fairly common practice to induce the heat promoted [4+2] cycloaddition reaction of SOY. The product from such process is most commonly referred to as heat-bodied SOY. Initially, heat-bodied SOY was suggested as an alternative for drying oils in coatings containing asphaltic components and other inorganic fillers for use in the construction industry (Ewart and Sieber 1959). More recently, heat-bodied SOY has been used as a precursor for the preparation of bio-based polyurethane (PU) foams (Lubguban et al. 2009). Latest advances on related systems include the enzymatic degradation of heat-bodied SOY into primary alcohols and their subsequent use in the preparation of high-density polymers (Kiatsimkul et al. 2008, 2007). Lastly, a study comparing the oxidative nature of bodied and blown SOY concluded that blown SOY exhibits a higher activation energy than bodied SOY (Kenar et al. 2002).

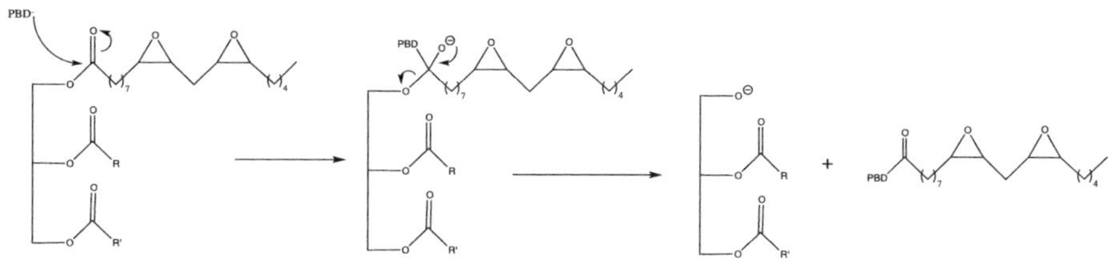

Fig. 13 Reaction scheme showing the attack of the PBD anion to the carbonyl of the ester in epoxidized soybean oil (ESO), where R and R′ are epoxidized carbon chains

The use of vegetable oils functionalized with azides and alkynes in click chemistry reactions leading to bio-based polymers dates back to 2001 (Hong et al. 2012). The optimization of reaction temperature leads to polymerizations under catalytic conditions without the use of a solvent, resulting in homogeneously cross-linked polymers (Hong et al. 2010). In non-catalyzed processes, thermal polymerization predominates. The absence of a catalyst is ideal for preparing encapsulating and insulating materials (Hong et al. 2012). Overall, alkyne-functionalized SOY has been reacted with other azide-containing vegetable oils, such as castor, canola, corn, soybean, and linseed oils, yielding fully cross-linked thermosets (Hong et al. 2010).

3 Polymer Composites

3.1 Organic Reinforcements

In order to prepare fully bio-based composites for structural applications, vegetable oil-based resins can be reinforced with a variety of natural fillers, including spent germ (Pfister et al. 2008), soybean hulls (Quirino and Larock 2009), corn stover (Pfister 2010), wheat straw (Pfister and Larock 2010), and rice hulls (Quirino and Larock 2011b). Free radical resins are typically chosen for the matrix of these composites over cationic systems. Indeed, hydrophilic fillers can strongly bind to Lewis acids used as initiators during cationic polymerization. Compression molding studies completed with soybean hull composites showed micro-fractures in the final material under scanning electron microscopy (SEM) imaging (Quirino and Larock 2009). In order to improve matrix–reinforcement interactions, recent studies have found maleic anhydride (MA) to be a suitable compatibilizer for the system, resulting in enhanced thermomechanical properties (Pfister 2010). Properties of tung oil-based composites reinforced with spent germ were also enhanced upon the use of shorter filler particle sizes until a filler loading of 60 wt% (Pfister et al. 2008). After that threshold, excessive reinforcement leads to filler agglomeration and formation micro-fractures (Bhuyan et al. 2010). More recent studies have suggested that it is feasible to compensate for the losses in properties due to

Table 1 Mechanical properties of rice hull composites prepared from conjugated soybean oil (CSO) and conjugated linseed oil (CLO) (Quirino and Larock 2011a)

Major matrix component (50%)[a]	Young's modulus (GPa)	Tensile strength (MPa)	Storage modulus at 130 °C (MPa)	T_g1 (°C)	T_g2 (°C)
Conjugated soybean oil	1.2 ± 0.3	4.3 ± 0.7	215	24	78
Conjugated linseed oil	2.3 ± 0.5	5.9 ± 0.6	220	52	–[b]

[a]The matrix composition is 50 wt% of vegetable oil, 35 wt% of n-butyl methacrylate, and 15 wt% of divinylbenzene. All composites contained 70 wt% of rice hulls
[b]Only one Tg was observed

excessive filler by raising molding pressure during the cure (Pfister et al. 2008), or by removing unreactive oil from the reinforcement prior to curing (Bhuyan et al. 2010).

In a similar fashion, CSO-based composites have been developed and reinforced with soybean hulls. Interestingly, the materials obtained exhibited two distinct T_gs, indicative of a phase separation in the matrix, that only appeared upon cure of the composite under compression molding conditions (Quirino and Larock 2009). When comparing the mechanical properties of rice hull composites made with CSO with those of composites made with CLO (Table 1), it becomes clear that CLO-based composites had superior properties, most likely due to their greater degree of unsaturation, allowing the polymer matrix to display an increased cross-link density (Quirino and Larock 2011a). Likewise, comprehensive studies have been conducted on the cationic polymerization of vegetable oil-based composites reinforced with natural fillers (Pfister and Larock 2012). Resins from CSO, CLO, and CCO have been reinforced with corn stover, wheat straw, and switchgrass, with the CLO-based materials displaying the best properties.

In a different approach, AESO free radical resins have been compression molded in the presence of several natural reinforcements, such as glass fiber, non-woven and woven jute fibers (Ramamoorthy et al. 2012). The woven jute fiber composites had the highest hygroscopic ability (22.2 wt% of water absorbed) while the glass fiber composites had the lowest absorption (2.4 wt%) (Ramamoorthy et al. 2012). This study suggests that the combined use of glass and natural fibers can increase the bio-composition without drastically compromising the composite's mechanical properties (Ramamoorthy et al. 2012).

3.1.1 Lignin as Reinforcement

Without modification, lignin exhibits an overall low reactivity and is not an ideal precursor for the preparation of advanced bio-based materials (Boudet et al. 2003). Lignin's hydroxyl, phenol, and ether groups can be used as potential sites for chemical modification in order to increase its reactivity (Gosselink 2011). For example, recent studies have shown lignin's potential to compete with current market's petroleum-based phenol-formaldehyde resins (Tejado et al. 2007; Hu et al. 2011; Gibbons et al. 2015). Also, due to lignin's aromatic units, it has been used as a substitute for phenol in the manufacturing of phenolic, polyester, PU, and epoxy resins (El Mansouri et al. 2011). Amino groups have been added to lignin in order to enhance the interfacial bonding in PVC/wood-flour composites (Yue et al. 2011). Lignin's affinity for condensation polymerization has been enhanced through demethylation of the aromatic methoxy structures, resulting in increased polarity, suitable for PU synthesis (Chung and Washburn 2012). Anhydrides have been used to esterify alkali lignin and enhance its solubility in nonpolar solvents for the manufacturing of saturated polyesters and vinyl esters (Thielemans and Wool 2005).

3.1.2 Cellulose as Reinforcement

The use of cellulose in composites often involves a pretreatment, which influences its solubility, and determines whether it can work as the matrix or as the reinforcement (Gindl et al. 2006) (Yang et al. 2013a, b). For example, acetylated cellulose nanowires are distributed in cellulose acetate to form a gel, which upon drying, results in a cellulose nanocomposite (Yang et al. 2013a, b). Another cellulose nanocomposites can also be prepared by drying cellulose in an ionic liquid after the addition of nanocrystalline cellulose (Ma et al. 2011). Silver/cellulose hybrid composites have been obtained by the microwave-assisted reduction of silver nitrate onto cellulose, yielding composites with high antimicrobial properties against Gram-positive bacteria (*Staphylococcus aureus*) and Gram-negative bacteria (*Escherichia coli*) (Li et al. 2011).

AESO resins have been reinforced with keratin fibers for printed circuit board applications with low dielectric constant (Zhan and Wool 2013). When keratin fibers are replaced with glass and flax fibers, AESO composites display increased flexural properties and impact resistance (Morye and Wool 2005). Along the same lines, resins reinforced with cellulose, pulp, and hemp exhibit properties with potential applications in the construction industry (O'Donnell et al. 2004). Cellulose is commonly used as a natural substitute for glass fibers due to their versatility and inherent properties. Cellulose fibers are often modified to enhance the adhesion between the fiber reinforcement and the matrix (Kalia et al. 2011; Shahzad 2012).

Compatibilizers can be used to enhance the adhesion between polar cellulose and a nonpolar matrix. More than half of asolectin's fatty acid chains are poly-unsaturated, with the potential to undergo free radical polymerization. Asolectin has also been used in the construction of membrane proteins and transporting medium associated with Cd^{+} and Pb^{+} (Park et al. 2007; Nozawa et al. 2010). It was shown that the addition of asolectin to the composite increased the storage modulus and the adhesion between fiber and matrix (Johns et al. 2015).

3.1.3 Carbon Nanotubes as Reinforcement

The reinforcement of an AESO-ST copolymer with multi-walled carbon nanotubes (MWCNTs) and soot resulted in an increase of the mechanical properties (Thielemans et al. 2005). At MWCNT loadings greater than 0.28 wt%, agglomeration of MWCNTs compromised the mechanical properties (Thielemans et al. 2005). In the biomedical field, MWCNTs have been mainly studied as reinforcement for polylactic acid (Zhang and Chen 2010; Laredo et al. 2013). In more traditional systems, CNTs have been used to reinforce epoxy resins, resulting in composites with enhanced dielectric constants in comparison with the non-reinforced resin (Rangri et al. 2011; Chang et al. 2012).

Due to CNT's exceptional ability to absorb microwaves and promote uniform and instantaneous heating through Joule effect, some studies have compared conventional heating with microwave heating of a tung oil-based thermoset reinforced with MWCNTs (Smith et al. 2015). It was observed that when CNT-containing samples were cured under microwave conditions, the cure time was highly dependent on the amount of CNTs loaded (Smith et al. 2015). With the shorter cure time under microwaves, the cross-link density was compromised, which had a negative overall effect on the mechanical properties of the composites (Smith et al. 2015).

3.2 Inorganic Fillers as Reinforcement

The PBD elastomers previously discussed in Sect. 2, while describing polymers formed through post-living anionic polymerization, have been reinforced with silica nanoparticles for application as synthetic shoe sole binders (Kim et al. 2014). The ionic polymerization with ESO increased the polarity of the resin, yielding a greater affinity between the elastomer and reinforcement, which was reflected in the 30% increase in tensile strength and thermal stability (Kim et al. 2014).

Glass fibers have been used as reinforcement to enhance the mechanical properties of SOY- and LSS-based composites (Lu and Larock 2007). Other oils, such as COR, have also been successfully reinforced with glass fibers (Lu and Larock 2006a). When COR is replaced with CCO, only slight improvements in properties of the composites are observed (Lu and Larock 2006a). Glass fibers have also been used to reinforce polymers synthesized via ROMP. It has been observed that the cross-link density of DCPD and Dilulin ROMP copolymers is not altered upon the addition of glass fibers, therefore not promoting any significant change in the thermal properties of the material (Henna and Larock 2009). Although the composites showed a substantial increase in mechanical properties with the addition of glass fibers, SEM analysis showed fragile interactions between the matrix and the fibers (Henna and Larock 2009).

Organo-functionalized clays have also been used in the preparation of vegetable oil-based composites. CSO and CLS cationic thermosets have been reinforced with organo-modified, montmorillonite clay (Lu and Larock 2006b). At 1–2 wt% clay loading, the reinforcement exhibited intercalated behavior and the most potential for the expansion of the basal interlayer space (Lu and Larock 2006b). In similar studies, montmorillonite clay was used in COR-based resins, revealing the same enhancement in mechanical properties with respect to the pure resin as studies with CSO and CLS (Lu and Larock 2006b).

4 Miscellaneous Systems

4.1 Vegetable Oil-Based Coatings and Emulsions

Unsaturated vegetable oils can be easily converted into polyols through epoxidation of carbon–carbon double bonds, followed by ring opening (Fig. 14) (Kong et al. 2013; Mosiewick and Aranguren 2013). Polyols can then react with diisocyanates by condensation reactions to yield polyurethanes (PU) (Thanh et al. 2014). The molar ratios of the comonomers may be adjusted to tailor the final PU polymers for the desired application (Bothe and Pretsch 2013; Mirtschin and Pretsch 2015). Studies have shown that in Jatropha oil-based PUs, the molar ratio of [NCO/OH] is directly proportional to the cross-link density of the final PU (Rios et al. 2013). Similar systems based on macauba oil (MO) have been studied (Ciconini et al. 2013). MO consists of approximately 55% oleic acid, 15% linoleic acid, 5% linolenic acid, and 25% other fatty acids. MO-based PUs have been reinforced with coconut husk fibers without the use of a compatibilizer (Quirino et al. 2015). As seen in previous literature, MO-based PU's thermal properties can be adjusted by altering the [NCO/OH] molar ratio (Quirino et al. 2015). Despite the absence of a compatibilizer between the nonpolar MO matrix and the polar fiber, the mechanical properties of MO-based PUs are still enhanced through the addition of coconut husk fibers (Quirino et al. 2015).

In recent research, PU foams derived from SOY have been studied. ESO was used to synthesize polyols and subsequently used for PU production (Suppes et al. 2007). ESO polyols can be obtained from the attack of the epoxide rings in ESO by

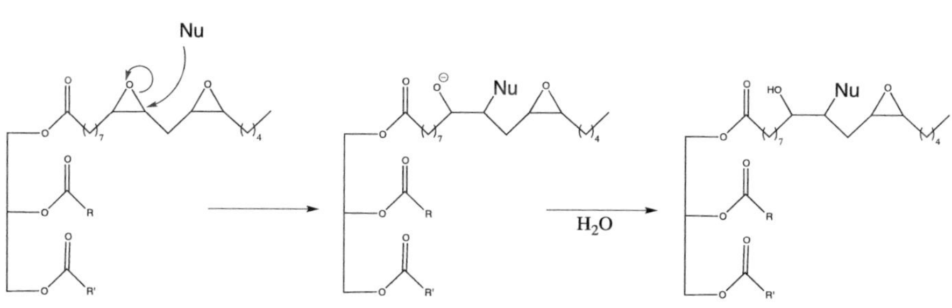

Fig. 14 Reaction scheme showing the nucleophilic attack of the epoxide by a nucleophile in soybean oil leading to polyols, where R and R′ are epoxidized carbon chains

ethylene glycol (Lubguban et al. 2009). When bodied SOY is used to prepare bio-based polyols prior to the formation of the PU, the oligomers formed allow for a need of lower amounts of diisocyanate, therefore resulting in PU foams with higher bio-based content and lower isocyanate content. Overall, PU foams from bodied SOY exhibited slightly enhanced properties in comparison with PUs derived from other sources (Lozada et al. 2009).

5 Conclusion

In recent years, bio-based polymer research has progressed rapidly, yielding promising results for bio-based thermosets with a wide range of applications. The goal behind this research trend is to replace current fossil fuel-based products with more environmentally friendly, sustainable products. Vegetable oils are made up of triglyceride units, which are glycerol esters connected to unsaturated fatty acids. Bio-based thermosets derived from regular and modified vegetable oils can be synthesized by free radical, cationic, ring-opening, and olefin metathesis reactions, leading to rigid, cross-linked polymers. Generally, in order to create homogeneous free radical or cationic copolymers, the similarity in the reactivity of the comonomers is necessary. Polyurethanes, polyester amides, and polyolefins are just some of the vegetable oil-based resins out in the current market. Bio-based polymers are often reinforced with organic or inorganic fibers and fillers to enhance their thermomechanical properties. Compatibilizers, such as MA and asolectin, have been used to enhance the adhesion between nonpolar polymer matrix and polar reinforcement. Most vegetable oil-based thermosets are only competitive for the substitution of petroleum-based materials when the oil content does not exceed 70 wt%. Future research efforts should focus on exploring other bio-based comonomers to replace petroleum-based monomers and cross-linkers without compromising the mechanical properties.

References

Ahn BK, Kraft S, Wang D, Sun XS (2011) Thermally stable, transparent, pressure-sensitive adhesives from epoxidized and dihydroxyl soybean oil. Biomacromolecules 12:1839–1843

Alam M, Alandis NM (2011) Development of ambient cured polyesteramide coatings from linseed oil: a sustainable resource. J Polym Environ 19:391–397

Andjelkovic DD, Larock RC (2006) Novel rubbers from cationic copolymerization of soybean oils and dicyclopentadiene. 1 Synthesis and characterization. Biomacromolecules 7:927–936

Andjelkovic DD, Valverde M, Henna P, Li FK, Larock RC (2005) Novel thermosets prepared by cationic copolymerization of various vegetable oils-synthesis and their structure-property relationships. Polymer 46:9674–9685

Andjelkovic DD, Min B, Ahn D, Larock RC (2006) Elucidation of structural isomers from the homogeneous rhodium-catalyzed isomerization of vegetable oils. J Agric Food Chem 54:9535–9543

Badrinarayanan P, Lu YS, Larock RC, Kessler MR (2009) Cure characterization of soybean oil-styrene divinylbenzene thermoseting copolymers. J Appl Polym Sci 113:1042–1049
Belgacem MN, Gandini A (2008) Monomers, polymers and composites from renewable resources. Elsevier, Amsterdam
Bhuyan S, Sundararajan S, Pfister D, Larock RC (2010) Effect of filler composition and crosslinker concentration on the tribological behavior of spent germ particle-based polymeric composites. Tribol Int 43:171–177
Biermann U, Metzger JO, Meier MAR (2010) Acyclic triene metathesis oligo- and polymerization of high oleic sunflower oil. Macromol Chem Phys 211:854–862
Bonnaillie LM, Wool RP (2007) Thermosetting foam with a high bio-based content from acrylated epoxidized soybean oil and carbon dioxide. J Appl Polym Sci 105:1042–1052
Boonkerd K, Moon BK, Kim MC, Kim JK (2014) Formation of a novel bio-based elastomer from polybutadiene and epoxidized soybean oil via post-living anionic polymerization. J Elastomers Plast 46:644–661
Bothe M, Pretsch T (2013) Bidirectional actuation of a thermoplastic polyurethane elastomer. J Mater Chem A 1:14491–14497
Boudet A, Kajita S, Grima-Pettenati J, Goffner D (2003) Lignins and lignocellulosics: a better control of synthesis for new and improved uses. Trends Plant Sci 8(12):576–581
Cakmakli B, Hazer B, Tekin IO, Kizgut S, Koksal M, Menceloglu Y (2004) Synthesis and characterization of polymeric linseed oil grafted methyl methacrylate or styrene. Macromol Biosci 4:649–655
Cakmakli B, Hazer B, Tekin IO, Comert FB (2005) Synthesis and characterization of polymeric soybean oil-g-methyl methacrylate (and n-butyl methacrylate) graft copolymers: biocompatibility and bacterial adhesion. Biomacromolecules 6:1750–1758
Can E, Kusefoglu S, Wool RP (2002) Rigid thermosetting liquid molding resins from renewable resources. II. Copolymers of soybean oil monoglyceride maleates with neopentyl glycol and bisphenol A maleates. J Appl Polym Sci 83:972–980
Can E, Wool RP, Kusefoglu S (2006a) Soybean and castor oil based monomers: synthesis and copolymerization with styrene. J Appl Polym Sci 102:2433–2447
Can E, Wool RP, Kusefoglu S (2006b) Soybean- and castor oil-based thermosetting polymers: mechanical properties. J Appl Polym Sci 102:1497–1504
Chang J, Liang G, Gu A, Cai S, Yuan L (2012) The production of carbon nanotube/epoxy composites with a very high dielectric constant and low dielectric loss by microwave curing. Carbon 50:689–698
Chung H, Washburn N (2012) Improved lignin polyurethane properties with Lewis acid treatment. ACS Appl Mat Interfaces 4:2840–2846
Ciconini G, Favaro SP, Roscoe R, Miranda CHB, Tapeti CF, Miyahira MAM, Bearari L, Galvani F, Borsato AV, Colnago LA, Naka MH (2013) Biometry and oil contents of Acrocomia aculeate fruits from the Cerrados and Pantanal biomes in Mato Grosso do Sul, Brazil. Ind Crops Prod 45:208–214
Cui HY, Kessler MR (2012) Glass fiber reinforced ROMP-based bio-renewable polymers: enhancement of the interface with silane coupling agents. Comp Sci Technol 72:1264–1272
Czub P (2009) Synthesis of high-molecular-weight epoxy resins from modified natural oils and Bisphenol A or Bisphenol A-based epoxy resins. Polym Adv Technol 20:194–208
Das G, Karak N (2009) Epoxidized Mesua ferrea L. seed oil-based reactive diluent for BPA epoxy resin and their green nanocomposites. Prog Org Coat 66:59–64
de Espinosa LM, Ronda JC, Galia M, Cadiz V (2009) A new route to acrylate oils: crosslinking and properties of acrylate triglycerides from high oleic sunflower oil. J Polym Sci A 47: 1159–1167
del Rio E, Lligadas G, Ronda JC, Galia M, Cadiz V, Meier MAR (2011a) Shape memory polyurethanes from renewable polyols obtained by ATMET polymerization of glyceryl triundec-10-enoate and 10-undecenol. Macromol Chem Phys 212:1392–1399

del Rio E, Lligadas G, Ronda JC, Galia M, Meier MAR, Cadiz V (2011b) Polyurethanes from polyols obtained by ADMET polymerization of a castor oil based diene: characterization and shape memory properties. J Polym Sci A 49:518–525
Deng Y, Fan XD, Waterhouse J (1999) Synthesis and characterization of soy-based copolyamides with different α-amino acids. J Appl Polym Sci 73:1081–1088
Dutta N, Karak N, Dolui SK (2004) Synthesis and characterization of polyester resins based on Nahar seed oil. Prog Org Coat 49:146–152
El Mansouri NE, Yuan Q, Huang F (2011) Characterization of alkaline lignins for use in phenol-formaldehyde and epoxy resins. Biores 6:2647–2662
Eren T, Kusefoglu SH, Wool R (2003) Polymerization of maleic anhydride-modified plant oils with polyols. J Appl Polym Sci 90:197–202
Ewart DC, Sieber EG (1959) Inventors. Airflo Roofing Compony of America, Inc., assignee; US2875080
Fokou PA, Meier MAR (2008) Acyclic triene metathesis polymerization with chain stoppers: molecular weight controlling the synthesis of branched polymers. Macromol Rapid Commun 29:1620–1625
Gandini A (2008) Polymers from renewable resources: a challenge for the future of macromolecular materials. Macromolecules 41:9491–9504
Gandini A (2011) The irruption of polymers from renewable resources on the scene of macromolecular science and technology. Green Chem 13:1061–1083
Gibbons L, Smith M, Quirino RL (2015) Modified lignin for composite and pellet binder applications. Int J Exp Comp Biomechanics 8:200–217
Gindl W, Schöberl T, Keckes J (2006) Structure and properties of a pulp fibre-reinforced composite with regenerated cellulose matrix. Appl Phys A 83:19–22
Goodman SH (1999) Handbook of Thermoset Plastics. William Andrew Publishing, Westwood
Gosselink R (2011) Ph.D. Dissertation, Wageningen University
Haman K, Badrinarayanan P, Kessler MR (2009) Cure characterization of the ring-opening metathesis polymerization of linseed oil-based thermosetting resins. Polym Int 58:738–744
Henna P, Larock RC (2007) Rubbery thermosets by ring-opening metathesis polymerization of a functionalized castor oil and cyclooctene. Macromol Mat Eng 292:1201–1209
Henna P, Larock RC (2009) Novel thermosets obtained by the ring-opening metathesis polymerization of a functionalized vegetable oil and dicyclopentadiene. J Appl Polym Sci 112:1788–1797
Henna PH, Andjelkovic DD, Kundu PP, Larock RC (2007) Biobased thermosets from the free-radical copolymerization of conjugated linseed oil. J Appl Polym Sci 104:979–985
Hong J, Luo Q, Shah BK (2010) Catalyst- and Solvent-Free "Click" chemistry: a facile approach to obtain cross-linked biopolymers from soybean oil. Biomacromolecules 11:2960–2965
Hong J, Luo Q, Wan XM, Petrovic ZS, Shah BK (2012) Biopolymers from vegetable oils via catalyst- and solvent-free "Click" chemistry: effects of cross-linking density. Biomacromolecules 13:261–266
Hu L, Pan H, Zhou Y, Zhang M (2011) Methods to improve lignin's reactivity as a phenol substitute and as a replacement for other phenolic compounds: a brief review. Bioresearch 6:3515–3525
Humar M, Lesar B (2013) Efficacy of linseed- and tung-oil-treated wood against wood-decay fungi and water uptake. Int Biodeterior Biodegrad 85:223–227
Jin FL, Park SJ (2008a) Impact-strength improvement of epoxy resins reinforced with a biodegradable polymer. Mat Sci Eng A 478:402–405
Jin FL, Park SJ (2008b) Thermomechanical behavior of epoxy resins modified with epoxidized vegetable oils. Polym Int 57:577–583
Johns A, Morris S, Edwards K, Quirino RL (2015) Asolectin from soybeans as a natural compatibilizer for cellulose-reinforced biocomposites from tung oil. J Appl Polym Sci 132:41833–41842
Kalia S, Dufresne A, Cherian BM, Kaith BS, Averous L, Njuguna J, Nassiopoulos E (2011) Cellulose-based bio- and nanocomposites: a review. Int J Polym Sci 2011:1–35

Kenar JA, McElligott J, Hwang H, Erhan SZ (2002) A DSC study of Z2–Z3 viscosity blown soybean oil. J Am Oil Chem Soc 79:1151–1155
Kiatsimkul P, Suppes GJ, Sutterlin WR (2007) Production of new soy-based polyols by enzyme hydrolysis of bodied soybean oil. Ind Crops Prod 25:202–209
Kiatsimkul P, Suppes GJ, Hsieh F, Lozada Z, Tu Y (2008) Preparation of high hydroxyl equivalent weight polyols from vegetable oils. Ind Crops Prod 27:257–264
Kim MS, Lee KW, Endo T, Lee SB (2004) Benzylpyrazinium salts as thermally latent initiators in the polymerization of glycidyl phenyl ether: substituent effect on the initiator activity and mechanistic aspects. Macromolecules 37:5830–5834
Kim K, Saha P, Kim S, Jo S, Kim JK (2014) Novel elastomer nanocomposite with uniform silica dispersion from polybutadiene rubber treated with epoxidized soybean oil. J Comp Mater 49:3005–3016
Kong X, Liu G, Qi H, Curtis JM (2013) Preparation and characterization of high-solid polyurethane coating systems based on vegetable oil derived polyols. Prog Org Coatings 76:1151–1160
Konwar U, Karak N, Mandal M (2009) Mesua ferrea L. seed oil based highly thermostable and biodegradable polyester/clay nanocomposites. Polym Deg Stab 94:2221–2230
Konwar U, Karak N, Jana T (2012) Vegetable oil-based highly branched polyester modified epoxy based low VOC high solid industrial paint. J Appl Polym Sci 125:E2–E9
Kundu PP, Larock RC (2005) Novel conjugated linseed oil-styrene-divinylbenzene copolymers prepared by thermal polymerization. 1. Effect of monomer concentration on the structure and properties. Biomacromolecules 6:797–806
Kundu PP, Larock RC (2009) Effect of drying catalysts on the properties of thermal copolymers from conjugated linseed oil-styrene-divinylbenzene. Prog Org Coat 65:10–18
La Scala J, Wool RP (2005) Property analysis of triglyceride-based thermosets. Polymer 46:61–69
Laredo E, Bello A, Diaz J, Grimau M, Martinez-Tong D, Wu D, Wu L (2013) Effect of cold-crystallization on the AC and DC conductive properties of polylactide biocomposites with carboxylic or neat large aspect ratio MWCNT. Polym Comp 34:67–76
Larock RC, Dong XY, Chung S, Reddy CK, Ehlers LE (2001) Preparation of conjugated soybean oil and other natural oils and fatty acids by homogeneous transition metal catalysis. J Am Oil Chem Soc 78:447–453
Li FK, Larock RC (2000) Thermosetting polymers from cationic copolymerization of tung oil: synthesis and characterization. J Appl Polym Sci 78:1044–1056
Li FK, Larock RC (2001) New soybean oil-styrene-divinylbenzene thermosetting copolymers. I. Synthesis and characterization. J Appl Polym Sci 80:658–670
Li FK, Larock RC (2002) New soybean oil-styrenedivinylbenzene thermosetting copolymers. V. Shape memory effect. J Appl Polym Sci 84:1533–1543
Li FK, Larock RC (2003) New soybean oil-styrene divinylbenzene thermosetting copolymers. VI. Time temperature-transformation cure diagram and the effect of curing conditions on the thermoset properties. Polym Int 52:126–132
Li FK, Hanson MV, Larock RC (2001) Soybean oil-divinylbenzene thermosetting polymers: synthesis, structure, properties and their relationships. Polymer 42:1567–1579
Li FK, Hasjim J, Larock RC (2003) Synthesis, structure, and thermophysical and mechanical properties of new polymers prepared by the cationic copolymerization of corn oil, styrene, and divinylbenzene. J Appl Polym Sci 90:1830–1838
Li SM, Jia N, Ma MG, Zhang Z, Liu QH, Sun RC (2011) Cellulose–silver nanocomposites: microwave-assisted synthesis, characterization, their thermal stability, and antimicrobial property. Carbohydr Polym 86:441–447
Liu ZX, Xu YD, Cao L, Bao CY, Sun HH, Wang L, Dai KR, Zhu LY (2012) Phosphoester cross-linked vegetable oil to construct a biodegradable and biocompatible elastomer. Soft Matter 8:5888–5895
Lozada Z, Suppes GJ, Hsieh F, Lubguban A, Tu Y (2009) Preparation of polymerized soybean oil and soy-based polyols. J Appl Polym Sci 112:2127–2135

Lu Y, Larock RC (2006a) Corn oil-based composites reinforced with continuous glass fibers: Fabrication and properties. J Appl Polym Sci 102:3345–3353
Lu Y, Larock RC (2006b) Novel Biobased Nanocomposites from Soybean Oil and Functionalized Organoclay. Biomacromolecules 7:2692–2700
Lu Y, Larock RC (2007) Fabrication, Morphology and Properties of Soybean Oil-Based Composites Reinforced with Continuous Glass Fibers. Macromol Mat Eng 292:1085–1094
Lu YS, Larock RC (2009) Novel polymeric materials from vegetable oils and vinyl monomers: preparation, properties, and applications. ChemSusChem 2:136–147
Lu J, Khot S, Wool RP (2005) New sheet molding compound resins from soybean oil. I. Synthesis and characterization. Polymer 46:71–80
Lubguban AA, Tu Y, Lozada ZR, Hsieh F, Suppes GJ (2009) Noncatalytic polymerization of ethylene glycol and epoxy molecules for rigid polyurethane foam applications. J Appl Polym Sci 112:2185–2194
Ma H, Zhou B, Li H, Li Y, Ou S (2011) Green composite films composed of nanocrystalline cellulose and a cellulose matrix regenerated from functionalized ionic liquid solution. Carbohydr Polym 84:383–389
Mauldin TC, Haman K, Sheng X, Henna P, Larock RC, Kessler MR (2008) Ring-opening metathesis polymerization of a modified linseed oil with varying levels of crosslinking. J Polym Sci A 46:6851–6860
Meier MAR, Metzger JO, Schubert US (2007) Plant oil renewable resources as green alternatives in polymer science. Chem Soc Rev 36:1788–1802
Mirtschin N, Pretsch T (2015) Designing temperature-memory effects in semicrystalline polyurethane. RSC Adv 5:46307–46315
Mol JC (2002) Application of olefin metathesis in oleocemistry: an example of green chemistry. Green Chem 4:5–13
Morye SS, Wool RP (2005) Mechanical properties of glass/flax hybrid composites based on a novel modified soybean oil matrix material. Polym Compos 26:407–416
Mosiewick MA, Aranguren MI (2013) A short review on novel biocomposites based on plant oil precursors. Europ Polym J 49:1243–1256
Mosiewicki M, Aranguren MI, Borrajo J (2005) Mechanical properties of linseed oil monoglyceride maleate/styrene copolymers. J Appl Polym Sci 97:825–836
Mustata F, Tudorachi N, Rosu D (2011) Curing and thermal behavior of resin matrix for composites based on epoxidized soybean oil/diglycidyl ether of bisphenol A. Compos B 42:1803–1812
Nayak PL (2000) Natural oil-based polymers: opportunities and challenges. J Macromol Sci Rev Macromol Chem Phys 40:1–21
Nozawa A, Nanamiya H, Tozawa Y (2010) Production of membrane proteins through the wheat-germ cell-free technology. Methods Mol Biol 607:213–218
O'Donnell A, Dweib M, Wool R (2004) Natural fiber composites with plant oil-based resin. Compos Sci Technol 64:1135–1145
Ozturk C, Mutlu H, Meier MAR, Kusefoglu SH (2011) 4-Vinylbenzenesulfonic acid adduct of epoxidized soybean oil: synthesis, free radical and ADMET polymerizations. Eur Polym J 47:1467–1476
Park SJ, Jin FL, Lee JR (2004) Effect of biodegradable epoxidized castor oil on physicochemical and mechanical properties of epoxy resins. Macromol Chem Phys 205:2048–2054
Park K, Berrier C, Lebaupain F, Pucci B, Popot J, Ghazi A, Zito F (2007) Fluorinated and hemifluorinated surfactants as alternatives to detergents for membrane protein cell-free synthesis. Biochem J 403:183–187
Petrovic ZS, Cvetkovic I, Hong D, Wan X, Zhang W, Abraham T, Malsam J (2008) Polyester polyols and polyurethanes from ricinoleic acid. J Appl Polym Sci 108:1184–1190
Pfister DP (2010) Ph.D. dissertation, Iowa State University
Pfister DP, Larock RC (2010) Green composites from a conjugated linseed oil-based resin and wheat straw. Compos A Appl Sci Manuf 41:1279–1288

Pfister DP, Larock RC (2012) Cationically cured natural oil-based green composites: effect of the natural oil and the agricultural fiber. J Appl Polym Sci 123:1392–1400

Pfister DP, Baker JR, Henna PH, Lu Y, Larock RC (2008) Preparation and properties of tung oil-based composites using spent germ as a natural filler. J Appl Polym Sci 108:3618–3625

Quirino RL, Larock RC (2009) Synthesis and properties of soy hull-reinforced biocomposites from conjugated soybean oil. J Appl Polym Sci 112:2033–2043

Quirino RL, Larock RC (2011a) Rice hull biocomposites, part 2: effect of the resin composition on the properties of the composite. J Appl Polym Sci 121:2050–2059

Quirino RL, Larock RC (2011b) Rice hull biocomposites. I. Preparation of a linseed-oil-based resin reinforced with rice hulls. J Appl Polym Sci 121:2039–2049

Quirino RL, da Silva T, Payne A, Lopes R, Paterno L, Sales M (2015) Synthesis and thermomechanical properties of polyurethanes and biocomposites derived from macauba oil and coconut husk fibers. Coatings 5:527–544

Ramamoorthy SK, Di Q, Adekunle K, Skrifvars M (2012) Effect of water absorption on mechanical properties of soybean oil thermosets reinforced with natural fibers. J Reinf Plast Comp 31:1191–1200

Rangri VK, Bhuyan MS, Jeelani S (2011) Comparative study of microwave and thermal curing of high temperature epoxy/Carbon nanotubes polymer nanocomposites and their properties. Int J Nanosci 10:1225–1230

Ratna D (2009) Handbook of thermoset resins. Smithers Rapra Publishing, Shropshire

Rios L, Echeverri D, Cardeño F (2013) Hydroxylation of vegetable oils using acidic resins as catalysts. Ind Crops Prod 43:183–187

Roy B, Karak N (2012) Modification of hyperbranched epoxy by vegetable oil-based highly branched polyester resin. Polym Bull 68:2299–2312

Rybak A, Meier MAR (2008) Acyclic diene metathesis with a monomer from renewable resources: control of molecular weight and one-step preparation of block copolymers. ChemSusChem 1:542–547

Sacristan M, Ronda JC, Galia M, Cadiz V (2009) Silicon containing soybean oil-based copolymers. Synthesis and properties. Biomacromolecules 10:2678–2685

Sacristan M, Ronda JC, Galia M, Cadiz V (2010) Synthesis and properties of boron containing soybean oil based thermosetting copolymers. Polymer 51:6099–6106

Samadi A, Martinez LA, Miranda MA, Morera IM (2001) Mechanism of lipid peroxidation photosensitized by tiaprofenic acid: product studies using linoleic acid and 1,4-cyclohexadienes as model substrates. Photochem Photobiol 73:359–365

Schrock RR, Czekelius C (2007) Recent advances in the syntheses and applications of molybdenum and tungsten alkylidene and alkylidyne catalysts for the metathesis of alkenes and alkynes. Adv Synth Catal 349:55–77

Shahzad A (2012) Hemp fiber and its composites – a review. J Comp Mater 46:973–986

Sharma V, Kundu P (2008) Condensation polymers from natural oils. Prog Polym Sci 33:1199–1215

Sharma V, Banait JS, Larock RC, Kundu PP (2008) Synthesis and characterization of styrene-co-divinylbenzene-graft-linseed oil by free radical polymerization. Express Polym Lett 2:265–276

Smith M, Payne A, Edwards K, Morris S, Beckler B, Quirino RL (2015) Effect of microwave cure on the thermo-mechanical properties of tung oil-based/carbon nanotube composites. Coatings 5:557–575

Stemmelen M, Pessel F, Lapinte V, Caillol S, Habas JP, Robin JJ (2011) A fully biobased epoxy resin from vegetable oils: from the synthesis of the precursors by thiol-ene reaction to the study of the final material. J Polym Sci A 49:2434–2444

Supanchaiyamat N, Shuttleworth PS, Hunt AJ, Clar JH, Matharu AS (2012) Thermosetting resin based on epoxidised linseed oil and bio-derived crosslinker. Green Chem 14:1759–1765

Suppes G, Hsieh F, Tu Y, Kiatsimkul P (2007) Inventors. University of Missouri, assignee; US0265459A1

Tallman KA, Roschek B, Porter NA (2004) Factors influencing the autoxidation of fatty acids: effect of olefin geometry of the nonconjugated diene. J Am Chem Soc 126:9240–9247

Tejado A, Peña C, Labidi J, Echeverria JM, Mondragon I (2007) Physico-chemical characterization of lignins from different sources for use in phenol-formaldehyde resin synthesis. Biores Technol 98:1655–1663

Tellez GL, Vigueras-Santiago E, Hernandez-Lopez S, Bilyeu B (2008) Synthesis and thermal cross-linking study of partially-aminated epoxidized linseed oil. Design Monom Polym 11:435–445

Thanh LT, Okitsu K, Maeda Y, Bandow H (2014) Ultrasound assisted production of fatty acid methyl esters from transesterification of triglycerides with methanol in the presence of KOH catalyst: optimization, mechanism and kinetics. Ultrason Sonochem 21:467–471

Thielemans W, Wool R (2005) Lignin esters for use in unsaturated thermosets: lignin modification and solubility modeling. Biomacromolecules 6:1895–1905

Thielemans W, McAninch IM, Barron V, Blau WJ, Wool RP (2005) Impure carbon nanotubes as reinforcements for acrylated epoxidized soy oil composites. J Appl Polym Sci 98:1325–1338

Tian QP, Larock RC (2002) Model studies and the ADMET polymerization of soybean oil. J Am Oil Chem Soc 79:479–488

Turley DB (2008) Introduction to chemicals from biomass. Wiley, New York

Valverde M, Andjelkovic D, Kundu PP, Larock RC (2008) Conjugated low-saturation soybean oil thermosets: free-radical copolymerization with dicyclopentadiene and divinylbenzene. J Appl Polym Sci 107:423–430

Van De Mark MR, Sandefur K (2009) Industrial uses of vegetable oil. AOCS Publishing, Peoria

Williams CK, Hillmyer MA (2008) Polymers from renewable resources: a perspective for a special issue of polymer reviews. Polym Rev 48:1–10

Xia Y, Larock RC (2010a) Castor oil-based thermosets with varied crosslink densities preparedby ring-opening metathesis polymerization (ROMP). Polymer 51:2508–2514

Xia Y, Larock RC (2010b) Vegetable oil-based polymeric materials: synthesis, properties, and applications. Green Chem 12:1893–1909

Xia Y, Henna PH, Larock RC (2009) Novel thermosets from the cationic copolymerization of modified linseed oils andicyclopentadiene. Macromol Mater Eng 294:590–598

Xia Y, Lu YS, Larock RC (2010) Ring-opening metathesis polymerization (ROMP) of norbornenyl-functionalized fatty alcohols. Polymer 51:53–61

Xia Y, Quirino RL, Larock RC (2013) Bio-based thermosetting polymers from vegetable oils. J Rev Mater 1:3–27

Yang X, Li S, Huang K, Li M, Xia J (2013a) Environment, energy and sustainable development. Taylor & Francis Group, London

Yang ZY, Wang WJ, Shao ZQ, Zhu HD, Li YH, Wang FJ (2013b) The transparency and mechanical properties of cellulose acetate nanocomposites using cellulose nanowhiskers as fillers. Cellulose 20:159–168

Yue X, Chen F, Zhou X (2011) Improved interfacial bonding of pvc/wood-flour composites by lignin amine modification. Biores 6:2022–2034

Zhan M, Wool RP (2013) Design and evaluation of bio-based composites for printed circuit board application. Compos A Appl Sci Manuf 47:22–30

Zhang H, Chen Z (2010) Fabrication and characterization of electrospun PLGA/MWNTs/ hydroxyapatite biocomposite scaffolds for bone tissue engineering. J Bioact Compat Polym 25:241–259

Chapter 7
Imprinted Polymeric Gels for Pharmaceutical and Biomedical Purposes

Monika Sobiech and Piotr Luliński

Abstract Molecular imprinting is a technology for the synthesis of tailor-made recognition materials by polymerization of suitable functional monomers in the presence of desired template. As a result, smart polymers with high specificity and selectivity are obtained. Hydrogels are defined as insoluble polymeric networks produced from hydrophilic polymer chains with capability to absorb water and preserve their shape without dissolving. The imprinting of hydrogels is more complicated comparing to rigid structure. However, the exponential increase in fabrication of imprinted hydrogels is observed. The novel, advanced applications in pharmacy and medicine are explored including areas such as drug delivery systems or sensors in different analytical methods utilized in diagnostics. In this chapter, a brief overview of hydrogel imprinting process together with various synthetic approaches is discussed followed by detailed description of possible applications of imprinted hydrogels in pharmaceutical and biomedical sciences. Finally, some current limits and future prospects for the imprinted hydrogels are pointed out.

Keywords Drug delivery · Imprinted hydrogel · Molecularly imprinted polymer Sensor · Separation

1 Introduction

The consideration of polymer-based gel could be started from a definition of gel. A gel is a soft, solid, or solid-like material composed from two or more components and one is a liquid component that is present in a substantial quantity. Gels resemble solid materials but they are characterized by high softness and they possess the capability to undergo a large deformation. Gels are widely encountered in nature, e.g., mammalian tissues are highly aqueous gels composed from protein

M. Sobiech · P. Luliński (✉)
Department of Organic Chemistry, Faculty of Pharmacy,
Medical University of Warsaw, Banacha 1, 02-097 Warsaw, Poland
e-mail: piotr.lulinski@wum.edu.pl

© Springer Nature Singapore Pte Ltd. 2018

V. K. Thakur et al. (eds.), *Polymer Gels*, Gels Horizons: From Science to Smart Materials, https://doi.org/10.1007/978-981-10-6080-9_7

and polyssacharide nets. Such composition facilitates transport of biomolecules or transduction of signals. These actions are possible because of an extraordinary ability of gels to swell and collapse as a response to various external stimuli. Behind mentioned abilities of gels is a gel volume transition that can be induced by the change of pH or temperature as well as the change of concentration of various salts or compounds. This phenomenon was used in a preparation of gels that act as actuators, sensors, sorbents, membranes, and drug carriers (Osada and Gong 1998). Based on above definition of gel, a polymeric gel can be defined as an elastic solid material composed from two (or more) components, one of which (polymer) forms a chemically or physically bonded three-dimensional matrix that occurs in a medium of another component, a liquid, wherein the volume of the liquid is sufficient for ensuring the elastic properties of the gel. The term of polymeric gel is also described as a polymer-solvent system involving a three-dimensional network that is composed from macromolecules or their aggregates. This network is capable for retaining the large volume of solvent preserving their shape under the action of its own weight and differs in this feature from the polymer solution. The polymeric gels are classified into two types depending on the strength of the cross-linking, viz. chemical, involving covalent bounds between macromolecular chains, and cross-linking agent or physical, involving various intermolecular interactions between macromolecular chains and cross-linking agent (Rogovina et al. 2008).

The polymeric gels that possess ability to adsorb a large volume of water are named as hydrogels. Hydrogels are defined as water-swollen and cross-linked polymeric networks that are produced from one or more monomers or are polymeric materials that exhibit the ability to swell and retain a significant fraction of water within their structure without dissolving. Recently, hydrogels have been defined as two- or multicomponent systems containing three-dimensional network of polymer chains and water that fills the regions between macromolecules. In term of the composition, the hydrogels can be divided into homopolymeric hydrogels which are derived from uniform monomer, copolymeric hydrogels which refer to at least two various monomers randomly located in the polymeric network (one should possess hydrophilic character) and multipolymer (interpenetrating) hydrogels that comprise of two independent cross-linkers (Ahmed 2015). The most important classification of hydrogels is based on the type interactions during the cross-linking (described above), and the molecularly imprinted hydrogels are mainly synthesized as chemically cross-linked polymers.

2 Principles of Imprinting Process

The molecularly imprinted polymers are characterized by the high level of selectivity due to the presence of specific recognition sites formed in the polymer network by the template-tailored synthesis. The synthesis of imprinted materials consists of three steps: formation of prepolymerization adduct, polymerization reaction, and template removal (Fig. 1).

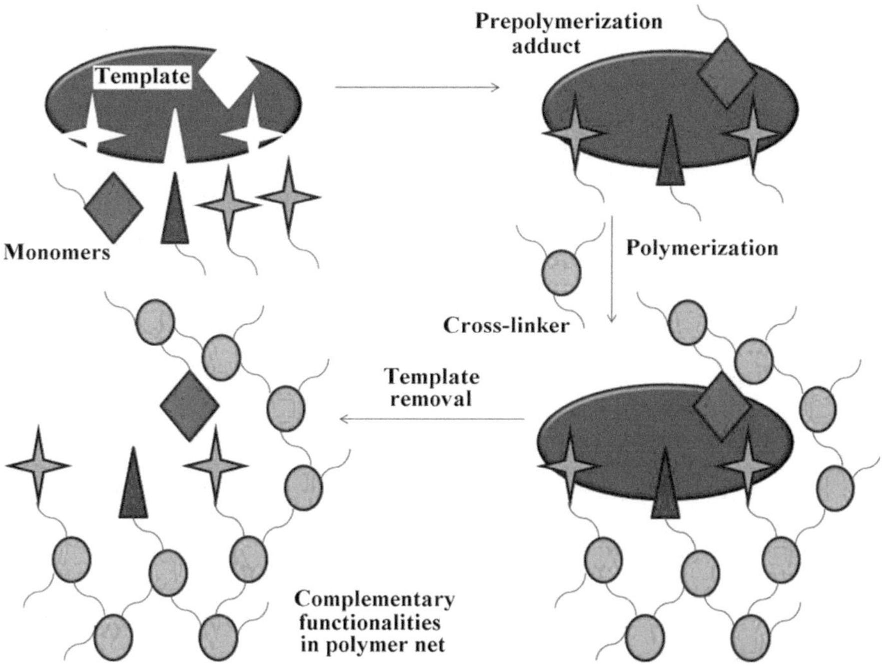

Fig. 1 Schematic idea of the imprinting process

There are four critical moments during the imprinting process that can strongly affect the property of imprinted materials and the effectiveness of imprinting. Those are the choice of template, the preselection of functional monomer, the conditions of polymerization reaction, and the effective template removal.

The choice of template is one of crucial moments during the synthesis of imprinted materials. The certain structural elements of the template molecule are required to facilitate the imprinting process. The presence of functional groups or/ and heteroatoms that could form covalent bonds or can interact non-covalently with functional monomers is the main factor for structural stabilization of prepolymerization adduct. The aromatic, heteroaromatic, or polyaromatic rings enhance the stabilization and together with cyclic systems as well as aliphatic chains are responsible for the formation of steric effects inside the polymeric network. The template molecule commonly acts as the target compound to which the imprinted polymer is dedicated. However, a majority of biocompounds or drugs is characterized by limited application in the synthesis of molecularly imprinted polymers because of insufficient solubility in organic solvents, limited availability, or high costs. Thus, the use of structural analogues of target molecule as the template during imprinting process is excellent approach (Alvarez-Lorenzo and Concheiro 2013).

The critical moment in the design of imprinted materials is the selection of functional monomer. It should interact with template molecule via complementary functionalities to stabilize the prepolymerization adduct. It could be obtained by covalent or non-covalent approach. The covalent approach assumes the chemical reaction between the template molecule and functional monomer to form functionalized compound prior to the polymerization. The non-covalent approach utilizes the range of weak intermolecular interactions such as ionic forces, hydrogen bonds, van der Waals forces or π–π interactions that can exist between the template molecule and functional monomer to form the prepolymerization complex prior to the polymerization (Alvarez-Lorenzo and Concheiro 2013). The selection of appropriate functional monomer could be a laborious process since only after the imprinted polymer is obtained it could be evaluated and the proper choice of the functional monomer could be validated. Here, the theoretical analyses have been employed as powerful tool in the preselection of functional monomer (Nicholls et al. 2015).

Next, the polymerization reaction is carried out mostly in the presence of cross-linker and solvent. Here, the choice of covalent or non-covalent strategy ought to be considered once again. The covalent approach allows the unrestricted choice of polymerization conditions due to high stability of the prepolymerization structure. In contrast, the harsh conditions of polymerization reaction could destabilize the prepolymerization complex formed by the non-covalent approach. This can result with failure of imprinting process. On the other hand, the ease of formation of the prepolymerization complex via non-covalent approach makes this strategy most commonly use to provide the imprinted material.

Finally, the template molecule has to be remove to leave three-dimensional cavities empty (Lorenzo et al. 2011). This is very important moment of synthesis because only total removing of the template molecule allows to obtain specific binding sites in the imprinted polymer. Thus, the imprinting process is considered as the process that enables to modify the surface of polymer by the template molecule.

The main parameter describing the efficiency of imprinting process is imprinting factor. In the simplest way, imprinting factor is defined as ratio of the binding capacity on imprinted polymer to the binding capacity on the reference non-imprinted polymer. Hence, the synthesis of non-imprinted polymer has to be carried out in the same conditions omitting the addition of template molecule.

The presented above brief overview of the synthesis of imprinted polymers outlined the most important moments of the imprinting process. The imprinting technique has gained widespread attention mainly due to its application in the fabrication of molecularly imprinted stationary phases. There are some excellent books and reviews that provide more detail information regarding to the synthesis and application of imprinted polymers (Wulff 1995, 2002; Alexander et al. 2006; Alvarez-Lorenzo and Concheiro 2013; Whitcombe et al. 2014).

3 Macromolecular Memory of Hydrogels

The idea of molecular imprinting involves the creation of three-dimensional sites in the mostly rigid, highly cross-linked polymeric network. In case of low cross-linked hydrogels that possess relatively high degree of flexibility of polymer chains, the process of imprinting can be more properly defined as the process of creation of the macromolecular memory in hydrogel network (Byrne and Salian 2008). Once the imprinting process is proceeded, the complementary interactions of various functionalities between the target molecule and polymer network as well as spatial or configurational specificity governs the recognition mechanism. It could be suspicious that the flexibility of polymeric chains in hydrogels can provide the macromolecular memory of the shape of chains since the relaxation of network can diminish or destroy the spatial effects of imprinting. However, multiple interactions during imprinting stage between the template and monomers evoke the incorporation of mer units into the polymeric chain in the conformation form which is characterized by low energy. When the polymerization occurs, the global energy minimum of growing polymer chains with template complexes is achieved promoting the memorization of the chain conformation. The molecular recognition of target molecule by the flexible structure of hydrogel is a dynamic process that comprises successive steps of mutual conformational adaptation (Chen et al. 2008). Such process allows to modify the specificity and affinity of interactions with target molecule by many trial-and-error steps until the proper interaction is amplified and compound is bound (Demchenko 2001).

The recognition of target molecule by the imprinted hydrogel takes place in the swollen state including water as the medium. The swollen volume at equilibrium has to be very similar to the relaxed polymer volume fraction during the imprinting stage. The hydrophilic groups in the polymer backbone and the hydrophobic regions of cross-linked junctions are responsible for the preservation of macromolecular memory and imprinting effect. Thus, the optimization of synthetic parameters is necessary to provide the imprinted hydrogel with controlled swelling.

The main factors that have to be considered during the synthesis of imprinted hydrogels are as follows: the ratio between the template and monomer, the diversity of monomers, the density of cross-linker, and the polymerization reaction conditions.

The very important factor is the optimal ratio between the template molecule and the functional monomers which is responsible for the enhancement of the interactions between the respective reagents. This is the major variable responsible for the effective imprinting of hydrogels. The large excess of monomers provides randomly incorporated monomer functionalities similarly to the non-imprinted hydrogel affecting the macromolecular memory of polymer chains and diminishing the imprinting effect. In contrary, the low ratio of monomer to the template results with the insufficient effectiveness of simultaneous interactions that have to occur in order to enhance the macromolecular memory in the polymer chain resulting with disappearing of imprinting effect, too. This phenomenon was observed during the optimization of prepolymerization conditions with timolol as the template. In the experiments, the various ratios of timolol to the functional monomer were employed. The optimized stoichiometry of the complexation was equal to one molecule of timolol to eight molecules of monomer, and it was confirmed by isothermal titration calorimetry. After the polymerization was finished, the binding capacities of the resulted imprinted hydrogels toward timolol were determined. It was proved that the increase of the amount of monomer functionalities (the stoichiometry of prepolymerization complex was more than one timolol to eight monomers) as well as decrease of the amount of monomer functionalities (the stoichiometry of prepolymerization complex was less than one timolol to six monomers) resulted with significant increase of the non-specific adsorption. In consequence, the disappearance of imprinting effect was observed due to similar binding ability of the imprinted and non-imprinted hydrogels (Yanez et al. 2011a). The optimization of template to monomer ratio is extremely important during the imprinting of macromolecular compounds such as proteins. The different monomer to cross-linker molar ratios was carefully investigated to ensure the formation of stable complexes between bovine serum albumin and monomer via multiple-point electrostatic interactions. Here, as higher as 64 molecules of monomers to one molecule of protein was established to be optimal (Hua et al. 2008, 2009).

The next important factor is the diversity of monomers that form heteropolymeric system. The employment of different monomers in the formation of the prepolymerization complexes enhances the strength of interactions with the template molecule. After the hydrogel is created, the diverse monomer functionalities play a crucial role in the recognition of the target molecule. This factor was analyzed during the determination of the diffusion coefficients of four ketotifen imprinted hydrogels possessing different compositions: poly(*N*-vinylpirrolidone-*co*-2-hydroxyethyl methacrylate-*co*-polyethylene glycol dimethacrylate), poly(acrylamide-*co*-2-hydroxyethyl methacrylate-*co*-polyethylene glycol dimethacrylate), poly(acrylic acid-*co*-2-hydroxyethyl methacrylate-*co*-polyethylene glycol dimethacrylate), and poly(*N*-vinylpirrolidone-*co*-acrylamide-*co*-acrylic acid-*co*-2-hydroxyethyl methacrylate-*co*-polyethylene glycol dimethacrylate). The following diffusion coefficients were determined: 80, 34, 54, and 0.7×10^{-9} cm^2/s, respectively. The slowest diffusion was noted for the ketotifen imprinted hydrogel with the most complex composition. It should be added that the simultaneous determination of structural parameters of the imprinted and non-imprinted hydrogels revealed no significant difference between them. It was concluded that imprinting technique resulted in the structural plasticity of polymeric chains and the imprinting process was responsible for the formation of macromolecular memory due to the organization of the diverse functionalities in the polymer architecture. The multiplicity and diversity of functionalities were responsible for binding events that affected the transport throughout the imprinted hydrogel and delayed the diffusion of drugs (Venkatesh et al. 2008).

Then, the density of cross-linker or the molar percentage of cross-linking agents in the total mass of polymer is very important. In the highly cross-linked rigid polymers, the movement of the functional residues is strongly restricted. In the low cross-linked swellable systems such as hydrogels, the density of the polymer chains ought to comprise between the ability to the free diffusion of the compound to the inner regions of the net, the sufficient binding capacity as well as the swelling parameters. The higher ratio of cross-linker affects the swelling property of imprinted hydrogel because it hampers the mobility of the polymer chains. In consequence, it prevents vast imbibe of water and counteracts against the destruction of imprinted sites. This problem was considered in the analysis of timolol imprinted hydrogel. The different molar concentrations of ethylene glycol dimethacrylate cross-linker were used in the synthesis of imprinted hydrogels. Then, Langmuir isotherms were determined and maximum binding capacities of hydrogels as well as the dissociation constants were calculated. It was found that the dissociation constants were similar for all imprinted hydrogels in all tested molar concentrations of cross-linker. The maximum binding capacities of all imprinted hydrogels were very similar and independent of the cross-linker molar concentration. In contrary, the non-imprinted hydrogels possessed ten times lower values of the maximum binding capacity. Moreover, the non-imprinted hydrogels revealed the decrease of maximum binding capacity with the increase of concentration of the cross-linker. It was explained that in the non-imprinted systems, the functionalities from monomer were randomly distributed in the hydrogel net and they were separated by the cross-linker chains which prevented interactions with timolol. In the imprinted systems, the macromolecular memory sites formed during synthesis imposed the architecture of cross-linker chains (Alvarez-Lorenzo et al. 2002).

The polymerization reaction can affect the sequence of imprinted hydrogel network. The polymerization is mainly carried out as a free radical process which is difficult to control. The resulted cross-linking polymeric structure could possess multiple defects. The formation of primary and secondary loops, entanglements, and dangling ends can affect the recognition process. Here, the interesting technique involving the living radical polymerization was introduced to overcome the problem. This technique provides better memorization of the hydrogel chain conformations (Salian et al. 2012).

The presence of solvent during polymerization plays also important role. The imprinted hydrogels that are prepared in the absence of solvent (or more precisely the largest volume of hydrophilic monomer plays a role of solvent) are characterized by the relatively high binding capacity and high imprinting effect. In contrary, the presence of solvent (mostly protic and polar one) requires to form the growing polymeric chains around the solvent molecules and provides a macro- or microstructure of hydrogel. The porous system is responsible for the transport of solutes inside the hydrogel network and is closely related to the water swellability.

Once the imprinted hydrogel is obtained, it should be characterized involving the determination of imprinting efficacy, swelling properties as well as the structural and mechanical properties. The effectiveness of imprinting process is estimated based upon the imprinting effect which was described above. The specific parameter devoted to hydrogels is the swelling ratio describing the ability of dry polymer to swell and retain water. In the absence of cross-linking points (junctions), the linear polymeric chains dissolve in water due to thermodynamic compatibility.

The cross-linking provides heterogeneous matrix with low water swelling (high cross-link density) hydrophobic regions, and the solubility is counterbalanced by the retractive forces of the elasticity of junctions in the polymer network. In the equilibrium swollen state, the thermodynamic forces of mixing and retractive forces of the polymeric chains are balanced. However, the swelling of hydrogels is a complex process. Firstly, the polar hydrophilic groups of imprinted hydrogel matrix are hydrated by water (primary bound water) followed by the swelling of the hydrophobic regions of hydrogel (secondary bound water). The primary and secondary bound water comprises the total bound water. Finally, the osmotic driving forces of the hydrogel network to imbibe additional water are hampered by chemical cross-linking providing equilibrium swelling state with additional bulk or free water which fills the regions between the cross-linked points (Hoffman 2012).

The structural determination of hydrogels could be done by the analysis of the polymer volume fraction in swollen state, the average molecular weight, and the average distance between two cross-linking junctions. The polymer volume fraction in the swollen state is defined as the amount of liquid phase imbibed and retained in the hydrogel network. The average molecular weight is a measure of the polymer cross-linking degree. The average distance (mesh size) describes the intervals between two neighboring junctions in the cross-linking net, and this value determines the space available between macromolecular chains that can be filled with water (Peppas et al. 2000).

4 Imprinted Polymeric Gels for Separation

The properties of the molecularly imprinted polymers such as preparation facility and low-costliness, high selectivity and unrestricted storage conditions make them a very good candidates for separation materials. They can be used for selective isolation or removal of target molecules from a mixture, solution, or complicated sample.

The molecularly imprinted hydrogels are often designed for separation of proteins. Proteins are not good candidates to serve as the templates because of their size, flexible structure, instability in organic solvents, and degradation under polymerization conditions. Moreover, additional problems can occur such as the purification of proteins, the restricted transfer of proteins within the polymer networks or the heterogeneity of recognition sites formed after imprinting of proteins (Lv et al. 2013). These problems could follow from poor reversibility and weak efficiency in adsorption or destruction of imprinting sites during post-polymerization grinding. The fabrication of hydrogel-form imprinted polymers with macroporous structure can help to overcome above limits. Surface imprinting is the common technique used for protein imprinting. In this method, the thin layer of imprinted hydrogel is grafted to the substrate surface (beads or flat surface) or the template protein is attached to the substrate and the polymerization is conducted around this system. The grafting method was employed for the imprinting process of following proteins: bovine serum albumin, human hemoglobin, bovine hemoglobin, and lysozyme. In one excellent series of articles, Hua and co-workers described synthesis of bovine serum albumin imprinted hydrogels using grafting technique (Hua et al. 2008, 2009). The preparation of hydrogels consisted of two stages. Firstly, the modification of silica gel was carried out by the immobilization of initiator on the surface of silica. Next, the preparation of surface imprinted hydrogel was proceeded. The imprinted hydrogel was composed from *N*-[3-(dimethylamino)propyl]methacrylamide as functional monomer and *N*-isopropylacrylamide as backbone monomer and thermosensitive element as well as *N,N*-methylenebisacrylamide, the cross-linker. The polymerization was carried out in a mixture of isopropanol and 2-amino-2-hydroxymethylpropane-1,3-diol hydrochloride buffer solution adjusted to pH 7. The evaluation of structural composition included differential scanning calorimetry which allows to measure the lower critical solution temperature. The results revealed a slightly higher value of lower critical solution temperature for imprinted hydrogel than for control one. It means that during the imprinting process tertiary amine groups were assembled around the cavity created by bovine serum albumin and after template removal these groups were exposed to the solvent resulting with a higher lower critical solution temperature. The adsorption parameters were analyzed for both imprinted and non-imprinted hydrogels. The maximum binding capacities were equal to 27.3 and 22.3 mg/g, respectively. It should be underlined that the values of the maximum binding capacities of imprinted hydrogels for macromolecules were relatively high (achieving micromolar values per gram of material) when compared to the maximum binding capacity of highly cross-linked rigid polymers obtained for low molecular weight templates (commonly achieving nanomolar values per gram of polymer). However, the specificity (expressed as imprinting factor) was generally lower for low cross-linked imprinted hydrogels than rigid imprinted polymer mainly due to the fact that the synthesis and analysis of hydrogels were carried out in aqueous medium. Kinetics study of adsorption and desorption revealed that the surface imprinted hydrogel possessed lower adsorption equilibrium time compared to the bulk imprinted hydrogel which was also synthesized for comparison (one minute to three hours, respectively). It confirmed that the sorption process occurs on the surface of particles quickly. The selectivity tests were provided with the group of various proteins, viz. ovalbumin, hemoglobin, lysozyme, myoglobin, and cytochrome c. The low binding capacities were observed for all tested compounds except ovalbumin and bovine serum albumin. The high adsorption of ovalbumin was observed on both imprinted and non-imprinted polymers. It was a proof that non-specific forces governed the adsorption of ovalbumin. In contrary, high adsorption of bovine serum albumin was observed only on imprinted material. The scanning electron micrographs showed the morphology of particles which were uniform and spherical with average diameter of 5 μm. This size is well accepted for particles used as the stationary phases in chromatographic columns. Therefore, the abilities of surface grafted imprinted particles were evaluated in the chromatographic resolution of bovine serum albumin and ovalbumin as a stationary phase in HPLC. The retention times of both proteins varied significantly confirming satisfactory resolution on imprinted material. Moreover, the changes in ionic strength of mobile phase revealed that the affinity of surface imprinted hydrogel with macromolecular memory of bovine serum albumin could be easily controlled. Nevertheless, the broad and tailing peaks of bovine serum albumin were observed on chromatograms limiting the applicability of analytical method. Another interesting example of surface bovine serum albumin imprinted hydrogel was described by Zhao et al. (2014). In this work, the non-woven polypropylene fiber was used as a matrix, acrylamide as the functional monomer, *N*,*N*-methylenebisacrylamide as the cross-linker, and bovine serum albumin acted as the template. The effects of functional monomer and cross-linker concentrations on the grafting rate, adsorption capacity, and imprinting factor of molecularly imprinted hydrogel were analyzed together with the impact of pH in bovine serum albumin solution on adsorption capacity. The thermogravimetric analysis was provided revealing four pyrolysis stages, and the differential scanning calorimetry was employed to assess the temperature of polymer thermal decomposition. The results of adsorption dynamics and isotherms showed high adsorption capacity of imprinted material. The selectivity tests were carried out with the group of macromolecules, viz. lysozyme, ovalbumin, γ-globulin, and hemoglobin. The results revealed high adsorption of lysozyme and low adsorption of γ-globulin on imprinted hydrogel but the adsorption of lysozyme was non-specific. It was explained by the size of tested compounds. The molecular volume of lysozyme was low enough to penetrate into the specific binding sites in the hydrogel but the functionalities from monomers were not complementary to these in lysozyme. The molecular volume of γ-globulin was larger than bovine serum albumin; thus, the access of γ-globulin molecules to specific binding sites in the imprinted material was restricted. In summary, it was stated that obtained imprinted hydrogel composite possessed high selectivity toward target protein. Moreover, this material could be used several times for sorption process after the subsequent regeneration without loss of its selectivity. The properties of fabricated imprinted material described above made them potentially applicable in the extraction field of required protein from solutions or cells. Ran et al. (2012) prepared thermosensitive bovine serum albumin imprinted hydrogels. The mixture of different monomers: 2-acrylamido-2-methylpropanosulfonic acid, acrylamide and 2-hydroxyethyl methacrylate together with *N*-isopropylacrylamide acting as the backbone monomer were used to synthesize the imprinted hydrogel toward bovine serum albumin in different temperatures, viz. 25 °C or −20 °C. The continuous shrinking with the increase of temperature was observed. The variations of swelling ratio as a function of temperature changes confirmed the intelligent response of hydrogel to the temperature. It was alleged that hydrogel obtained during sub-zero temperature polymerization presented higher adsorption capacity for the template than polymer prepared in room temperature. The phenomenon was explained by the formation of more regularly macroporous structure.

The naturally occurred polysaccharide, chitosan is the example of another material that was used for the preparation of composite imprinted hydrogel for separation of proteins. Fu et al. (2007, 2008) investigated imprinting effect of bovine serum albumin imprinted gel composite of chitosan and acrylamide. The composite material was synthesized by graft copolymerization of both components, and the polymerization conditions such as initiator, monomer, and chitosan concentrations were optimized. The differential scanning calorimetry analysis was provided to determine glass transition temperature of composite hydrogel revealing the information about its thermal stability. In the first part of the studies, it was concluded that the obtain composite hydrogel could be a good material for selective separation of bovine serum albumin but later re-examination studies revealed no imprinting effect. This fact was explained by the inappropriate washing conditions that were applied to remove the template after synthesis. Another type of polysaccharide materials used for preparation of imprinting hydrogel composite are alginates. Li et al. (2015) prepared bovine serum albumin molecularly imprinted polyurethane-grafted calcium alginate hydrogel. The synthesis of composite polyurethane-grafted alginate imprinted hydrogel was carried out in aqueous medium in the presence of 2-hydroxyethyl methacrylate (functional monomer) and bovine serum albumin. The composite microspheres were obtained by dropping the reaction mixture into the aqueous solution of calcium chloride. The non-imprinted composite was also obtained. The impact of different grafting ratios on adsorption capacity and imprinting effect was investigated revealing the optimum percentage of grafted alginate at 20%. The maximum binding capacity was equal to 4.5 and 2.5 mg/g, respectively, for imprinted and non-imprinted hydrogel showing nearly twofold higher adsorption capacity on imprinted hydrogel. This result confirmed that the imprinting effect was achieved. The satisfactory adsorption capacity and high specificity of optimized composite imprinted hydrogel could allow for application in the bioseparation of target protein. Ying et al. (2011, 2013) fabricated different bovine serum albumin imprinted alginate microspheres for protein separation. Bovine serum albumin imprinted calcium alginate microspheres were prepared by the inverse phase suspending gelling method in the presence of hydroxyl ethyl cellulose. This compound was used to modify the swelling property of composite but its impact on the size of particles was minimal. The presented method allowed to obtain uniform microspheres with diameter of ca 160 μm. It was confirmed that better protein adsorption properties were found in hydrogel with greater swelling capacity. In the following study, Ying and co-workers evaluated bovine serum albumin imprinted polyvinyl acetate grafted calcium alginate core–shell hydrogel microspheres. Two different preparation schemes were proposed: the prepolymerization gelatin and the gelating in situ grafting. The influence of monomer concentration for grafting ratio was analyzed prior to the analysis of grafting yield on swelling ratio. This parameter was important because the reduction of swelling ratio affects imprinting effect. The investigation of various external factors on adsorption properties, imprinting effect, and swelling ratio were tested. The impact of external factors such as ionic strength, temperature, and pH was also analyzed. The adsorption dependence on swollen state was discussed and glass transition temperature of composite hydrogel with respect to control alginate materials was determined revealing difference of about 20 °C between both materials. It was summarized that the selective stimuli-responsive bovine serum albumin imprinted hydrogel for potential application to protein separation was prepared.

Another polysaccharide used in the fabrication of composite imprinted hydrogels is agarose. Bovine serum albumin imprinted agarose microspheres were prepared by the inverse suspension gelating method (Ying et al. 2010). Depth analysis of different factors on the synthesis and adsorption properties of imprinted hydrogel was provided. The influence of temperature, surfactant content, and agitation frequency on diameter distribution of imprinted hydrogels was observed. Moreover, the effect of paraffin droplets used during the polymerization on the porous agarose microspheres morphology was analyzed together with the influence of particle diameters and pore size on sorption ability of obtained material. Additional selectivity studies revealed that selective agarose-based material was obtained proofing its utility in the separation of proteins.

The very important protein form is hemoglobin which has to be separated from human blood. Hemoglobin was used as a template to obtain molecularly imprinted polymers. Derazshamshir et al. (2010) prepared imprinted cryogel for depletion of hemoglobin from human blood. The fabrication of imprinted cryogel was preceded by synthesis of *N*-methacryloyl-L-histidine methyl ester, a functional ligand for the interaction with hemoglobin. The complex of ligand and hemoglobin was polymerized in the presence of 2-hydroxyethyl methacrylate and *N*,*N*-methylenebisacrylamide. The free radical polymerization was initiated at 0 °C and then kept in −16 °C to obtain imprinted cryogel. The effect of pH, flow rate, and ionic strength on the hemoglobin adsorption process was investigated. The isotherm adsorption data fitted into Freundlich model revealing quite heterogeneous population of binding sites with heterogeneity index equal to 0.33 (this index is ranged between zero and one where zero or one means respectively that the system is characterized by the heterogeneous or homogeneous population of binding sites). The selectivity studies were carried out with myoglobin and bovine serum albumin. It was concluded that the selectivity was dependent on shape and size of imprinted cavity. To prove the applicability of novel cryogel, the adsorption of hemoglobin from human blood was provided. The main advantage of cryogel was related to its supermacroporous morphology. It allowed to pass through the sorbent very complex samples such as whole blood without any blockade or stoppage. The maximum binding capacity for hemoglobin was equal to 167.4 mg/g cryogel for diluted hemosylate. The percentage of hemoglobin removal from human blood was 95%. Finally, the reusability study revealed that even after several extraction cycles, the sorbent possessed high affinity to hemoglobin. There are also another examples of the synthesis of hemoglobin imprinted hydrogels (Guo et al. 2004, 2005; Xia et al. 2005; Uysal et al. 2008) as well as bovine hemoglobin (Hawkins et al. 2005; Janiak et al. 2010), lysozyme (Adrus and Ulbricht 2012; Tov et al. 2010), myoglobin (Turan et al. 2009), fibrinogen (Cimen et al. 2009), maltose binding protein (Zayats et al. 2014). As a result of these works, the imprinted materials for selective separation of target proteins were fabricated. In an interesting study described by El-Sharif and co-workers, the impact of buffer solution composition and pH during preparation, template removal, conditioning, and adsorption was investigated for three hemoglobin imprinted hydrogel build up from different monomers, viz. acrylamide, *N*-isopropylacrylamide, and *N*-hydroxymethyl acrylamide (El-Sharif et al. 2014). It was demonstrated that use of certain buffers during conditioning caused the loss of specificity during the adsorption from water. It was found also that too strong acidicity of such buffers had negative effect on the sorption efficacy on hemoglobin imprinted hydrogel. However, the optimized solvent conditions allowed to reduce significantly the impact of interference from biological samples. Nematollahzadeh et al. (2014) fabricated interpenetrating polymer network modified hierarchically protein imprinted hydrogel for the separation of human serum albumin and immunoglobulin G. The preparation of composite material was carried out in the following sequence: first, the template protein was used to modify the porous solid silica support inside the pores, then the pores were filled with the polymerization mixture of acrylamide monomer and *N*,*N*-methylenebisacrylamide cross-linker, next the penetration of the secondary cross-linker (ethylene glycol dimethacrylate) into the reticulated polymer matrix was carried out. Finally, the solid silica support and the template were removed providing highly porous interpenetrating imprinted hydrogel system. The application of interpenetrating polymer network allowed to produce the composite imprinted hydrogel with satisfactory mechanical strength. It was important advantage because previous hydrogel-based stationary phases for chromatographic resolution evoked significant elevation of column pressure due to low mechanical resistance or compressibility. After the structural and morphological characterization, the composite material was successfully used as a stationary phase in solid-phase extraction as well as in columns for high-performance liquid chromatography analyses. The very bright work was described by Chen et al. (2008). Here, the stimuli-responsive protein imprinted polymers were obtained via combination of molecular imprinting and reversible stimuli-responsive polymer using lysozyme or cytochrome c as templates, *N*-isopropylacrylamide as backbone monomer, methacrylic acid and acrylamide as functional monomers, and *N*,*N*-methylenebisacrylamide as cross-linker. The obtained imprinted hydrogel could respond not only to external stimuli such as temperature but also to the corresponding template protein with significant specific volume shrinking. This specific shrinking behavior of imprinted material was attributed to the synergistic effect of multiple-site weak interactions in the macromolecular memory site such as electrostatic force, hydrogen bonding, and hydrophobic interaction as well as the size and shape of this site. In conclusion, it was stated that the imprinted hydrogel possessed ability to change the shape in order to accommodate the conformation of the template protein leading to the formation of the shape complementary cavity or complementary macromolecular memory site. In this sense, the adsorption of the template protein on imprinted hydrogel mimicked the natural processes where sequential steps of adaptation of mutual conformation occur during recognition.

Another group of templates used during preparation of molecularly imprinted hydrogels are glycoproteins. Here in one example, carcinoembryonic antigen was used as a template for preparation of imprinted poly(allylamine hydrochloride) hydrogel (Casey and Kofinas 2008). This imprinted hydrogel possessed the ability for the selective binding of sugar residues. The cationic polymer had high affinity to carcinoembryonic antigen and could selectively adsorb template molecule what would be employed in the separation of glycoprotein. Miyata et al. (2012) applied lectins and antibodies as ligands in the biomolecular imprinting of human α-fetoprotein, a target tumor marker glycoprotein. In the synthetic protocol, acrylamide was used as the monomer and the impact of various cross-linkers: low-(*N*,*N*-methylenebisacrylamide) and high molecular weight (polyethylene glycol acrylate or polyethylene glycol dimethacrylate) was analyzed. It was concluded that imprinted hydrogel fabricated from high molecular weight cross-linker shrank more drastically as the response to target tumor marker than counterpart synthesized from low molecular weight cross-linker. The compressive modulus of obtained hydrogels was measured to find the relationship between the compressive stress and strain of obtained hydrogel. As a result, the selective hydrogels from high molecular weight cross-linker which yielded an optimal molecular weight were fabricated and evaluated as the separation material. The literature survey reveals also the application of peptides as the templates in the molecular imprinting process. In one interesting paper, Abbate and co-workers prepared hepcidin (peptide hormone) selective imprinted hydrogels using so-called epitope approach (Abbate et al. 2010). The epitope is a short amino acid sequence complementary to binding site of antibody. In the field of molecular imprinting, the so-called epitope approach was developed for imprinting of proteins and it used a short peptide sequence as the template instead of the whole protein. Once the polymerization was finished, the resultant imprinted material should be able to recognize and bind the whole protein. Here only the amino-terminus of hepcidin (six amino acids) was used as the template together with acrylamide, *N*-isopropylacrylamide, acrylic acid, and 2-(dimetylamino)ethyl methacrylate as the functional monomers. The water/oil suspension polymerization technique was employed for synthesis of imprinted hydrogel. The imprinted material possessed high affinity to *N*-amino-terminus of hepticidin but the adsorption of whole peptide was similar on both imprinted and non-imprinted hydrogel revealing practically no imprinting effect. Carnosine was used as another peptide template to produce imprinted material composite with chitosan (Singh et al. 2014). The production of composite imprinted hydrogel consisted of the synthesis of a zwitterionic form of chitosan, polybetaine (during the reaction of chitosan amine group with dihydrofuran-2(3H)-one) prior to gelation of betainized chitosan in the presence of the template and *N*,*N*-methylenebisacrylamide. The zwitterionic form of chitosan increased the stability of prepolymerization complex with carnosine resulting with highly specific imprinted hydrogel (imprinting factor was equal to 3.64). The swelling and thermogravimetric (for water content determination) measurements were provided revealing that hydrogel is thermosensitive, pH-sensitive and responsive to ionic strength. The experimental results demonstrated that novel material could be used for selective isolation of carnosine.

The very interesting and challenging example of imprinting process was described aiming to obtain lipopolysaccharide imprinted materials. Lipopolysaccharides similarly to proteins are high molecular weight compounds and possess amphiphilic nature. Ogawa et al. (2012) prepared lipid A—imprinted polymer hydrogels that selectively recognized lipopolysaccharides. Lipid A was derived from *Escherichia coli*. The new functional monomers, viz. acryloyllysine and acryloylphenylalanaline, were obtained in order to effectively interact with lipopolysaccharide. The polymerization was carried out with the presence of acrylamide and *N,N*-methylenebisacrylamide in 4-(2-hydroxyethyl)-1-piperazineethanesulfonic acid buffer with addition of *n*-octane to promote the formation of lipopolysaccharide accessible sites on the surface of hydrogel. The selectivity analysis was performed for various lipopolysaccharides derived from *Helicobacter pylori* and *Porphyromonas gingivalis*. The results suggested certain degree of selectivity mainly due to the strong interactions of monomers with two phosphate groups of lipid A but non-specific adsorption was also high.

Molecularly imprinted gels may be used as matrices for electrophoresis. Ogiso et al. (2006) prepared DNA imprinted polyacrylamide gel employing 2-vinyl-4,4-diamino-1,3,5-triazine as a functional monomer. During imprinted gel electrophoresis, the migration of target DNA was hindered by catching effect of binding sites. The detection of target sequence of DNA from DNA mixture was achieved and selectivity was confirmed by analyzing the base-pair-substituted DNA.

Another interesting example of gel-based imprinting material is ATP binding peptide hydrogel (Takata et al. 2015). In this study, vinyl-end-capped on beads peptide was used as a functional monomer to produce hydrogel for selective ATP recognition. The affinity and selectivity of the hydrogel toward ATP were analyzed and confirming satisfactory properties of obtained hydrogel.

The advanced research examples were dedicated to the imprinting of viruses. The virus imprinted hydrogel was obtained from poly(allylamine hydrochloride) in the presence of *Tobacco mosaic virus* acting as the template cross-linked by epichlorohydrin. The choice of poly(allylamine hydrochloride) was dictated by the high amount of amine groups that could interact with viral surface. The particular attention was paid for the appropriate concentration of template because the excess of viruses resulted with their spontaneous aggregation. The resulted imprinted hydrogel was characterized by the specific adsorption of *Tobacco mosaic virus* with nearly twofold higher binding capacity than on non-imprinted hydrogel. The selectivity test was also carried out toward *Tobacco necrosis virus* revealing lower affinity to the imprinted hydrogel (Bolisay et al. 2006; Bolisay and Kofinas 2010).

The another example of the employment of poly(allylamine hydrochloride) as the matrix for fabrication of imprinted hydrogel was demonstrated by Divya et al. (2015) who produced molecularly imprinted hydrogel for the separation of lactate from milk samples. During the analysis, it was found that the hydrogel did not distinguish between L- and D-lactic acid enantiomers revealing similar binding capacity. There is also an example of the preparation of poly(allylamine hydrochloride)-based imprinted hydrogels toward glucose (Wizeman and Kofinas 2001). Glucose is a very important molecule because of their role in diet of type 2 diabetes. Molecularly imprinted hydrogels can serve to adsorb glucose in the stomach and small intestine and would be simply ingested with a high glucose content food. Wizeman and Kofianas prepared glucose phosphate mono-sodium imprinted polymer with good glucose rebinding properties but the re-examination of their work revealed no imprinting effect observed during further studies impugning the imprinting process in poly(allylamine hydrochloride) and epichlorohydrin system (Fazal and Hansen 2007).

There are also reports related to fabrication of imprinted hydrogels dedicated for separation of small molecules of biomedical interest as well as various ions. Thermoresponsive hydrogels based on *N*-isopropylacrylamide and methacrylic acid or itaconic acid were prepared for selective separation of dopamine (Korytkowska-Wałach 2013). Stoichiometry in the prepolymerization complexes of dopamine with monomers was determined by ^{1}H NMR method to ensure the optimal arrangement of binding sites in the final product.

The efficient removal of heavy ions from water and environmental samples is interesting from the academic and important from industrial point of view. Additionally, the level of ions of great biological importance like zinc(II) or potassium needs to be controlled in biological fluids or water samples. Different techniques were developed for the isolation of ions but searching of new, easier, faster, and better is still continued and some excellent papers are described below.

The interpenetrated polymer network was fabricated by free radical/cationic hybrid polymerization for copper(II) ion-imprinted hydrogel. The composition of hydrogel consisted of acrylamide and triethylene glycol vinyl ether as monomers (Wang et al. 2014a). The adsorption capacity of imprinted material depended on pH values and temperature; however, the imprinted hydrogel was thermally stable due to the interpenetrating polymer structure. In further work, Wang and Li obtained polyethylenimine-functionalized copper(II) ion-imprinted hydrogel by cobalt isotope-γ-induced copolymerization (Wang and Li 2015). It was concluded that this hydrogel is a promising sorbent for the selective removal of copper(II) ions from wastewater. In another study, the copper(II) ion-imprinted polymer was prepared as a Congo Red-impregnated chitosan hydrogel (Liu et al. 2012). The analysis of binding properties and the influence of different factors on adsorption process revealed that the obtained adsorbent possessed better adsorption performance than the adsorbent prepared by the traditional method. The interpenetrating polymer network was also introduced for the preparation of uranyl-imprinted hydrogel composite of chitosan and poly(vinyl alcohol). The material was successfully prepared as a selective sorbent for removing uranyl ions from aqueous solutions (Liu et al. 2010). The similar method involving the interpenetrated polymer network was employed by Wang and co-workers to fabricate thermosensitive zinc(II)-imprinted hydrogel. This structure based on *N*,*N*-dimethylaminoethyl methacrylate and bis-(3,4-epoxycyclohexyl)adipate as functional monomers (Wang et al. 2014b). Chitosan tripolyphosphate gel beads were tested as bioadsorbents for iron ions from aqueous solution and from human blood samples (Yalinca et al. 2012). The removal of iron ions as well as hemoglobin from blood samples was observed after application of chitosan gel as bioadsorbent. Wu and co-workers prepared imprinted hydrogel with recognition sites compatible with potassium ions (Wu et al. 2011). Before the polymerization, "host-guest" complexes of potassium ions with benzo-15-crown-5-acrylamide acting as the monomer were produced. Then, the synthesis of hydrogel with *N*-isopropylacrylamide as the backbone monomer was carried out. The fabricated material was characterized by faster response rate to potassium ions than reference hydrogel prepared in the same conditions but without prepolymerization complex involving crown ether complex.

5 Imprinted Polymeric Gel-Based Detectors

The quick detection of biological signals or dangerous chemicals relies on the application of expensive specific antibodies. Molecularly imprinted hydrogels in contrast to natural antibodies are marked by stability, mechanical and thermal resistance as well as relatively low manufacturing cost. Thus, the properties of molecularly imprinted hydrogels allow to use them commonly as the recognition element or receptor part of biosensors. Several implementations of molecularly imprinted hydrogels for detection of different analytes relevant to medical diagnostics or environmental monitoring were reported. Apart from the advantages of imprinted hydrogels used as sensors, the response is sometimes limited to binding of specified molecule. This fact leads to further responses using transduction ways. In the biosensor area, usually the binding event is converted into a detectable signal. The two of the main transducer techniques are optical and mass-sensitive detectors.

Optical techniques are commonly used in the fabrication of biosensors. Many different kinds of optical detectors can be chosen to produce appropriate sensor such as fluorescence, absorption, reflectometric interference, Raman, surface plasmon resonance spectroscopies. Hydrogel-based imprinted materials adsorb target analyte inducing shrinking or swelling of the hydrogel network. This transition is associated with the refractive index or other changes of optical properties and can be detected.

One of the simplest techniques of optical detection is colorimetry and spectrophotometry. In these methods, light absorbance, transmittance, or reflectance of analyzed material is measured. Here, molecularly imprinted photonic hydrogels used as label-free chemosensors in described techniques can be an example. Photonic crystals are periodically arranged materials built with regularly shaped blocks. Refractive index can be modulated by photonic crystal on a length scale which is the light wavelength of interest. It is governed by the specific distance between blocks (layers or spheres) and appeared as a specific color. When the external stimulus occurs, the wavelength of reflectance and periodicity of the crystal will change providing visible as a color changes that could be easily observed by the naked eye. Those properties make photonic crystals very promising candidates for application as signal transducers for optical detection accurate even for unskilled operators (Fenzl et al. 2014). Photonic crystals based on hydrogels are interesting in the sensing area because they can respond to a broad spectrum of different stimuli and they can change their volume by swelling or shrinking after analyte binding what enhances the optical readout. However, it is necessary to know that hydrogel photonic crystals often suffer from slow response times. This limitation arises from thick hydrogel structure leading to hinder diffusion of the analyte molecule (Shin et al. 2010; Wang et al. 2007). The combination of molecular imprinting and photonic crystals techniques gives a highly ordered, interconnected, and porous structure with thin wall composed from hydrogel. The molecular imprinting provides nanocavities which make the hydrogel specific to analytes. The recognition events enhance the optical signal through changing the hydrogel optical properties (Wu et al. 2008a). Li group fabricated molecularly imprinted photonic polymer hydrogels for detection of various molecules (Hu et al. 2006, 2007, 2008; Wu et al. 2008b; Peng et al. 2012). Three-step procedure which was used for the fabrication of material consisted of the preparation of silica colloidal crystals, the polymerization of template-functional monomer complex in the interspacers of the colloidal crystal, and the removal of the templates (colloid particle and the template molecule). Methacrylic acid or acrylic acid were used as the monomers and ethylene glycol dimethacrylate as the cross-linker in optimized synthesis to produce imprinted photonic hydrogel-based sensor for specific recognition of vanillin, theophylline, 1(*R*),2(*S*)-ephedrine, L-DOPA, bovine serum albumin, egg albumin or lysozyme. One of interesting finding in above-cited reports was related with appropriate choice of functional monomer and solvent and the proper molar ratio between the monomer and cross-linker. High molar ratio of cross-linker led to less diffraction peak shift. This is the result of lower capacity of hydrogel and weaker hydrogel response to template connected with reduced swelling and shrinking capability. In contrast, the insufficient amount of cross-linker caused problems with the construction of integrated and ordered macroporous array and the creation of selective and specific three-dimensional cavities providing to disappearance of Bragg diffraction peak.

Another optical analytical technique which was reported as imprinted hydrogel sensor is surface plasmon resonance spectroscopy. This method allows to determine the molecule concentration based on a measuring of change in the refractive index of medium in the close contiguity of metal surface. The glass plate covered by thin metal layer is used during the measurement. This plate contacts with a solution containing molecules which absorb on the metal surface. Surface plasmon resonance occurs when a polarized light hits the metal layer, and under certain condition, free electrons of the metal adsorb light photons and convert them into surface plasmon waves. The disturbance of metal surface changes the resonance conditions what is seen as a different reflectivity and can be measured. There are interesting papers describing fabrication of molecularly imprinted hydrogels for surface plasmon resonance detection of vitamin B_3 or macromolecular compounds (Wang et al. 2011; Verma and Gupta 2013; Patching 2014; Naraprawatphong et al. 2016). Verma and co-workers constructed surface plasmon resonance sensor for vitamin B_3 detection. The sensor consisted of silver-coated optical fiber probe covered by molecularly imprinted photonic hydrogel. Control sensors were carried out to confirm the role of each component: probe without colloidal crystal templating and without molecular imprinting, probe without colloidal crystal templating but with template imprinted polymerization, and probe with colloidal crystal templating but without template during polymerization process. After surface plasmon resonance analysis of different template concentrated solutions, it was found that colloidal crystal-templated-vitamin B_3 imprinted optical fiber surface plasmon resonance probe gave the best results for the template concentration range of interest (Verma and Gupta 2013). Another example of the application of imprinted hydrogel in the surface plasmon resonance spectroscopy is the sensor for glucose detection (Wang et al. 2011). Here, in the first step of fabrication, the optimization process of relative amounts of compounds used during the preparation was carried out in order to obtain the highest sensitivity of sensor. Gold nanoparticles were added pending the synthetic process to enhance the surface plasmon resonance sensitivity due to the impact on the refractive index. Selectivity and interferences studies were provided revealing good sensor selectivity toward glucose. Nevertheless, high ionic strength of analyzed sample was responsible for decreasing sensitivity. During development of protein recognition surface plasmon resonance devices, three different kinds of products were constructed: IgG-imprinted and non-imprinted hydrogel-based sensor, imprinted and non-imprinted polymer brush-based sensor as well as lectins and anti-IgG-immobilized sensor (Naraprawatphong et al. 2016). The last chip was fabricated to reveal the effect of polymer brush and gel layers on the surface plasmon resonance response for target biomolecules. Hydrogel and polymer brush were produced respectively with and without the cross-linker. The characterization of materials revealed that imprinted and non-imprinted polymer brush layer chip possessed comparable affinity, whereas the imprinting effect was clearly showed in hydrogel-based sensors confirming the important role of cross-linker in the creation of recognition cavities in the polymer matrix.

Reddy et al. (2013) analyzed protein detection using imprinted hydrogels integrated with dual polarization interferometry. Dual polarization interferometry is an analytical technique that can explore the molecular layers adsorbed to the surface of a waveguide occurring within an evanescent field made by a laser beam confined to the waveguide. It is typically used to measure the refractive index, the thickness of thin layers, and the conformational change in biomolecules. The experimental results showed that only imprinted hydrogel bound to the preadsorbed protein layer significantly in contrast to non-imprinted polymer and was selective against the bovine hemoglobin which acted as template.

The second class of detectors is mass-sensitive devices. The most popular mass-sensitive sensor is quartz crystal microbalance. Quartz crystal microbalance device principle consists in sending an electrical signal through a gold-plated quartz crystal. That fact causes a vibration at some resonant frequency. Detection mechanism of mass system devices depends on analyte binding to the surface of measuring tool and results in a resonance frequency decline what enables to quantify the analyte incorporation. While the imprinted polymer with quartz crystal microbalance is used, the imprinted polymer films are deposited onto the capped quartz crystal microbalance crystals (Schirhagl 2014). El-Sharif et al. (2015a) and Reddy et al. (2012) prepared quartz crystal microbalance sensors to differentiate the imprinted and control hydrogel behavior in the present cognate and non-cognate protein. Three different functional monomers, viz. acrylamide, *N*-hydroxymethylacrylamide, and *N*-isopropylacrylamide, were used in synthetic protocol.

It was observed that quartz crystal microbalance sensor with *N*-hydroxymethylacrylamide-based imprinted hydrogel was characterized by the best response for the template protein of bovine hemoglobin. Authors constructed quartz crystal microbalance device which designated sensor surface activity and supplied the physical interpretation in terms of hydrophilicity.

The final examples of molecularly imprinted hydrogel-based biosensors are ultrasonic detectors. Imprinted hydrogel microspheres have a specific resonance frequency after perturbation by the ultrasound pressure waves. The frequency depends on size and stiffness of material. When imprinted hydrogel adsorbs the molecule, the physical properties changed resulting in the modification of resonance frequency. Troiani et al. (2011) analyzed two theophylline imprinted hydrogels constructed on the basis of two different materials: hydroxypropyl cellulose and *N*-isopropylacrylamide as potential ultrasonic sensors. The characterization of hydrogels revealed greater selectivity of *N*-isopropylacrylamide-based polymer and quantification of the interfering molecules by observing different ultrasonic response frequencies of the sensor.

There is also another example of specific imprinted material which has a potential for application in electrochemical measurements. In one report, El-Sharif and co-workers fabricated material as a polyacrylamide metal-coded imprinted hydrogel for bovine serum albumin. Hemin, vinylferrocene, protoporphyrin and zinc(II) protoporphyrin along with cobalt(II) complex were introduced as a co-monomers for metal coding of template selectively recognized imprinted cavities. It was concluded that metal-coded imprinted hydrogel possessed higher bovine serum albumin adsorption and selective capacities than pure polyacrylamide imprinted material (El-Sharif et al. 2015b).

6 Drug Delivery Imprinted Polymeric Gel Carriers

Apart from the use of imprinted polymers for separation and detection purposes, the new promising area of application has been explored. Molecularly imprinted polymers could serve as the drug delivery devices because they can provide high selectivity toward the particular drug. Moreover, they show also high stereoselectivity which is extremely important in drug delivery forms. A majority of the drugs has chiral atoms and their pharmacological activity is demonstrated only by one isomer or only one isomer shows a significantly higher activity. The ability to maximize the delivery of given eutomer (the isomer of interest) and reduce or even eliminate the delivery of the distomer (the undesirable isomer) is the enormous advantage of imprinted polymers. Molecularly imprinted polymers as the drug delivery devices have yet not found any commercial applications because there are still a lot of challenges that should be overcome. Some of the problems are pointed out in numerous books and reviews (Alvarez-Lorenzo and Concheiro 2004; Alvarez-Lorenzo et al. 2013; Luliński 2013). However, the interesting investigations are in progress to apply molecularly imprinted hydrogels as the drug dosage forms. The most advanced experiments and recent results are discussed below behind the short characterization of imprinted hydrogels.

The release of drug from pharmaceutical form is performed in different routes (Byrne et al. 2002; Hilt and Byrne 2004; Serra et al. 2006; Hoare and Kohane 2008). In the imprinted hydrogels, the numerous complementary or sterically oriented functionalities interact with the drug. The interactions in the macromolecular memory sites delay the release of the drug from the hydrogel despite the swelling degree of the polymer. The strength of those interactions can be responsible for delayed release as well as the so-called tumbling effect related to the migration of the drug from one macromolecular memory site to another. The functionally diverse hydrogel system is able to multiple contact points with the drug or target molecule to create efficient complexation due to the reduction of the entropically unfavorable translational and rotational free energies. In the analyzed timolol imprinted poly(*N*-vinylpirrolidone-*co*-acrylamide-*co*-acrylic acid-*co*-2-hydroxyethylmethacrylate-*co*-polyethylene glycol dimethacrylate) hydrogel that consisted of three various functional monomers (Venkatesh et al. 2008; Yanez et al. 2011a), it was proposed that the imprinting effect was responsible for delay transport and sustained release of drug due to the random interactions of drug through the network imprinted by it. The drug itinerated from one macromolecular memory site to another high-affinity region where the complexation took place and then escaped from it and itinerated to another one and so on. This behavior increased the mean residence time of the drug in the hydrogel net.

In another mechanism, the competitive binding of structurally related compound present in the intracorporeal environment will promote the release of the drug. This mechanism was investigated with hydrocortisone imprinted polymer. This imprinted system showed the ability to adsorb a considerable amount of testosterone due to the structural similarity of both compounds. The release of testosterone in water was very slow and even after 24 h only 45% of testosterone was released. However, in the presence of hydrocortisone (which was also the template molecule during synthesis of polymer), the competitive binding promoted a rapid release of testosterone from the binding sites completing the total release of testosterone within 4 h (Sreenivasan 1999). Those imprinting systems are extremely interesting as the drug carriers, because they can respond or release the drugs when the concentration of important bioanalyte or biomarker increases.

There are also mechanisms that involve a hydrolytically induced releasing of drug or release in the physical stimuli-responsive manner. The first approach requires the use of the erodible imprinting system from which the drug cannot be released unless the whole (or a part of) the polymer degrades. The last one describes the drug release after the response to shift of the physicochemical properties of external fluids (Li et al. 2010).

The imprinted hydrogels were most intensively evaluated as the drug carriers for ophthalmic purposes. This is because the available ocular pharmaceutical formulations such as eye drops are characterized by low effectiveness of ophthalmic drug delivery. The protective mechanisms of lacrimal apparatus and fast removal of drugs from anterior segments of eye by nasolacrimal drainage are responsible for low bioavailability of ophthalmic drugs. The application of ophthalmic ointments and gels slightly improves the therapeutic effects. The bioavailability of ophthalmic drugs significantly increased when the contact lenses ophthalmic drug delivery systems were introduced into the treatment. However, there are limitations arisen from a low loading capacity of such devices (Kirchhof et al. 2015).

Molecularly imprinted hydrogels were extensively investigated as the ocular drug delivery forms with a hope to overcome existing problems. Alvarez-Lorenzo et al. (2002) prepared the imprinted soft contact lenses able to delivery of timolol in sustained mode. Timolol is a non-selective beta adrenergic receptor antagonist that is widely applies in the treatment of glaucoma. The molecule of timolol possesses chiral carbon atom but only *S*-enantiomer possesses high pharmacological activity. Therefore, the design of drug delivery form to deliver necessary eutomer is important scientific challenge. The authors evaluated the influence of composition of the matrix and the loading capacity of hydroxyethyl methacrylate hydrogels weakly cross-linked with ethylene glycol dimethacrylate and with the presence or the absence of methacrylic acid and methyl methacrylate. The dry hydrogels were clear and smooth with their poreless surface and presented appropriate properties required for the contact lenses. The diffusion of *S*-timolol into the physiological saline from the both, hydroxyethyl methacrylate and combined hydroxyethyl methacrylate-methacrylic acid hydrogels, was slow. The latter hydrogel presented the highest loading capacity equal to 12 μg of *S*-timolol in 1 mg of dry hydrogel. In the following study, the effects of different monomers on the binding capacity as well as the impact of the stoichiometry of reagents and polymerization conditions were investigated (Hiratani and Alvarez-Lorenzo 2002, 2004; Hiratani et al. 2005a; Yanez et al. 2011a). The prepolymerization systems including methacrylic acid (functional monomer), *N*,*N*-diethylacrylamide (backbone monomer as well as solvent), ethylene glycol dimethacrylate (cross-linker, various concentrations used in the synthesis), and timolol (template) were investigated. The polymerization was proceeded in-mold forming, finally contact lenses with diameter of 16 mm and thickness of 0.3 mm. The adsorption of the target drug (*S*-timolol) was carried out after the template removal step. The soaking procedure was used for loading until the equilibrium state was achieved. The binding capacities of imprinted hydrogels were satisfied and more than two times higher than non-imprinted ones. The dissociation constants were between 5.0 and 8.9 μmol/L (depending on the concentration of cross-linker) for imprinted hydrogels and between 6.3 and 29.4 μmol/L for non-imprinted hydrogels.

The mechanical properties of hydrogel are very important considering the application of imprinted material as drug carrier. The mechanical properties are mainly related to the viscoelastic behavior of hydrogels which play significant role when hydrogels are considered to the pharmaceutical formulations. Here, the standard mechanical tests can be carried out and the parameters such as the elastic modulus, viscous modulus, and loss modulus should be determined (Koetting et al. 2015). The set of rheological analyses was performed on timolol imprinted poly (methacrylic acid-*co*-2-hydroxyethyl methacrylate) and poly(methyl methacrylate-*co*-2-hydroxyethyl methacrylate) hydrogels. The dependence between the elastic and viscous moduli and the angular frequency showed similar pattern for imprinted and non-imprinted hydrogels. It was concluded that timolol did not modify the mechanical properties. The plot of dissipation factor (tan δ) against temperature for dry hydrogels at fixed angular frequency revealed peaks that indicate the glass transition temperature of imprinted hydrogel. It was also stated that the rheological analysis can indicate modification of the inter- and intrachain interactions due to the presence of the template molecule. The presence of timolol in the prepolymerization mixture affected the relative positions of hydroxy or ester groups in the functional monomers confirming the imprinting process (Alvarez-Lorenzo et al. 2002).

The determination of the release behavior of above imprinted hydrogel was carried out in 0.9% NaCl aqueous medium imitating the physiological conditions of the ocular surface. The profiles of *S*-timolol release were analyzed. The Higuchi model was employed for analysis indicating Fickian diffusion. Moreover, the presence of physiological saline strongly affected the diffusion of drug retarding the release process. However, the composition of imprinted hydrogels has to ensure appropriate permeability of oxygen and carbon dioxide recommended for safe day or month-wearing soft contact lenses. Therefore, the nature of monomers on the profiles of *S*-timolol release was investigated and 3-[tris(trimethylsiloxy)silyl]propyl methacrylate was selected as the functional monomer for analysis. This monomer provides hydrogel with very good permeability of oxygen and is used as the component of month-wearing soft contact lenses. The imprinted hydrogel matrix was build up from 3-(tris(trimethylsiloxy)silyl)propyl methacrylate as well as methacrylic acid (both used as functional monomers), *N*,*N*-diethylacrylamide (backbone monomer), and ethylene glycol dimethacrylate (cross-linker). The diffusion coefficient obtained for imprinted hydrogel confirmed that *S*-timolol was released in sustained mode. Thus, the modulation of the composition of hydrogel was possible to adapt the drug release behavior for imprinted soft contact lenses of better oxygen permeability. Finally, the optimization of dissolution of timolol in a mixture of monomers was performed. This step allowed to exclude organic solvents and to apply as minimum of cross-linking agent as necessary to secure macromolecular memory sites in the hydrogel network. The detailed analysis of the imprinted hydrogels was performed and in vivo experiments of releasing *S*-timolol from the imprinted contact lenses instilled in cornea were carried out (Hiratani et al. 2005b). *S*-timolol was detected in the tear fluid for a period of 180 min, which was twofold longer than that from the non-imprinted hydrogel. The comparative analysis of *S*-timolol released from the eye drops containing 0.068 and 0.25% of *S*-timolol, showed the presence of the drug in the tear fluid for only 60 min. It was also found that the ocular bioavailability of *S*-timolol from the eye drops only slightly depended on the dose, but the corresponding bioavailability from the presoaked imprinted soft contact lenses increased significantly (for 34 μg of timolol the AUC—area under curve values were 1.24 and 10.76 mmol min/L, respectively). However, the capacity of the imprinted soft contact lenses to in vivo release was lower probably due to the effect of the release medium and the continuous renovation of the lacrimal fluid as well as the blinking. In order to improve the loading capacity, an innovative method using a supercritical fluid technology was employed (Braga et al. 2010; Yanez et al. 2011b). The selected commercial soft contact lenses (Hilafilcon B®) were used. They were impregnated using the supercritical carbon dioxide technology. The sequential impregnation-imprinting procedure and extraction steps were performed with a non-steroid anti-inflammatory drug *R*/*S*-flurbiprofen as the template. The rearrangement of polymeric regions of soft contact lenses in the presence of the template as well as the effect of supercritical carbon dioxide on the plasticization of hydrogel was responsible for imprinting effect. It was observed that each consecutive impregnation step resulted with the increase of binding capacity of the imprinted soft contact lenses. The comprehensive analysis of the structure and binding experiments with some structurally related compounds revealed the recognition ability and a higher affinity for *R*/*S*-flurbiprofen in aqueous solutions. The release profiles were investigated. The release of *R*/*S*-flurbiprofen was sustained for six hours in aqueous medium. The study of *S*-ibuprofen release from soft contact lenses prepared using supercritical fluid technology was also provided (da Silva et al. 2011). The imprinted hydrogel was build up only from 2-(dimethylamino)ethyl methacrylate (functional as well as backbone monomer) and ethylene glycol dimethacrylate (cross-linker). The release behavior showed strong dependence on pH of releasing medium. In the neutral pH, the anomalous transport was observed and release of *S*-ibuprofen after 24 h was equal to 181 mg/g for imprinted hydrogel and only 65 mg/g for non-imprinted one. In the acidic pH, the release was close to kinetic of zero order according to Korsmeyer-Peppas model which was employed to analysis. The lower diffusion coefficient of *S*-ibuprofen from imprinted hydrogel in acidic pH (0.0357 h^{-1} vs. 0.1337 h^{-1} in neutral pH) was explained by the lower solubility of *S*-ibuprofen in acidic medium and enhancement of hydrophobic interactions between the drug and hydrogel. It was concluded that different interactions (hydrogen bonds and hydrophobic forces) are involved in the stabilization of imprinted hydrogel and *S*-ibuprofen complex. Nevertheless, it was found that the procedure was not effective for monthly wearing soft contact lenses probably due to different composition of hydrogel. The imprinted soft contact lenses were also investigated for various anti-inflammatory drugs as well as antibiotics carriers (Alvarez-Lorenzo et al. 2006; Maryam et al. 2007; Malaekeh-Nikouei et al. 2012; Tieppo et al. 2012; Malakooti et al. 2015).

The investigations of imprinted hydrogels as the drug delivery forms for other administrating routes are rather scarce and limited to selected important pharmacological agents. There is also insufficiency in the analysis of release profiles in vivo which indisposes further development or commercialization of imprinted hydrogels. However, there are a few interesting examples that can be found in literature survey describing imprinted hydrogel systems dedicated to delivery of selected anticancer drugs, antibiotics, or insulin. 5-Fluorouracil is an anticancer drug that is used in treatment of breast, colorectal, liver, or brain carcinoma. This compound undergoes fast transformation due to the high rate of metabolism in the human body. However, the high concentration of this agent in serum is necessary to achieve sufficient therapeutic effect. Such concentration has to be maintained for a prolonged time but cannot exceed a certain value to avoid severe side effects. Hence, the imprinted hydrogel was design for sustained release of 5-fluorouracil (Singh and Chauhan 2008; Cirillo et al. 2009). The hydrogel was build up from 2-hydroxyethyl methacrylate, acrylic acid, and *N,N*-methylenebisacrylamide. Different concentrations of 5-fluorouracil (template) were applied to observe the effect of the number of recognition sites in the imprinted net on entrapment of drug and release profile. As it was expected, the higher binding capacity was noted for the hydrogel prepared in the presence of higher concentration of 5-fluorouracil. However, in vitro release profiles revealed a non-Fickian diffusion of the drug from hydrogel with n values below 0.5 for both imprinted hydrogels. It was supposed that the release of drug was also controlled by the degradation of the covalent linkage of the hydrogel network. In the further study, uniform, spherical hydrophilic nanoparticles (diameter ca 250 nm) were obtained for delivery of 5-fluorouracil. The hydrogel was a copolymer of methacrylic acid and ethylene glycol dimethacrylate with nearly equimolar amount of each reagent. The in vitro release experiments revealed that non-imprinted hydrogel completely desorbed the drug within 5 h but in case of imprinted hydrogel the release was continued for more than 50 h. In another interesting work, the metal coordinate bond strategy was employed to prepare imprinted hydrogel with pH-responsive property and controlled release of the antibiotic, doxorubicin. The hydrogel was built up from 4-vinylpyridine (functional monomer), 2-hydroxyethyl methacrylate (backbone monomer), and *N,N*-methylenebisacrylamide (cross-linker). Additionally, the cupric ions were used to serve as the bridge between the functional monomer and the template. The strong coordinate interaction enhanced the stability of the prepolymerization complex of 4-vinylpyridine and doxorubicin. The strong pH-dependence of the release of doxorubicin from hydrogel was observed. At pH 6 or 7.2, approximately 10% of the loaded drug was released to medium, but at pH 5, the release was accelerated and nearly 60% of the loaded drug was released. The pH value affected the coordinate interactions between the cupric ions and doxorubicin which were steady in the basic or neutral range. In lower pH values, the protonation of 4-vinylpyridine residues proceeds resulting with cleavage of the coordinate bonds and release of doxorubicin. The in vitro release analysis revealed that the drug was released from imprinted hydrogel for seven days and only for one day from non-imprinted counterpart. Finally, low cross-linker and also pH-responsive insulin delivery system was proposed (Li et al. 2010) based on methacrylic acid and poly(ethylene glycol dimethacrylate) as functional and backbone monomers as well as *N,N*-methylenebisacrylamide as the cross-linker. The imprinted hydrogel showed the sustained release of insulin only in acidic medium.

7 Current Limits and Future Prospects

The unique properties of polymeric gels to swell and to respond for stimuli from external environment are the main factor for extensive interest of this class of materials. In other hand, the exceptional selectivity of polymers provided by the imprinting technology attracted attention of many scientific groups to produce novel, advanced materials. The merge of both advantages resulted in the fabrication of selective and stimuli-responsive imprinted hydrogels with potential application as the sorbents, biosensors, or drug carriers. The huge practical capability of imprinted polymeric hydrogels was shown in above chapter. However, there are few significant limitations that have to be overcome to progress commercialization of imprinted hydrogels. The biomedical and clinical utilization of imprinted hydrogel-based sorbents required further improvement in order to minimize the impact of interferences that coexist in real sample. In biosensing, the betterment of transduction of signals from the receptor part of sensor to detector is still a challenge. The particular attention should be notice to imprinted hydrogel-based drug delivery forms where in vivo release profiles, cytotoxicity tests, long-term compatibility, and clinical performance should become a standard in the future studies. Nevertheless, the significant progress in the area of molecularly imprinted hydrogels could contribute to the mercantile introduction of new and advanced materials dedicated to separation, sensing, or drug delivery in near future.

References

Abbate V, Frascione N, Bansal SS (2010) Preparation, characterization, and binding profile of molecularly imprinted hydrogels for the peptide hepcidin. J Polym Sci A 48:1721–1731

Adrus N, Ulbricht M (2012) Molecularly imprinted stimuli-responsive hydrogels for protein recognition. Polymer 53:4359–4366

Ahmed EM (2015) Hydrogel: preparation, characterization and applications: a review. J Adv Res 6:105–121

Alexander C, Andersson HS, Andersson LI, Ansell RJ, Kirsch N, Nicholls IA, O'Mahony J, Whitcombe MJ (2006) Molecular imprinting science and technology: a survey of the literature for the years up to and including 2003. J Mol Recognit 19:106–180

Alvarez-Lorenzo C, Concheiro A (2004) Molecularly imprinted polymers for drug delivery. J Chromatogr B 804:231–245

Alvarez-Lorenzo C, Concheiro A (2013) Handbook of molecularly imprinted polymers. Smithers Rapra, Shawbury

Alvarez-Lorenzo C, Hiratani H, Gomez-Amoza JL, Martinez-Pacheco R, Souto C, Concheiro A (2002) Soft contact lenses capable of sustained delivery of timolol. J Pharm Sci 91:2182–2192

Alvarez-Lorenzo C, Yanez F, Barreiro-Iglesias R, Concheiro C (2006) Imprinted soft contact lenses as norfloxacin delivery systems. J Control Release 113:236–244

Alvarez-Lorenzo C, González-Chomón C, Concheiro A (2013) Molecularly imprinted hydrogels for affinity-controlled and stimuli-responsive drug delivery. In: Alvarez-Lorenzo C, Concheiro A, Schneider HJ, Shahinpoor M (eds) Smart materials for drug delivery. RSC, Cambridge, pp 228–260

Bolisay LD, Kofinas P (2010) Imprinted polymer hydrogels for the separation of viruses. Macromol Symp 291–292:302–306
Bolisay LD, Culver JN, Kofinas P (2006) Molecularly imprinted polymers for tobacco mosaic virus recognition. Biomaterials 27:4165–4168
Braga MEM, Yanez F, Alvarez-Lorenzo C, Concheiro C, Duarte CMM, Gil MH, de Soussa HC (2010) Improved drug loading/release capacities of commercial contact lenses obtained by supercritical fluid assisted molecular imprinting method. J Control Release 148:e102–e104
Byrne ME, Salian V (2008) Molecular imprinting within hydrogels II: progress and analysis of the field. Int J Pharm 364:188–212
Byrne ME, Park K, Peppas NA (2002) Molecular imprinting within hydrogels. Adv Drug Deliv Rev 54:149–161
Casey BJ, Kofinas P (2008) Selective binding of carcinoembryonic antigen using imprinted polymeric hydrogels. J Biomed Mater Res 87A:359–363
Chen Z, Hua Z, Xu L, Huang Y, Zhao M, Li Y (2008) Protein-responsive imprinted polymers with specific shrinking and rebinding. J Mol Recognit 21:71–77
Cimen D, Akbulut E, Demirel G, Caykara T (2009) Molecularly imprinted hydrogels for fibrinogen recognition. React Funct Polym 69:655–659
Cirillo G, Iemma F, Puoci F, Parisi OI, Curcio M, Spizzirri UG, Picci N (2009) Imprinted hydrophilic nanospheres as drug delivery systems for 5-fluorouracil sustained release. J Drug Target 17:72–77
da Silva MS, Viveiros R, Morgado PI, Aguiar-Ricardo A, Correia IJ, Casimiro T (2011) Development of 2-(dimethylamino)ethyl methacrylate-based molecular recognition devices for controlled drug delivery using supercritical fluid technology. Int J Pharm 416:61–68
Demchenko AP (2001) Recognition between flexible protein molecules: induced and assisting folding. J Mol Recognit 14:42–61
Derazshamshir A, Baydemir G, Andac M, Say R, Galaev IY, Denizli A (2010) Molecularly imprinted PHEMA-based cryogel for depletion of hemoglobin from human blood. Macromol Chem Phys 211:657–668
Divya MP, Rajput YS, Sharma R, Singh G (2015) Molecularly imprinted polymer for separation of lactate. J Anal Chem 10:1213–1217
El-Sharif HF, Phan QT, Reddy SM (2014) Enhanced selectivity of hydrogel-based molecularly imprinted polymers (HydroMIPs) following buffer conditioning. Anal Chim Acta 809:155–161
El-Sharif HF, Aizawa H, Reddy SM (2015a) Spectroscopic and quartz crystal microbalance (QCM) characterisation of protein-based MIPs. Sensor Actuat B 206:239–245
El-Sharif HF, Yapati H, Kalluru S, Reddy SM (2015b) Highly selective BSA imprinted polyacrylamide hydrogels facilitated by a metal-coding MIP approach. Acta Biomater 28: 121–127
Fazal FM, Hansen DE (2007) Glucose-specific poly(allylamine) hydrogels—a reassessment. Bioorg Med Chem Lett 17:235–238
Fenzl C, Hirsch T, Wolfbeis OS (2014) Photonic crystals for chemical sensing and biosensing. Angew Chem Int Ed 53:3318–3335
Fu G, Zhao J, Yu H, Liu L, He B (2007) Bovine serum albumin-imprinted polymer gels prepared by graft copolymerization of acrylamide on chitosan. React Funct Polym 67:442–450
Fu G, Yu H, Zhu J (2008) Imprinting effect of protein-imprinted polymers composed of chitosan and polyacrylamide: a re-examination. Biomaterials 29:2138–2142
Guo TY, Xia YQ, Hao GJ, Song MD, Zhang BH (2004) Adsorptive separation of hemoglobin by molecularly imprinted chitosan beads. Biomaterials 25:5905–5912
Guo TY, Xia YQ, Wang J, Song MD, Zhang BH (2005) Chitosan beads as molecularly imprinted polymer matrix for selective separation of proteins. Biomaterials 26(5737):5745
Hawkins DM, Stevenson D, Reddy SM (2005) Investigation of protein imprinting in hydrogel-based molecularly imprinted polymers (HydroMIPs). Anal Chim Acta 542:61–65
Hilt JZ, Byrne ME (2004) Configurational biomimesis in drug delivery: molecular imprinting of biologically significant molecules. Adv Drug Deliv Rev 56:1599–1620

Hiratani H, Alvarez-Lorenzo C (2002) Timolol uptake and release by imprinted soft contact lenses made of *N*,*N*-diethylacrylamide and methacrylic acid. J Control Release 83:223–230
Hiratani H, Alvarez-Lorenzo C (2004) The nature of backbone monomers determines the performance of imprinted soft contact lenses as timolol drug delivery system. Biomaterials 25:1105–1113
Hiratani H, Mizutani Y, Alvarez-Lorenzo C (2005a) Controlling drug release from imprinted hydrogels by modifying the characteristics of the imprinted cavities. Macromol Biosci 5: 728–733
Hiratani H, Fujiwara A, Tamiya Y, Mizutani Y, Alvarez-Lorenzo C (2005b) Ocular release of timolol from molecularly imprinted soft contact lenses. Biomaterials 26:1293–1298
Hoare TR, Kohane DS (2008) Hydrogels in drug delivery: progress and challenges. Polymer 49:1993–2007
Hoffman AS (2012) Hydrogels for biomedical application. Adv Drug Deliv Rev 64:18–23
Hu X, An Q, Li G, Tao S, Liu J (2006) Imprinted photonic polymers for chiral recognition. Angew Chem Int Ed 45:8145–8148
Hu X, Li G, Huang J, Zhang D, Qiu Y (2007) Construction of self-reporting specific chemical sensors with high sensitivity. Adv Mater 19:4327–4332
Hu X, Li G, Li M, Huang J, Li Y, Gao Y, Zhang Y (2008) Ultrasensitive specific stimulant assay based on molecularly imprinted photonic hydrogels. Adv Funct Mater 18:575–583
Hua Z, Chen Z, Li Y, Zhao M (2008) Thermosensitive and salt-sensitive molecularly imprinted hydrogel for bovine serum albumin. Langmuir 24:5773–5780
Hua Z, Zhou S, Zhao M (2009) Fabrication of a surface imprinted hydrogel shell over silica microspheres using bovine serum albumin as a model protein template. Biosens Bioelectron 25:615–622
Janiak DS, Ayyub OB, Kofinas P (2010) Effects of charge density on the recognition properties of molecularly imprinted polyampholyte hydrogels. Polymer 51:665–670
Kirchhof S, Goeperich AM, Brandl FP (2015) Hydrogels in ophthalmic applications. Eur J Pharm Biopharm 95:227–238
Koetting MC, Peters JT, Steichen SD, Peppas NA (2015) Stimuli-responsive hydrogels: theory, modern advances and applications. Mater Sci Eng R 93:1–49
Korytkowska-Wałach A (2013) Molecularly imprinted hydrogels for application in aqueous environment. Polym Bull 70:1647–1657
Li S, Tiwari A, Ge Y, Fei D (2010) A pH-responsive, low crosslinked, molecularly imprinted insulin delivery system. Adv Mater Lett 1:4–10
Li L, Ying X, Liu J, Li X, Zhang W (2015) Molecularly imprinted polyurethane grafted calcium alginate hydrogel with specific recognition for proteins. Mater Lett 143:248–251
Liu Y, Cao X, Hua R, Wang Y, Liu Y, Pang C, Wang Y (2010) Selective adsorption of uranyl ion on ion-imprinted chitosan/PVA cross-linked hydrogel. Hydrometallurgy 104:150–155
Liu Y, Kang Y, Huang D, Wang A (2012) Cu^{2+} removal from aqueous solution by modified chitosan hydrogels. J Chem Technol Biotechnol 87:1010–1016
Lorenzo RA, Carro AM, Alvarez-Lorenzo C, Concheiro A (2011) To remove or not to remove? The challenge of extracting the template to make the cavities available in molecularly imprinted polymers (MIPs). Int J Mol Sci 12:4327–4347
Luliński P (2013) Molecularly imprinted polymers as the future drug delivery devices. Acta Pol Pharm 70:601–609
Lv Y, Tan T, Svec F (2013) Molecular imprinting of proteins in polymers attached to the surface of nanomaterials for selective recognition of biomacromolecules. Biotech Adv 31:1172–1186
Malaekeh-Nikouei B, Ghaeni FA, Motamedshariaty VS, Mohajeri SA (2012) Controlled release of prednisolone acetate from molecularly imprinted hydrogel contact lenses. J Appl Polym Sci 126:387–394
Malakooti N, Alexander C, Alvarez-Lorenzo C (2015) Imprinted contact lenses for sustained release of polymyxin B and related antimicrobial peptides. J Pharm Sci 104:3386–3394

Maryam A, Horikawa S, Venkatesh S, Saha J, Hong JW, Byrne ME (2007) Zero order therapeutic release from imprinted hydrogel contact lenses within in vitro physiological ocular tear flow. J Control Release 124:154–162

Miyata T, Hayashi T, Kuriu Y, Uragami T (2012) Responsive behavior of tumor-marker imprinted hydrogels using macromolecular cross-linkers. J Mol Recognit 25:336–343

Naraprawatphong R, Kawanaka G, Hayashi M, Kawamura A, Miyata T (2016) Development of protein-recognition SPR devices by combination of SI-ATRP with biomolecular imprinting using protein ligands. Mol Impr 4:21–30

Nematollahzadeh A, Lindemann P, Sun W, Stute J, Lütkemeyer D, Sellergren B (2014) Robust and selective nano cavities for protein separation: an interpenetrating polymer network modified hierarchically protein imprinted hydrogel. J Chromatogr A 1345:154–163

Nicholls IA, Chavan S, Golker K, Karlsson BCG, Olsson GD, Rosengren AM, Suriyanarayanan S, Wiklander JG (2015) Theoretical and computational strategies for the study of the molecular imprinting process and polymer performance. In: Mattiasson B, Ye L (eds) Molecularly imprinted polymers in biotechnology. Springer, Berlin-Heidelberg, pp 25–50

Ogawa K-I, Hyuga M, Okada T, Minoura N (2012) Development of lipid a-imprinted polymer hydrogels that selectively recognize lipopolysaccharides. Biosens Bioelectron 38:215–219

Ogiso M, Minoura N, Shinbo T, Shimizu T (2006) Detection of a specific DNA sequence by electrophoresis through a molecularly imprinted polymer. Biomaterials 27:4177–4182

Osada Y, Gong J-P (1998) Soft and wet materials: polymer gels. Adv Mater 10:827–837

Patching SG (2014) Surface plasmon resonance spectroscopy for characterisation of membrane protein–ligand interactions and its potential for drug discovery. Biochim Biophys Acta 1838:43–55

Peng H, Wang S, Zhang Z, Xiong H, Li J, Chen L, Li Y (2012) Molecularly imprinted photonic hydrogels as colorimetric sensors for rapid and label-free detection of vanillin. J Agric Food Chem 60:1921–1928

Peppas NA, Bures P, Leobandung W, Ichikawa H (2000) Hydrogels in pharmaceutical formulations. Eur J Pharm Biopharm 50:27–46

Ran D, Wang Y, Jia X, Nie C (2012) Bovine serum albumin recognition via thermosensitive molecular imprinted macroporous hydrogels prepared at two different temperatures. Anal Chim Acta 723:45–53

Reddy SM, Phan QT, El-Sharif H, Govada L, Stevenson D, Chayen NE (2012) Protein crystallization and biosensor applications of hydrogel-based molecularly imprinted polymers. Biomacromolecules 13:3959–3965

Reddy SM, Hawkins DM, Phan QT, Stevenson D, Warriner K (2013) Protein detection using hydrogel-based molecularly imprinted polymers integrated with dual polarisation interferometry. Sens Actuat B 176:190–197

Rogovina LZ, Vasil'ev VG, Braudo EE (2008) Definition of the concept of polymer gel. Polym Sci C 50:85–92

Salian VD, Vaughan AD, Byrne ME (2012) The role of living/controlled radical polymerization in the formation of improved imprinted polymers. J Mol Recognit 25:361–369

Schirhagl R (2014) Bioapplications for molecularly imprinted polymers. Anal Chem 86:250–261

Serra L, Domenech J, Peppas NA (2006) Drug transport mechanism and release kinetics from molecularly imprinted designed poly(acrylic acid-g-ethylene glycol) hydrogels. Biomaterials 27:5440–5451

Shin J, Braun PV, Lee W (2010) Fast response photonic crystal pH sensor based on templated photo-polymerized hydrogel inverse opal. Sensor Actuat B 150:183–190

Singh B, Chauhan N (2008) Preliminary evaluation of molecular imprinting of 5-fluorouracil within hydrogels for use as drug delivery systems. Acta Biomater 4:1244–1254

Singh LK, Singh M, Singh M (2014) Biopolymeric receptor for peptide recognition by molecular imprinting approach—synthesis, characterization and application. Mater Sci Eng C 45: 383–394

Sreenivasan K (1999) On the application of molecularly imprinted poly(HEMA) as a template responsive release system. J Appl Pol Sci 71:1819–1821

Takata A, Usui K, Matsui J (2015) ATP-binding peptide-hydrogel composite synthesized by molecular imprinting on beads. Mol Impr 2:93–98
Tieppo A, Pate KM, Byrne ME (2012) In vitro controlled release of an anti-inflammatory from daily disposable therapeutic contact lenses under physiological ocular tear flow. Eur J Pharm Biopharm 81:170–177
Tov OY, Luvitch S, Bianco-Peled H (2010) Molecularly imprinted hydrogel displaying reduced non-specific binding and improved protein recognition. J Sep Sci 33:1673–1681
Troiani D, Dion JR, Burns DH (2011) Ultrasonic quantification using smart hydrogel sensors. Talanta 83:1371–1375
Turan E, Ozcetin G, Caykara T (2009) Dependence of protein recognition of temperature-sensitive imprinted hydrogels on preparation temperature. Macromol Biosci 9:421–428
Uysal A, Demirel G, Turan E, Çaykara T (2008) Hemoglobin recognition of molecularly imprinted hydrogels prepared at different pHs. Anal Chim Acta 625:110–115
Venkatesh S, Saha J, Pass S, Byrne ME (2008) Transport and structural analysis of molecularly imprinted hydrogels for controlled drug delivery. Eur J Pharm Biopharm 69:852–860
Verma R, Gupta BD (2013) Fiber optic SPR sensor for the detection of 3 pyridinecarboxamide (vitamin B_3) using molecularly imprinted hydrogel. Sensor Actuat B 177:279–285
Wang J, Li Z (2015) Enhanced selective removal of Cu(II) from aqueous solution by novel polyethylenimine-functionalized ion imprinted hydrogel: behaviors and mechanisms. J Hazard Mater 300:18–28
Wang J, Cao Y, Feng Y, Yin F, Gao J (2007) Multiresponsive inverse-opal hydrogels. Adv Mater 19:3865–3871
Wang J, Banerji S, Menegazzo N, Peng W, Zou Q, Booksh KS (2011) Glucose detection with surface plasmon resonance spectroscopy and molecularly imprinted hydrogel coatings. Talanta 86:133–141
Wang J, Ding L, Wei J, Liu F (2014a) Adsorption of copper ions by ion-imprinted simultaneous interpenetrating network hydrogel: thermodynamics, morphology and mechanism. Appl Surf Sci 305:412–418
Wang J, Li J, Li H, Ding L (2014b) Modulated ion recognition by thermosensitive ion imprinted hydrogels with IPN structure. Mater Lett 131:9–11
Whitcombe MJ, Kirsch N, Nicholls IA (2014) Molecular imprinting science and technology: a survey of the literature for the years 2004–2011. J Mol Recognit 27:297–401
Wizeman WJ, Kofinas P (2001) Molecularly imprinted polymer hydrogels displaying isomerically resolved glucose binding. Biomaterials 22:1485–1491
Wu Z, Hu X, Tao C, Li Y, Liu J, Yang C, Shen D, Li G (2008a) Direct and label-free detection of cholic acid based on molecularly imprinted photonic hydrogels. J Mater Chem 18:5452–5458
Wu Z, Tao C, Lin C, Shen D, Li G (2008b) Label-free colorimetric detection of trace atrazine in aqueous solution by using molecularly imprinted photonic polymers. Chem Eur J 14:11358–11368
Wu H-G, Ju X-J, Xie R, Liu Y-M, Deng J-G, Niu CH, Chu L-Y (2011) A novel ion-imprinted hydrogel for recognition of potassium ions with rapid response. Polym Adv Technol 22:1389–1394
Wulff G (1995) Molecular imprinting in cross-linked materials with the aid of molecular templates—a way towards artificial antibodies. Angew Chem Int Ed 34:1812–1832
Wulff G (2002) Enzyme-like catalysis by molecularly imprinted polymers. Chem Rev 102:1–27
Xia YQ, Guo TY, Song MD, Zhang BH, Zhang BL (2005) Hemoglobin recognition by imprinting in semi-interpenetrating polymer network hydrogel based on and chitosan. Biomacromolecules 6:2601–2606
Yalinca Z, Yilmaz E, Bullici FT (2012) Evaluation of chitosan tripolyphosphate gel beads as bioadsorbents for iron in aqueous solution and in human blood in vitro. J Appl Polym Sci 125:1493–1505
Yanez F, Chauhan A, Concheiro A, Alvarez-Lorenzo C (2011a) Timolol-imprinted soft contact lenses: influence of the template: functional monomer ratio and the hydrogel thickness. J Appl Polym Sci 122:1333–1340

Yanez F, Martikainen L, Braga MEM, Alvarez-Lorenzo C, Concheiro C, Duarte CMM, Gil MH, de Soussa HC (2011b) Supercritical fluid-assisted preparation of imprinted contact lenses for drug delivery. Acta Biomater 7:1019–1030

Ying X, Cheng G, Zhao K (2010) Preparation and characterization of protein imprinted agarose microspheres. Polym Bull 65:245–263

Ying X, Cheng G, Li X (2011) The imprinting induce-fit model of specific rebinding of macromolecularly imprinted polymer microspheres. J Appl Polym Sci 122:1847–1856

Ying X, Qi L, Li X, Zhang W, Cheng G (2013) Stimuli-responsive recognition of BSA imprinted poly vinyl acetate grafted calcium alginate core-shell hydrogel microspheres. J Appl Polym Sci 127:3898–3909

Zayats M, Brenner AJ, Searson PC (2014) Protein imprinting in polyacrylamide-based gels. Biomaterials 35:8659–8668

Zhao K, Lin B, Cui W, Feng L, Chen T, Wei J (2014) Preparation and adsorption of bovine serum albumin-imprinted polyacrylamide hydrogel membrane grafted on non-woven polypropylene. Talanta 121:256–262

Chapter 8
Applications of Biopolymeric Gels in Agricultural Sector

Sumit Mishra, Nandkishore Thombare, Mohd Ali and Saurabh Swami

Abstract Hydrogels are three-dimensional cross-linked polymeric network having substantial affinity for water. Classes of hydrogels which are derived from biopolymers have been widely used in number of industries because of their biocompatibility and environmental safety. In agricultural sector, they are extensively used as soil conditioners, water retainers, and bio-remediating agents. Their recent application claims customized diffusion of different materials such as fertilizers and pesticides into surrounding soil matrix. They are proving very useful for crops in efficient distribution of water with minimum wastage, and utilization of fertilizer nutrients and pesticides in targeted zones. This chapter covers the recent advances on biopolymeric gels and different aspects of their usage with respect to agricultural sector.

Keywords Polysaccharide · Gum · Hydrogel · Soil conditioner
Pesticide · Water retainer

1 Introduction

Biopolymers are polymeric materials that are obtained from living beings. According to Dr. Pat Smith, "Biopolymers are not only materials of 'green birth' but polymers with 'green death' as well." The examples of biopolymers include carbohydrates, nucleic acids, proteins, lipids, peptides, and polysaccharides. Among these, polysaccharides have large popularity owing to their manifold uses, particularly in the field of agriculture, food, pharmacotherapy and pharmacy, cosmetic, and mining industry. Polysaccharides have high molecular weight and are

S. Mishra (✉)
Department of Chemistry, Birla Institute of Technology, Mesra, Ranchi 835215, India
e-mail: sumitmishra1@gmail.com

N. Thombare · M. Ali · S. Swami
Processing and Product Development Division, ICAR-Indian Institute of Natural Resins and Gums, Namkum, Ranchi 834010, India

© Springer Nature Singapore Pte Ltd. 2018

V. K. Thakur et al. (eds.), *Polymer Gels*, Gels Horizons: From Science to Smart Materials, https://doi.org/10.1007/978-981-10-6080-9_8

composed of one or different types of monosaccharide units arranged in linear or branched fashion via glycosidic linkages. On hydrolysis, polysaccharides give their constituent oligosaccharides or monosaccharides. Cellulose, starch, gum arabic, guar gum, alginate, chitosan, xanthan, etc., are commercially exploited polysaccharides which find application in various industries. Polysaccharides are generally heterogeneous, containing slight modifications of the repeating units. Depending on the molecular structure and chemical composition, polysaccharides have different physicochemical properties derived from their building blocks. They may be amorphous or even insoluble in water (Varki et al. 1999, 2008). When a polysaccharide is composed of same units of monosaccharides, the polysaccharide is called a homoglycan or homo-polysaccharide, but when different units of monosaccharides are present, the polysaccharides are called heteroglycans or heteropolysaccharides.

1.1 Advantages of Natural Polysaccharide

Biodegradability: Biopolymers are naturally available and are produced by living organisms. They represent truly renewable source, and they are easily degraded by microbes; hence, they do not have any unfavorable effect on the environmental well-being.

Non-toxic and biocompatible: Natural polysaccharides being biocompatible are non-toxic and can be used in different commodities, ranging from foods to pharmaceuticals and cosmetics to drug delivery.

Environment-friendly processing: Polysaccharides from various natural sources are effortlessly gathered or harvested in various seasons in substantial amounts because the processes involved in their production are mostly natural and simple.

Low cost: As there is no specialized set up or infrastructure required to produce natural polysaccharide, the cost of production is also much lower compared to synthetic material.

Easy availability: Most of the natural polysaccharides are obtained either from cropping or plants or sea; hence, they are readily available. Also, for most of them production can be increased as per demand, e.g., guar gum.

Gums are important class of biopolymers which are heterogeneous, water soluble or water swellable, high molecular weight polysaccharides extracted from terrestrial or marine plants or from microorganisms. They have gelling capability or ability to contribute viscosity to their dispersions (Abu Baker et al. 2007). Generally, gums are insoluble in nonpolar or organic solvents such as hexane, alcohols, ether, and other hydrocarbons. On hydrolysis, depending on their chemical composition, gums yield mannose, dextrose, rhamnose, xylose, arabinose, galactose, glucuronic acid, galacturonic acid, etc. Due to their unique physicochemical properties, gums have broad applications in both food and non-food industries. All applications rely on the properties given by macromolecules in various states of hydration, but mostly depend on the properties they impart to solutions and gels.

2 Classification of Natural Polysaccharides

Natural polysaccharides are available in large amounts as they are obtained from large varieties of animals, plants, fungi and microbes, and seaweed sources; where they perform various structural and metabolic functions. Naturally obtained polysaccharides from various sources can be classified as follows (Jani et al. 2009) (Figs. 1, 2 and 3).

Based on their origin/sources, the natural gums are differentiated into four major groups. Out of these four groups, polysaccharides with plant origin are mostly utilized commercially and found in numerous day-to-day life applications.

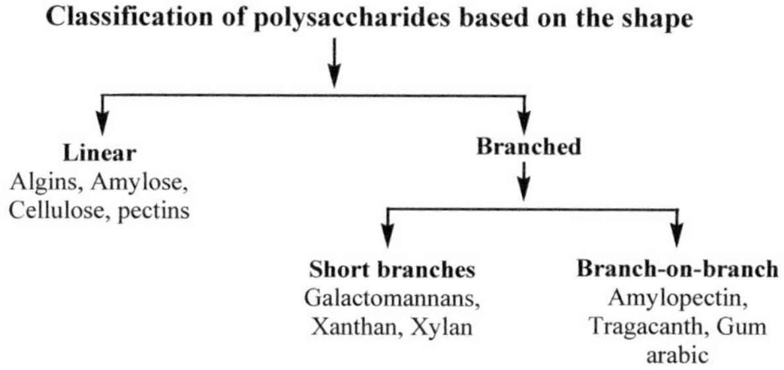

Fig. 1 Classification of polysaccharides based on the shape

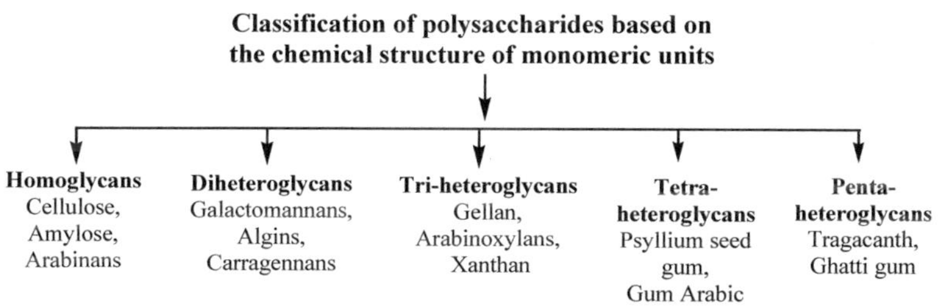

Fig. 2 Classification of polysaccharides based on the chemical structure of monomeric unit

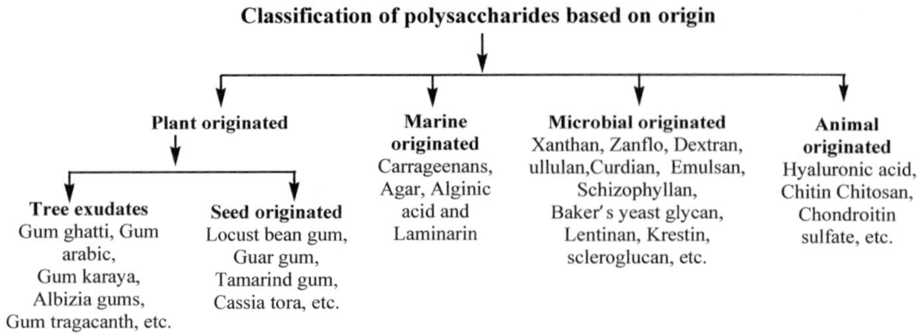

Fig. 3 Classification of polysaccharides based on origin

2.1 *Polysaccharides with Plant Origin*

2.1.1 Tree Exudates

Exudate gums are one of the oldest natural polysaccharides, which were already being used 5000 years ago as a thickening and stabilizing agents. Exudate gums are formed from the breakdown of internal plant tissues (mainly cellulose) in a process called gummosis. They are exudated naturally from stems, in response to mechanical wounding or after an insect, bacterial, or fungal attack. Though large numbers of gums are available, only few of them could find commercial applications and are produced or collected in substantial quantity. Some of them are discussed below:

Gum Arabic:

Gum arabic is an edible gum and approved as food additive by European Food Safety Authority with E-Number 414. This is obtained from the stems and branches of diverse subspecies of trees and shrubs of genus Acacia, in the form of dried, gummy exudates (Nishinari and Doi 2012). Gum arabic is predominantly collected from *Acacia nilotica* in India and from *Acacia senegal* and *Acacia seyal* in different parts of the world. *A. senegal* and *A. seyal* grow naturally in the semiarid sub-Saharan regions of Africa. Major gum arabic-producing countries are Sudan, Nigeria, Mali, Niger, Burkina Faso, Chad, India, Tanzania, and Kenya. Among these countries, Sudan alone produces 80% of gum arabic, followed by Nigeria which is the second largest producer (Iqbal 1993). *A. senegal* is found in some parts of India mainly in dry rocky hills of south east Punjab, in north Aravalli hills, and in other drier parts of Rajasthan and Gujarat.

Gum arabic has been extensively used as a stabilizer (Chung et al. 2016; Kong et al. 2014), thickening agent and emulsifier (Hosseini et al. 2015), micro-encapsulant (Ramakrishnan et al. 2007), in fruit coating (Addai et al. 2013), corrosion inhibitor of aluminum (Umoren et al. 2006), antioxidant (Gamal el-din et al. 2003; Al-Majed et al.2002, 2003; Abd-Allah 2002) as an adsorbent for heavy metal (Banerjee and Chen 2007), and also used in lithography, textile, pottery, cosmetics, and pharmaceutical industries (Verbeken et al. 2003).

Karaya Gum:

The Joint Expert Committee on Food Additives (JECFA/FAO) defines gum karaya as the dried exudates obtained from *Sterculia urens* Roxburgh and other related species of Sterculia (family Sterculiaceae) or *Cochlospermum gossypium* (Mbuna and Mhinzi 2003) or other species of *Cochlospermum kunth* (family Bixaceae). The other two contributing species are *Sterculia setigera* and *Sterculia villosa*. Karaya gum is acetylated polysaccharide which on acid or base hydrolysis gives galactose, rhamnose, galacturonic acid with little quantity of glucuronic acid. *S. urens* is indigenous to India having wide distribution. It is abundantly found in the dry deciduous forests of Madhya Pradesh, Chhattisgarh, and central India. Conventionally, India is the largest producer and exporter of karaya gum, while Europe is its largest importer (Verbeken et al. 2003). Gum karaya trees are also found in Australia, Pakistan, Panama, Philippines, Indonesia, Sudan, and Vietnam. Owing to its high viscosity, acid stability, and suspension properties, it is widely used in both non-food and food applications (Verbeken et al. 2003). It is also used for the removal of heavy metals (Vinod et al. 2011), dyes, cationic dyes (Mittal et al. 2016), methylene blue (Mittal et al. 2015a, b), as an excipient for muco-adhesive drug delivery systems (Bahulkar et al. 2015), as a thickening agents (Ibrahim et al. 2010). Other major exudate gums are gum ghatti, albizia gums, salai, jhingan, moringa gum, etc.

2.1.2 Seed Derived

Seeds are also a source of polysaccharides. Most seeds contain starch as the principal food stored for use by the embryonic plant in its initial growth. Most of the plants which are used as food produce seeds with starch as a carbohydrate reserve. Few plant species produce seeds without starch food reserves, which can be harvested to produce the seed gums. More ancient seed gums were extracted from quince, psyllium, flax, and locust seeds, and some of these are still quite important.

Those gum-producing seeds that are amenable to normal agricultural production are lower in price. These are the seeds from annual plants with a normal growing season, which can be grown on agricultural land by normal methods, and which can be planted and harvested by standard agricultural machinery. In the last few decades, guar gum has emerged as very important seed gum.

Guar gum:

Guar gum is one of the outstanding representatives of green and eco-friendly biopolymers. It is a water-soluble nonionic polysaccharide isolated from the seeds of *Cyamopsis tetragonolobus* (Family leguminosae) (Whistler and BeMiller 1993). This plant is cultivated for centuries in semiarid and subtropical areas of India and

Pakistan and in some areas of North Africa and South America. Guar was introduced into the USA from India in 1903 (Whistler and BeMiller 1993). Guar is also known as "Black Gold" because its demand supply pattern has turned it into cash crop and hence a precious commodity. The main guar gum-producing states in India are Rajasthan, Uttar Pradesh, Gujarat, Tamil Nadu, Karnataka, Haryana, Punjab, Madhya Pradesh, and Andhra Pradesh. India is the largest exporter of guar gum to the world and has exported 3388.4 thousand tons of guar gum in 2013–14 (Yogi et al. 2015) to USA, Canada, China, Russia, and Germany (APEDA).

Due to its unique gelling properties and rheology, it is being widely used across various industries such as oil well drilling (Robinson et al. 1982), pharmaceuticals (Celkan et al. 2016; Vollmer 2003), textiles (Aggarwal and Sharma 2010), cement (Blackburn 2004), cosmetics (Vijn et al. 2002), food (Cretois et al. 2000), paper (Anderson et al. 1993), paint, explosives (Thombare et al. 2016), agriculture (Chandrika et al. 2014).

Tamarind gum:

Tamarind seed polysaccharide (TSP) derived from endosperm of tamarind kernels is an important natural polysaccharide. Tamarind (*Tamarindus indica*) is also known as "Indian date." Tamarind is long-lived, medium growth, bushy evergreen tree belonging to the family Fabaceae. Tamarind tree grows well in clayey, loamy, sandy, and acidic soil types, with a high drought and aerosol salt resistance (Joseph et al. 2012). This is liberally found in dry tracks of central and south Indian states, furthermore in other Southeast Asian nations.

Tamarind seed polysaccharide is a multifunctional polymer, which plays the role of stabilizer, thickener, binder (Kulkarni et al.1998), release retardant (Srinivasan et al. 2011), modifier (Kulkarni et al. 2005), emulsifying agent, and suspending agent (Deveswaran et al. 2009), as a carrier for novel drug delivery systems for oral (Alka et al. 2011; Srinivasan et al. 2011), buccal (Patel et al. 2009; Bangale et al. 2011; Jana et al. 2010), colon (Mishra and Khandare 2011), ocular systems (Rolando and Valente 2007; Mehra et al. 2010), nanofabrication, wound dressing (Patil et al. 2011; Burgalassi et al. 2000), food (Shirakawa and Yamatoya 2003; Glicksman 1996), cosmetics, confectionery, bakery, etc.

2.2 *Marine Origin*

Gum extraction from seaweeds which was originally practiced in oriental countries has spread to many parts of the world where shallow waters and seaweeds are abundant. The cost of production of seaweed extracts are soared up due to tedious harvesting and processing practices used in the extraction processes, which remove a large portion of the dry weight of the weeds. When harvesting is done manually, as with agar weed, or when the seaweeds are picked from beaches where they are deposited by the tides, labor costs become high. These days, advance engineering practices are being used for harvesting and collection of seaweeds, e.g., harvesting of giant kelp (*Macrocystis pyrifera*) in California. Cut kelp is collected and lifted by a rake to the barge top where it is stacked with a claw on a drag line. By such simple mechanical means, many tons of trimmings can be harvested rapidly and brought to the processing plant on the coast. Rapid growth of the kelp beds permits retrimming within a few months. Such standard mechanical practices lower harvesting cost and tend to stabilize weed cost at the extraction point. Few important marine polysaccharides are discussed here.

Carrageenans:

Carrageenans are straight sulphated polysaccharides extracted from cell wall of red sea weed of the Rhodophyceae class. Seaweeds commonly used for the carrageenans are *Kappaphycus alvarezii* and *Eucheuma denticulatum* (McHugh 2003). These seaweeds are harvested, washed several times with seawater to remove sediments, and sun-dried until they acquire 30–40% moisture content. Major countries producing carrageenans include Indonesia, Tanzania, Malaysia, Philippines, Kenya, Kiribati, Fiji, and Madagascar. Overall business sector volume now surpasses 140,000 tons per annum with an estimation of more than $70 million (Bixler and Porse 2011). Primarily, wild-collected genera, for example, Gigartina, Chondrus, Mazzaella, Furcellaria, Sarcothalia, Iridaea, Mazzaella, Tichocarpus, and Mastocarpus, are additionally delivered as carrageenan crude materials. Carrageenan-producing nations also include Japan, North Korea, Canada, Argentina, Russia, Chile, Spain, Denmark, South Korea, France, Mexico, Portugal, Spain, USA, and Morocco. Carrageenan has numerous applications in both food and non-food industries (Hambleton et al. 2009; Necas and Bartosikova 2013) and can be utilized as stabilizer (Hsu and Chung 1999) in dairy items, for example, flavored products (Varela and Fiszman 2011), pet nourishment (McHugh 2003), newborn child sustenance, and dietary supplement refreshments. Carrageenan have been utilized to postpone microbial development in gels containing antimicrobial agent (Varela and Fiszman 2011). The capacity of suspending cocoa in chocolate milk at low concentration is special in carrageenan (Necas and Bartosikova 2013). Mostly, they are utilized as a part of pharmaceuticals, beautifying agents, printing and as a material for commercial enterprises (Cosenza et al. 2014).

Alginate:

Alginate is a water-soluble straight anionic polysaccharide, isolated from cell wall of brown algae *Ascophyllum nodosum* and *Laminaria digitata,* where it as present as magnesium, sodium, and calcium salt of alginic acid (Vu and Won 2013; McHugh 2003; Hambleton et al. 2011). Microorganisms can also produce alginate (Blanco-Pascual et al. 2014; Alboofetileh et al. 2014).

Among the world edible seaweed producers, China ranks first with a production of about five million tons and most of this is for Kombu, obtained from *Laminaria japonica* grown on hundreds of hectares on suspended ropes in the oceans. Korea mainly grows three species and produces eight million tons, and almost 50% of this is for Wakame obtained from *Undaria pinnatifida* in a similar way to that of Laminaria in China. Japan produces about six lakh tons mainly from three species, among these species 75% of this for Nori, obtained from *Porphyra* sp.

Alginate has wide range of applications and can be used as a colloidal stabilizer, gel forming agent, and thickening agent in beverage industries (Liakos et al. 2013), as an anti-dehydrating agent on natural products such as fruits, meat, and fish (Hambleton et al. 2011; Varela and Fiszman 2011), for encapsulation of protein, DNA, drugs, cell, etc. (Ashikin et al. 2010), as a binder in fish feed, welding rods, paper, releasing agents, and immobilizing catalyst and also used in medical and pharmaceuticals, and material printing (Vu and Won, 2013).

2.3 Polysaccharides with Microbial Origin

Microbial polysaccharides are extracellular polysaccharides produced by certain microorganisms. Such gums are fashioned by selected, and perhaps carefully mutated, organisms growing on low-cost energy sources, such as grains or molasses. A variety of microbial origin gums having different physicochemical properties are already available and many more can be found. Thus, a class of fermentation gums may be developed to cater to the numerous industrial needs. Homoglycans comprising of one kind of sugar unit can be made, as exemplified in the generation of dextran. However, synthesis of enzyme-catalyzed polysaccharides from simple sugars and enzyme modification of existing polysaccharides will eventually develop and become common technique in future. Important microbial polysaccharides include:

Xanthan:

China is the world's largest producer of xanthan and exports about 66% of its produce to the world. Countries such as USA, Australia, Japan, and France are other producers and exporters of xanthan gum. Xanthan is complex extracellular bacterial exo-polysaccharide produced by the yellow-pigmented gram-negative bacteria *Xanthomonas campestris* (Ielpi et al. 1981; Ashraf et al. 2008). Xanthan is an anionic, acidic polymer produced by microbial fermentation of glucose. It is chemically composed of repeating units of pentasaccharide having two units of mannose, two units of glucose and a glucuronic acid (Becker et al. 1998). The anionic nature of xanthan is because of the presence of two acidic groups, i.e., glucuronic acid and pyruvic acid in the side chain (Sandford and Baird 1983). Xanthan is highly stable over wide range of pH and temperature and also resistant to enzymatic hydrolysis. It is highly water-soluble gum and also shown synergistic interaction with other gums. These properties are very unique to xanthan gum, and it makes it versatile hydrocolloid with applications in many industries such as pharmaceuticals as a stabilizer, film forming, thickening, gelling agent, and emulsifier. It is also used in agriculture, paint, oil, paper, cosmetics, and textile industries. Xanthan gum finds applications in petroleum production, oil well drilling fluids, fracturing, pipeline cleaning, enhanced oil recovery (EOR), textile printing and dyeing, ceramic glazes, cleaners, slurry explosives. In food industry, it is used in dressing, dry mixing, beverages, dairy products, and baked foods. Besides, it is used in animal feed, agricultural chemicals, pharmaceuticals, and cosmetics.

2.4 Polysaccharides with Animal Origin

Chitin and chitosan:

Chitin is the second most ubiquitous natural polysaccharide after cellulose on earth. It is a hard and inelastic polysaccharide, found in invertebrate exoskeleton and internal structure. Chitin and chitosan are very promising biomaterials. The deacetylated chitin derivative, chitosan, is more useful and interesting bioactive polymer. It has many reactive amino side groups, which offer possibilities of chemical modifications, formation of a large variety of useful derivatives that are commercially available or can be made available via graft reactions and ionic interactions.

Due to its unique physicochemical properties, it is being used in number of industries such as cosmetics (Libio et al. 2016), textiles (Dutta et al. 2002), food processing (Klein et al. 2010), agriculture (Kashyap et al. 2015), photography (Dutta et al. 2002), chromatographic separations (Rhee et al. 1998), and biomedical applications such as tissue engineering (Pangon et al. 2016; Suh and Matthew 2000), burn treatment (Sohrabi et al. 2016), ophthalmology (Cheng et al. 2016), wound healing/wound dressing (Dragostin et al. 2016), and drug delivery systems (Soares et al. 2016; Pathania et al. 2016).

3 Chemical Modification of Polysaccharides

The natural raw polysaccharides have wide applications, yet there is a lot of scope to enhance their applications by improving their physicochemical properties. Properties of natural raw gum can be improved or modified by chemical methods such as grafting, derivatization, and cross-linking by least affecting its inherent properties. This is because chemical modification yields the hybrid derivatives of raw polysaccharides, which can fit into various applications (Zhang et al. 2005). For example, by synthesizing the hybrid derivatives of guar gum such as hydroxypropyl or carboxymethyl, its properties such as solubilization time, viscosity, and clarity of solution can be significantly improved (Dumitriu 2002).

3.1 Grafting

Graft copolymerization is one of the powerful tools for the modification of biopolymers as it functionalizes these natural polymers with improved and desirable properties. Therefore, these days much attention has been paid to grafting method as a tool for chemical modifications (Mishra et al. 2010; Tripathy et al. 2009). Grafting of natural gums can be done by various methods such as microwave irradiation (Pal et al. 2011; Adhikary et al. 2011), chemical initiator, or γ-radiation (Abdel-Halim and Al-Deyab 2011; Srivastava et al. 2007). The grafted polymers can be utilized in drug delivery, pharmaceuticals, and agriculture for controlled release of nutrient and agrochemicals.

3.2 Derivatization

Another method for the improvement of physicochemical properties of polysaccharides is derivatization. A large number of natural polysaccharide-based derivatives have already been synthesized and many more will be added in the future. Some of derivatized products of natural polysaccharides which are synthesized, characterized, and evaluated for their applications are dodecenyl succinic anhydride gum arabic (Wang et al. 2014) hydroxymethyl guar gum (Lapasin et al. 1991), hydroxypropyl guar gum (Lapasin et al. 1995), o-carboxymethyl-o-hydroxypropyl guar gum (CMHPG) (Shi and Zhang 2007), quaternary ammonium chitosan derivatives (De-Oliveira-Pedro et al. 2016), triazolyl-functionalized chitosan derivatives (Li et al. 2015), amphiphilic alginate-amide derivatives (Vallée et al. 2009), thiolated karaya gum (Bahulkar et al. 2015), carboxymethyl cellulose derivatives (Monier et al. 2016), metallo-terpyridine carboxymethyl cellulose derivatives, etc.

3.3 Cross-linking

Natural gums are biopolymers having numbers of free hydroxyl groups in their linear or branched long chain, which on dispersion in water form viscous solution. This is due to interaction of free hydroxyl groups of gum with water molecules by forming intra- and intermolecular hydrogen bonding, which increases viscosity of solution. Natural raw gum as such cannot be used as hydrogel due to low swelling and water-holding capacity, but it can be improved by using the synthetic cross-linkers. The cross-linker binds the chains of biopolymers by chemical or physical means and increases the stability, swelling, and water-holding capacity of hydrogels. This is because the cross-linkers undergo intra-molecular bonding with hydroxyl group of biopolymers and form a three-dimensional structure (Fig. 4).

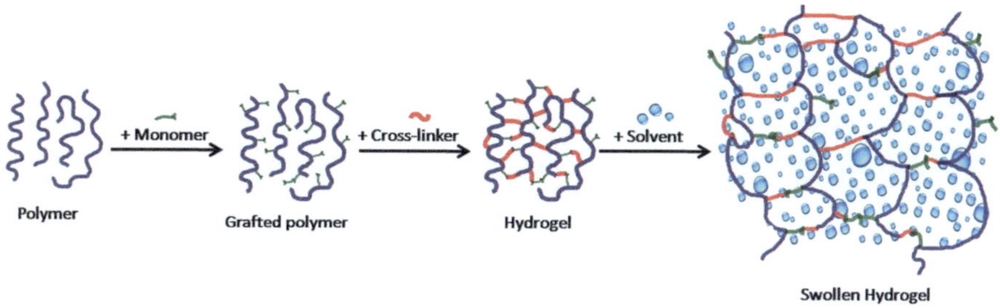

Fig. 4 Schematic diagram of synthesis of hydrogel through grafting and cross-linking

When water is added, the molecules of water are entrapped in this three-dimensional structure and cannot escape easily. Commonly used cross-linkers are methylene-bis-acrylamide, divinyl-benzene, glutaraldehyde, derivatives of ethylene-glycol-di(meth)acrylate, etc.

Hydrogels are three-dimensional matrix constituted by linear or branched hydrophilic polymers that are cross-linked chemically or physically, with the ability to absorb large quantity of water or biological fluids (Chang et al. 2010; Pourjavadi et al. 2004). Further, even in the swollen state hydrogels keep their network stable because of their cross-linked structure which imparts stability in different environments. The final properties and applications of hydrogels depend on the type of cross-linking methods used for hydrogel preparation. Important properties such as water uptake, swelling, kinetics, rheological properties, porosity, degradation rate, and toxicity are closely related to cross-linking methods (Bordi et al. 2013; Ahmed 2013). Therefore, hydrogels are synthesized to possess properties such as fast swelling, porosity, degradability. Due to these properties, they find applications in many fields such as biomedical (Pangon et al. 2016), agriculture (Pourjavadi et al. 2007), cosmetics, tissue engineering (Kim et al. 2007), drug delivery (Rodríguez et al. 2003; Zhang et al. 2002), biosensors (Adhikari and Majumdar 2004; Pourjavadi et al. 2007), and sorbents for the removal of heavy metals (Guilherme et al. 2007).

The structural strength of hydrogels depends upon the nature of bonds (Chemical or physical) between the biopolymers and the cross-linker (Kamath and Park 1993). Hydrogels can be classified on the following basis:

- Physical structure: hydrogen bonded or supramolecular, amorphous, or semicrystalline;
- Electric charge: neutral or ionic (charged);
- Cross-link: chemically or physically cross-linked;
- Responses to external stimuli: sensitive and insensitive;
- Origin: natural, semisynthetic, and synthetic.

After the first polymerization of acrylic acid and divinylbenzene in 1938, it took a decade for the first group of hydrogels to become commercially viable. These hydrogels were made of hydroxyalkyl methacrylate and other monomers having similar structures. However, they had swelling capacity of only 40–50% and were used for production of contact lenses. HSPAN (hydrolyzed starch-polyacrylonitrile) was developed in the 1970s by USDA as first commercial superadsorbent hydrogel by alkaline hydrolysis of starch-grafted polyacrylonitrile (Buchholz and Peppas 1994). The product could not succeed because of its poor mechanical properties and high cost. The polymeric material used for hydrogel preparation affects its hydrophilicity and biodegradability. Polyacrylamides and acrylates have been extensively used over the years for synthesis of highly hydrophilic hydrogels (Laftah et al. 2011; Rodrigues et al. 2014).

However, increasing environmental concerns have led to the development of hydrogels based on natural polymers. Among natural polymers, polysaccharides have been widely used to develop hydrogels due to their high hydrophilicity, compatibility, low cost, and biodegradability (Wang and Wang 2010a, b; Wang et al. 2013). The interconnected porous structure of hydrogel materials leads to high water absorption capacity by capillary action (Kuang et al. 2011; Hemvichian et al. 2014). Various natural polymers such as starch, chitosan, chitin, pectin, gum arabic, cashew gum, and others have been used to develop hydrogels for specific use in biomedical, agriculture, biotechnological, and wastewater treatment applications (Omidian et al. 2005; Kazanskii and Dubrovskii 1992; Mekonnen et al. 2013; Rinaudo 2006; Heinze et al. 2006).

4 Applications of Hydrogels in Agriculture

In recent years, the use of biopolymer-based hydrogels in agriculture has been widely studied because they are ecologically and economically viable alternatives for soil conditioning and water and nutrient retention (Kazanskii and Dubrovskii 1992). Also, these biopolymer gels are biodegradable, non-toxic, and abundantly available and have great application potential (Sinha and Kumria 2001; Thakur et al. 2015; Thakur and Kessler 2015). Hydrogels are suitable for use in agricultural fields because of their susceptibility to degradation by physical, chemical, and microbial agents (Baldrian and Valášková 2008; Villay et al. 2012). These materials retain water and nutrients and release them over a long period of time. The rate of release of nutrients throughout the degradation of hydrogel can be made harmonious to the plant nutrient requirements.

The polysaccharides in their native form may not produce hydrogels stable enough for use as pesticide or nutrient carriers in agriculture. These hydrogels are prepared using physical or chemical methods of cross-linking or both. Various polysaccharides such as gum arabic, cashew gum, starch, and pectin have been modified by introducing vinyl groups to develop hydrogels (Fajardo et al. 2013; Sannino and Nicolais 2005; Lionetto et al. 2005; Marcì et al. 2006; Sannino et al. 2003).

Though large numbers of hydrogels have been developed at laboratory stage, only very few satisfy the requisite environment safety parameters such as biodegradability and non-toxicity for their use in agriculture (Pillai 2010). Economic considerations also play a major role in final release of the product in the market. The use of hydrogel for water retention and as soil conditioner has been investigated long back in 1966. The commercial sale of these hydrogels started in the 1980s mainly as disposable diapers (Castel et al. 1990; Kataja et al. 1992). In later years, various reports on the structure and properties of such hydrogels were published owing to the growing interest in commercialization of these materials for use in agriculture (Buchholz and Graham 1998). The most common application method of these hydrogels in agriculture is by mixing granular hydrogel particles in soil at required concentration.

4.1 Water Retention

Globally, agriculture is predominantly dependent on rainfall. Around 65–95% of cultivated land in sub-Saharan Africa, Latin America, North Africa, East Asia, and South Asia is under rainfed agriculture (IWMI 2010). Uncertainties in frequency and pattern of rainfall in arid areas result in crop losses every year. Climate change has further aggravated the problem of water scarcity. Various methods are being employed to increase the water use efficiency in agriculture. One of the strategies is to use water retainers to grab and preserve limited irrigation or rainwater for a prolonged period. Owing to their water imbibing property, hydrogel materials are being widely investigated for water retention in agriculture. For example, in sandy areas the use of hydrogels may help in improving the water-holding capacity and thus the growth and quality of crops (Wang and Wang 2010a, b). The hydrogel particles act as miniature reservoirs through which water is drawn when required by the osmotic pressure difference. The use of hydrogels in agriculture is showing very good results. Some of the advantages of hydrogels can be listed as follows (Lee and Mooney 2001; Shalviri et al. 2010; Ulery et al. 2011):

- Reduction in requirement of water for irrigation purpose,
- Increase in availability of soil water which results in longer survival of plants under stress conditions,
- Improved fertilizer use efficiency and decreased contamination of ground water,
- Decrease in plant evapotranspiration rate,
- Improved soil physical properties such as reduced compaction and better soil aeration,
- Enhanced microbial activity,
- Prevention of excess runoff and thus reduction in soil erosion,
- Adsorption of heavy metals and reducing their effect on plants,
- Maintenance of soil moisture that helps in reducing the effects of salinity, and
- Better germination and establishment of seedlings.

Hydrogels are known for their capacity of absorbing large amounts of water. Materials commonly used as absorbents exhibit absorption capacity of around hundred times of their weight; in case of hydrogels, it increases up to thousand times. For example, Guilherme et al. (2005) synthesized a hydrogel having water absorption capacity of 1500 times. The hydrogel was prepared by copolymerization of cashew gum with acrylamide followed by partial hydrolysis of the acrylamide repeat units. Hydrogels have such high water absorption capacity due to the thermodynamic compatibility found between functional groups of hydrogel matrix and water molecules.

Most of the times, the electrically charged groups (ions) of the hydrogel material are responsible for electrostatic affinity toward water molecules during the swelling phase of absorption. Also, the hydrophobic units of the network structure interact with water molecules by weak van der Walls forces. The free water present in soil is absorbed into hydrogel by osmosis. When all the hydrophilic and hydrophobic sites are occupied, the water molecules fill the empty spaces present in hydrogel matrix. Therefore, the porosity of the hydrogel material, as well as polymer chain density and extent of cross-linking, affects the water absorption capacity of the hydrogel. On the other hand, the mechanical strength and rheological properties of the hydrogel are dependent on the degree of swelling. A high swelling capacity may significantly reduce the mechanical strength. The swelling of hydrogels by absorption of high amounts of water is the characteristic property for their use in water retention, nutrient delivery, and maintenance of various soil properties (Ramezani et al. 2013; Campos et al. 2015). Poor mechanical strength becomes significant drawback of such materials when higher and higher amount of water is absorbed (Omidian et al. 2005). Recently works have been done to overcome this problem by using materials such as nanofibrils and nanowhiskers as fillers for better mechanical strength (Rodrigues et al. 2014; Cheng et al. 2012; Spagnol et al. 2012). The use of filler material helps in obtaining hydrogels having high absorption capacity as well as mechanical strength.

Among natural biopolymers, starch has advantage for use in hydrogel preparation because it is abundant, cheaper, and suitable for chemical modification. Starch also has better plasticity and mechanical resistance. Guilherme et al. (2012) chemically modified starch with glycidyl methacrylate using 4-(*N*,*N*-dimethylamino)pyridine and *N*,*N*,*N*′,*N*′-tetramethylethylenediamine as catalysts. The resultant product undergoes hydrogelation by free-radical reaction. The free-radical polymerization reaction was carried out in the presence of acrylic acid and acrylamide assisted by ultrasound. The material displayed 150 times absorption of its dry weight in a duration of 200 min.

Similarly, hydrogels based on gum arabic were prepared by using glycidyl methacrylate (GMA) for modification (Guilherme et al. 2007). Organic solvents and toxic reagents are no longer used for modification process and are replaced by GMA. Vinylated gum arabic has been obtained without catalyst by using water as solvent. As GMA is insoluble in water, the system was stirred at high speed at 60–65 °C. The modification reaction occurs at the interface layers of GMA and water. The hydrogel cross-linking was done by reaction with sodium acrylate and acrylamide. The product absorbed water 500 times its dry weight and showed good mechanical strength.

Pourjavadi and coworkers used agar for development of superabsorbent hydrogels. Graft copolymerization of acrylic acid and 2-acrylamido-2-methylpropane sulfonic acid was done over agar in aqueous medium. A persulfate initiator was used along with a bifunctional hydrophilic cross-linker. The swelling capacity of the final product varied with changes in reaction parameters. Maximum of 1100 g/g water absorption capacity at optimized reaction conditions was reported by the authors (Pourjavadi et al. 2007).

4.2 Soil Conditioners

A soil conditioner is defined as any synthetic organic chemical or chemically modified natural substance that stabilizes soil aggregates, and/or favorably modifies the structural or physical properties (Aslam 1990). Synthetic polymers when used as soil conditioners improved the physical properties of soil, increased crop growth, and reduced soil erosion (Boodt 1975). The use of hydrogels in agriculture as such has not been prevalent because of high cost. Scarcity of water and desertification of soils are one of the most severe anthropogenic problems in about one-third of lands around the world. To feed the ever-growing population, it is necessary to restore these degrading lands. As these lands are also low in organic matter content, hydrogel materials when added to these soils can act as humus like substance because of their hydrophilicity and free carboxylic groups. Therefore, along with water retention, these hydrogels also increase cation exchange capacity and overall physical properties of the soils (Hüttermann et al. 2009). Hydrogels have been successfully utilized as soil conditioners in horticultural crops for increasing water and nutrient retention in sandy soils (Bouranis et al. 1995). Hydrogels affect various soil properties such as soil structure, porosity, density, texture, permeability, and water infiltration. They reduce evaporation and irrigation requirement, reduce erosion, and enhance aeration and micro-flora activity (Abd El-Rehim et al. 2004).

Hydrogel can act as reclamation agent for light sandy soils and for substrates in hydroponics as it imparts various soil properties which are present in normal arable land (Azzam 1985). The optimum concentration of application of hydrogel depends on various factors such as age and nature of the plant as well as soil properties and environmental conditions. Generally, 0.05–0.1% dry hydrogel is applied with seeds during planting (Zohuriaan-Mehr 2006). In case of forestry, hydrogel can be used during transplantation. Hydrogel can be applied over tree roots when they are transported for transplantation to prevent them from drying. Hydrogel composites have been used in dry areas of China to grow rice, soybean, sugar beet, etc. It was found that the hydrogels increased the yield of rice, soybean, and sugar beet crops (Gao 2003).

Several workers have investigated on application of hydrogels as soil conditioners. Saponified cassava-based starch-graft-poly(acrylamide) hydrogels were evaluated for their effect on physical, chemical, and biological properties of soil and growth-related parameters of chilly (*Capsicum annum* L.) at different irrigation intervals (Parvathy and Jyothi 2012). It was reported that the moisture retained in the soil was dependent on the concentration of hydrogel which provided a controlled release of absorbed water. These hydrogels can be utilized to combat climate change mainly in moisture stress conditions as they improve the soil moisture levels and physicochemical properties. Agaba et al. (2011) reported that moisture retention by using hydrogel is effective for plantation trees in forest establishment and influences the plant growth and various soil properties such as temperature, aeration, nutrient uptake, transport, and transformation. Demitri et al. (2013) studied the feasibility of carbodiimide cross-linked cellulose hydrogels in arid areas. Three formulations of the hydrogel were used for controlled release of nutrients along with water. It was reported that the water stored in hydrogel is released as the soil dries, and thus, moisture levels are maintained for longer durations.

Hydrogels also increase the soil porosity and provide better aeration to plant roots. Chemically modified pectin-based hydrogels were studied for release of urea, phosphate, and potassium (Guilherme et al. 2010). Swelling capacity of hydrogels was measured in saline and distilled water at different pressures. It was concluded that these hydrogels can conserve moisture in a pressure range in which a large variety of horticultural crops can absorb water. Therefore, such hydrogels can be used as soil conditioners. Effect of hydrogel type and concentration on germination and growth of maize (*Zea mays*) was investigated (Abd El-Rehim et al. 2004). The plant growth parameters such as plant height, dry weight, and leaf width were increased with concentration of hydrogel. It was reported that the polyacrylate-based hydrogels improved soil physical properties and reduced the wilting period. Optimum concentration of hydrogel was investigated for its use in soils of Haouz, Morocco. Apart from the water retention capacity of the hydrogel, the study also focused on effect of pH and ions present in soil. The polymer was found to increase the water retention in soil and reduce the irrigation requirement (Bakass et al. 2001).

Hydrogels were evaluated for use as conditioners to help the establishment and growth of plant in limited irrigation conditions. Starch copolymer and polyacrylamide copolymer were studied for their effect on growth of barley and lettuce in sandy soil medium (Woodhouse and Johnson 1991). The use of hydrogel increased the period between field capacity and wilting by 300%. Total dry matter produced and the water use efficiency were also increased by hydrogel use. Effect of hydrogel on emergence and growth of seedling were studied. The starch-based hydrogels prepared by graft copolymerization with acrylic acid and acrylamide were studied. It was reported that the water-absorbing capacity of hydrogels depends on water conductivity. The use of these hydrogels increased the overall water retention capacity of soil (Chen et al. 2004). A commercial hydrogel (Stockosorb K 400) was evaluated for growth of *Pinus halepensis* seedlings in water stress conditions (Hüttermann et al. 1999). Maximum survival of plants was observed when the hydrogel concentration was highest.

Poly(AA-ethyl acrylate-vinyl acetate) hydrogel was found to increase the growth and germination of soyabean (Knypl and Knypl 1993). It has been found useful in decreasing the evaporation loss from soil and water consumption of the crop as well. When plants were inoculated with Frankia and hydrogel polymer, the root nodules showed better growth (Kohls et al. 1999). Polyacrylamide-based hydrogels were prepared by Raju and Raju (2001) and were evaluated for their soil conditioning properties. The hydrogels showed better water retention and growth in sunflower and bean crop. Hydrogels derived from guar gum have also been reported for their use as soil additive (Lokhande and Varadarajan 1992). Chu and coworkers prepared composite hydrogel made up of Polyacrylic acid and sodium humate (Chu et al. 2006). It was synthesized by acrylic acid graft copolymerization on sodium humate. The composite hydrogel was evaluated and found to have positive effect on the growth of maize crop.

Abedi-Koupai and coworkers studied the effect of hydrogel on water absorption in soil and other plant growth parameters (Abedi-Koupai and Sohrab 2004). Application of hydrogel and its effect on water retention properties of three different soils of Iran was investigated. Plant growth parameters of *Cupressus arizonica*, an ornamental plant in water-stressed conditions were also determined (Abedi-Koupai and Asadkazaemi 2006). The residual water and saturated water contents were reported to increase by use of hydrogel. Application of 6 g/kg hydrophilic polymer in sandy loam soil resulted in 2.3 times increase in available water.

Apart from crop plants, hydrogels have also been evaluated for use in cultivation of mushrooms. Sook and coworkers evaluated the effect of hydrogel as medium for growth of edible mushrooms, *Pleurotus sajor-caju* and *Hericium erinaceus* (Sook and Jae-Sik 2000). The hydrogel improved the mycelial growth and production of sporophores of mushrooms. The optimum hydrated level concentration of hydrogel was determined to be 2–2.5 g/cm^3.

Hydrogels have also been studied for use in landscaping and turf grasses (Quinn 1990). The use of hydrogel reduced the amount of irrigation required for maintenance of turf grass especially in hot summer conditions. Hydrogel also improved the density, color, and coverage of turf.

4.3 Nutrient Delivery

Plant nutrients when applied to soil are subject to various forms of losses such as leaching, volatilization, runoff. Therefore, only a portion, about 20–25% of applied nutrients is available to crops and the loss of nutrients in leaching, chemical processes, excess rains, and runoff also results in contamination of groundwater and eutrophication of surface water bodies. Out of these, nutrient loss by leaching is high in porous sandy soils. An alternative approach that has been more recently investigated involves the controlled release of nutrients from the fertilizer-loaded hydrogels (Ni et al. 2009, 2011; Guilherme et al. 2010; Davidson and Gu 2013; Aouada et al. 2008; Zhou et al. 2014).

A controlled release system is aimed at protecting the reserve of active ingredient for releasing it in a slow controlled rate so that the concentration in the target system is maintained at optimum levels for extended period of time without affecting the efficiency. Controlled release application of agrochemicals is helpful in maintaining their concentration in the soil at optimum level and also reduces runoff losses (Aouada et al. 2011). A variety of biopolymers such as cellulose, chitin, tragacanth gum, guar gum have been used for controlled release application of fertilizers (Jamnongkan and Kaewpirom 2010; Guilherme et al. 2010; Buchholz and Graham 1998; Saruchi et al. 2014).

Various controlled nutrient release hydrogels based on natural polysaccharides have been found to enhance the efficiency of agrochemicals by reducing their cost, toxicity, and environmental pollution (Noppakundilograt et al. 2015). Another advantage is that a sustained release of optimum level of nutrients can be achieved in one application.

The nutrients in hydrogels are loaded by two approaches, viz. post-synthesis loading and in situ loading. Post-synthesis loading is done after the processing of hydrogel, while in situ incorporation occurs during the hydrogel processing itself. In the post-synthesis approach, the hydrogel is swelled together with active ingredient which diffuses inside the swollen polymer matrix by absorption. The effectiveness of this method depends upon the physical and chemical affinity of the active ingredient for the polymeric network of the hydrogel. In case of in situ loading, the nutrient is incorporated in hydrogel material during synthesis and remains in dried form before adding to the soil. The hydrogel swells by irrigation or rainwater, and the release of nutrient is activated. The water absorbed in hydrogel dissolves the nutrient which can diffuse out through the polymer matrix (Fig. 5).

The release of nutrients outside the hydrogel matrix is affected by the swelling rate (Ruvalcaba et al. 2009; Gil et al. 2007). The whole quantity of nutrients present in the matrix is not released, and a portion of it remains as reserve during drier periods. When irrigation or rainwater appears, the release mechanism is activated again, thereby providing a prolonged supply of nutrients with minimum leaching losses. The in situ method is better as it has higher loading efficiency when compared to post-loading method (Zheng et al. 2007). More than one active ingredient

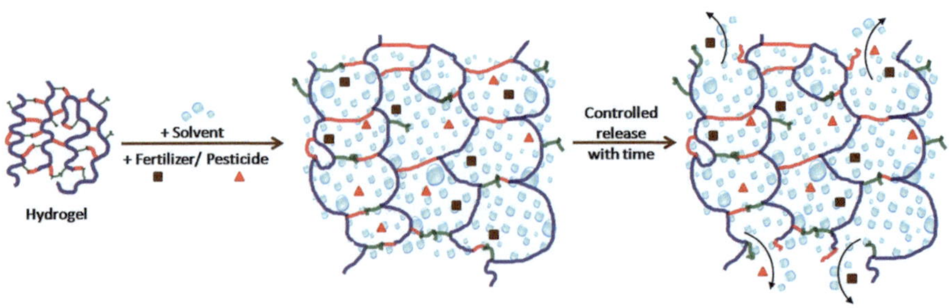

Fig. 5 Entrapping of fertilizer/pesticide and their controlled release through hydrogels

can be added to single hydrogel, each one of them having separate specific rates of release. This way, the cost of application can be reduced.

Optimum availability of water and nutrients in soil is important for growth of agricultural crops. The levels of essential nutrients such as nitrogen, phosphorus, potassium, calcium, sulfur, copper, iron, and boron are often depleted and are supplemented by addition of fertilizers and manures (Saruchi et al. 2014). There are leaching losses from 40 to 70% of applied nitrogen and 50 to 70% of potassium which require application of large quantities of fertilizers (Wu and Liu 2008), and consequently, this results into environmental pollution. Therefore, the use of polysaccharide-based chemical cross-linked hydrogels has been studied for controlled release of fertilizers in soil (Wu and Liu 2008; Wang et al. 2014; Shaviv 2001). Chemically cross-liked hydrogels for nutrient release have been more reliable than coated polymers (Zheng et al. 2009). The nutrient delivery in chemically cross-linked hydrogels is dependent on the concentration gradient of nutrient from inside of hydrogel to external medium and therefore can provide nutrient as per the requirement of the plant (Zheng et al. 2009). Other than the nutrient flow by concentration gradient, processes such as diffusion, convective flow, and chain relaxation also take place. These processes can be described by Fickian or non-Fickian mathematical models. There may be combination of diffusion and convective flows resulting in controlled nutrient release (Shaviv 2001; Wang et al. 2011; Shavit et al. 1997). Macromolecular chain relaxation of polymeric hydrogels occurs by swelling and deswelling (Brazel and Peppas 1999). However, other than these processes, the absorption of water and release of nutrients depends upon the type of polymer and density of cross-linking during the synthesis of polymeric hydrogel as well as the pH and ionic strength of the solution.

Guilherme and coworkers reported that swelling properties of superabsorbents based on modified pectin in saline solutions were of same order that of distilled water and were not affected by presence of salts (Guilherme et al. 2009). These hydrogels showed controlled release of urea, phosphate, and potassium. Xu and coworkers prepared acrylic acid and diallyl-dimethyl-ammonium-chloride based amphoteric hydrogels by solution polymerization for controlled release of ammonium nitrate (Xu et al. 2005). The ratio of anionic groups to cationic groups was varied to obtain different hydrogels. The effect of salt concentration, pH of solution, and temperature was also investigated. The hydrogels having lower ratio of anionic group to the cationic groups had better swelling capacity and tolerance to high salt concentration. These hydrogels were found to be effective for release of ammonium nitrate.

Methylcellulose- and polyacrylamide-based hydrogels were evaluated by Bortolin and coworkers for controlled release of two fertilizers, viz. ammonium sulfate and potassium phosphate (Bortolin et al. 2012). The addition of methyl cellulose with polyacrylamide increased the amount of total fertilizer loaded and prolonged their release. Loading amount as well as release of fertilizers was affected by the hydrophilicity of the polymer decided by the ratio of acrylamide to methylcellulose. Recently controlled release fertilizers based on graphene oxide were prepared by Zhang and coworkers (2014a, b). The fertilizers encapsulated with films of graphene oxide had their release prolonged by 8 h.

4.4 Pesticide Carriers

Pesticides in soil are also subject to leaching and runoff losses which may contaminate groundwater and surface water bodies causing serious hazards. These losses can be minimized by using slow release pesticide systems. Controlled release of pesticides benefits crops for longer periods and reduces the dosage and number of application. In conventional agriculture, application of excessive quantity of agrochemicals is being practiced to get quick results. But actually, the use of pesticides beyond recommended doses in the greed to get quick results leads to their discharge in the environment affecting non-target organisms and causes environmental pollution (Bajpai and Giri 2003; Thakur and Thakur 2014). Due to overuse of the chemicals, resistance can be developed in the target pests toward the pesticide. By adopting controlled release of agrochemicals, pesticide-related health hazards can be minimized and residues on food stuffs can also be controlled which eases handling of the harvested product (Tsuji 2001).

The delivery of agrochemicals using controlled release polymer matrix offers several advantages by avoiding the use of surplus amounts of active substances and also delivers active ingredient slowly over a period of time (Wang et al. 2007). It also reduces quantity of active ingredients required for obtaining same results over a particular time span due to which other plant or animal species are least affected (Aouada et al. 2011).

Based on mode of functioning, controlled release polymer systems are divided into two groups (Mitrus et al. 2009). The first is one where active ingredient is dissolved, dispersed, or encapsulated within the polymeric matrix. Here, the release takes place by diffusion or through biological or chemical breakdown of the releasing polymer. In the second category the active ingredient either constitutes a part of the macromolecular backbone, or is chemically or physically attached to it. After biological or chemical cleavage of the bond with the polymer, the bioactive agent is released in the surrounding matrix.

The advantages of the controlled pesticides release systems include reduced toxicity, increased efficacy, lesser environmental impact from pesticides and their applications, and reduced potential transportation hazards. It also addresses new product development through which advanced pesticide delivery technologies can be facilitated (Aouada et al. 2011; Abd-El-Rehim et al. 2005). Therefore, it is economical and reduces the environmental load of pesticides. Several biopolymers have been tried as matrices for controlled release of agrochemicals.

Kenawy in 1998 synthesized cross-linked polyacrylamide gels by using the free-radical polymerization technique (Kenawy 1998). Further, their derivatives were prepared by trans-amidation with different diamines such as ethylenediamine, hydrazine hydrate. The synthesized gels were studied for release of

2,4-D (2,4-dichlorophenoxyacetic acid) herbicide. The release from the matrices was examined and estimated at 25 °C in water solution buffered at pH 4, 7, and 9 by UV–Vis spectrophotometer. Results revealed the pH-dependent release of 2,4-D, which was lower at pH 4 than in neutral or alkaline medium. Cross-linked polyacrylamide hydrogels trans-amidated with bis-(3-aminopropyl) poly(tetrahydrofuran)-1100 showed the best release rate.

Kulkarni et al. (2000) studied the encapsulation and release rate of a natural liquid pesticide "neem seed oil (NSO)" derived from seeds of *Azadirachta indica*, using sodium alginate as a vehicle carrier and glutaraldehyde as a cross-linker. The absence of any chemical interactions between active ingredients and polymer as well as cross-linking agent was confirmed by FTIR. With increase in the degree of cross-linking of the sodium alginate by glutaraldehyde, a significant decrease of NSO release from the beads was observed. The empirical parameter "n" and the kinetic constant "k" values calculated for the release of NSO from the beads indicated that the diffusion deviates slightly from Fickian transport and showed a decreased release with the increase in cross-linking.

Işıklan (2004) studied the release of carbaryl insecticides through polymeric beads. They investigated the effect of various factors during bead preparation such as percent of carboxymethylcellulose, ratio of carbaryl insecticide to carboxymethylcellulose, concentration of cross-linker as well as effect of addition of filler material (kaolin clay). The carbaryl release was increased when the ratio of carbaryl to carboxymethyl cellulose was low or carboxymethyl cellulose concentration was higher or the quantity of filler added was more.

To study the release of thiram, a dithiocarbamate fungicide, Singh and coworkers synthesized starch–alginate–clay beads with different compositions by varying the amount of kaolin and bentonite clays (Singh et al. 2009a, b). The beads showed good loading capacity of thiram fungicide. The integration of kaolin and bentonite in starch–alginate beads was found to be effective in controlling the release of thiram. Bentonite-based formulations showed slower release than kaolin-based formulations. The release followed non-Fickian diffusion mechanism. The decrease in the release of thiram from 10 mg in control, to 6.9 and 6.3 mg in the presence of kaolin and bentonite, respectively, was due to differences in the intercalation ability of bentonite clay mineral, whereas no such intercalation of thiram was found with kaolin. Also, the presence of kaolin and bentonite in starch–alginate bead further reduced the release of the thiram from the formulation. Likewise, other starch-based hydrogels have also been used for agrochemical delivery system (Baur 1980; Schreiber et al. 1988; Jana et al. 2001; Frederiksen et al. 2002).

Roy and coworkers prepared biopolymer microspheres of sodium alginate and starch by $CaCl_2$ cross-linking. A series of such microspheres was prepared with different compositions by varying the amounts of sodium alginate, starch, and $CaCl_2$ (Roy et al. 2009). The prepared beads were loaded with pesticide, chlorpyrifos, and studied for its release pattern. The microspheres beads delivered optimum swelling at 57.3: 42.7 wt% of alginate: starch composition. The cross-linked beads showed great potential for the release of chlorpyrifos. It was observed that the fractional release of chlorpyrifos increases with increasing wt% of alginate and decreases with increasing content of starch. The sustained and controlled release was given by the beads with more alginate and less starch with cumulative release up to 14 days.

Chevillard et al. (2011) introduced organically modified nanoclay, montmorillonite (C30B) in wheat gluten (WG)-based formulation, and a model pesticide, and ethofumesate was imbibed in it in order to obtain slow release pattern. The aim was to use nanoclay in modulating transfer and biodegradation properties of active ingredient in bio-sourced polymers. Controlled release properties were examined through release experiments in water in comparison with the commercial formulations. Degradation study of ethofumesate in soil by respirometric experiments confirmed its non-biodegradable behavior, whereas after addition of 0.26% of ethofumesate, biodegradation of WG-based formulation was slightly but significantly delayed. This ecotoxic effect of ethofumesate which is responsible for delaying biodegradation of formulation was reduced after introducing C30B in the formulation. This also resulted in slower release of pesticide in water that could be further enhanced by adding organoclays in the materials. Binding of ethofumesate with nanoclays reduces its hazards to microorganisms and also probably less subjected to leaching, making this delivery system eco-friendly.

Alemzadeh and Vossoughi (2001) prepared hydrogel systems based on polyvinyl alcohol polymeric network and membranes with glutaraldehyde as cross-linking agent. The product was studied for release of paraquat herbicide. It was reported that higher concentration of the cross-linking agent decreased the release of active ingredient from the system. The hydrogel showed higher adsorption at lower temperature.

Aouada and coworkers synthesized polyacrylamide- and methylcellulose-based biodegradable hydrogels for the controlled delivery of paraquat herbicide (Aouada et al. 2009, 2010). The hydrogels were synthesized using *N,N*-methylene-bis-acrylamide as cross-linker, *N,N,N′,N′*-tetramethylethylenediamine as catalyst agent, and sodium persulfate as initiator. Hydrogels were loaded with paraquat herbicide by soaking into its aqueous solution for 30 h. Around 82% of the paraquat from the solution was loaded in the matrix. The adsorption of paraquat was higher with methylcellulose as compared to the hydrogels without methylcellulose. Each hydrogel was removed from the solution, and the quantity of paraquat left in the remaining solution was determined. It was reported that the rate of release was fast initially indicating the release of active ingredient from surface while swelling. The release rate was slower in later stages, and the herbicide was released in a controlled manner at a constant concentration for a period of 15–46 days.

Singh et al. (2011a, b) have developed polysaccharide-based controlled release beads of herbicide atrazine. The controlled release formulations were made using alginate, neem leaf powder, kaolin, and bentonite clays. The use of neem leaves powder in ionotropic gelation of alginate had added advantage of its natural pesticidal activity. Bead size (diameter), entrapment efficiency, and amount of beads formed were taken as reaction parameters for the synthesis of various bead formulations. It was reported that increase in clay content resulted in increase in size and yield of beads. Various release characteristics of the formulation such as entrapment efficiency, gel characteristic, diffusion mechanism, and bead size were evaluated. The amount of herbicide atrazine released from beads based on calcium alginate was found to be 14.8 $\pm$ 1.2 mg. The release occurred slowly for 300 h, and the amount was further increased with addition of neem leaf powder.

In vitro release studies of atrazine were conducted by using dry and loaded formulations in water (Bergaya et al. 2006). The amount of atrazine released from calcium alginate beads was found to increase when neem leaf powder was added to the bead. Among clays, bentonite incorporated beads resulted in slower release of atrazine as compared to those of kaolinite. The rate of release of atrazine was also higher initially as compared to later stages. Therefore, the overall release of atrazine occured for a prolonged period and thus was helpful in minimizing environment hazards. The study concluded that the presence of clays in neem leaf powder-alginate beads has resulted in slower atrazine release from formulations.

Alginate-based formulations were also evaluated for the release of herbicides metribuzin and isoproturon. The release of the active ingredients was slower when compared to conventional formulations (Pepperman and Kuan 1993; Villafranca-Sánchez et al. 2000). Also, the effectiveness of these delivery systems were reported to be better when two or more herbicides are used (Johnson and Pepperman 1998). Herbicides such as atrazine, monolinuron, simazine, chloridazon, chloroxuron, and desmetryn were formulated in beads based on alginate. The beads were evaluated for controlled release and were found to slower down the release of herbicides (Pfister et al. 1986).

In case of alginate- and gelatin-based beads, it has been reported that increase in proportion of gelatin decreases the release of pesticides such as cypermethrin and neem seed oil (Roy et al. 2009; Kulkarni et al. 2000). Both starch and alginate are biodegradable in soil. The pesticides are released from these beads by both swelling and degradation processes. Also, various soil factors such as enzymes like amylase and alginate lysases affect the release (Trimnell et al. 1985; Wong et al. 2000). Zhu and coworkers prepared controlled release systems based on gelatin and acacia gum (Zhu et al. 2009). The delivery systems were prepared by coacervation for insecticide release.

Polyvinyl chloride, carboxymethyl cellulose, and carboxymethyl cellulose-kaolinite-based composite delivery system were prepared for metribuzin herbicide. The formulation showed better results by controlling 75% of weeds than conventional product (57.14%) in wheat crop (Kumar et al. 2010). Singh and coworkers reported starch- and acrylamide-based slow release hydrogel system for the release of fungicide thiram (Singh et al. 2007, 2008). Apart from the release, the hydrogels showed good water-holding capacity. Therefore, such materials can be used for pesticide delivery as well as for water retention.

Lignin and lignin-based natural polymers have been investigated for the release of agrochemicals (Thakur and Thakur 2015). The release of pesticides and water absorption was dependent on the interactions between the functional groups present in pesticides with that of the polymer. The size of the polymeric granule also affected the release. Smaller granules resulted in higher release of pesticides (Chowdhury 2014). Singh et al. (2015) prepared agar-, starch-, and polyacrylamide-based hydrogels for the release of atrazine herbicide. The hydrogel showed maximum 551% swelling. Release studies conducted for 144 h showed that hydrogel was suitable for slower release of pesticide in agriculture.

Recently, polymeric delivery systems having in situ gelling properties have been studied for drug delivery applications. It is a new technique in which the solution containing the drug or bioactive agent converts into a gel when it applied to the target site. Such gel-based delivery systems can also be explored for their applications in pesticide delivery. The in situ formation of gels depends on various factors such as temperature, pH, ions, radiations. These gels forming formulation shall be able to release the product in controlled manner (Hari et al. 2015). These gels are polymer-based colloidal solutions, which undergo phase transition from sol to gel. The physical and chemical changes responsible for release of product are affected by physiological environment (Geethalakshmi et al. 2012, 2013). Many researchers have been working on the agricultural applications of biopolymer-based products, mainly hydrogels. Significant contributions in this field are listed in Table 1.

4.5 Other Applications

Agriculture relies on an adequate supply of good quality irrigation water. Until now, irrigation water quality concerns have often been neglected because good quality water supplies have been plentiful and readily available (Kitila et al. 2014). But today, the situation is changing in many areas and due to contaminated irrigation water toxic chemicals and heavy metals are ruining soil health and also leaving their residues in the crops. In most of the cases, water is taken from water bodies such as river, pond, and dam or pumped up from the soil subsurface. Recently due to increasing industrialization, fair quality water is used in the industry and its wash water/wastewater is being released in the same water body which adds to many toxic pollutants in it. Key pollutants in such contaminated water include heavy metals, industrial dyes, dissolved salts, industrial sludge, organic pollutants, oils, and excess nutrients due to runoff or leaching down from cultivated lands. Though there is infrastructure available to recycle and purify this water for domestic and drinking purpose, no such care is being taken for pre-treatment of water before irrigation. This leads to accumulation of these pollutants in agricultural land and subsequently entering into the food chain via crops grown on that area. Irrigation with such contaminated water can lead to accumulation of heavy metals such as cadmium, chromium, mercury, lead, and arsenic, all of which appear in the World Health Organization's list of 10 chemicals of major public concern.

In recent researches, lot of emphasis is being given on the use of biopolymer-based sorbents for removal of the heavy metals and dyes from the

Table 1 Application of biopolymers in agriculture

Source	Polymer	Applications	Reference
Cellulose	Carboxymethyl cellulose with polyvinyl chloride and Kaolinite composites	Pesticide delivery	Kumar et al. (2010)
	Methylcellulose and polyacrylamide hydrogel	Pesticide delivery	Aouada et al. (2010)
	Carboxymethyl cellulose beads with kaolin clay by ionotropic cross-linking with copper ions	Pesticide delivery	Işıklan (2004)
	Carboxymethyl cellulose copolymer with poly(acrylamide)	Nutrient delivery	Bortolin et al. (2012)
	Cross-linked with carbodiimide	Water retention and nutrient delivery	Demitri et al. (2013)
	Carboxymethyl cellulose hydrogel with starch using aluminumsulfate octadecahydrate as cross-linker	Soil conditioner	Nnadi and Brave (2011)
	Carboxymethyl cellulose ionically cross-linked with iron and calcium salts	Nutrient delivery	Davidson and Gu (2013)
	Poly(*N*-vinyl-pyrrolidone) hydrogels with outer coating of ethylcellulose	Nutrient delivery	Ni et al. (2009)
	Hydroxyethyl cellulose and carboxymethyl cellulose cross-linked with citric acid	Nutrient delivery	Ni et al. (2009)
Starch	Starch-based hydrogel prepared by modification with glycidyl methacrylate	Water retention	Guilherme et al. (2012)
	Acrylamide/methyl acrylamide/methyl acrylicacid	Pesticide delivery	Singh et al. (2007, 2008, 2009a, b)
	Graft copolymer with acrylic acid and acrylamide	Soil conditioner	Chen et al. (2004)
	Microencapsulation	Pesticide delivery	Frederiksen et al. (2002)
	Biodegradable film	Pesticide delivery	Jana et al. (2001)
	Microsphere with sodium alginate	Pesticide delivery	Roy et al. (2009)
	Copolymer with acrylamide	Soil conditioner	Woodhouse and Johnson (1991)
Cassava starch	Saponified cassava starch grafted with polyacrylamide	Soil conditioner	Parvathy and Jyothi (2012)
Corn starch	Copolymer with acrylamide	Soil conditioner	Kohls et al. (1999)

(continued)

Table 1 (continued)

Source	Polymer	Applications	Reference
Alginate	Sodium alginate encapsulation	Pesticide delivery	Kulkarni et al. (2000)
	Calcium alginate gel	Pesticide delivery	Pfister et al. (1986)
	Beads with neem leaf powder, kaolin, and bentonite clays	Pesticide delivery	Singh et al. (2011a, b)
	Copolymer with acrylamide by electron beam irradiation	Soil conditioner	Abd El-Rehim et al. (2005)
Chitosan	Polyvinyl alcohol, polyvinyl alcohol/chitosan, and chitosan-based hydrogel cross-linked by glutaraldehyde	Water retention and nutrient delivery	Jamnongkan and Kaewpirom (2010)
	Poly(vinyl alcohol) and chitosan hydrogel by cross-linking with glutaraldehyde vapor deposition	Nutrient release and water absorption	Noppakundilograt et al. (2015)
	Fertilizer with chitosan as inner coating and external coating of poly(acrylic acid-co-acrylamide) hydrogel	Nutrient delivery and water retention	Wu and Liu (2008)
Lignin	Radical polymerization with corn starch, acrylamide, and acrylic acid by using potassium persulfate as insulator	Soil conditioner	Lee et al. (2013)
	Polyurethane-based hydrogel with isocyanate-terminated polyurethane ionomer (IPUI) as cross-linker	Nutrient delivery	Peng and Chen (2011)
Guar gum	Graft copolymer with acrylonitrile	Soil additive	Lokhande and Varadarajan (1992)
	Graft copolymer with acrylic acid	Water retention	Wang and Wang (2010a, b)
	Graft copolymer with acrylic acid	Soil conditioner and nutrient delivery	Chandrika et al. (2014)
Tragacanth gum	Graft copolymer with acrylic acid	Water retention and nutrient delivery	Saruchi et al. (2014)
	Graft copolymer with acrylic acid	Nutrient delivery and water retention	Kaith et al. (2013)
Agar	Copolymer with starch and acrylamide	Pesticide delivery	Singh et al. (2015)
	Copolymer with acrylic acid and 2-acrylamido-2-methylpropane sulfonic acid (AMPS)	Water retention	Pourjavadi et al. (2009)
Gum arabic	Hydrogel synthesis by modification with glycidyl methacrylate	Water retention	Guilherme et al. (2007)
	Coacervation with gelatin	Pesticide delivery	Zhu et al. (2009)
Cashew gum	Copolymer with acrylamide	Water retention	Guilherme et al. (2005)

(continued)

Table 1 (continued)

Source	Polymer	Applications	Reference
Gluten	Extrusion granules with montmorillonite	Pesticide delivery	Chevillard et al. (2011)
Pectin	Pectin hydrogels with acrylamide and sodium acrylate modified with glycidyl methacrylate	Water retention, nutrient delivery, soil conditioner	Guilherme et al. (2009)
Humate (humic acid)	Sodium humate graft copolymer with acrylic acid	Soil conditioner	Chu et al. (2006)
Heteropolysaccharide resin from *Cochlospermum*species	Graftco polymerization with acrylamide, persulfate initiator by microwave-assisted free-radical polymerization	Soil conditioner	Singh et al. (2009a, b)

Table 2 Application of biopolymers for water decontamination

Gum	Modified adsorbents	Use in removal of-	References
Alginate	Post-cross-linking sodium alginate (SA) beads	Cu^{2+}, Ag^{+}, and Fe^{3+}	Lu et al. (2015)
	Nanocomposites of glucose oxidase/$MnFe_2O_4$/calcium alginate, laccase/$MnFe_2O_4$/calcium alginate, and $MnFe_2O_4$/calcium alginate	Methylene blue	Shojaat et al. (2016)
	Cross-link copolymerization of acrylic acid (AA) and hydroxy ethyl methacrylate and sodium alginate	Congo red and methyl violet	Mandal and Ray (2013)
	Semi-IPN hydrogels of Acrylic copolymers and sodium alginate	Basic fuschin (BF) and methyl violet (MV) dyes	Bhattacharyya and Ray (2015)
	Macroporous alginate substrate-bound growth of Fe nanoparticles (Fe NPs/MAS)	Nitrates	Lee et al. (2016)
	Magnetic ferrite nanoparticle–alginate composite	Basic Blue 9 (BB9), Basic Blue 41 (BB41), and Basic Red 18	Mahmoodi (2013)
	Mesoporous titania spheres derived from sodium alginate–gum acacia composite beads	Reactive blue H5G dye	Singh et al. (2015)
	Superparamagnetic nanocomposite of sodium alginate (Fe_3O_4@TSTC[4]AS-s-SA) fabricated sodium alginate-supported tetra-sodium thiacalix[4]arene tetrasulfonate	Cu^{2+}, Cd^{2+}, Pb^{2+}, Co^{2+}, Ni^{2+}, and Cr^{3+}	Lakouraj et al. (2014)
	Magnetic alginate beads	Methylene blue and methyl orange	Rocher et al. (2008)
	Porous alginate aerogel beads	Cu^{2+}, Cd^{2+},	Deze et al. (2012)
Chitosan	Magnetic polydopamine (PDA) cross-linked chitosan (CS) hybrid nano-biosorbent	Methylene blue and malachite green Cr(VI), Pb^{2+}, Hg^{2+}	Wang et al. (2016)
	Magnetic chitosan composites	Metal and dyes	Reddy and Lee (2013)
	Chitosan–lignin composites	Anthraquinonic dye, Remazol Brilliant Blue R (RBBR), and Cr(VI)	Nair et al. (2014)
	Ethyl acrylate-grafted chitosan	Pb^{2+}, Cd^{2+}, and Zn^{2+}	Maleki et al. (2015)
	Nanochitosan and its acetophenone derivative	Cu^{2+}, Pb^{2+}, Cd^{2+}, and Hg^{2+}	Mahmoud et al. (2015)

(continued)

Table 2 (continued)

Gum	Modified adsorbents	Use in removal of-	References
	Chitosan/organic rectorite–Fe_3O_4 composite microspheres (CS/χOREC-Fe_3O_4)	Cu^{2+}, Cd^{2+}	Xie et al. (2015)
	PAMAM-grafted chitosan	Pb^{2+}	Zarghami et al. (2016)
	Natural polymer chitosan, modified chitosan, and chitosan composites	Heavy metals and radio nuclides	Wang et al. (2014)
	O-carboxymethyl functionalization of chitosan	Cr^{2+}, Cd^{2+}	Borsagli et al. (2015)
Cellulose	Carboxy methyl cellulose incorporated acrylic hydrogels	Safranine T and brilliant cresyl blue dyes	Mandal and Ray (2015)
	Epichlorohydrin cross-linked carboxymethyl cellulose fiber	Cd^{2+}	Wei et al. (2015)
	Acrylic acid grafting polymerization carboxymethyl cellulose	Methyl orange, disperse Blue 2BLN, and malachite green chloride	Zhang et al. (2014a, b)
Guar gum	Groundnut husk modified with guar gum	Pb^{2+}, Cu^{2+}, Ni^{2+}	Ahmad and Haseeb (2015)
	Poly(acrylic acid)-grafted guar gum	Pb^{2+}	Pal et al. (2016)
	Guar gum–cerium (IV) tungstate hybrid cationic exchanger	Methylene blue	Gupta et al. (2014)
	Cross-linked polyacrylamide/guar gum graft copolymer	Cr(VI)	Abdel-Halim and Al-Deyab (2011)
	Polyacrylamide-grafted carboxymethyl guar gum (CMG-g-PAM)	Methylene blue	Pal et al. (2011)
	Poly(methylacrylate)-functionalized guar gum	Cr(VI)	Singh et al. (2009a, b)
Gum arabic	Gum arabic-modified magnetic nanoadsorbent	Cu^{2+}	Banerjee and Chen (2007)
	Gum acacia-inspired silica hybrid xerogels	Hg^{2+}	Singh et al. (2011a, b)
	Sulfonated form of gum arabic-powdered	Zn^{2+}, Hg^{2+}	Singh et al. (2011a, b) Das et al. 2015

(continued)

Table 2 (continued)

Gum	Modified adsorbents	Use in removal of-	References
Gum ghatti	Gum ghatti with a copolymer mixture of acrylamide and methacrylic acid	Methylene blue and Methyl violet	Mittal et al. (2015a, b)
	TiO_2 nanoparticle (TiO2NP)-containing hydrogel nanocomposite (HNC) of polyacrylamide-grafted gum ghatti (PAAm-g-Gg).	Methylene blue	Mittal et al. (2016)
	Poly(acrylic acid-aniline)-grafted gum ghatti-based conducting hydrogel	Dyes	Sharma et al. (2014)
Gum kondagogu	–	Cd^{2+}, Cu^{2+}, Fe^{2+}, Se^{2+}, Pb^{2+}, total Cr, Ni^{2+}, Zn^{2+}, Co^{2+}, As^{2+}	Vinod et al. (2010)
	Gum kondagogu-modified magnetic iron oxide nanoparticles	Cd^{2+},Cu^{2+},Pb^{2+}, Ni^{2+}, Zn^{2+}, Hg^{2+}	Saravanan et al. (2012)
	–	Cd^{2+}, Pb^{2+}	Vinod et al. (2009)
Tragacanth gum	Fe_3O_4/poly(methyl methacrylate)-grafted tragacanth gum nanocomposite	Cr (VI)	Sadeghi et al. (2014)
	Tragacanth gum-g-polyamidoxime nano hydrogel	Co^{2+}, Zn^{2+},Cr^{3+}, Cd^{2+}.	Masoumi and Ghaemy (2014)
Karaya gum	Gum karaya grafted with poly(acrylic acid − acrylamide) (GK-cl-P(AA-co-AAM))	Methylene blue	Mittal et al. (2015a, b)
	–	Hg^{2+}	Vinod et al. (2011)
Xanthan gum	Xanthan gum-g-poly(ethylacrylate)	Pb2+	Pandey and Mishra (2012)
	Nanocomposite based on nanosilica modified with xanthan gum grafted with polyacrylamide	Pb2+	Ghorai et al. (2012)
Agar	Agar-based bimetallic nanoparticles	Methylene blue and rhodamine B	Patra et al. (2016)

wastewater, which can be used for the treatment of industrial wastewater, before being released into the water bodies. The biopolymer-based adsorbents/hydrogels showed potential to remove heavy metals and dyes from wastewater and can also be used repetitively in most of the cases. Important research works in this line carried out in the recent past are summarized in Table 2.

5 Conclusion

This chapter discusses in nut shell the biopolymers, their classification, modification by different techniques, and application in various fields of agriculture. Due to their biodegradability, renewability, ease to blend into different products, cost-effectiveness, easy handling, and storage, biopolymers can be used in agricultural applications such as soil conditioners, moisture retainers, controlled nutrient, and pesticide release, and also in other applications such as remediation of heavy metals, dyes form irrigation water. All these properties make biopolymers a unique natural material and provide an edge over its other synthetic counterparts.

The present discussion on biopolymers signifies that the biopolymers and their chemically modified derivatives are yet to be fully explored in agriculture industry. Natural biopolymer-based superabsorbent hydrogels have tremendous potential to be used as culture media in nursery and tissue culture for growing high value seedlings. It can be used for multipurpose operations such as moisture retainer and controlled release of nutrients and pesticides imbibed in it, and being biodegradable, it will improve soil's physical properties after decomposition. Modified biopolymers also have potential to be used as drift control agents in sprays and mists, in liquid flowable pesticides as a stabilizers and emulsifying agents. Recently emerged concept of in situ gels can be utilized in pesticide formulations. Plants and soil systems having various specific ions and pH conditions can activate the formation of gel. The in situ gel-based pesticide formulation would have added advantage of sticking over leaf surface for longer period as compared to conventional foliar spray. For targeting soil application of hydrogels, they need to be cheaper, available in bulk, and easy to handle. As multistep synthesis and using high value reagents such as acrylates, increase in cost, the challenge is to prepare the cheaper hydrogels, which at the same time show good absorption and mechanical properties. The use of nanotechnology in synthesis or incorporating nanofillers to improve mechanical strength and performance of the hydrogels can be one of the important areas to work upon.

Owing to consequences of the issues such as climate change, degrading land, decreasing cultivable area, scarce resources, population rise, the ever-mounting food demand has been a greatest challenge for achieving adequate food production. Biopolymer-based eco-friendly hydrogels ensure potential and high future prospects in building sustainability in agriculture.

References

Abd-Allah AR, Al-Majed AA, Mostafa AM, Al-Shabanah OA, Din AGE, Nagi MN (2002) Protective effect of arabic gum against cardiotoxicity induced by doxorubicin in mice: a possible mechanism of protection. J Biochem Mol Toxicol 16(5):254–259

Abd El-Rehim HA, Hegazy EA, Abd El-Mohdy HL (2004) Radiation synthesis of hydrogels to enhance sandy soils water retention and increase plant performance. J Appl Polym Sci 93:1360–1371

Abd El-Rehim HA, Hegazy EA, Abd El-Mohdy HL (2005) Properties of polyacrylamide-based hydrogels prepared by electron beam irradiation for possible use as bioactive controlled delivery matrices. J Appl Polym Sci 98:1262–12670

Abdel-Halim ES, Al-Deyab SS (2011) Hydrogel from crosslinked polyacrylamide/guar gum graft copolymer for sorption of hexavalent chromium ion. Carbohydr Polym 86:1306–1312

Abedi-Koupai J, Asadkazemi J (2006) Effects of a hydrophilic polymer on the field performance of an ornamental plant (*Cupressus arizonica*) under reduced irrigation regimes. Iran Polym J 15:715

Abedi-Koupai J, Sohrab F (2004) Evaluating the application of superabsorbent polymers on soil water capacity and potential on three soil textures. Iran J PolymSciTechnol 17:163–173

Abu Baker A, Tahir A, Sabah Elkheir MK (2007) Effect of tree and nodule age on some physicochemical properties of gum from *Acacia senegal* (L.) Willd Sudan. Res J Agric and Bio Sci 3:866–870

Addai ZR, Abdullah A, Mutalib SA, Musa KH (2013) Effect of gum arabic on quality and antioxidant properties of papaya fruit during cold storage. Int J Chem Tech Res 5:2854–2862

Adhikari B, Majumdar S (2004) Polymers in sensor applications: progress in polymer science. Polym Sci 29:699–766

Adhikary P, Krishnamoorthi S, Singh RP (2011) Synthesis and characterization of grafted carboxymethyl guar gum. J Appl Polym Sci 120:2621–2626

Agaba H, Orikiriza LJ, Obua J, Kabasa JD, Worbes M, Hüttermann A (2011) Hydrogel amendment to sandy soil reduces irrigation frequency and improves the biomass of *Agrostis stolonifera*. Agri Sci 2:544

Aggarwal G, Sharma A (2010) Controlled release of a poorly water soluble drug prednisolone from heterodisperse hydrogel system. Int J Pharma Biol Sci 2:1–8

Ahmad R, Haseeb S (2015) Absorptive removal of Pb^{2+}, Cu^{2+} and Ni^{2+} from the aqueous solution by using groundnut husk modified with Guar Gum (GG): kinetic and thermodynamic studies. Groundw Sustain Dev 1:41–49

Ahmed EM (2013) Hydrogel: preparation, characterization, and applications. J Adv Res 6:105–121

Alboofetileh M, Rezaei M, Hosseini H, Abdollahi M (2014) Antimicrobial activity of alginate/clay nanocomposite films enriched with essential oils against three common foodborne pathogens. Food Control 36:1–7

Alemzadeh I, Vossoughi M (2001) Controlled release of paraquat from poly vinyl alcohol hydrogel. Chem Eng Process 41:707–710

Alka A, Laxmi B, Suman R (2011) Formulation development and evaluation of sustained release matrix tablet of Lamivudine using tamarind seed polysaccharide. Der Pharmacia Lettre 3:250–2508

Al-Majed AA, Mostafa AM, Al-Rikabi AC, Al-Shabanah OA (2002) Protective effects of oral arabic gum administration on gentamicin-induced nephrotoxicity in rats. Pharmacol Res 46:445–451

Al-Majed AA, Abd-Allah AR, Al-Rikabi AC, Al-Shabanah OA, Mostafa AM (2003) Effect of oral administration of arabic gum on cisplatin-induced nephrotoxicity in rats. J Biochem Mol Toxicol 17:146–153

Anderson SA, Fisher KD, Talbot JM (1993) Evaluation of the health aspects of using partially hydrolyzed guar gum as a food ingredient. FASEB J, 1–61

Aouada FA, Moura MR, Menezes ED, Nogueira AR, Mattoso LH (2008) Hydrogel synthesis and kinetics of ammonium and potassium release. Rev Bras De Cienc Do Solo 32:1643–1649
Aouada FA, Pan Z, Orts WJ, Mattoso LH (2009) Removal of paraquat pesticide from aqueous solutions using a novel adsorbent material based on polyacrylamide and methylcellulose hydrogels. J Appl Polym Sci 114:2139–2148
Aouada FA, de Moura MR, Orts WJ, Mattoso LH (2010) Polyacrylamide and methylcellulose hydrogel as delivery vehicle for the controlled release of paraquat pesticide. J Mater Sci 45:4977–4985
Aouada FA, Mattoso LH, de Moura MR (2011) Biodegradable hydrogel as delivery vehicle for the controlled release of pesticide. INTECH Open Access Publisher
Ashikin WH, Wong TW, Law CL (2010) Plasticity of hot air-dried mannuronate-and guluronate-rich alginate films. Carbohydr Polym 81:104–113
Ashraf S, Soudi MR, Sadeghizadeh MA (2008) Isolation of a novel mutated strain of *Xanthomonas campestris* for xanthan production using whey as the sole substrate. Pak J Biol Sci 11:438–442
Aslam M (1990) Polymers as soil conditioners and sealing agents. Pakistan Academy of Sciences, Islamabad
Azzam RA (1985) Tailoring polymeric gels for soil reclamation and hydroponics. Commun Soil Sci Plant Anal 16:1123–1138
Bahulkar SS, Munot NM, Surwase SS (2015) Synthesis, characterization of thiolated karaya gum and evaluation of effect of pH on its mucoadhesive and sustained release properties. Carbohydr Polym 130:183–190
Bajpai AK, Giri A (2003) Water sorption behaviour of highly swelling (carboxy methylcellulose-g-polyacrylamide) hydrogels and release of potassium nitrate as agrochemical. Carbohydr Polym 53:271–279
Bakass M, Mokhlisse A, Lallemant M (2001) Absorption and desorption of liquid water by a superabsorbent polyelectrolyte: role of polymer on the capacity for absorption of a ground. J Appl Polym Sci 82:1541–1548
Baldrian P, Valášková V (2008) Degradation of cellulose by basidiomycetous fungi. FEMS Microbiol Rev 32:501–521
Banerjee SS, Chen DH (2007) Fast removal of copper ions by gum arabic modified magnetic nano-adsorbent. J Hazard Mater 147:792–799
Bangale GS, Shinde GV, Umalkar DG, Rajesh KS (2011) Natural mucoadhesive material based buccal tablets of nitrendipine-formulation and in-vitro evaluation. J Pharm Res 4:33–38
Baur JR (1980) Release characteristics of starch xanthide herbicide formulations. J Environ Qual 9:379–382
Becker A, Katzen F, Pühler A, Ielpi L (1998) Xanthan gum biosynthesis and application: a biochemical/genetic perspective. Appl Microbiol Biotechnol 50(2):145–152
Bergaya F, Theng BKG, Legaly G (2006) Handbook of clay science. Elsevier, Amsterdam, pp 693–702
Bhattacharyya R, Ray SK (2015) Adsorption of industrial dyes by semi-IPN hydrogels of Acrylic copolymers and sodium alginate. J Ind Eng Chem 22:92–102
Bixler HJ, Porse H (2011) A decade of change in the seaweed hydrocolloids industry. J Appl Phycol 23:321–335
Blackburn RS (2004) Natural polysaccharides and their interactions with dye molecules: applications in effluent treatment. Environ Sci Technol 38:4905–4909
Blanco-Pascual N, Montero MP, Gómez-Guillén MC (2014) Antioxidant film development from unrefined extracts of brown seaweeds *Laminaria digitata* and *Ascophyllum nodosum*. Food Hydrocolloids 37:100–110
Boodt DM (1975) Use of soil conditioners around the world. Soil Conditioners. SSS A Special Publication. Soil Science Society of America, Madison, pp 1–12
Bordi F, Paradossi G, Rinaldi C, Ruzicka B (2013) Chemical and physical hydrogels: two case systems studied by quasi elastic light scattering. Phys A 304:119–128

Borsagli FG, Mansur AA, Chagas P, Oliveira LC, Mansur HS (2015) O-carboxymethyl functionalization of chitosan: complexation and adsorption of Cd (II) and Cr (VI) as heavy metal pollutant ions. React Funct Polym 97:37–47

Bortolin A, Aouada FA, de Moura MR, Ribeiro C, Longo E, Mattoso LH (2012) Application of polysaccharide hydrogels in adsorption and controlled-extended release of fertilizers processes. J Appl Polym Sci 123:2291–2298

Bouranis DL, Theodoropoulos AG, Drossopoulos JB (1995) Designing synthetic polymers as soil conditioners. Commun Soil Sci Plant Anal 26:1455

Brazel CS, Peppas NA (1999) Mechanisms of solute and drug transport in relaxing, swellable, hydrophilic glassy polymers. Polymer 40:3383–3398

Buchholz FL, Graham AT (1998) Modern superabsorbent polymer technology. Wiley, New York, pp 279

Buchholz FL, Peppas NA (1994) Superabsorbent polymers: science and technology. American Chemical Society

Burgalassi S, Raimondi L, Pirisino R, Banchelli G, Boldrini E, Saettone MF (2000) Effect of xyloglucan (tamarind seed polysaccharide) on conjunctival cell adhesion to laminin and on corneal epithelium wound healing. Eur J Ophthalmol 10:71–76

Campos EVR, de Oliveira JL, Fraceto LF, Singh B (2015) Polysaccharides as safer release systems for agrochemicals. Agron Sustain Dev 35:47–66

Castel D, Ricard A, Audebert R (1990) Swelling of anionic and cationic starch-based superabsorbents in water and saline solution. J Appl Polym Sci 39:11–29

Celkan A, Acartürk F, Tuğcu-Demiröz F, Gökçora N, Akkaş BE, Güner LA (2016) Gamma scintigraphic studies on guar gum-based compressed coated tablets for colonic delivery of theophylline in healthy volunteers. J Drug Deliv Sci Technol 32(3):1–37

Chandrika KS, Singh A, Sarkar DJ, Rathore A, Kumar A (2014) pH sensitive crosslinked guar gum based superabsorbent hydrogels: swelling response in simulated environments and water retention behavior in plant growth media. J Appl Polym Sci 131:1–12

Chang C, Duan B, Cai J, Zhang L (2010) Superabsorbent hydrogels based on cellulose for smart swelling and controllable delivery. Eur Polym J 46:92–100

Chen P, Zhang WA, Luo W, Fang YE (2004) Synthesis of superabsorbent polymers by irradiation and their applications in agriculture. J Appl Polym Sci 93:1748–1755

Cheng Y, Lu L, Zhang W, Shi J, Cao Y (2012) Reinforced low density alginate-based aerogels: preparation, hydrophobic modification and characterization. Carbohydr Polym 88:1093–1099

Cheng YH, Tsai TH, Jhan YY, Chiu AW, Tsai KL, Chien CS, Chiou SH, Liu CJ (2016) Thermosensitive chitosan-based hydrogel as a topical ocular drug delivery system of latanoprost for glaucoma treatment. Carbohydr Polym 144:390–399

Chevillard A, Angellier-Coussy H, Guillard V, Bertrand C, Gontard N, Gastaldi E (2011) Biodegradable herbicide delivery systems with slow diffusion in soil and UV protection properties. Pest Manag Sci 70:1697–1705

Chowdhury MA (2014) The controlled release of bioactive compounds from lignin and lignin-based biopolymer matrices. Int J Biol Macromol 65:136–147

Chu M, Zhu SQ, Li HM, Huang ZB, Li SQ (2006) Synthesis of poly (acrylic acid)/sodium humate superabsorbent composite for agricultural use. J Appl Polym Sci 102:5137–5143

Chung C, Rojanasasithara T, Mutilangi W, McClements DJ (2016) Enhancement of colour stability of anthocyanins in model beverages by gum arabic addition. Food Chem 201:14–22

Cosenza VA, Navarro DA, Fissore EN, Rojas AM, Stortz CA (2014) Chemical and rheological characterization of the carrageenans from *Hypnea musciformis* (Wulfen) Lamoroux. Carbohydr Polym 102:780–789

Cretois I, Samain H (inventors) (2000) L'oreal, assignee. Hair styling. United States patent US 6,060,044

Das D, Vimala R, Das N (2015) Removal of Ag (I) and Zn (II) ions from single and binary solution using sulfonated form of gum arabic-powdered mushroom composite hollow semispheres: equilibrium, kinetic, thermodynamic and ex-situ studies. Ecol Eng 75:116–122

Davidson DWVM, Gu FX (2013) Controlled root targeted delivery of fertilizer using an ionically crosslinked carboxymethyl cellulose hydrogel matrix. Springerplus 2:1–9

De Oliveira Pedro R, Schmitt CC, Neumann MG (2016) Syntheses and characterization of amphiphilic quaternary ammonium chitosan derivatives. Carbohydr Polym 147:97–103

Demitri C, Scalera F, Madaghiele M, Sannino A, Maffezzoli A (2013) Potential of cellulose-based superabsorbent hydrogels as water reservoir in agriculture. Int J Polym Sci

Deveswaran R, Abraham S, Bharath S, Basavaraj BV, Furtado S, Madhavan V (2009) Design and characterization of diclofenac sodium tablets containing tamarind seed polysaccharide as release retardant. Int J PharmTech Res 1:191–195

Deze EG, Papageorgiou SK, Favvas EP, Katsaros FK (2012) Porous alginate aerogel beads for effective and rapid heavy metal sorption from aqueous solutions: effect of porosity in Cu^{2+} and Cd^{2+} ion sorption. Chem Eng J 209:537–546

Dragostin OM, Samal SK, Dash M, Lupascu F, Pânzariu A, Tuchilus C, Ghetu N, Danciu M, Dubruel P, Pieptu D, Vasile C (2016) New antimicrobial chitosan derivatives for wound dressing applications. Carbohydr Polym 141:28–40

Dumitriu S (2002) Polymeric biomaterials, revised and expanded, 2nd edn. MarcelDekker Inc., NY, USA, p 480

Dutta PK, Ravikumar MN, Dutta J (2002) Chitin and chitosan for versatile applications. JMS Polym Rev C 42:307–354

Fajardo AR, Fávaro SL, Rubira AF, Muniz EC (2013) Dual-network hydrogels based on chemically and physically crosslinked chitosan/chondroitin sulfate. Funct Polym 73:1662–1671

Frederiksen HK, Hansen HCB, Borggaard O, Pedersen M (2002) Starch encapsulated chlorpyrifos: release rate, insecticidal activity and degradation in soil. J Microencapsul 19:319–331

Gamal el-din AM, Mostafa AM, Al-Shabanah OA, Al-Bekairi AM, Nagi MN (2003) Protective effect of arabic gum against acetaminophen-induced hepatotoxicity in mice. Pharmacol Res 48:631–635

Gao D (2003) Superabsorbent polymer composite (SAPC) materials and their industrial and high-tech applications. Ph.D. dissertation, Bergakademie, Technische Universitat, Freiberg, Germany

Geethalakshmi A, Karki R, Jha SK, Venkatesh DP, Nikunj B (2012) Sustained ocular delivery of brimonidine tartrate using ion activated in situ gelling system. Curr Drug Deliv 9:197–204

Geethalakshmi A, Karki R, Sagi P, Jha SK, Venkatesh DP (2013) Temperature triggered in situ gelling system for betaxolol in glaucoma. J Appl Pharm Sci 3:153

Ghorai S, Sinhamahpatra A, Sarkar A, Panda AB, Pal S (2012) Novel biodegradable nanocomposite based on XG-g-PAM/SiO_2: application of an efficient adsorbent for Pb^{2+} ions from aqueous solution. Bioresour Technol 119:181–190

Gil EC, Colarte AI, Bataille B, Caraballo I (2007) Estimation of the percolation thresholds in lobenzarit disodium native dextran matrix tablets. Pharm Sci Tech 8:1–7

Glicksman M (ed) (1996) Tamarind seed gum. Food Hydrocolloids III:191–202. CRC Press Inc, Boca Raton, Florida, USA

Guilherme MR, Reis AV, Takahashi SH, Rubira AF, Feitosa JPA, Muniz EC (2005) Synthesis of a novel superabsorbent hydrogel by copolymerization of acrylamide and cashew gum modified with glycidyl methacrylate. Carbohydr Polym 61:464–471

Guilherme MR, Reis AV, Paulino AT, Fajardo AR, Muniz EC, Tambourgi EB (2007) Superabsorbent hydrogel based on modified polysaccharide for removal of Pb^{2+} and Cu^{2+} from water with excellent performance. J Appl Polym Sci 105:2903–2909

Guilherme MR, Moia TA, Reis AV, Paulino AT, Rubira AF, Mattoso LH, Muniz EC, Tambourgi EB (2009) Synthesis and water absorption transport mechanism of a pH-sensitive polymer network structured on vinyl-functionalized pectin. Biomacromolecules 10:190–196

Guilherme MR, Reis AV, Paulino AT, Moia TA, Mattoso LH, Tambourgi EB (2010) Pectin-based polymer hydrogel as a carrier for release of agricultural nutrients and removal of heavy metals from wastewater. J Appl Polym Sci 117:3146–3154

Guilherme MR, Oliveira RS, Mauricio MR, CelletTSP Pereira GM, Kunita MH (2012) Albumin release from a brain-resembling superabsorbent magnetic hydrogel based on starch. Soft Matter 8:6629–6637
Gupta VK, Pathania D, Singh P, Kumar A, Rathore BS (2014) Adsorptional removal of methylene blue by guar gum–cerium (IV) tungstate hybrid cationic exchanger. Carbohydr Polym 101:684–691
Hambleton A, Fabra MJ, Debeaufort F, Dury-Brun C, Voilley A (2009) Interface and aroma barrier properties of iota-carrageenan emulsion-based films used for encapsulation of active food compounds. J Food Eng 93:80–88
Hambleton A, Voilley A, Debeaufort F (2011) Transport parameters for aroma compounds through i-carrageenan and sodium alginate-based edible films. Food Hydrocoll 25:1128–1133
Hari BNV, Lobo FJ, Devi DR (2015) Polymeric in-situ gels as smart carriers for pesticide delivery in agricultural practice. Int J Pharm Clin Res 7:2
Heinze T, Liebert T, Heublein B, Hornig S (2006) Functional polymers based on dextran. Adv Polym Sci 205:199–291
Hemvichian K, Chanthawong A, Suwanmala P (2014) Synthesis and characterization of superabsorbent polymer prepared by radiation-induced graft copolymerization of acrylamide onto carboxymethyl cellulose for controlled release of agrochemicals. Radiat Phys Chem 103:167–171
Hosseini A, Jafari SM, Mirzaei H, Asghari A, Akhavan S (2015) Application of image processing to assess emulsion stability and emulsification properties of Arabic gum. Carbohydr Polym 126:1–8
Hsu SY, Chung HY (1999) Comparisons of 13 edible gum-hydrate fat substitutes for low fat Kung-wan (an emulsified meatball). J Food Eng 40:279–285
Hüttermann A, Zommorodi M, Reise K (1999) Addition of hydrogels to soil for prolonging the survival of *Pinus halepensis* seedlings subjected to drought. Soil Till Res 50:295–304
Hüttermann A, Orikiriza LJ, Agaba H (2009) Application of superabsorbent polymers for improving the ecological chemistry of degraded or polluted lands. Clean-Soil, Air, Water 1:517–526
Ibrahim NA, Abo-Shosha MH, Allam EA, El-Zairy EM (2010) New thickening agents based on tamarind seed gum and karaya gum polysaccharides. Carbohydr Polym 81:402–408
Ielpi L, Couso R, Dankert M (1981) Lipid-linked intermediates in the biosynthesis of xanthan gum. FEBS Lett 130:253–256
International Water Management Institute (2010) Issue 10
Iqbal M (1993) International trade in non-wood forest products: an overview. http://www.fao.org/docrep/x5326e/x5326e08.htm#1.%20gum%20arabic
Işıklan N (2004) Controlled release of insecticide carbaryl from crosslinked carboxymethylcellulose beads. Fresen Environ Bull 13:537–544
Jamnongkan T, Kaewpirom S (2010) Potassium release kinetics and water retention of controlled-release fertilizers based on chitosan hydrogels. J Polym Environ 18:413–421
Jana T, Roy BC, Maiti S (2001) Biodegradable film: 6 modification of the film for control release of insecticides. Eur Polym J 37:861–864
Jana S, Lakshman D, Sen KK, Basu SK (2010) Development and evaluation of epichlorohydrin cross-linked mucoadhesive patches of tamarind seed polysaccharide for buccal application. Int J Pharm Sci Drug Res 2:193–198
Jani GK, Shah DP, Prajapati VD, Jain VC (2009) Gums and mucilages: versatile excipients for pharmaceutical formulations. Asian J Pharm Sci 4:309–323
Johnson RM, Pepperman AB (1998) Release of atrazine and alachlor from clay-oxamide controlled-release formulations. PesticSci 53:233–240
Joseph J, Kanchalochana S, Rajalakshmi G, Hari V, Durai R (2012) Tamarind seed polysaccharide: a promising natural excipient for pharmaceuticals. Int J Green Pharm 6(4):270
Kaith BS, Jindal R, Kapur GS (2013) Enzyme-based green approach for the synthesis of gum tragacanth and acrylic acid cross-linked hydrogel: its utilization in controlled fertilizer release and enhancement of water-holding capacity of soil. Iran Polym J 22:561–570

Kamath KR, Park K (1993) Biodegradable hydrogels in drug delivery. Adv Drug Deliver Rev 11:59–84

Kashyap PL, Xiang X, Heiden P (2015) Chitosan nanoparticle based delivery systems for sustainable agriculture. Int J Biol Macromol 77:36–51

Kataja M, Rybin A, Timonen J (1992) Permeability of highly compressible porous medium. J Appl Phys 72:1271–1274

Kazanskii KS, Dubrovskii SA (1992) Chemistry and physics of agricultural hydrogels. Polyelectrolytes hydrogels chromatographic materials. Springer, Berlin, pp 97–133

Kenawy ER (1998) Biologically active polymers: controlled-release formulations based on crosslinked acrylamide gel derivatives. React Funct Polym 36:31–39

Kim J, Lee KW, Hefferan TE, Currier BL, Yaszemski MJ, Lu L (2007) Synthesis and evaluation of novel biodegradable hydrogels based on poly (ethylene glycol) and sebacic acid as tissue engineering scaffolds. Biomacromolecules 9(1):149–157

Kitila G, Gebrekidan H, Alamrew T (2014) Assessment of irrigation water quality and suitability for irrigation in the Fincha'a Valley Sugar Estate, Nile Basin of Western Ethiopia. Sci Technol Arts Res J 3(1):64–73

Klein M, Aserin A, Svitov I, Garti N (2010) Enhanced stabilization of cloudy emulsions with gum Arabic and whey protein isolate. Colloids Surf B: Biointerfaces 77(1):75–81

Knypl JS, Knypl M (1993) Increased chilling stress tolerance in soybean seeds treated with poly (acrylic acid-ethyl acrylate-vinyl acetate, a superabsorbent). Acta Physiol Plant 15:3

Kohls SJ, Baker DD, Kremer DA, Dawson JO (1999) Water-retentive polymers increase nodulation of actinorhizal plants inoculated with Frankia. Plant Soil 214:105–115

Kong H, Yang J, Zhang Y, Fang Y, Nishinari K, Phillips GO (2014) Synthesis and antioxidant properties of gum arabic-stabilized selenium nanoparticles. Int J Biol Macromol 65:155–162

Kuang J, Yuk KY, Huh KM (2011) Polysaccharide-based superporous hydrogels with fast swelling and superabsorbent properties. Carbohydr Polym 83:284–290

Kulkarni D, Dwivedi AK, Singh S (1998) Performance evaluation of tamarind seed polyose as a binder and in sustained release formulations of low drug loading. Indian J Pharm Sci 60:50–53

Kulkarni AR, Soppimath KS, Aminabhavi TM, Dave AM, Mehta MH (2000) Glutaraldehyde crosslinked sodium alginate beads containing liquid pesticide for soil application. J Control Release 63:97–105

Kulkarni GT, Gowthamarajan K, Dhobe RR, Yohanan F, Suresh B (2005) Development of controlled release spheriods using natural polysaccharide as release modifier. Drug Deliv 12:201–206

Kumar J, Nisar K, Shakil NA, Sharma R (2010) Residue and bio-efficacy evaluation of controlled release formulations of metribuzin against weeds in wheat. Bull Environ ContamToxicol 85:357–361

Laftah WA, Hashim S, Ibrahim AN (2011) Polymer hydrogels: a review. Polym Plast Technol Eng 50:1475–1486

Lakouraj MM, Mojerlou F, Zare EN (2014) Nanogel and superparamagnetic nanocomposite based on sodium alginate for sorption of heavy metal ions. Carbohydr Polym 106:34–41

Lapasin R, Pricl S, Tracanelli P (1991) Rheology of hydroxyethyl guar gum derivatives. Carbohydr Polym 14(4):411–427

Lapasin R, De Lorenzi L, Pricl S, Torriano G (1995) Flow properties of hydroxypropyl guar gum and its long-chain hydrophobic derivatives. Carbohydr Polym 28(3):195–202

Lee KY, Mooney DJ (2001) Hydrogels for tissue engineering. Chem Rev 101:1869–1879

Lee SS, Chang SX, Chang YY, Ok YS (2013) Commercial versus synthesized polymers for soil erosion control and growth of Chinese cabbage. Springer Plus 2:534

Lee CS, Gong J, Huong CV, Oh DS, Chang YS (2016) Macroporous alginate substrate-bound growth of Fe^0 nanoparticles with high redox activities for nitrate removal from aqueous solutions. Chem Eng J 298:206–213

Li Q, Tan W, Zhang C, Gu G, Guo Z (2015) Novel triazolyl-functionalized chitosan derivatives with different chain lengths of aliphatic alcohol substituent: design, synthesis, and antifungal activity. Carbohydr Res 418:44–49

Liakos I, Rizzello L, Bayer IS, Pompa PP, Cingolani R, Athanassiou A (2013) Controlled antiseptic release by alginate polymer films and beads. Carbohydr Polym 92:176–183
Libio IC, Demori R, Ferrão MF, Lionzo MI, da Silveira NP (2016) Films based on neutralized chitosan citrate as innovative composition for cosmetic application. Mater Sci Eng C 67:115–124
Lionetto F, Sannino A, Maffezzoli A (2005) Ultrasonic monitoring of the network formation in superabsorbent cellulose based hydrogels. Polymer 46:1796–1803
Lokhande HT, Varadarajan PV (1992) A new Guargum-based superabsorbent polymer synthesised using gamma radiation as a soil additive. Bioresour Technol 42:119–122
Lu T, Xiang T, Huang XL, Li C, Zhao WF, Zhang Q, Zhao CS (2015) Post-crosslinking towards stimuli-responsive sodium alginate beads for the removal of dye and heavy metals. Carbohydr Polym 133:587–595
Mahmoodi NM (2013) Magnetic ferrite nanoparticle–alginate composite: synthesis, characterization and binary system dye removal. J Taiwan Inst Chem Eng 44:322–330
Mahmoud ME, Kana MT, Hendy AA (2015) Synthesis and implementation of nano-chitosan and its acetophenone derivative for enhanced removal of metals. Int J Biol Macromol 81:672–680
Maleki A, Pajootan E, Hayati B (2015) Ethyl acrylate grafted chitosan for heavy metal removal from wastewater: equilibrium, kinetic and thermodynamic studies. J Taiwan Inst Chem Eng 51:127–134
Mandal B, Ray SK (2013) Synthesis of interpenetrating network hydrogel from poly (acrylic acid-co-hydroxyethyl methacrylate) and sodium alginate: modeling and kinetics study for removal of synthetic dyes from water. Carbohydr Polym 98:257–269
Mandal B, Ray SK (2015) Removal of safranine T and brilliant cresyl blue dyes from water by carboxy methyl cellulose incorporated acrylic hydrogels: isotherms, kinetics and thermodynamic study. J Taiwan Inst Chem Eng 60:313–327
Marcì G, Mele G, Palmisano L, Pulito P, Sannino A (2006) Environmentally sustainable production of cellulose-based superabsorbent hydrogels. Green Chem 8:439–444
Masoumi A, Ghaemy M (2014) Removal of metal ions from water using nanohydrogel tragacanth gum-g-polyamidoxime: isotherm and kinetic study. Carbohydr Polym 108:206–215
Mbuna JJ, Mhinzi GS (2003) Evaluation of gum exudates from three selected plant species from Tanzania for food and pharmaceutical applications. J Sci Food Agr 83(2):142–146
McHugh DJ (2003) A guide to the seaweed industry FAO Fisheries Technical Paper 441. Food and Agriculture Organization of the United Nations, Rome
Mehra GR, Manish M, Rashi S, Neeraj G, Mishra DN (2010) Enhancement of miotic potential of pilocarpine by tamarind gum based in-situ gelling ocular dosage form. Acta Pharma Sci 52:145–154
Mekonnen T, Mussone P, Khalil H, Bressler D (2013) Progress in bio-based plastics and plasticizing modifications. J Mater Chem A 1:13379–13398
Mishra MU, Khandare JN (2011) Evaluation of tamarind seed polysaccharide as a biodegradable carrier for colon specific drug delivery. Int J Pharm Pharm Sci 3:139–142
Mishra MM, Yadav M, Sand A, Tripathy J, Behari K (2010) Water soluble graft copolymer (κ-carrageenan-g-N-vinyl formamide): preparation, characterization and application. Carbohydr Polym 80:235–241
Mitrus M, Wojtowicz A, Moscicki L (2009) Biodegradable polymers and their practical utility. In: Thermoplastic starch: a green material for various industries. Wiley, Weinheim
Mittal H, Maity A, Ray SS (2015a) Effective removal of cationic dyes from aqueous solution using gum ghatti-based biodegradable hydrogel. Int J Biol Macromol 79:8–20
Mittal H, Maity A, Ray SS (2015b) Synthesis of co-polymer-grafted gum karaya and silica hybrid organic–inorganic hydrogel nanocomposite for the highly effective removal of methylene blue. Chem Eng J 279:166–179
Mittal H, Maity A, Ray SS (2016) Gum karaya based hydrogel nanocomposites for the effective removal of cationic dyes from aqueous solutions. Appl Surf Sci 364:917–930
Monier M, Abdel-Latif DA, Ji HF (2016) Synthesis and application of photo-active carboxymethyl cellulose derivatives. React Funct Polym 102:137–146

Nair V, Panigrahy A, Vinu R (2014) Development of novel chitosan–lignin composites for adsorption of dyes and metal ions from wastewater. Chem Eng J 254:491–502
Necas J, Bartosikova L (2013) Carrageenan: a review. Vet Med 58:187–205
Ni B, Liu M, Lü S (2009) Multifunctional slow-release urea fertilizer from ethylcellulose and superabsorbent coated formulations. Chem Eng J 155:892–898
Ni B, Liu M, Lu S, Xie L, Wang Y (2011) Environmentally friendly slow-release nitrogen fertilizer. J Agric Food Chem 59:10169–10175
Nishinari K, Doi E (eds) (2012) Food hydrocolloids: structures, properties, and functions. Springer Science & Business Media
Nnadi F, Brave C (2011) Environmentally friendly superabsorbent polymers for water conservation in agricultural lands. J Soil Sci Environ Manag 2:206–211
Noppakundilograt S, Pheatcharat N, Kiatkamjornwong S (2015) Multilayer coated NPK compound fertilizer hydrogel with controlled nutrient release and water absorbency. J Appl Polym Sci 132:41249–41250
Omidian H, Rocca JG, Park K (2005) Advances in superporous hydrogels. J Control Release 102:3–12
Pal S, Ghorai S, Dash MK, Ghosh S, Udayabhanu G (2011) Flocculation properties of polyacrylamide grafted carboxymethyl guar gum (CMG-g-PAM) synthesized by conventional and microwave assisted method. J Hazard Mater 192:1580–1588
Pal A, Giri A, Bandyopadhyay A (2016) Influence of hydrodynamic size and zeta potential of a novel polyelectrolyte poly (acrylic acid) grafted guar gum for adsorption of Pb (II) from acidic waste water. J Environ Chem Eng 4:1731–1742
Pandey S, Mishra SB (2012) Microwave synthesized xanthan gum-g-poly (ethylacrylate): an efficient Pb^{2+} ion binder. Carbohydr Polym 90:370–379
Pangon A, Saesoo S, Saengkrit N, Ruktanonchai U, Intasanta V (2016) Hydroxyapatite-hybridized chitosan/chitin whisker bionanocomposite fibers for bone tissue engineering applications. Carbohydr Polym 144:419–427
Parvathy PC, Jyothi AN (2012) Synthesis, characterization and swelling behaviour of superabsorbent polymers from cassava starch-graft-poly (acrylamide). Starch-Stärke 64:207–218
Patel B, Patel P, Bhosale A, Hardikar S, Mutha S, Chaulang G (2009) Evaluation of Tamarind Seed Polysaccharide (TSP) as a mucoadhesive and sustained release component of nifedipine buccoadhesive tablet & comparison with HPMC and Na CMC. Int J Pharm Tech Res 1:404–410
Pathania D, Gupta D, Kothiyal NC, Eldesoky GE, Naushad M (2016) Preparation of a novel chitosan-g-poly (acrylamide)/Zn nanocomposite hydrogel and its applications for controlled drug delivery of ofloxacin. Int J Biol Macromol 84:340–348
Patil BS, Mastiholimath V, Kulkarni AR (2011) Novel wound dressing material derived from crosslinked tamarind seeds polysaccharide. In vitro characterization and wound healing activity. Trends Biomater Artif Organs 25(149):153
Patra S, Roy E, Madhuri R, Sharma PK (2016) Agar based bimetallic nanoparticles as high-performance renewable adsorbent for removal and degradation of cationic organic dyes. J Ind Eng Chem 33:226–238
Peng Z, Chen F (2011) Synthesis and properties of lignin-based polyurethane hydrogels. Int J Polym Mat 60:674–683
Pepperman AB, Kuan J-CW (1993) Slow release formulations of metribuzin based on alginate-kaolin-linseed oil. J Control Release 26:21–30
Pfister G, Bahadir M, Korte F (1986) Release characteristics of herbicides from Ca alginate gel formulations. J Control Release 3:229–233
Pillai CK (2010) Challenges for natural monomers and polymers: novel design strategies and engineering to develop advanced polymers. Designed Monomers Polym 13:87–121
Pourjavadi A, Harzandi AM, Hosseinzadeh H (2004) Modified carrageenan synthesis of a novel polysaccharide-based superabsorbent hydrogel via graft copolymerization of acrylic acid onto kappa-carrageenan in air. Eur Polym J 40:1363–1370

Pourjavadi A, Ghasemzadeh H, Soleyman R (2007) Synthesis, characterization, and swelling behavior of alginate-g-poly (sodium acrylate)/kaolin superabsorbent hydrogel composites. J Appl Polym Sci 5:2631–2639

Quinn JJ (1990) 22nd congress of national agricultural plastics association, Montreal, Canada, 1990

Raju KM, Raju MP (2001) Synthesis of novel superabsorbing copolymers for agricultural and horticultural applications. Polym Int 50:946–951

Ramakrishnan A, Pandit N, Badgujar M, Bhaskar C, Rao M (2007) Encapsulation of endoglucanase using a biopolymer Gum Arabic for its controlled release. Bioresour Technol 98:368–372

Ramezani M, Jafari M, Tavili A, Javadi SA, Amiri QZ (2013) The effect of calcium and sodium chloride levels on the dehydration power of hydrophilic polymer. J Food Agric Environ 11:896–898

Reddy DH, Lee SM (2013) Application of magnetic chitosan composites for the removal of toxic metal and dyes from aqueous solutions. Adv Colloid Interface Sci 201:68–93

Rhee JS, Jung MW, Paeng KJ (1998) Evaluation of chitin and chitosan as a sorbent for the preconcentration of phenol and chlorophenols in water. Anal Sci 14:1089–1092

Rinaudo M (2006) Chitin and chitosan: properties and applications. Progr Polym Sci 31:603–632

Robinson G, Ross-Murphy SB, Morris ER (1982) Viscosity-molecular weight relationships, intrinsic chain flexibility, and dynamic solution properties of guar galactomannan. Carbohydr Res 107:17–32

Rocher V, Siaugue JM, Cabuil V, Bee A (2008) Removal of organic dyes by magnetic alginate beads. Water Res 42:1290–1298

Rodrigues FH, Spagnol C, Pereira AG, Martins AF, Fajardo AR, Rubira AF, Muniz EC (2014) Superabsorbent hydrogel composites with a focus on hydrogels containing nanofibers or nanowhiskers of cellulose and chitin. J Appl Polym Sci 131(2)

Rodríguez R, Alvarez-Lorenzo C, Concheiro A (2003) Cationic cellulose hydrogels: kinetics of the cross-linking process and characterization as pH-/ion-sensitive drug delivery systems. J Control Release 86:253–265

Rolando M, Valente C (2007) Establishing the tolerability and performance of tamarind seed polysaccharide (TSP) in treating dry eye syndrome: results of a clinical study. BMC ophthalmology 7:1

Roy A, Bajpai J, Bajpai AK (2009) Dynamics of controlled release of chlorpyrifos from swelling and eroding biopolymeric microspheres of calcium alginate and starch. Carbohydr Polym 76:222–231

Ruvalcaba AM, Sanchez-Diaz JC, Becerra F, Cruz-Barba LE, Gonzalez-Alvarez A (2009) Swelling characterization and drug delivery kinetics of polyacrylamide-co-itaconic acid/chitosan hydrogels. Express Polym Lett 3:25–32

Sadeghi S, Rad FA, Moghaddam AZ (2014) A highly selective sorbent for removal of Cr (VI) from aqueous solutions based on Fe_3O_4/poly (methyl methacrylate) grafted Tragacanth gum nanocomposite: optimization by experimental design. Mater Sci Eng C 45:136–145

Sandford PA, Baird J (1983) In: Aspinall GO (ed) The polysaccharides. Academia Press, Prague, pp 470–473

Sannino A, Nicolais L (2005) Concurrent effect of microporosity and chemical structure on the equilibrium sorption properties of cellulose-based hydrogels. Polymer 46:4676–4685

Sannino, A Maffezzoli, L Nicolais (2003) Introduction of molecular spacers between the crosslinks of a cellulose-based superabsorbent hydrogel: effects on the equilibrium sorption properties. J Appl Polym Sci 90(1):168–174

Saravanan P, Vinod VT, Sreedhar B, Sashidhar RB (2012) Gum kondagogu modified magnetic nano-adsorbent: an efficient protocol for removal of various toxic metal ions. Mater Sci Eng C 32:581–586

Saruchi KBS, Jindal R, Kapur GS, Kumar V (2014) Synthesis, characterization and evaluation of Gum tragacanth and acrylic acid based hydrogel for sustained calcium chloride release–enhancement of water-holding capacity of soil. J Chin Adv Mat Soc 2:40–52

Schreiber MM, White MD, Wing RE (1988) Bioactivity of controlled release formulations of starch-encapsulated EPTC. J Control Release 7:237–242

Shalviri A, Liu Q, Abdekhodaie MJ, Wu XY (2010) Novel modified starch–xanthan gum hydrogels for controlled drug delivery: synthesis and characterization. Carbohydr Polym 79:898–907

Sharma K, Kaith BS, Kumar V, Kalia S, Kumar V, Swart HC (2014) Water retention and dye adsorption behavior of Gg-cl-poly (acrylic acid-aniline) based conductive hydrogels. Geoderma 232:45–55

Shavit U, Shaviv A, Shalit G, Zaslavsky D (1997) Release characteristics of a new controlled release fertilizer. J Control Release 43:131–138

Shaviv (2001) Advances in controlled-release fertilizers. Adv Agronomy 71:1–49

Shi HY, Zhang LM (2007) New grafted polysaccharides based on O-carboxymethyl-O-hydroxypropyl guar gum and N-isopropylacrylamide: synthesis and phase transition behavior in aqueous media. Carbohydr Polym 67(3):337–342

Shirakawa M, Yamatoya K (2003) Xyloglucan: its structure and function. Foods and food ingredients. J Japan 208:922–929

Shojaat R, Saadatjoo N, Karimi A, Aber S (2016) Simultaneous adsorption–degradation of organic dyes using $MnFe_2O_4$/calcium alginate nano-composites coupled with GO_x and laccase. J Environ chem Eng 4:1722–1730

Singh B, Sharma DK, Gupta A (2007) Controlled release of thiram fungicide from starch-based hydrogels. J Environ Sci Health B 42:677–695

Singh B, Sharma DK, Gupta A (2008) In vitro release dynamics of thiram fungicide from starch and poly(methacrylic acid)-based hydrogels. J Hazard Mater 154:278–286

Singh B, Sharma DK, Kumar R, Gupta A (2009a) Controlled release of the fungicide thiram from starch–alginate–clay based formulation. Appl Clay Sci 45:76–82

Singh V, Kumari P, Pandey S, Narayan T (2009b) Removal of chromium (VI) using poly (methylacrylate) functionalized guar gum. Bioresour Technol 100:1977–1982

Singh A, Sarkar DJ, Singh AK, Parsad R, Kumar A, Parmar BS (2011a) Studies on novel nanosuperabsorbent composites: swelling behavior in different environments and effect on water absorption and retention properties of sandy loam soil and soil-less medium. J Appl Polym Sci 120:1448–1458

Singh B, Sharma DK, Dhiman A, Gupta A (2011b) Applications of natural polysaccharide-based beads for slow release herbicide formulation. Toxicol Environ Chem 93:616–622

Singh B, Sharma DK, Negi S, Dhiman A (2015) Synthesis and characterization of agar-starch based hydrogels for slow herbicide delivery applications. Int J Plastics Technol 19:263–274

Sinha VR, Kumria R (2001) Polysaccharides in colon-specific drug delivery. Int J Pharma 224:19–38

Soares PI, Sousa AI, Silva JC, Ferreira IM, Novo CM, Borges JP (2016) Chitosan-based nanoparticles as drug delivery systems for doxorubicin: optimization and modelling. Carbohydr Polym 147:304–312

Sohrabi S, Haeri A, Mahboubi A, Mortazavi A, Dadashzadeh S (2016) Chitosan gel-embedded moxifloxacin niosomes: an efficient antimicrobial hybrid system for burn infection. Int J Biol Macromol 85:625–633

Sook Y, Jae-Sik H (2000) Application of super water absorbent for edible mushroom production. J Korean Soc Agr Eng 28:16

Spagnol C, Rodrigues FHA, Pereira AGB, Fajardo AR, Rubira AF, Muniz EC (2012) Superabsorbent hydrogel composite made of cellulose nanofibrils and chitosan-graft-poly (acrylic acid). Carbohydr Polym 87:2038–2045

Srinivasan B, Ganta A, Rajamanickam D, Veerabhadraiah BB, Varadharajan M (2011) Evaluation of tamarind seed polysaccharide as a drug release retardant. Int J Pharm Sci Rev Res 9:27–31

Srivastava A, Tripathy J, Mishra MM, Behari K (2007) Modification of guar gum through grafting of 4 vinyl pyridine using potassium peroxymonosulphate/ascorbic acid redox pair. J Appl Polym Sci 106:1353–1358

Suh JK, Matthew HW (2000) Application of chitosan-based polysaccharide biomaterials in cartilage tissue engineering: a review. Biomaterials 21(24):2589–2598
Thakur VK, Kessler MR (2015) Self-healing polymer nanocomposite materials: a review. Polymer 69:369–383
Thakur VK, Thakur MK (2014) Recent trends in hydrogels based on psyllium polysaccharide: a review. J Clean Prod 82:1–15
Thakur VK, Thakur MK (2015) Recent advances in green hydrogels from lignin: a review. Int J Biol Macromolec 72:834–847
Thakur MK, Thakur VK, Gupta RK, Pappu A (2015) Synthesis and applications of biodegradable soy based graft copolymers: a review. ACS Sustain Chem Eng 4(1):1–17
Thombare N, Jha U, Mishra S, Siddiqui MZ (2016) Guar gum as a promising starting material for diverse applications: a review. Int J Biol Macromol 88:361–372
Trimnell D, Shasha BS, Otey FH (1985) The effect of α-amylases upon the release of trifluralin encapsulated in starch. J Control Release 1:183–190
Tripathy J, Mishra DK, Yadav M, Sand A, Behari K (2009) Modification of κ-carrageenan by graft copolymerization of methacrylic acid: synthesis and applications. J Appl Polym Sci 114:3896–3905
Tsuji K (2001) Microencapsulation of pesticides and their improved handling safety. J Microencapsul 18:137–147
Ulery BD, Nair LS, Laurencin CT (2011) Biomedical applications of biodegradable polymers. J Polym Sci Part B Polym Phys 49:832–864
Umoren SA, Obot IB, Ebenso EE, Okafor PC, Ogbobe O, Oguzie EE (2006) Gum arabic as a potential corrosion inhibitor for aluminium in alkaline medium and its adsorption characteristics. Anti-Corros Methods Mater 53:277–282
Vallée F, Müller C, Durand A, Schimchowitsch S, Dellacherie E, Kelche C, Cassel JC, Leonard M (2009) Synthesis and rheological properties of hydrogels based on amphiphilic alginate-amide derivatives. Carbohydr Res 344:223–228
Varela P, Fiszman SM (2011) Hydrocolloids in fried foods. A review. Food Hydrocolloids 25:1801–1812
Varki A, Cummings R, Esko J, Jessica Freeze, Hart G, Marth J (1999) Essentials of glycobiology. Cold Spring Harbor Laboratory Press
Varki A, Cummings R, Esko J, Freeze H, Stanley P, Bertozzi C, Hart G, Etzler M (2008) Essentials of glycobiology, 2nd edn. Cold Spring Harbor Laboratory Press
Verbeken D, Dierckx S, Dewettinck K (2003) Exudate gums: occurrence, production, and applications. Appl Microbiol Biotechnol 63:10–21
Vijn JP, Leiden JB, Bach D (2002) Delayed thixotropic cement compositions and methods. US Patent, 6,610,140 B2
Villafranca-Sánchez M, González-Pradas E, Fernández-Pérez M (2000) Controlled release of isoproturon from an alginate–bentonite formulation: water release kinetics and soil mobility. Pest Manag Sci 56:749–756
Villay A, de Filippis FL, Picton L, Le Cerf D, Vial C, Michaud P (2012) Comparison of polysaccharide degradations by dynamic high-pressure homogenization. Food Hydrocoll 27:278–286
Vinod VT, Sashidhar RB, Sreedhar B, Rao BR, Rao TN, Abraham JT (2009) Interaction of Pb^{2+} and Cd^{2+} with gum kondagogu (*Cochlospermum gossypium*): a natural carbohydrate polymer with biosorbent properties. Carbohydr Polym 78:894–901
Vinod VT, Sashidhar RB, Sukumar AA (2010) Competitive adsorption of toxic heavy metal contaminants by gum kondagogu (*Cochlospermum gossypium*): a natural hydrocolloid. Colloids Surf B 75:490–495
Vinod VT, Sashidhar RB, Sivaprasad N, Sarma VU, Satyanarayana N, Kumaresan R, Rao TN, Raviprasad P (2011) Bioremediation of mercury (II) from aqueous solution by gum karaya (*Sterculia urens*). A natural hydrocolloid. Desalination 272:270–277
Vollmer DP (inventor) (2003) Well treatment fluids and methods for the use thereof. United States patent US 6,509,301

Vu CH, Won K (2013) Novel water-resistant UV-activated oxygen indicator for intelligent food packaging. Food Chem 140:52–56

Wang W, Wang A (2010a) Nanocomposite of carboxymethyl cellulose and attapulgite as a novel pH sensitive superabsorbent: synthesis, characterization and properties. Carbohydr Polym 82:83–91

Wang WB, Wang AQ (2010b) Preparation, swelling and water-retention properties of crosslinked superabsorbent hydrogels based on guar gum. Adv Mater Res 96:177–182

Wang M, Xu L, Hu H, Zhai M, Peng J, Nho Y, Li J, Wei G (2007) Radiation synthesis of PVP/CMC hydrogels as wound dressing. Nucl Instrum Methods Phys Res Sect B: Beam Interact Mater Atoms 265:385–389

Wang S, Alva AK, Li Y, Zhang M (2011) A rapid technique for prediction of nutrient release from polymer coated controlled release fertilizers. Open J Soil Sci 1:40–44

Wang Y, Shi X, Wang W, Wang A (2013) Synthesis, characterization, and swelling behaviors of a pH-responsive CMC-g-poly(AA-co-AMPS) superabsorbent hydrogel. Turkish J Chem 37:149–159

Wang X, Lü S, Gao C, Xu X, Wei Y, Bai X, Feng C, Gao N, Liu M, Wu L (2014) Biomass-based multifunctional fertilizer system featuring controlled-release nutrient, water-retention and amelioration of soil. RSC Adv 4:18382–18390

Wang Y, Zhang Y, Hou C, Liu M (2016) Mussel-inspired synthesis of magnetic polydopamine–chitosan nanoparticles as biosorbent for dyes and metals removal. J Taiwan Inst Chem Eng 61:292–298

Wei W, Kim S, Song MH, Bediako JK, Yun YS (2015) Carboxymethyl cellulose fiber as a fast binding and biodegradable adsorbent of heavy metals. J Taiwan Inst Chem Eng 57:104–110

Whistler RL, BeMiller JN (1993) Industrial gums: polysaccharides and their derivatives, 2nd edn. Academic Press Inc, pp 145–181

Wong TY, Preston LA, Schiller NL (2000) Alginate Lyase: review of major sources and enzyme characteristics, structure-function analysis, biological roles, and applications. Annu Rev Microbiol 54:289–340

Woodhouse J, Johnson MS (1991) Effect of superabsorbent polymers on survival and growth of crop seedlings. Agric Water Manag 20:63–70

Wu L, Liu M (2008) Preparation and properties of chitosan-coated NPK compound fertilizer with controlled-release and water-retention. Carbohydr Polym 72:240–247

Xie M, Zeng L, Zhang Q, Kang Y, Xiao H, Peng Y, Chen X, Luo J (2015) Synthesis and adsorption behavior of magnetic microspheres based on chitosan/organic rectorite for low-concentration heavy metal removal. J Alloys Compd 647:892–905

Xu S, Wu R, Huang X, Cao L, Wang J (2005) Effect of the anionic-group/cationic-group ratio on the swelling behavior and controlled release of agrochemicals of the amphoteric, superabsorbent polymer poly (acrylic acid-co-diallyldimethylammonium chloride). J Appl Polym Sci 102:986–991

Yogi RK, Bhattacharya A, Jaiswal AK, Kumar A (2015) Lac, plant resin and gums statistics 2014: at a glance. ICAR-Indian Institute of Natural Resins And Gums, Ranchi (Jharkhand), India. Bulletin (technical) No. 27/2015.01–68 pp

Zarghami Z, Akbari A, Latifi AM, Amani MA (2016) Design of a new integrated Chitosan-PAMAM Dendrimer biosorbent for heavy metals removing and study of its adsorption kinetics and thermodynamics. Bioresour Technol 205:230–238

Zhang XZ, Yang YY, Chung TS (2002) The influence of cold treatment on properties of temperature-sensitive poly (N-isopropylacrylamide) hydrogels. J Colloid Interface Sci 246:105–111

Zhang LM, Zhou JF, Hui PS (2005) A comparative study on viscosity behavior of water soluble chemically modified guar gum derivatives with different functional lateral groups. J Sci Food Agric 85:2638–2644

Zhang G, Yi L, Deng H, Sun P (2014a) Dyes adsorption using a synthetic carboxymethyl cellulose-acrylic acid adsorbent. J Environ Sci 26:1203–1211

Zhang M, Gao B, Chen J, Li Y, Creamer AE, Chen H (2014b) Slow-release fertilizer encapsulated by graphene oxide films. Chem Eng J 255:107–113

Zheng Y, Li P, Zhang J, Wang A (2007) Study on superabsorbent composite XVI synthesis, characterization and swelling behaviors of poly(sodium acrylate)/vermiculite superabsorbent composites. Eur Polym J 43:1691–1698

Zheng T, Liang Y, Ye S, He Z (2009) Superabsorbent hydrogels as carriers for the controlled-release of urea: experiments and a mathematical model describing the release rate. Biosyst Eng 102:44–50

Zhou W, Lou Y, Ren L, Han Y, Meng Y, Wu L (2014) Application of controlled-release nitrogen fertilizer decreased methane emission in transgenic rice from a paddy soil. Water Air Soil Pollut 225:2–7

Zhu Y, An X, Li S, Yu S (2009) Nanoencapsulation of β-cypermethrin by complex coacervation in a microemulsion. J Surfactant Deterg 12:305–311

Zohuriaan-Mehr MJ (2006) Super-absorbents. Iran Polymer Society, Tehran

Chapter 9
Agro-Based Polymer Gel and Its Application in Purification of Industrial Water Wastes

Altaf Halim Basta and Houssni El-Saied

Abstract A polymer gel swells when exposed to water significantly when imbibing a large amount of solvent (e.g., water). In this contribution chapter, we present the literature reported polymer gels and their applications in different fields. Moreover, we highlight on the upgrading the utilization of agro-wastes in the production of hydrogels in agronomic purposes. Moreover, our published work deals with our published article on the beneficial effect of rice straw-based hydrogels in purifying the local industrial water wastes from metals, to be suitable for agricultural purposes.

Keywords Polymers hydrogels · Agro-based hydrogels · Economical adsorbents Municipal and industrial wastewater treatment

1 General Introduction

The word "gel" was defined by Scottish chemist Thomas Graham, at nineteenth century. While, according to Wikipedia, it is defined as the non-fluid network "colloidal or polymer," that is swelled throughout its whole volume by a fluid. There are several types of gels, e.g., hydrogels, organogels, xerogels, and nanocomposites hydrogels. In this chapter, we will concern on polymer hydrogels, as well as its sources and general applications.

A. H. Basta (✉) · H. El-Saied
Cellulose and Paper Department, National Research Centre, Dokki, 12622 Cairo, Egypt
e-mail: altaf_basta2004@yahoo.com

© Springer Nature Singapore Pte Ltd. 2018

V. K. Thakur et al. (eds.), *Polymer Gels*, Gels Horizons: From Science to Smart Materials, https://doi.org/10.1007/978-981-10-6080-9_9

1.1 Hydrogels Preparation and Applications

A hydrogel is expressed on polymer "gel" or colloidal "gel" when water is used as swelling agent or dispersing medium. Hydrogels are characterized by the high affinity to absorb liquid "water," and first investigated by researchers at the "Northern Regional Research Laboratories," and subjected as the application of the grafted starch (Burrg et al. 1976). The literature reported many methods for preparing the synthetic- and natural-synthetic-based hydrogels, such as:

- Hydrogels from synthetic polymers, e.g., acryloxypropionic acid. This polymer was prepared by using ethylene glycol dimethacrylate monomer (Ansel 1995).
- Hydrogels from modification of agro-wastes by concentrated nitric acid, or phosphorylating agents (Saito et al. 1991, 1995; Buchholz and Peppas 1994). Moreover, graft copolymerization approach of agro-wastes and carbohydrate derivatives with vinyl monomers was reported. The monomers used must be contained carboxyl, sulfonic, or nitrile groups. In the case of using hydrophobic monomer (acrylonitrile), the additional step was provided (alkali hydrolysis), to convert the hydrophobic nitrile groups to hydrophilic groups (Kennedy et al. 1993; Athawale and Lele 2000; El-Saied et al. 2000, 2004, 2007, 2013; Fanta et al. 2006). Copolymerization of starch by binary hydrophobic and hydrophilic monomers provided hydrogel, without an alkali hydrolysis stage (Fanta and Doane 1990).
- Recently, numerous literature reported psyllium-based hydrogels via grafting polymerization by monomer, e.g., acrylamide to prepare Psy-cl-poly (AAm) hydrogels (Kumar et al. 2010), or dual monomers, e.g., acrylamide (AAm)/methacrylamide (Singh and Chauhan 2010), acrylic acid/polyvinyl alcohol (Singh et al. 2011), or acrylamide/2-acrylamido-methyl propane sulfonic acid (Singh and Bala 2012). The role of the method of preparation psyllium-based hydrogels on its swelling behavior and mechanism of solvent diffusion was reported in review article (Thakur and Thakur 2014).
- Lignin is also used as inexpensive material for production of green hydrogels (Hofman 2012; Thakur et al. 2014). The schematic diagram for synthesizing the hydrogels by physical and chemical approaches was illustrated by Hoffman 2012 (Fig. 1).

Based on hydrogel behavior, the literature specified its application, e.g.,

- Due to its flexibility behavior like natural tissue, it has applied in different fields, e.g., scaffolds in tissue engineering (Drury and Mooney 2003).
- Due to its ability to sense changes of pH, temperature, or the concentration of metabolite, it has applied as environmentally sensitive hydrogels (also known as "Smart Gels" or "Intelligent Gels") (Tomatsu et al. 2011; Techawanitchai et al. 2012).
- Due to its stimuli-responsive changes in their structural network, it has used in controlling release drug delivery system (Ko et al. 2002).

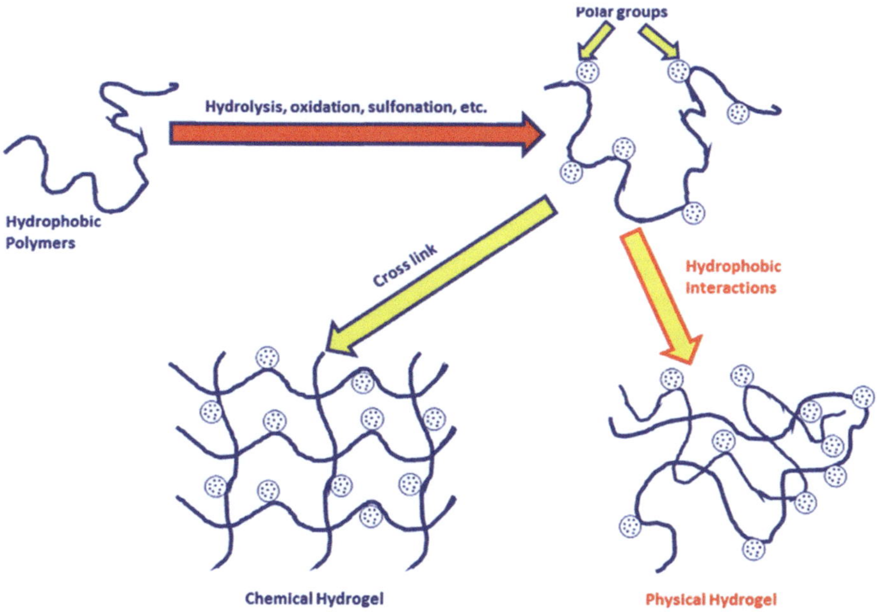

Fig. 1 Schematic diagram of formation lignin-based hydrogels (Hoffman 2012)

- Due to its responsive to specific material, such as glucose or antigens, it can be used as biosensor (Luo et al. 2005).
- Due to its affinity to adsorb liquids and pollutants, e.g., urine, water, and dyes, it can be used in sanitary napkins and sandy soil conditioners (El-Saied et al. 2007; Bajpai and Shrivastava 2011; Gupta 2014).
- Due to the synthetic polymer (e.g., polyacrylamide) in gel has the high affinity to absorb the water, it can be considered as one of the most widely used materials as soil conditioning sandy soils, when mixed with such soils (Zhao et al. 2010).

2 Evaluation the Adsorption Behavior of Hydrogels

2.1 Measurement of Swelling

Equilibrium swelling experiments were carried out to investigate the swelling behavior of the gels prepared. In this respect, the dry gels were weighed and then soaked in distilled water at 25 ± 0.1 °C. Swollen gels were removed from water at predetermined times, blotted dry, and weighed in air. The weight swelling ratios were determined. (q), according to the following equation.

$$q = \frac{W}{W_o} \quad (1)$$

where W is the weight of a swollen sample and W_o is the initial dry weight (Kosmala et al. 2000).

2.2 Water Absorption Capacity

In this test, about one gram of air-dried gel was placed in a 300-ml conical flask. A 150-ml water was added, and the flask subjected to cover with plastic or aluminum foil to control water evaporation. Each water treatment was replicated twice with each gel. The moisture was allowed to equilibrate overnight at room temperature (25 °C). Then the contents were filtered through a filter paper in Buchner funnel attached to a vacuum pump. The unabsorbed was then measured by a graduated cylinder to calculate the amount of water absorbed per gram of gel-forming soil conditioners to the irrigation of water quality. The data were subjected to appropriate statistical techniques for comparing the soil conditioners (Hussain et al. 1989).

2.3 Sorption Equilibrium Experimental

Freundlich and Langmuir models were applied for assessing the adsorption of the dyes or metals onto hydrogels.

Freundlich model (Freundlich 1907) describes adsorption over a heterogeneous surface through a multilayer adsorption mechanism. This model is represented as follows:

$$q_e = k_f \cdot C_e^{1/n} \quad (2)$$

where q_e is the amount of adsorbate per unit mass of adsorbent (mg/g), k_f is the Freundlich constant related to the adsorption capacity (mg/g·(mg/L)^n), C_e is the equilibrium concentration of adsorbate in the solution (mg/L), and n is the empirical parameter representing the energetic heterogeneity of the adsorption sites (dimensionless).

The Langmuir adsorption model (Langmuir 1918) is related to maximum uptake exists, and it considers several assumptions (adsorption is localized, all the surface active sites have similar energies, no interaction between adsorbed molecules exists, and the limiting reaction step is the surface reaction as in the heterogeneous catalytic reaction).

In this model, all the adsorption sites have the same sorption activation energy. Langmuir's equation can be written as:

$$Q = \frac{Q_m b C_e}{1 + bC_e} \tag{3}$$

where C_e and q_e have the same meaning as before, Q_O is the maximum uptake per unit mass of adsorbent (mg/g), and β is the Langmuir constant related to the adsorption energy (L/mg).

3 Scope for Agro-Wastes as Precursors for Hydrogels

Agricultural residues considered to be an important and available biopolymer source with useful applications in various industries, especially for countries lake forests. The application of such types of biopolymers includes the paper, artificial wood, furfural, activated carbons, and cellulose derivatives for desalination of salted water (El-Saied et al. 2003; Fierro et al. 2010; Li et al. 2011; Basta and El-Saied 2003, Basta 2003, Basta et al. 2011a, b). In the previous investigations deals with carbohydrate-based hydrogels (Fanta and Doane 1990; Makoto et al. 2005; Vasquez and Tumolva 2015), hydroxyethyl cellulose (HEC) and other polysaccharides (i.e., konjac mannan, alginic acid, agar, pectin, starch, and chitin) were selected as the precursors. These hydrogels were prepared by graft copolymerization with some monomers (e.g., partially hydrolyzed polyacrylamide (PAM) or polymethyl acrylate (PMA), poly-(2-dimethylamino) ethylmethacrylate (PDM) or polyacrylic acid sodium and potassium salts (PAANa)), to give highly water absorbents.

For the application of synthetic polymers-based hydrogels, e.g., polyacrylamide (Zhao et al. 2010) it is very wide, especially in desert area and water purification. But unfortunately, their use in practice is not spread as expected due to high cost. The agro-based wastes represent the most important cellulosic source because it represent inexpensively. Moreover, it is one of the pollution sources, due to their behavior to microorganisms grow on the humid agro-wastes, and also as fire sources. The utilization of agro-wastes, as precursor for economical preparation of hydrogels, was studied by numerous authors (Allen and Mckay 1988; Abdel-Aal et al. 2006, El-Saied et al. 2000, 2004, 2007, 2013; Neto et al. 2011; Cheng et al. 2015). These hydrogels were prepared from chemical modification of agro-wastes (peat, sunflower, rice husks, coconut bagasse, bagasse fly ash, cotton stalks, waste paper, rice straw, and corn straw).

Egypt, likes many other countries, will not be able to cover the increasing freshwater demand with increasing the populations; therefore, reusing of wastewater is very necessary to utilize in other human activities, for the trial to preserve the fresh water. Because of the chemical, physical, and biological nature of that water, there are potential problems associated with its reuse in agriculture and others. In this chapter, we present our published article (Basta et al. 2013), deals with preparation of rice straw-based hydrogels, as adsorbent for different transition metals loaded municipal and industrial wastewater, for purifying these water wastes, and to be suitable for agricultural purposes.

4 RS-Based Hydrogels in Wastewater Treatments

Due to the economical and environmental performance of the prepared hydrogel for wastewater treatments, therefore in this subtitle, we present the results of our trial to utilize rice straw as undesirable bio-waste in the production of hydrogel and removing the metals loaded aqueous medium, municipal and industrial wastewater.

4.1 Examined Hydrogels

Four RS-based hydrogels were used as metal absorbents. These hydrogels were prepared from rice straw as the methods reported elsewhere (El-Saied et al. 2013). The RS-based hydrogels which were provided relatively higher absorption capacities for both distilled and Nile water (i.e., Sat/Dry $\times$ 100 (w/w) were 7610, 3318, 3008 and 7061 in distilled water; while 6107, 2025, 3128 and 3789 in Nile water, respectively) were candidate for removing of metals. The conditions of hydrogels preparation were:

- Gel #1, it was prepared by using free-radical initiator (ferrous ammonium sulfate/H_2O_2 initiation system) and acrylonitrile monomer, followed by KOH hydrolysis and AcOH for neutralizing the pH-value.
- Gel #2, it was prepared under the same conditions of Gel #1, but by using HNO_3 for neutralizing the pH.
- Gel #3, it was prepared by using initiation system (persulfate/bisulfite) and acrylonitrile as monomer, followed by KOH hydrolysis, using HNO_3 for neutralizing the pH-value to $\sim$6–7.
- Gel #4, it was prepared by grafting the partially cyanoethylated RS, under the same conditions of hydrogel #3, using AcOH for neutralize the pH-value to $\sim$6–7.

These prepared hydrogels were coded as G1, G2, G3, and G4.

4.2 Examined Metal Ions Containing Waters

- The investigated hydrogels were preliminarily tested, their adsorption capacities toward four heavy metal salts, e.g., $CuSO_4{\cdot}5H_2O$, $FeSO_4{\cdot}5H_2O$, $ZnSO_4{\cdot}5H_2O$, and $MnSO_4{\cdot}5H_2O$, with concentration $\sim$50 ppm in bidistilled water. These metals were selected based on their always loading the wastewaters.
- Effluents (wastewaters) were:

- Municipal wastewater from El-Saff municipal waste of pH = 8.58 and EC = 1.76 dSm^{-1}. The concentrations of Cu(II), Fe(II), Zn(II), and Mn(II) in this wastewater were 160, 1332, 160, and 232 μg/l, respectively.
- Industrial wastewater from iron and steel industrial factory (Helwan governorate) of pH = 8.11 and EC = 0.83 dSm^{-1} The concentrations of Cu(II), Fe(II), Zn(II), and Mn(II) in this wastewater were 606.7, 1628, 740, and 310 μg/l, respectively.

4.3 Removal Efficiency of Metal Ions Loaded Laboratory-Prepared Solutions

In this evaluation, different amounts of hydrogel (0.05–0.25 g) were added to 25 ml of metal salt solutions with concentration of ~50 mg L^{-1} in bidistilled water. The hydrogel–metal systems were kept in a shaker at a fixed temperature of ~23 °C for 24 h. This time is enough for guaranteeing adsorption equilibrium of the studied ions over hydrogel. Then separating the suspended hydrogel gel from the filtrate, and determining the unadsorbed metal ions, by using atomic absorption spectroscopy. The efficiency of the investigated hydrogels to adsorb the metal ions was calculated by the following equation

$$\text{Efficiency, \%} = \frac{C_o - C}{C_o} \times 100 \quad (4)$$

where C_o is the initial concentration of metal ions, C is the concentration of the residual metal ions retained in aqueous solution after treating with adsorbent gel.

Freundlich and Langmuir equations were also applied for adsorption of the metal ions onto RS-based hydrogels.

The Freundlich and Langmuir equations (Eqs. 2 and 3) were also applied to determine the maximum adsorption capacities of metal ions onto hydrogels.

4.4 Evaluating the Removal Efficiency of Metal Ions in Effluents

A completely randomized columns experiment, with three replications, was conducted. For each examined hydrogel, 0.5 g was used, and fifty milliliters of either the municipal or the industrial wastewaters were passed on the hydrogels six times, every three days interval, The retained amounts of Fe(II), Cu(II), Zn(II), and Mn(II) on the hydrogels were calculated from determining the metal content in the eluted aqueous solution, by atomic adsorption.

For examining the durability of these investigated hydrogels, the foregoing hydrogels loaded with metal ions were washed with 0.1 M EDTA solution at 80 °C for 1 h, followed by hot distilled water twice, then rewashed another time with EDTA solution and twice with hot bidistilled water. This was carried out for complete desorption of metal ions from the hydrogel. The efficiency of the washed hydrogels for removing the heavy metal ions from wastewaters was compared with original ones.

4.5 Findings

For evaluating the application of RS-based hydrogels in the purification of municipal and industrial water wastes loaded by Cu(II), Fe(II), Zn(II), and Mn(II) ions, two series of experiments were carried out. First, the preliminary evaluation was performed on adsorption behavior of hydrogels for removing the foregoing metal ions, as individually, in laboratory-prepared metal solutions. The optimum dosage of applied hydrogels which provided higher removal efficiency (0.5 g/50 ml adsorbate) will be applied in the second evaluation on their affinity to purify the municipal and industrial water wastes.

4.5.1 Removal Efficiency of Metal Ions in Aqueous Solutions Versus Hydrogel Type

From Fig. 2, it is clear that, as expected, increasing the amount of hydrogel provided the increase in removal efficiency of metal ions. The relatively higher removal efficiency was noticed in the case of hydrogel derived from grafting initiation by $H_2O_2/(NH4)_2Fe(SO_4)_2$ and using AcOH for neutralization (G1). Applying 0.5 g of G1 provided higher removal of Cu(II), Fe(II), Zn(II), and Mn(II), where the removal efficiency of these metal ions reached 60%, 40%, 83% and 64%, respectively. As can be seen that, using $H_2O_2/(NH4)_2Fe(SO_4)_2$ initiation system provided hydrogel with higher metal removal efficiency than using persulfate/bisulfite initiation system for grafting the raw RS (G3) or cyanoethylated RS (G4). The changes in removal efficiencies of G3 and G4 are not significant. The role of grafting initiation system maybe probably related to the degree of reducing the nitrogen content during hydrolyzing the nitrile groups to amide or/and carboxyl groups. Whereas, in the case of G1 and G2, the degree of reduction is 19.25%; while in the later hydrogels (G3 and G4) are 54 and 48%, respectively (Tables 1 and 2).

On applying the Langmuir's and Freundlich models, the calculated parameters (Table 1) show that the maximum adsorption capacity "Q_0" which was calculated by Langmuir's is varied between 100 and 8.795 mg/g in case of Cu(II) ions, between 144.9 and 5.29 mg/g in case of Fe(II), between 119 and 57.8 mg/g for Zn (II), and between 70.9 and 8.9 mg/g of Mn(II) ions on applying G1–G4 hydrogels, respectively. While the calculated Freundlich parameters (k_f and n, as well as the

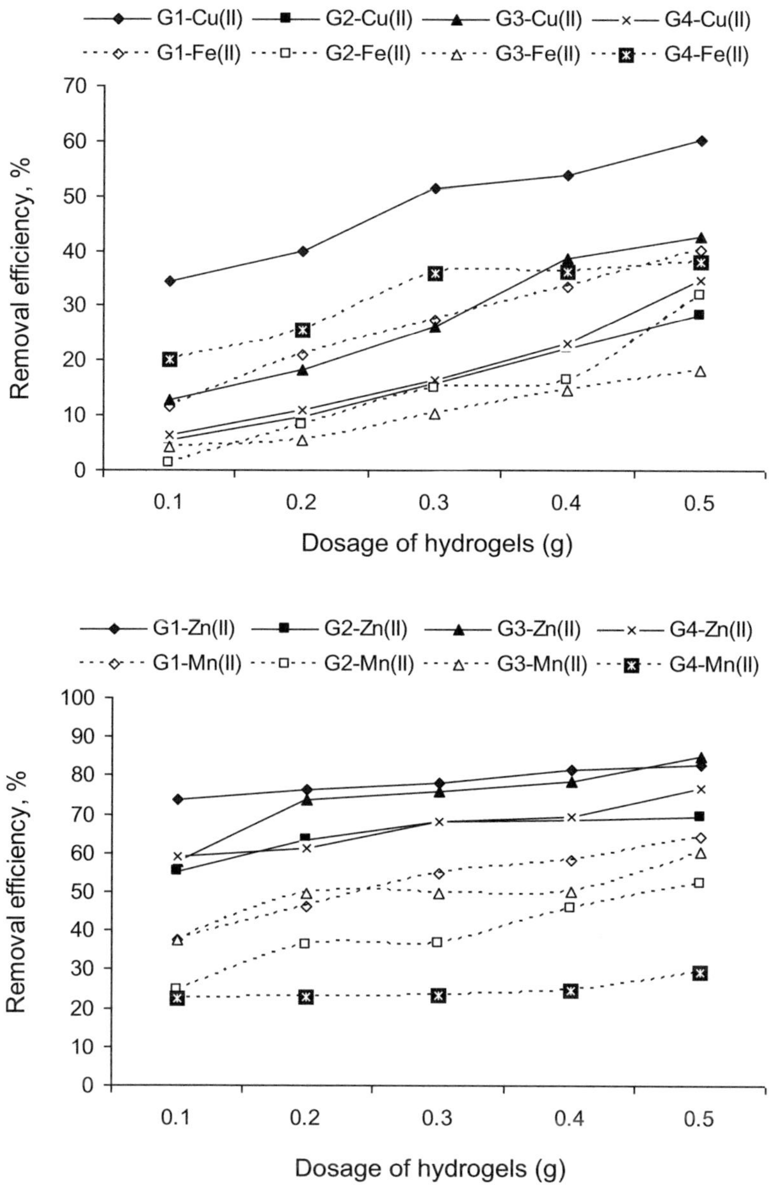

Fig. 2 Adsorption efficiencies of RS-based hydrogels toward different metal ions

Table 1 Freundlich and Langmuir parameters for metal ions removal by RS-based hydrogels

AC precursor	Freundlich			Langmuir		
	k_f (mg/g)(mg/L)n	n	R^2	Q_o (mg/g)	β (L/mg)	R^2
Cu(II) ions						
G1	380	4.0	0.82	100	0.08	0.99
G2	46	18.5	0.83	15	0.07	0.98
G3	92	5.0	0.78	23	8.7×10^{-6}	0.83
G4	48	1.3	0.83	9	0.12	0.93
Fe(II) ions						
G1	111	11.5	0.84	145	0.019	0.94
G2	33	6.7	0.86	5	0.14	0.93
G3	27	11.3	0.71	5	0.11	0.81
G4	160	9.6	0.80	23	0.20	0.96
Zn(II) ions						
G1	892	2.8	0.84	119	0.16	0.95
G2	683	3.0	0.87	58	0.78	0.99
G3	711	3.3	0.86	75	0.28	0.94
G4	681	3.1	0.82	61	0.73	0.94
Mn(II) ions						
G1	431	3.8	0.84	71	0.18	0.94
G2	280	4.6	0.71	37	0.21	0.94
G3	453	3.0	0.734	25	1.134×10^{-5}	0.97
G4	251	3.1	0.81	9	1.0703	0.91

Table 2 Nitrogen content of PAN-g-RS and HPAN-g-RS prepared in pilot scale

Samples	Nitrogen (%)		Polymer loading (%)	Reduction in N (%)
	Before hydrolysis	After hydrolysis		
HPAN-g-RS (G1 & G2)	10.39	8.39	66.88	19.25
HPAN-g-RS (G3)	15.48	7.11	148.19	54.07
HPAN-g-CE-RS G4	15.51	8.11	148.91	47.71

corresponding correlation coefficient, R^2) revealed the poor adsorption behavior of hydrogels G3 and G4 toward adsorption of Cu(II) and Fe(II), where "n" was higher than 10.

For comparing the adsorption affinity of G1 versus metal ions (Table 1), it was observed that the adsorption capacity in the order Fe(II) > Zn(II) > Cu(II) > Mn(II) ions. The higher adsorption capacity for Fe(II) ions than Mn(II) is mainly related to both ionic radius and electronegativity of metal ions.

4.5.2 Evaluating the Adsorption Behavior of RS-Hydrogels in Wastewaters

The promising affinity of the investigated RS-based hydrogels for removing the metal ions, this persuades us to study their behavior on purification of municipal and industrial wastewaters, for the trial to reuse these wastewaters in irrigation purposes. In this respect, column experiments were studied using half gram hydrogels and passing 50 ml of the industrial wastewater through it. Leached ions through six successive leachates are illustrated in Figs. 3, 4, 5, and 6 for Cu(II), Fe(II), Zn(II), and Mn(II), using municipal wastewater and industrial wastewater, respectively. For studying the durability of hydrogels, the leached ions through six successive leachates after washing the hydrogels with EDTA solution were also evaluated and the data are also illustrated in Figs. 3, 4, 5, and 6. Efficiencies of RS-based hydrogels for removing the studied heavy metal ions from municipal and industrial wastewaters either before or after washing with EDTA solution are shown in Table 3, taking into consideration that removing efficiency for any metal ion is calculated as a percentage with respect to the original concentration of metal ion in wastewaters.

For municipal wastewater, Figs. 3, 4, 5, 6, and Table 3 show that the four examined hydrogels are considered as excellent adsorbents, where their removal efficiencies exceed 95% for Cu(II) and Mn(II) ions and in the range of 85–90% for Fe(II) and Zn(II) ions. Moreover, the type of hydrogel is not significant on removal efficiency. After washing the hydrogels with EDTA solution to be reused for removing the heavy metal ions, their efficiencies were decreased. G1 and G2 (grafting using Fe^{2+}/H_2O_2 initiation system) provided lower change (insignificant) in the efficiencies before and after washing with EDTA; however, higher changes are noticed on using G4 (pre-cyanoethylated before grafting and using persulfate/bisulfite initiation system). In other words, only one exception was noticed in case of washing the G4, where its decrement the removal efficiency of heavy metals, it in the range of 5–10% for Cu(II) and Fe(II), 10–15% for Mn(II), and 10–20% for Zn(II).

For industrial wastewater, Figs. 3, 4, 5, 6, and Table 3 show that although the concentration of heavy metal ions in industrial wastewater is higher than that of the municipal waste one, the efficiency of removing heavy metal ions from this water using the examined hydrogels is still excellent. The examined industrial wastewater has Cu(II) as 3.38 folds, Fe(II) as 1.22 folds, Zn(II) as 4.63 folds, and Mn(II) as 1.35 folds that of municipal wastewater. In most examined hydrogels, the removing efficiency of Cu(II) and Mn(II) from industrial wastewater after passing on the examined hydrogels exceeded 95%; while for Fe(II) and Zn(II) ions, this efficiency reaches or exceeds 90%. Similar to that observed in case of municipal wastewater; washing the hydrogels with EDTA solution to remove the retained heavy metal ions on the hydrogels are possible to be reused for purifying the wastewater another time since the maximum reduction in their removal efficiency is not exceeded 10%.

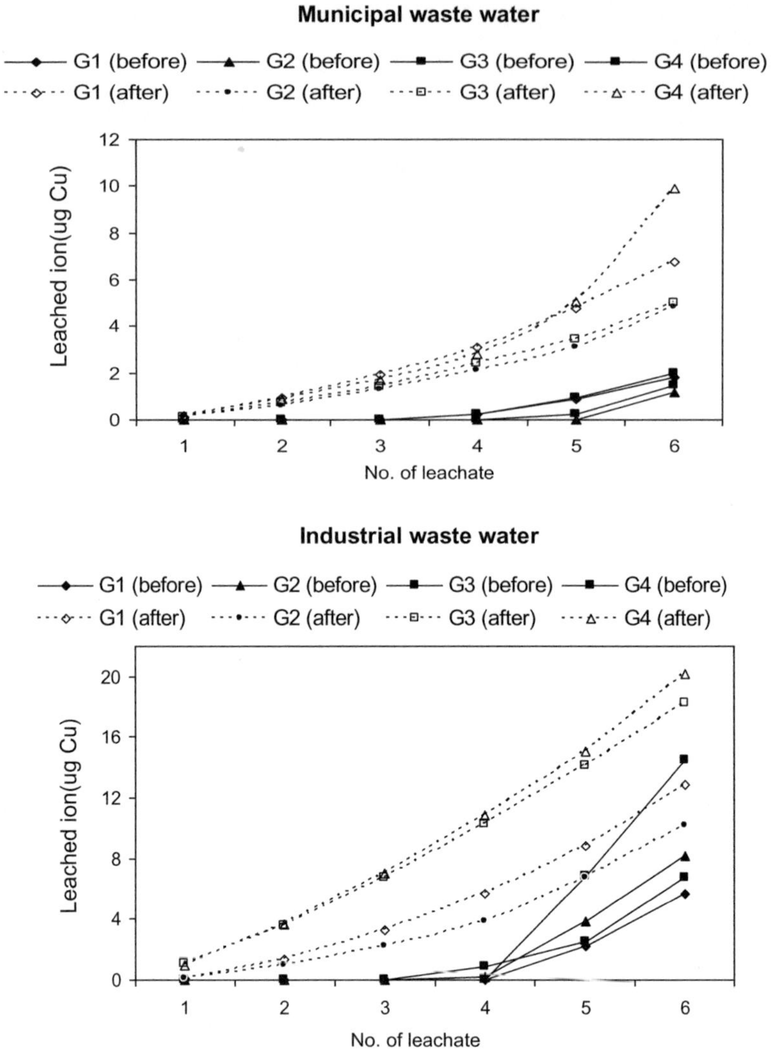

Fig. 3 Effect of hydrogels on the efficiency of removing Cu(II) from wastewaters before and after washing the hydrogels with EDTA (Leached ion/hydrogel)

It is interesting to note that the concentrations of metal ions of the purified municipal and industrial wastewater comply that reported in the specification of water quality for agriculture (Ayers and Westeot 1994).

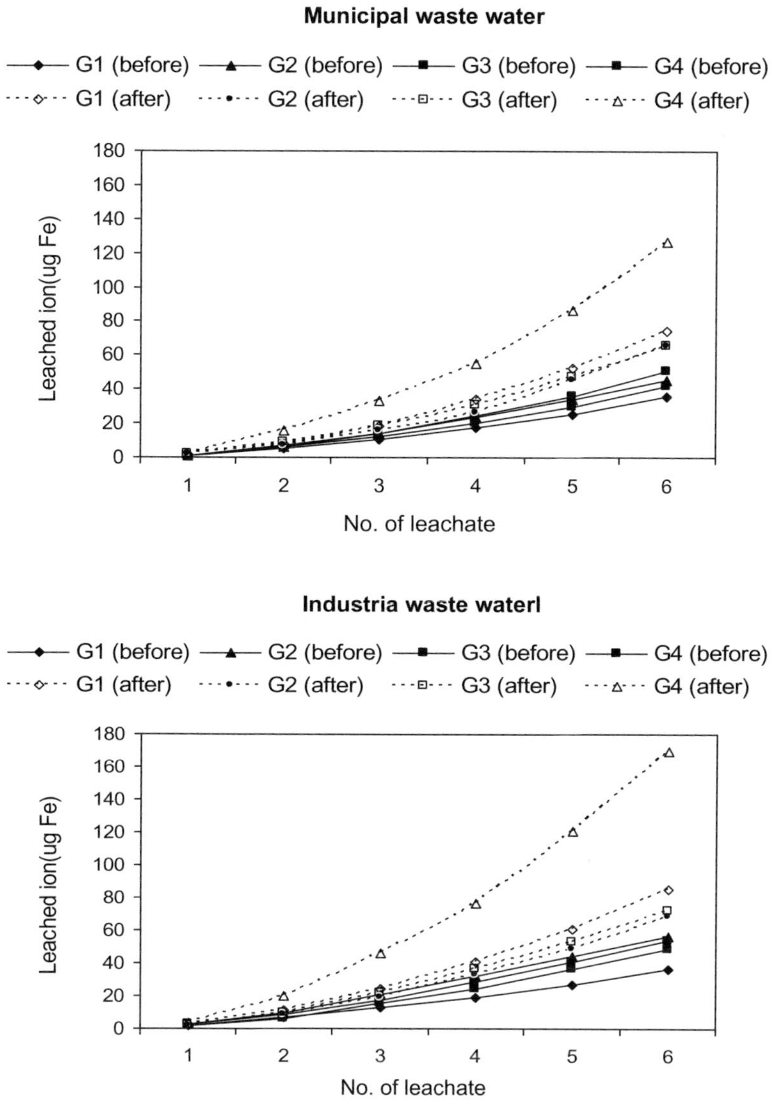

Fig. 4 Effect of hydrogels on the efficiency of removing Fe(II) from wastewaters before and after washing the hydrogels with EDTA (Leached ion/hydrogel)

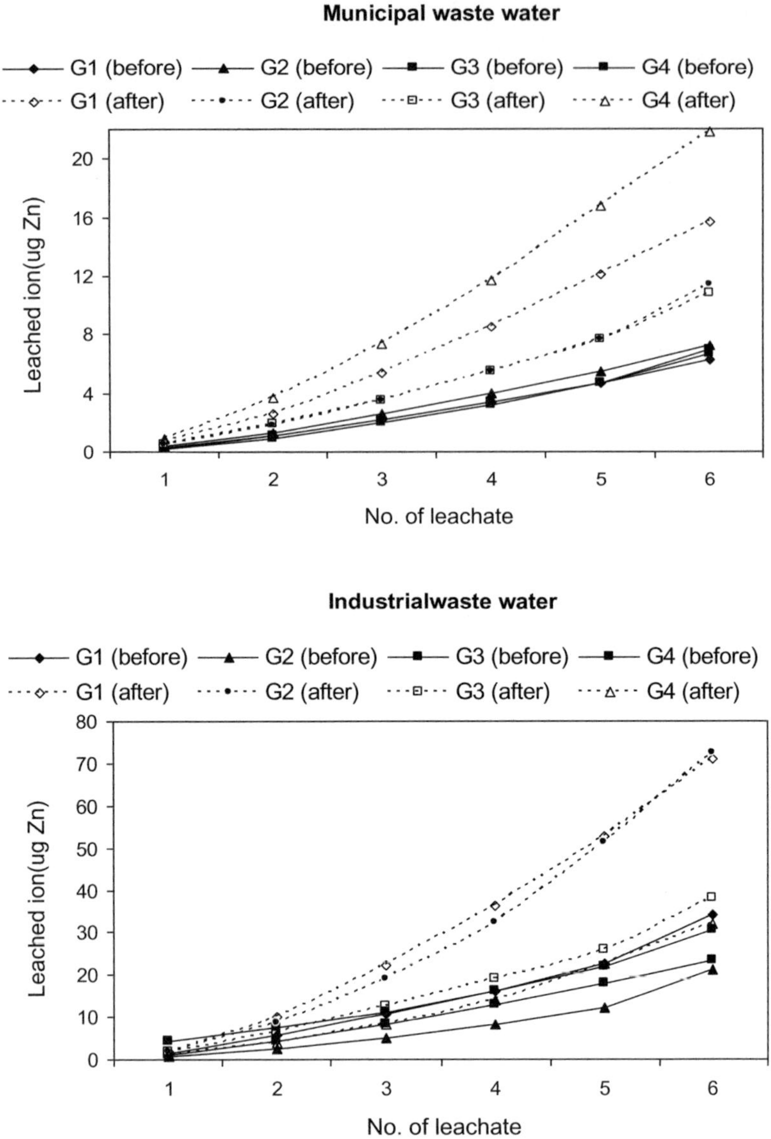

Fig. 5 Effect of hydrogels on the efficiency of removing Zn(II) from wastewaters before and after washing the hydrogels with EDTA (Leached ion/hydrogel)

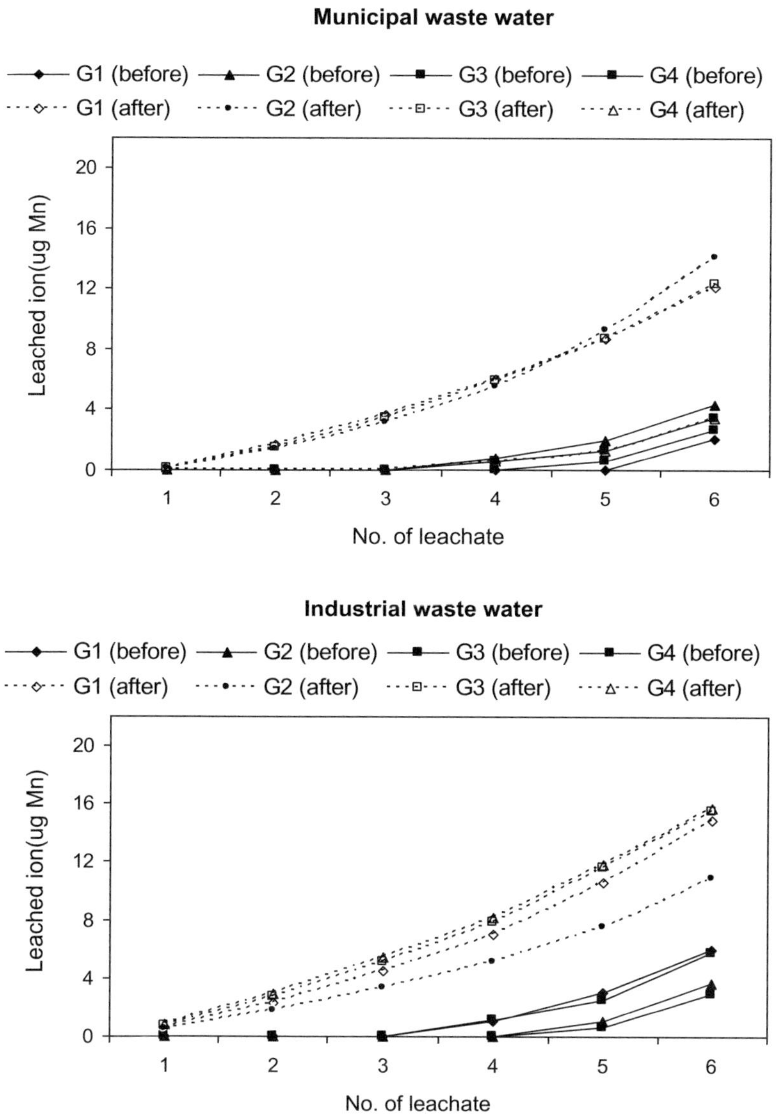

Fig. 6 Effect of hydrogels on the efficiency of removing Mn(II) from wastewaters before and after washing the hydrogels with EDTA (Leached ion/hydrogel)

Table 3 Efficiencies of RS-based hydrogels for removing some heavy metals from municipal and industrial wastewaters

Metal ions	Type of hydrogel	Removal efficiency in municipal wastewater		Removal efficiency in industrial wastewater	
		Before washing	After washing	Before washing	After washing
Cu(II)	G1	96.4	85.6	96.5	92.0
	G2	97.5	89.7	95.0	93.7
	G3	95.9	89.3	95.9	88.7
	G4	97.0	79.0	91.1	87.6
Fe(II)	G1	91.1	81.5	92.6	82.5
	G2	88.6	83.5	88.4	86.0
	G3	89.5	83.8	89.0	85.3
	G4	87.2	68.4	90.1	67.1
Zn(II)	G1	87.1	67.4	84.6	67.9
	G2	84.8	75.8	90.5	67.4
	G3	85.5	77.6	89.5	82.9
	G4	86.2	54.6	87.7	85.7
Mn(II)	G1	97.0	82.4	93.6	84.0
	G2	93.7	79.5	96.1	88.3
	G3	96.2	82.1	93.8	83.3
	G4	94.9	78.4	96.9	83.1

5 Conclusions and Future Perspective

For economical and environmental purposes agricultural wastes were used as low-cost biomaterials for preparation of hydrogels, as adsorbent materials for wastewater treatment. In this respect, we presented our trial in preparing RS-based hydrogels for purifying the local municipal and industrial wastewaters. The preparation of hydrogels was carried out by free-radical grafting process, using ferrous sulfate/hydrogen peroxide and persulfate/bisulfite initiation systems, followed by alkali hydrolysis. The former initiation system provided a high-performance hydrogel with maximum adsorption capacities (Lanqmuir constant; Q_o) for metal ions from laboratory-prepared metal solutions. Where the adsorption capacities of Fe(II), Zn(II), Cu(II), and Mn(II) ions were 145, 119, 100, and 71 mg/g, respectively. Moreover, these investigated hydrogels have significant efficiencies for purifying the municipal and industrial wastewaters. The promising results were cleared from examining the durability of hydrogels, where their possible to be reused for purifying the wastewater another time since the maximum reduction in their removal efficiency is not exceeded the range 5–10%. The concentrations of metal ions of the purified municipal and industrial wastewater comply that reported in the specification of water quality for agriculture. Now we undertaken on applying these investigated hydrogels in combination with agro-based carbon materials for reclamation of Sinai sandy soil (field experiment).

References

Abdel-Aal SE, Gad YH, Dessouki AM (2006) Use of rice straw and radiation-modified maize starch/acrylonitrile in the treatment of wastewater. J Hazard Mater 129(1–3):204–215

Allen SJ, Mckay G (1988) Diffusion model for the sorption of dyes on peat. J Sepn Process Technol 9:18–25

Ansel CWG (1995) New synthetic hydrogels. Macromol Rep A32:733–739

Athawale VD, Lele V (2000) Syntheses and characterisation of graft copolymers of maize starch and methacrylonitrile. Carbohydr Polym 41:407–416

Ayers RS, Westeot SW (1994) Water quality for agriculture. FAO irrigation and drainage paper 29 Rev. 1, FAO, Rome

Bajpai SK, Shrivastava S (2011) Sorptive removal of methylene blue from aqueous solutions by polymer/activated charcoal composites. J Appl Polym Sci 119(5):2525–2532

Basta AH, El-Saied H (2003) Furfural production and kinetics of pentosans hydrolysis in corn cobs. Cellulose Chem Technol 37(1–2):79–94

Basta AH (2003) The role of chitosan in improving the ageing resistance of rosin sized paper. Restaurator 24(2):106–117

Basta AH, Fierro V, Saied H, Celzard A (2011a) Effect of deashing rice straws on their derived activated carbons produced by phosphoric acid activation. Biomass Bioenergy 35(5):1954–1959

Basta AH, El-Saied H, Winandy JE, Sabo R (2011b) Preformed amide-containing biopolymer for improving the environmental performance of synthesized urea-formaldehyde in agro-fiber composites. J Polym Environ 19:405–412

Basta AH, El-Saied H, El-Hadi O, El-Dewiny C (2013) Evaluation of rice straw-based hydrogels for purification of wastewater. Polym-Plast Technol Eng 52(11):1074–1080

Buchholz FL, Peppas NA (1994) Superabsorbent polymer science and technology. American Chemical Society, Washington, DC, p 123

Burrg RC, Fanta GF, Doane W, Russell C (1976) Graft copolymers of starch and mixtures of acrylamide and acrylic acid. Appl Polym Sci 20:3201–3204

Cheng WM, Hu XM, Wang DM, Liu GH (2015) Preparation and characteristics of corn straw-Co-AMPS-Co-AA superabsorbent hydrogel. Polymers 7(11):2431–2445

Drury JL, Mooney DJ (2003) Hydrogels for tissue engineering: scaffold design variables and applications. Biomaterials 24:4337–4351

El-Saied H, Basta AH, Barsoum BN, Elberry MM (2003) Cellulose membranes for reverse osmosis Part I. RO cellulose acetate membranes including a composite with polypropylene. Desalination 159(2):171–181

El-Saied H, Waly AI, Basta AH (2000) Highly water absorbents from lignocelluloses. Effect of reaction variables on the water absorbency of polymerized lignocelluloses. Polym Plastic Technol Eng 39:905–926

El-Saied H, Waley AI, Basta AH, El-Hadi O (2004) Water absorbents from lignocelluloces: novel soil conditioners of sandy soil from lignocellulosic wastes. Polym Plastic Technol Eng 43 (3):777–795

El-Saied H, Basta AH, El-Hadi O, Waley AI (2007) High water absorbents from lignocelluloses. Pt. III: upgrading the utilization of old news paper (ONP) in agronomic applications. Plast Polym Technol Eng 46(3):311–319

El-Saied H, Basta AH, Waly AI, El-Hady OA, El-Dewiny CY (2013) Evaluating the grafting approaches for utilizing the rice straw as environmental friendly and potential low cost hydrogels. Emir J Food Agri 25(3):211–224

Fanta GF, Doane WM (1990) Advances in carbohydrates chemistry. ACS Sym Ser 433:288–303

Fanta GF, Burr RC, Doane WM (2006) Factors influencing absorbent properties of saponified starch-g-polyacrylonitrile isolated by methanol precipitation Starch-Stärke 36(12):416–419
Fierro V, Muñiz G, Basta AH, El-Saied H, Celzard A (2010) Rice straw as precursor of activated carbons: activation with ortho-phosphoric acid. J Hazards Mater 181(1):27–34
Freundlich H (1907) Ueber die adsorption in loesungen. Z Phys Chem 57:385–470
Gupta A (2014) Design of an absorbent and comfortable sanitary napkin for applications in developing countries Ms thesis, Massachusetts Institute of Technology
Hoffman AS (2012) Hydrogels for biomedical applications. Adv Drug Deliv Rev 64:18–23
Hussain G, Al-Gosaibi AM, Al-Ajaji AN (1989) Effect of water quality on water absorption by gel-forming soil conditioners. Arab J Sci Eng 14(3):393–401
Kennedy JK, Philips GO, Williams PA (1993) Cellulosics: chemical biochemical and material aspects. Ellis Horwood, London, p 391
Ko JA, Park HJ, Hwang SJ, Park JB, Lee JS (2002) Preparation and characterization of chitosan microparticles intended for controlled drug delivery. Int J Pharm 249:165–174
Kosmala JD, Henthorn DB, Brannon-Peppse L (2000) Preparation of inter-penetrating networks of gelatin and dextran as degradable biomaterials. Biomaterials 21:2019–2023
Kumar K, Kaith BS, Mittal H (2010) Utilization of acrylamide and natural saccharide based polymeric networks in Ph controlled release of 5-amino icylic acid. J Chil Chem Soc 55:522e–526
Langmuir I (1918) The adsorption of gases on plane surfaces of glass, mica and platinum. J Am Chem Soc 40:1361–1368
Li X, Cai Z, Winandy JE, Basta AH (2011) Effect of oxalic acid and steam pretreatment on the primary properties of UF-bonded rice straw particleboards. Ind Crops Prod 33(3):665–669
Luo X, Xu J, Zhang Q, Yang G, Chen H (2005) Electrochemically deposited chitosan hydrogel for horseradish peroxidase immobilization through gold nanoparticles self-assembly. Biosens Bioelectron 21:190–195
Makoto T, Kouji K, Ayako S, Ping X, Ken-Ichi F, Toshio T, Yutaka T, Eiji H (2005) Physicochemical properties of cross-linked poly-γ-glutamic acid and its flocculating activity against kaolin suspension. J BioSci Bioenergy 99(2):130–135
Neto VOS, Oliveira AG, Teixeira RNP, Silva MAA, Freire PTC, Keukeleire DD, Nascimento RF (2011) Use of coconut bagasse as alternative adsorbent for separation of copper (II) ions from aqueous solutions: isotherms, kinetics, and thermodynamic studies. BioResources 6(3):3376–3395
Saito N, Seki K, Aoyama M (1991) Superabsorbent materials prepared from lignocellulosic materials by phosphorylation Sen-1Gakkaishi 47:255–258
Saito N, Aoyama M, Shimizu Y, Takai M, Hayashi J (1995) Super-absorbent materials prepared from lignocellulosic materials by phosphorylation VI. Swelling and morphological changes Mokuzai Gakkaishi 41:179–183
Singh B, Chauhan N (2010) Release dynamics of tyrosine from dietary fiber psyllium based hydrogels for use in Parkinson's disease Food Res Int 43:1065–1072
Singh B, Sharma V, Kumar S (2011) Synthesis of smart hydrogels by radiation polymerisation for use as slow drug delivery devices. Can J Chem Eng 89:1596–1605
Singh B, Bala R (2012) Design of dietary polysaccharide and binary monomer mixture of acrylamide and 2-acrylamido-2-methylpropane sulphonic acid based antiviral drug delivery devices. Chem Eng Res Des 90:346–358
Tomatsu I, Peng K, Kros A (2011) Photoresponsive hydrogels for biomedical applications. Adv Drug Del Rev 63:1257–1266. http://dx.doi.org/10.1016/j.addr.2011.06.009
Techawanitchai P, Idota N, Uto K, Ebara M, Aoyagi T (2012) A smart hydrogel-based time bomb triggers drug release mediated by pH-jump reaction. Sci Technol Adv Mat 13:064–202

Thakur VK, Thakur MK (2014) Recent trends in hydrogels based on psyllium polysaccharide: a review. J Clean Prod 82:1–15

Thakur VK, Thakur MK, Raghavan P, Kessler MR (2014) Progress in green polymer composites from lignin for multifunctional applications: a review. ACS Sustain Chem Eng 2:1072–1092

Vasquez J-MG, Tumolva TP (2015) Synthesis and characterization of a self-assembling hydrogel from water-soluble cellulose derivatives and sodium hydroxide/thiourea solution. Am J Chem 5(2):60–65

Zhao Q, Sun J, Lin Y, Zhou Q (2010) Study of the properties of hydrolyzed polyacrylamide hydrogels with various pore structures and rapid pH-sensitivities. React Funct Polym 70: 602–609

Chapter 10
Pharmaceutical Polymer Gels in Drug Delivery

Geeta Aggarwal and Manju Nagpal

Abstract Polymer gels are being developed at prolific pace as prominent delivery systems to circumvent obstacles in drug delivery and impart easier application of therapeutics. A large number of molecules migrate into an elastic cross-linked network of long polymers inflated with a solvent such as water, causing the swelling of network and resulting in the formation of an aggregate known as a polymer gel. Polymer gels are utilized in different technologies, including drug delivery, biomedical devices, biosensors, tissue engineering, actuators, and even in cosmetics and food. Recently, there has been growing interest in smart polymer gels constituting a new generation of biomaterials that respond to significant changes due to little changes in the environment. The polymers that form smart gels are referred as "smart" polymers or "stimuli-sensitive" polymers or "environmentally" sensitive polymers, and they can be classified on the basis of their structure (superporous gels, shape memory gels, artificial protein gels) and stimuli (pH-sensitive gels, temperature-sensitive or thermoresponsive gels, analyte-sensitive gels). The gel can be prepared by covalent and non-covalent cross-linking (chemical or thermally irreversible gels and physical or thermoreversible gels, respectively), and it can be evaluated for morphology, rheological behavior, adhesion, spreadability, in vitro diffusion/permeation, in vivo behavior, and stability of the formulation. Polymer gels that are sensitive to the presence of various biomarkers have been extensively used in targeting specific disease conditions. This chapter will be organized into four separate sections: defining the concept of polymer gels; describing types, formulation, and evaluation of gels; illustrating applications especially in the context of drug delivery; and including recent advances and future perspectives.

G. Aggarwal (✉)
Delhi Pharmaceutical Sciences and Research University,
Pushp Vihar, Sector III, New Delhi 110017, India
e-mail: geetaaggarwal17@gmail.com

M. Nagpal
Chitkara College of Pharmacy, Chitkara University, Rajpura, Punjab, India

© Springer Nature Singapore Pte Ltd. 2018

V. K. Thakur et al. (eds.), *Polymer Gels*, Gels Horizons: From Science to Smart Materials, https://doi.org/10.1007/978-981-10-6080-9_10

Keywords Polymer · Smart gels · Nanogels · Pharmaceutical gels
Transdermal · Topical

1 Introduction

Polymer gels usually represent cross-linked solvent systems (in most case, a binary system, although in principle the presence of several solvents in the system is possible) in which there exists a three-dimensional network composed of macromolecules and is capable of staining a large amount of solvent (times to hundreds of times that of the polymers itself). This internal network structure is due to physical bonding (physical gel) or chemical bonding (chemical gel) within the extending fluid. The fluids which can be used as extender include water (hydrogel) (Omidian et al. 2005, 2006), oil (organogel), or air (aerogel).

Polymeric gels are viscoelastic solids that exhibit different properties varying from soft and weak to hard and tough (Hong et al. 2008). They tend to shrink or swell reversibly (up to 1000 times of volume) by little changes in the environment (e.g., temperature, pH, and electric field) (Thakur and Thakur 2014a, b, 2015). Microsized gels may contract in less than a second, whereas thick polymers layers need many minutes to respond. The polymeric gels can be chemical or thermally irreversible gels and physical or thermoreversible gels. The thermoreversible gel is formed when the solution is cooled down due to the conformational transition of polymer chains, and chemical gels are formed by cross-linking of polymer chains. The cross-linking is caused by certain additives, which are added to aqueous solution. In general, physical gels are observed at high temperature range than the chemical gels (Dolbow et al. 2004, 2005).

1.1 Chemical Gels

Chemical gels are covalently cross-linked network of polymers swollen in a solvent. These gels can be prepared by swelling of cross-linked polymer and polymerization or polycondensation or cross-linking of polymer in solvent medium.

These gels are characterized by equilibrium modulus of elasticity. The factors which affect the elasticity of gels are the solvents and the flexibility of polymer chains. The higher flexibility of polymer chains along with small concentration of polymer results in larger distances within the network, leading to higher elasticity.

When chemical gels have high chain rigidity and dense network or it swells high in a solvent with poor thermodynamic quality with respect to the polymer, it results in fragile gels. These gels can be destroyed by thermal degradation or chemical flow of polymer, because junctions in chemical gels network are formed by covalent bond between cross-linking agent and macromolecules.

1.2 Physical Gels

The difference in microscopic properties of cross-linked network in physical and chemical gels endows these gels with variation in macroscopic characteristics. Since the cross-linked network in chemical gels prevents them from dissolving in surrounding solvent, they appear macroscopically as solids. However, due to weaker bonds in cross-linking, the cross-links undergo dissolution and create physical gels. During short time periods, the physical cross-links are not able to dissolve, because of less time available and then gel shows the solid-like behavior of chemical gels. Similarly during long time periods, the distruction in bonds tends to release shear stress or anisotropic stress. It reveals that at long time intervals, physical gels may liquid. Thus, physical gels exhibit thermally driven properties. The phase diagram in these gels shows reversible sol–gel transition by changing temperature of the host fluid (mostly fluid is water). These changes are useful in various applications, where the material exhibits lower critical solution temperature. In these systems, the polymer occurs in solution form at lower temperatures, which transforms into a gel by increasing temperatures. These types of materials can be used as injectable polymers in biomedical applications, which are liquid at the time of drug administration (at room temperature), but form gel inside the body (at physiological temperature). So, physical gels also known as thermoreversible gels, in which creation as well as dissolution of physical cross-linked network or sol–gel transition is reversible and temperature-dependent (Dong et al. 2006).

2 Pharmaceutical Polymer Gels

Pharmaceutical gel is a semisolid network system, which consists of dispersion made up of large organic molecules or small particles enclosing and interpenetrated with solvent. Various advantages of pharmaceutical gels include formulation of gastro-retentive vehicle (Park et al. 2006) enhancing gastrointestinal (GI) drug absorption and avoids difficulties created by gastrointestinal pH, drug interactions and metabolizing enzymatic activity; colon targeted drug delivery; bypass first-pass metabolism to enhance bioavailability of drug and to reduce undesirable action of metabolites; used for cutaneous and percutaneous drug delivery to enhance penetration of drug through skin; controlled drug delivery through nasal and ocular route and enhanced patient compliance (Chaterji et al. 2007).

Disadvantages of pharmaceutical polymer gels

- Gels may cause allergic reactions.
- Enzyme in gastrointestinal tract or epidermis may denature the drug molecule in gels.

- Selection of area where the gel is applied affects the bioavailability of drug through skin.
- Gels which are applied into eyes or other body cavity should be sterilized.

3 Classification of Gels

Pharmaceutical polymer gels are classified on the basis of colloidal phase, type of solvent, rheology, and the type of drug delivery.

3.1 On the Basis of Colloidal Phase

Polymer gels can be classified either as two-phase system (inorganic gels) or single-phase system (organic gels) on the basis of the presence of colloidal phase. In the two-phase system, if particles form three-dimensional structures throughout the gel due to their relatively large particle size in dispersed phase, then gel mass is known as magma (e.g., bentonite magma). Generally, two-phase system has floccules of small particles rather, which results in unstable gel structure. Both magma and gel may exhibit thixotropic forming semisolids characteristics on standing, which on agitation becomes liquid (Bhasha et al. 2013).

In single-phase system, large organic molecules are dissolved in the liquid phase. These polymer gels may be formed of semisynthetic polymer (e.g., cellulose derivatives) or synthetic polymer (e.g., carbomer) or natural gums (e.g., tragacanth).

3.2 On the Basis of Solvent

Gels can be classified as hydrogels (water-based) or organogels (non-aqueous solvent) or aerogels/xerogels on the basis of type of solvent used as continuous phase in the formulation.

3.2.1 Hydrogel

Hydrogel or aquagel is a cross-linked polymer chain network, which is water-insoluble and may occur as colloidal gel, in which dispersion medium is water. Hydrogel absorbs substantial amounts of aqueous solutions due to cross-linked polymer network. Because of the presence of large water content, the gels behave like natural living tissue than another type of semisynthetic or synthetic biomaterial (Laftah et al. 2011). Hydrogels are also characterized as superabsorbent as they may contain over 99% water. These gels have the special ability to sense changes in stimuli in the environment, i.e., pH, temperature, or any change in concentration of analyte or metabolite, causing gels to release their active constituent as a result of any type of change. Hydrogels can be of different types depending on the type of polymer, cross-linking, configuration, stimuli, etc., as indicated in Fig. 1. They are commonly used for pharmaceuticals as (Venugopalarao et al. 2013; Sabale and Vora 2012; Stamatialis et al. 2008; Kashyap et al. 2005; Hamidi et al. 2009; Hoffman 2012):

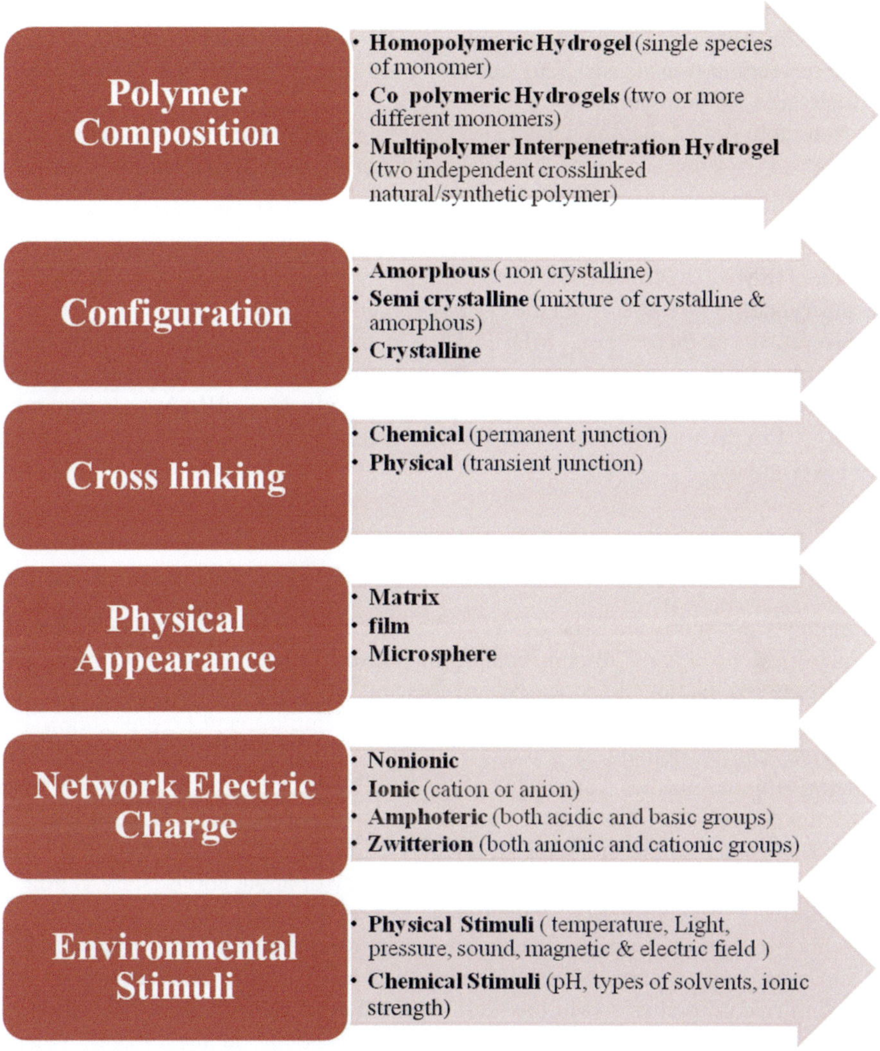

Fig. 1 Classification of hydrogels on the basis of different parameters

- Topical/transdermal drug delivery vehicles.
- As sustained/controlled release delivery vehicle.
- Hydrogels responsive to specific moieties, such as antigens or glucose can be used as stimuli-responsive biosensors in drug delivery systems.
- Hydrogels are used in medical electrodes, which are composed of cross-linked polymers (e.g., polyvinylpyrrolidone).
- Used in contact lenses.
- Used in disposable diapers or in sanitary napkins.
- Hydrogel explosives are formulated.
- Can be used as implants for breast.
- Dressings for burns or for difficult-to-heal burns or wounds. These gels are superior to achieve a moist environment, when required in wounds.
- Reservoirs in transdermal drug delivery used mainly for ionic drugs and can be delivered by iontophoresis.
- Natural hydrogel materials have also been fabricated for tissue engineering (e.g., agarose, hyaluronan, methylcellulose, and other natural polymers) (Caló and Khutoryanski 2015).

Examples of monomers used to synthesize hydrogels for pharmaceutical use include HEMA (hydroxyethyl methacrylate), HDEEMA (hydroxydiethoxyethyl methacrylate), MEMA (methoxyethyl methacrylate), HEEMA (hydroxyethoxyethyl methacrylate), MDEEMA (methoxydiethoxyethyl methacrylate), MEEMA (methoxyethoxyethyl methacrylate), EGDMA (ethylene glycol dimethacrylate), AA (acrylic acid), EG (ethylene glycol), NIPAAm (*N*-isopropyl AAm), PEGMA (PEG methacrylate) and HPMA (*N*-(2-hydroxypropyl) methacrylamide).

3.2.2 Organogels

Organogels are semisolid type of networks, in which organic continuous phase is immobilized by a three-dimensional (3D) network made up of intertwined and self-assembled gelator fibers. In spite of their major liquid part, organogels exhibit the rheological behavior and appearance of solids (Vintiloiu and Leroux 2008). The organogel can be defined as a non-glassy, non-crystalline thermoreversible solid material (thermoplastic) made up of a liquid organic phase entrapped in a cross-linked polymer network. Examples of continuous phase include organic solvent, vegetable oil, or mineral oil (Jadhav et al. 2009).

Organogels have potential to use in diverse array of applications, such as topical drug delivery for cosmetics, opthalmics, and ointments, dermal drug delivery (eczema), transdermal drug delivery, parenteral drug delivery, oral drug delivery, in situ organogels as mucoadhesives, medicated pastilles and suppositories and for culturing microorganisms (Murdan 2005; Anand et al. 2001; Garg et al. 2011).

Different types of organogels are formed depending on the type of gelator (Table 1).

Table 1 Types of organogels

Types of gel	Composition	Properties	Applications
ALS-based organogel	Gelator molecules have three components: an aromatic moiety linked to steroidal molecules with linker atoms. Example: butanoate (CAB) cholesteryl-4-(2-anthryloxy)	Stable	Sustained drug delivery
Phenol-based organogel	Hydrogen bonding between phenol and the head group of twin-tailed anionic surfactant, sodium bis (2-ethylexyl) sulfosuccinate	Moisture-sensitive	Drug delivery
Lecithin-based organogel	Lecithin extracted from different plant tissues as well as animal tissues other than egg yolk	Thermodynamically stable, thermoreversible, transparent, non-irritant, stable for long period	Controlled delivery vehicles
Pluronic lecithin organogel	Soya lecithin-based organogel and consists of isopropyl palmitate or isopropyl myristat, water and pluronic F27 (poloxamer 407)	Thermostable, viscoelastic biocompatible	Topical and transdermal drug delivery
Gelatin stabilized organogel	Gelatin solubilized with water in oil microemulsion, which results in gelatin	Thermostable	Enzyme immobilization, transdermal and topical drug delivery
Non-ionic surfactant-based gel	Sorbitan—monostearate, nonionic surfactant gels a variety of solvents	Stable	Delivery of hydrophilic vaccines
Eudragit organogel	Mixture of Eudragit L or S and polyhydric alcohols (propylene glycol, like glycerol and liquid polyethylene glycol)	High gel rigidity and stability	pH-sensitive sustained release
Poly ethylene organogel	Polyethylene dissolved in mineral oil	Colorless	Ointment bases
L-alanine derivative organogel	*N*-lauroyl-L- alanine methyl ester polymer gels in organic solvent (soybean oil and triglycerides)	Gel state at room temperature	Sustained-release implants

3.2.3 Xerogels and Aerogels

Xerogels are solid gels with low solvent concentration, frequently formed by evaporation of solvent and leaving gel framework behind. They may be reverted to gel state by addition of an agent that imbibes and then swells the gel matrix. Examples of xerogels are acacia tears, dry gelatin, dry cellulose tragacanth ribbons, and polystyrene. A xerogel is a solid form of gel, which retains high porosity (25%), high surface area (150–900 m^2/g), and small pore size (1–10 nm). The xerogels can retain its original shape, but often crack due to the extreme shrinkage that is experienced while being dried. Heat treatment of xerogel by elevating temperature causes shrinkage of xerogel because of small amount of viscous flow and efficiently changes the porous gel network into a dense glass.

An aerogel is obtained when the liquid phase of a gel is replaced with gas while retaining the solid network, with only a minor or no shrinkage of the gel. It was firstly observed under supercritical conditions but now it is possible under ambient drying conditions as well. When solvent removal from the gel occurs in supercritical conditions, the cross-linking network does not shrink and a highly porous and a low-density aerogel is formed. Therefore, the method of drying will dictate whether an aerogel or xerogel will be formed.

Aerogels can be used in pharmaceuticals as a matrix base for delivery of enzymes and proteins (Ulker and Erkey 2014). Xerogels are widely used biomaterials as matrix systems for extended and controlled release of various biologically active agents administered by different routes. The drug release from xerogels could be customized by adjusting several parameters, such as (i) type of precursor (ii) concentration of the catalyst (iii) drying temperature (Quintanar-Guerrero et al. 2009; Munusamy et al. 2009; Kortesuo et al. 2000).

3.3 On the Basis of Rheology

The flow properties of fluids (liquids and semisolids) control each and every step in pharmaceutical process involved in the development of pharmaceutical dosage forms, such as mixing, filling, packaging, and removal from container during the application to action site. The flow properties of fluids also influence their in vivo behavior. In general cases, gels exhibit properties of non-Newtonian flow. On the basis of rheology, they can be classified as plastic, pseudo-plastic, or thixotropic gels.

3.3.1 Plastic Gels

These polymer gels exhibit non-Newtonian plastic flow, in which rheogram plot shows the yield value above which the elastic gel disturbs and begins to flow.

Example includes flocculated form of suspensions, e.g., aluminum hydroxide, and bingham bodies exhibit plastic flow.

3.3.2 Pseudo-plastic Gels

The viscosity of these gels reduces with increase in rate of shear, but shows no yield value. The rheogram plot results from the application of shear stress on the long-chain polymer molecules. As the shear stress is applied and increased, the disturbed long-chain polymer molecules start to align along the direction of flow, which results in the release of solvent from the gel matrix and increase in viscosity, e.g., dispersion of sodium alginate, sodium carboxymethyl cellulose, and tragacanth.

3.3.3 Thixotropic Gels

The gels are generally thick and viscous under static conditions, but become thin, less viscous and will flow over time when agitated, shaken or stressed, i.e., time-dependent viscosity. The bonds in between molecules in the gel networks are weak and thus can be easily broken down by shaking or stirring. The resultant sol again forms gel because of collision in the particles and again linking together. This reversible isothermal sol–gel change is known as thixotropy. This occurs mainly in non-spherical particles containing colloidal system, e.g., kaolin, bentonite, and agar solution.

3.4 On the Basis of Drug Delivery

Pharmaceutical gels are generally used as vehicles in drug delivery. On the basis of drug delivery, these gels may be classified as:

1. Sustained/controlled release gels
2. Fast release gels
3. Bioadhesive gels
4. Smart gels: thermosensitive gels, pH-sensitive gels, analyte-sensitive gels
5. In situ gels

3.4.1 Sustained/Controlled Release Gels

Controlled drug delivery systems offer considerable advantages such as reduced frequency of dosing thereby improved patient compliance, maintenance of blood levels within a desired range by minimized fluctuation of concentrations, localized/

targeted drug delivery, and remarkably decreased adverse effects. Gels act as controlled drug delivery vehicles by diffusion mechanism, in which drug is delivered by polymer matrix system (water-insoluble matrix) or reservoir system (water-insoluble polymeric membrane). Hydrogels, aerogels, and organogels act as sustained or controlled drug delivery vehicles in pharmaceuticals.

3.4.2 Fast Release Gels

The gels which have fast swelling and superabsorbent properties may be utilized for fast disintegrating tablets. Fast release gels can be used for the treatment of acute diseases, in which rapid onset of action is required.

3.4.3 Bioadhesive Gels

The bioadhesive systems increase the residence time of the drug in the oral cavity and localize the drugs in a region. Bioadhesive gels are used for topical, transdermal, vaginal, mucoadhesive, buccoadhesive, and ophthalmic drug delivery. Gel-forming bioadhesive polymers are cross-linked polyacrylic acid which can adhere to the mucosal surfaces for longer periods of time and exhibit controlled release of drug at absorption area. Various hydrogel-based bioadhesive systems for controlled drug release has been reported in the literature. These kinds of polymer gel systems act as controlled drug delivery device and bioadhesive system, which can release drug at a particular site with enhanced bioavailability (Reddy et al. 2011).

3.4.4 Smart Polymer Gels

These are also known as environmental-sensitive or intelligent gels. Interest of researchers in stimuli-responsive gel systems is steadily increasing, particularly in controlled or self-regulated delivery systems. Stimuli-responsive or "smart" gels show a remarkable physiochemical change in response to little changes in their environment including pH, temperature, light, magnetic field, and ionic factors. These are reversible changes and so the system is capable of returning to its initial state when the signal/trigger is removed (Almeida et al. 2012a, b). Mostly hydrogels respond to environmental changes, which act as smart gels (Table 2).

Table 2 Smart polymer gels for drug delivery

Signal	Polymers	Applications
Temperature	PNIPAAm, PDEAAm, Poloxamers, Prolastin	On–off drug release, squeezing device
pH	Dendrimers ester poly(hydroxyproline), poly(L-lisine), poly(propyl acrylic acid), polysilamine, poly(ethacrylic acid), carbopol, Eudragit S-100 and Eudragit L-100	pH-dependent oral drug delivery
Magnetic field	Ferromagnetic material PNIPAAm-co-acrylamide	Targeted drug delivery
Organic solvent	Eudragit S100	Modified drug release
Electric signal	Chitosan, polyelectrolytes (pH-sensitive)	Actuator, on–off drug release, artificial muscle
Light	Copolymer of PNIPAAm and light-sensitive chromophore, such as triphenylmethane and leuco derivatives	Optical switches, ophthalmic drug delivery
Glucose	pH-sensitive hydrogels; concanavalin A-grafted polymers; polymers containing phenylborate groups	Self-regulated insulin delivery
Antigen	Semi-IPN with grafted antibodies or antigens	Regulated drug release by the presence of a specific antigen or sensor for immunoassay

4 Properties of Gels

- **Swelling**: After the gelling agent comes in contact with solubilizing solvent, it absorbs a large amount of solvent and its volume enhances. This phenomenon is known as swelling. The process occurs due to significant amount of penetration of solvent into the matrix. The degree of swelling usually depends upon the strength and number and type of linkages between the individual molecules of the gelling agent.
- **Syneresis**: Some gels contract on standing and squeeze out some fluid instantaneously. This process is known as syneresis. This process of syneresis indicates that the polymer gel was thermodynamically unstable. The degree to which syneresis occurs gets increased by decreasing the concentration of gelling agent in gel. The contraction mechanism is due to the relaxation of the elastic stress which is formed during the settling of the gels. As the stress is removed, the interstitial space between the molecules available for solvent is reduced, which forces the liquid out.
- **Structure**: The interlinking of molecules of the gelling agents leads to network structure and causes the rigidity of the gel.

- **Aging**: Aging is very slow but spontaneous aggregation. Because of the presence of gelling agent, aging causes gradual formation of a dense network in gel.
- **Rheology**: Solutions of the gelling agents exhibit non-Newtonian flow behavior, which may be plastic flow or pseudo-plastic flow. It is characterized by decrease in viscosity with increase in shear rate.

5 Gel-Forming Substances/Gelling Agents

A variety of polymers can be used to provide cross-linked network. The polymers or gelling agents in gels are ingredients that can turn water, or oil, phase into a gel, which is thickened but is not stiffed. These gelling agents make gels thick products, which can be stirred or shaken under shearing stress, for easier packing or for spraying (Kadajji and Betageri 2011).

Natural polymer Examples of natural gelling agents are gums, agar, alginates, tragacanth, gellan gum, pectin, xanthan gum, guar gum, carrageenan, and chitosan.

Semisynthetic polymers Examples of semisynthetic polymers are cellulose derivatives, methylcellulose, hydroxy propyl methyl cellulose, carboxymethyl cellulose, hydroxypropyl cellulose, and hydroxyethyl cellulose.

Synthetic polymers Examples of synthetic gel-forming polymers are carbomer, poloxamer/surfactants, and polyacrylamide.

Miscellaneous gel-forming polymers Examples of miscellaneous gel-forming polymers are microcrystalline silica, cetyl ester wax, clays, beeswax, and aluminum stearate.

Though each gelling agent has its own unique properties, they have some generalizations, which include:

1. If we add gelling agent in the solvent in a haphazard manner, it can result in the formation of clump. In this case, the gelling agent molecules make contact with the medium and form surface layer which creates difficulty for solvent medium to penetrate. The clumps hydrate after some time, but it takes more time. In an alternative manner, gelling agent is sieved on the surface of the solvent/medium with the continuous stirring. The presence of wetting agent always helps to solve the problem of clump formation.
2. The solubility of gelling agent also depends on the temperature of the medium. Some polymers are more soluble in hot water than in cold water and vice-versa. For example, poloxamers and methylcellulose are more soluble in cold water, whereas bentonite, gelatin and sodium carboxymethyl cellulose are more soluble in hot water.
3. Some of the gelling agents may require a pH adjustment by chemical or a neutralizer to form gel after addition of a gelling agent, e.g., carbomers.

4. Gelling agents mostly require one to two days to completely hydrate in the medium and acquire required viscosity and clarity.
5. They may be used up to 0.5–10% of concentrations, depending on the gel and nature of agent.
6. It is better to add the crude drug before the formation of gel, but in this case drug should not interfere with the process of gel formation.
7. Somc gelling agents, viz., carbopol 934P, sodium carboxymethyl cellulose, hydroxypropyl methylcellulose, and methylcellulose, are recommended to use in oral formulations.

5.1 Natural Polymers

Plant gums or natural gums are hydrophilic polymers of carbohydrates with high molecular weights. They are generally made up of monosaccharide units, which are joined with glucosidic bonds. The natural gums generally do hold various advantages over other synthetic polymers, because they are less expensive, non-toxic, and easily available (Bhardwaj et al. 2000). Gencrally, these gums are insoluble in organic solvents and oils. They may be water soluble or may absorb water or swell up in cold water and form a viscous solution or gel. Gums are produced by members belonging to a large number of different families, but commercially obtained from a few members of Leguminosae, Combretaceae, and Sterculiaceae. They can be extracted from seeds, seaweeds, microorganisms and aloe, seed coats or barns of corn, oats, barley, wheat, rice, and soybean (hemicellulose). These gums have diverse application in pharmaceuticals like gelling agent, tablets binder, disintegrating agent, suspending agent, emulsifier, thickener, and stabilizing agent. They are also used as adjuvant in various pharmaceutical formulations (Goswami and Naik 2014).

Most of the gums are anionic (i.e., negatively charged in aqueous dispersion) while some of them, such as gaur gum, are neutral molecules. For the formation of gel, gums are used either alone or in combination with other. Gelling occurs as a result of inter- and intramolecular associations which results in the formation of three-dimensional network entrapped with water molecules. These kinds of associations occur by physical change (pH, temperature) or chemical (addition of chemical reagent) treatment (Dionísio and Grenha 2012). Aqueous solutions or dispersions containing natural gums must contain a suitable amount of preservative. But cationic antimicrobial agents should not be used with anionic gums. Although most of the well-known gums are extracts of plant exudates, other sources can be used.

Alginates are polysaccharide and are composed of D-mannuronic and L-guluronic acids extracted from the alginic salts found in seaweed. They are ideal as a stabilizing, thickening or gelling agent. Among the alginates, sodium alginate is most commonly used. Gelation by alginates takes place by reduction in pH or any reaction with divalent cations. The carboxylate ions are converted to free carboxyl groups by reduction in pH. This causes reduction in hydration of polymer and repulsion between the molecules. Practically 0.5% alginate concentration is sufficient for gel formation.

Carrageenan is a anionic hydrocolloid which is extracted from red seaweeds belonging to the species of *Hypnea, Eucheuma, Gigartina, Chondrus,* and *Iridaea.* Carrageenan has unique ability to form a different type of gel textures at room temperature including rigid or elastic, tough or tender, clear or turbid, stable to heat or thermally reversible, low or high gelling temperatures. It can be used for gelling, suspending, emulsifying, and stabilizing the pharmaceutical applications.

Pectin is the polysaccharide, which is extracted from citrus fruit or apple pomace, and inner skin of these fruits is generally used to extract pectin. Pectin is used as pharmaceutical gel. The pectin gel forms at acidic pH in the aqueous solutions in the presence of calcium or other reagents.

Chitosan is biopolymer generally obtained from outer shells of crustaceans. For producing chitosan, chitin is used after extraction and deacetylation. Chitosan exhibits positive charge and therefore attracted to biological tissues and surfaces carrying negative charge. Concentrated aqueous solutions of chitosan result in gel formation. Alginates can be combined with chitosan to produce gel system (Khong et al. 2013).

Gums Natural gums like tragacanth, xanthan, gellan, guar gums are used as gelling agents. Tragacanth gum in the concentration of 2% or above produces gel. Xanthan gum is generally used as stabilizer in emulsions as well as suspensions systems at less than 0.5 of concentrations. At higher concentration, xanthan gum in aqueous solution turns into gel. Thermally reversible gels can be obtained by the combination of xanthan gum with gaur gum or locust bean gum.

Gellan gum is a polysaccharide which is produced after fermentation and is highly potent. It can produce gel formulation at concentration of 0.05%. But gels will only be formed in the presence of free cations. It is also important to note that divalent ions are required in very less concentrations compared to monovalent ions to form gel.

Guar gum is a type of nonionic polysaccharide which is obtained from seeds. The aqueous solutions of guar gum are cross-linked by the presence of polyvalent cations which yields gels via the process of chelation.

5.2 Semisynthetic Polymers

Cellulose derivatives They are natural polymers derived from cellulose found in plants. Cellulose is treated with different active substances, which lead to breakdown of the cellulose backbone and substitution of hydroxyl moieties.

Carboxymethyl cellulose or sodium carboxymethyl cellulose or cellulose gum is available in various grades that differ in degree of substitution and molecular weight. Generally, they are anionic polymers and gelation needs addition of a polyvalent cation in the polymer solution, e.g., aluminum salts.

Methylcellulose Gel formation in methylcellulose results in the increase in temperature of its aqueous solution, which results in the increase in viscosity, decrease in solubility and formation of gel network. This process is also known as thermal gelation.
Other derivatives of cellulose like hydroxypropyl cellulose are soluble in polar organic solvents along with water. Therefore, some grades of hydroxypropyl cellulose and hydroxyethyl cellulose having high molecular weight are very viscous, but occur as fluids.

5.3 Synthetic Polymers

Carbomer is a synthetic polymer of family known as Carbopol®. It is the most commonly used polymer for the formation of gels. Carbopols® were first introduced and used in the mid-1950s for gels. They are in powdered form and have high bulk densities and used to form acidic solution of pH 3 in water. They become thickened at higher pH (5–6). They can swell up to ten times in volume in the presence of aqueous solution. Carbopol solutions have viscosity from 0 to 80,000 centipoise. The examples are given in Table 3.

Table 3 Types of carbopol

Name of the polymer	Viscosity range	General properties
Carbopol® 910	3,000–7,000	• Used in lower concentrations • Produce less viscous formulation
Carbopol® 934P	29,400–39,400	• Used in thick formulation systems such as suspensions, emulsions, sustained-release formulations, transdermal as well as topical formulations • Produce clear gels in water • "P" = highly purified product
Carbopol® 934	30,500–39,400	• Same properties as 934P, but not highly purified
Carbopol® 940	40,000–60,000	• Effective in viscous formulations • Produce clear gels in hydroalcoholic solution
Carbopol® 941	4,000–11,000	• Yields low viscosity gels • High clarity

6 Method of Preparation of Gels

In industrial scale, gels can be prepared at room temperature, but some gel-forming agents require special treatments for the development of cross-linking during gel formation. These gels can be prepared by various ways.

6.1 By Thermal Change

Temperature affects the formulation of gel, and it depends on the nature of gelling agents. Some gelling agents (e.g., lyophilic (solvent loving) colloids like agar and gelatine) show decrease in solubility with decrease in temperature, which results in increase in viscosity of solution and gel formation, while in case of other polymers (e.g., cellulose ethers) solubility decreases with increase in temperature, which results in the disruption of hydrogen bonding, and reduced solubility causes gelation.

6.2 By Flocculation with Neutralizers

In this method, gelation occurs by addition of just sufficient precipitation (but not complete precipitation) to produce the gel state. During the gel formation, rapid mixing is required which can avoid high concentrations of precipitant at one point. Example includes solutions of polystyrene in benzene and ethyl cellulose, which are gelled by fast mixing with non-solvent, e.g., petroleum ether.

6.3 By Chemical Reaction

Gels are prepared by precipitation of sols from solutions such as aluminum hydroxide (sol is produced by interaction of aqueous solution with aluminum salt and sodium carbonate). In this case higher concentration of reactants results in a gel structure.

7 Evaluation Parameters of the Formulated Gels

Homogeneity After the gels are set, they are visually tested for homogeneity. Along with it, the gels are tested for the presence of any aggregate and their appearance.

Grittiness Gels are observed microscopically for the presence of any particulate matter.

Physicochemical characterization The gels are characterized spectroscopically by various techniques like Fourier transform infrared spectroscopy (FTIR) and transmission scanning electron microscopy (TEM), SEM DLS, i.e., dynamic and static light scattering (elastic and quasi-elastic light scattering technique), and small-angle neutron scattering (SANS).

Determination of pH The pH is determined with the help of digital pH meter. After dissolving the gel with distilled water, its pH is measured.

Drug content A specified quantity of gel is added in suitable solvent. After filtering, a number of different concentrations are prepared by suitable dilutions and absorbance is measured to calculate drug content.

Viscosity and Rheology The viscosity of gel is determined using Brookfield Viscometer.

Phase transition temperature Phase transition temperature (i.e., gel to sol and sol to gel) indicates the nature of microstructures that form cross-linked gelling networks. It is determined by accurate as well as sensitive techniques like differential hot stage microscopy and scanning calorimetry. Turbidimetry method is use to determine the gelation kinetics.

Spreadability It determines the extent of area up to which gel can readily spread on applying to skin and affected part. It is generally measured as the time in seconds taken as two slides slip off when gel is placed in between the slides in the direction of load. So if less time is taken for separation of slides, then spreadability is better. It is calculated by using the formula:

$$S = M.L/T$$

where

M weight attached on upper slide
L length of slide
T time used to separate the slides

Extrudability After gel is set in container, the formulation is filled in collapsible tube. The extrudability of gel is defined in terms of weight in grams which is required to extrude a 0.5 cm ribbon of gel in 10 s from the tube.

In vitro Diffusion studies They are carried out in Franz diffusion cell. The release and permeation rate is determined through a cellophane membrane. Gel sample is applied to the surface of membrane and diffusion studies are carried out at 37 $\pm$ 0.5 °C using suitable dissolution medium. Aliquots of 5 ml are withdrawn periodically at predetermined time intervals and analyzed spectrophotometrically for the drug content. Sink conditions are maintained through out the in vitro diffusion studies.

Stability Stability studies are carried out by freeze–thaw cycle. The gel formulation is stored at 4 °C for a month followed by 25 °C a month and then followed by 40 °C for a month. The occurrence of syneresis is observed. Then, the gel is exposed at ambient room temperature, and separation of liquid exudate (if any) is noted down formulation.

Safety and skin compatibility studies If the gel has to be used for applying to skin for topical or transdermal drug delivery, the skin retention studies and skin irritation studies are performed. The skin irritation potential of gel can be determined by skin irritation studies in animals (Draize et al.). Mice/rat/Guinea pigs of either sex can be used for skin irritation studies. Animals are kept on standard animal feed with free access to water. Hair is shaved from back of animals and area is marked for both the sides. Out of which one side serves as control and other side is test. Gel is applied twice a day for 7 days. The site is observed for any reaction, and it is graded as 0, 1, 2, 3 [for no reaction (0), slight patchy erythema (1), slight but confluent or moderate but patchy erythema (2), and severe erythema with or without edema (3)].

In vivo studies In vivo studies are done to determine pharmacodynamic and pharmacokinetic parameters.

8 Recent Approaches in Polymer Gel Formulation

8.1 Nanostructured Gels

The applications of nanotechnology in medicine have enabled the development of nanovesicular and nanoparticulate drug delivery carriers that can be loaded with drugs/ genetic material, which can be released or targeted with a controlled mechanism in specific organs/tissues of the organism. Many types of nanocarriers for drug delivery have been developed within the past 30 years, which includes liposomes, niosomes, transferosomes, dendrimers, quantum dots, and solid lipid nanoparticles. Nanostructured gels utilizing vesicular or particulate carriers deserve a special attention.

The advantageous properties of nanostructured gels which include ability to retain a large amount of water, biocompatibility with the polymers, ability to incorporate, encapsulate, and protect drugs within nanocarriers, higher stability, stimuli-responsive behavior (if required), and the modified release of drugs in a prolonged and controlled manner make them potential carrier systems in drug delivery. A large variety of materials are used for the formulation of nanogels starting from natural polymers including methacrylated chondroitin sulfate, pullulan, hyaluronic acid, and chitosan, to synthetic polymers including poly (N-isopropylacrylamide-co-acrylic acid), poly (*N*-isopropylacrylamide), and poly (ethylene glycol)-b-poly(methacrylic acid).

These nanostructured gels can be utilized as reservoirs for anticancer drugs as well as imaging agents. Smart nanostructured gels respond to environmental stimuli including temperature and pH, etc., and are utilized for various therapeutic and diagnostic applications. These gels can be surface functionalized with specific ligands which help to target the drugs.

Nanostructured gels can be designed to facilitate the incorporation of variety of bioactive compounds, and following applications of these gels show their utility as a potential nanomedicine carrier:

1. Used for transdermal and topical drug delivery.
2. Upon intravenous injection, nanostructured gels can reach to the areas which are not easily accessed by hydrogels.
3. Nanogels are ideal candidates for intracellular delivery and can be safely delivered into the cytoplasm of the cell.
4. Large surface area of gels helps in better in vivo applications.
5. They have high dispersion stability in aqueous media.
6. The gels enhance efficacy of nucleosides.
7. These gels can incorporate drugs for targeted drug delivery.
8. Drug delivery across cellular barriers can be increased.
9. The nanoscale dimensions of nanogels make them to respond rapidly to any stimuli or change in pH and temperature.

The nanostructured gels can be classified into following types:

a. Lecithin-based nanostructured gels
    a. Liposomal gels
    b. Modified liposomal gels
        i. Transferosomal gels
        ii. Ethosomal gels
        iii. Proliposomal gels
        iv. Pluronic lecithin organogels
b. Microemulsion-/nanoemulsion-based gels
c. Nanoparticulate gels
d. Niosomal gels

1. **Lecithin-based nanostructured gels**:

Phospholipid (lecithin)-based nanocarrier systems are found intriguing among different types of lipid-based systems. Lipids are chemical compounds (e.g., oils and waxes) and occur naturally in living organisms. They are sparingly soluble in aqueous medium. The phospholipids are special lipids, which contain phosphate. Generally, lipids are hydrophobic and are nonpolar. However, phosphate group in phospholipids is able to mix with water and hydrophilic. Phospholipids are the building blocks of lecithin-based nanostructured gels and cell membranes. They are natural and biocompatible molecules. Because of their similarity to biomembrane composition, phospholipids are found to be non-allergic, biofriendly, and permeation enhancers. The amphiphilic nature combines the benefits of aqueous and oily vehicles in drug delivery, while circumventing the drawbacks against both of them. The most commonly used phospholipid-based nanocarrier systems for drug delivery are liposomes and lecithin-based microemulsions, which further can be converted into gels particularly for transdermal or topical drug delivery.

Gels based on liposomes and lecithin microemulsions as drug carriers, are known as traditional phospholipid based vesicles systems and widely used as effective and safe drug delivery vehicles in transdermal/topical treatment of various diseases due to their effective potential in improving skin penetration and therapeutic efficacy of various drugs. These nanostructured carriers could hold different drugs, in both the aqueous and lipid phases, depending on their solubility characteristics. Modified liposomes such as transferosomes, ethosomes, and proliposomes can be utilized to impart deeper permeation compared to traditional liposomal formulation. The microemulsions have been found to have advantages over liposomes, such as easier method of preparation, lower cost, absence of organic solvents, and more storage stability. In addition, microemulsions are thermodynamically stable and can be spontaneously formed by mixing aqueous phase with lipophilic phase in the presence of a surfactant or cosurfactant mixture.

The drug delivery through lecithin-based nanocarrier system has significant disadvantage of its liquid state and so has low contact time with the application area in addition to drug leakage upon storage (Elnaggar et al. 2014).

These nanocarrier systems can be incorporated into gel matrices which circumvent drawbacks of liquid vehicles upon drug delivery. In addition to it, gel formulation helps to modify the drug release at the application site (Tian et al. 2010; Xuan et al. 2012). It is well known that liposomes are unstable and may form aggregates with time, but gelling can improve the stability along with modifying the drug release kinetics.

8.1.1 Liposomal Gels

Phospholipids are amphiphilic in nature; i.e., one part is hydrophilic (water-loving) while the other part is hydrophobic (water-hating). Phospholipids on coming in contact with water, the hydrophobic part of phospholipid interacts with the solvent and causes self-assembly of phospholipids, generally in the form of layered structures called liposomes.

Liposomes can be formed from a variety of phospholipids. The most widely used lipids are phosphatidyl choline, phosphatidyl ethanolamine, and phosphatidlylserine either alone or in combination with other substance, which affects liposome's physical, chemical, and biological properties, liposome size, charge, drug loading capacity, and permeability.

Cholesterol is generally added to condense the packing of phospholipids in bilayer above transition temperature, which reduce their permeability to encapsulated compounds. *Stearylamine* can be used to give positive charge to the liposome surface.

Topical application of liposome formulation has drawback of short residence time on the application site due to lower viscosity and drug leakage upon administration. Along with it, the drug can leak from the liposomal formulations during storage due to diffusion and erosion of the vesicles into the surrounding dispersion buffer. Various efforts have been done in order to incorporate liposomes into a gel matrix to avoid their problems (Cereda et al. 2016; Singh et al. 2016). Liposomal formulations have been found to be compatible with different gel-forming polymers such as methyl cellulose, carbopol, and hydroxyethyl cellulose. Carbopol is most commonly used gelling agent with concentrations ranging from 1 to 2%.

Many researchers have prepared and evaluated liposomal gels for their therapeutic efficacy and compared them to conventional drug-incorporated gels or creams. The results indicated that liposomal gels can enhance the skin retention and absorption of the drugs through skin.

Kumar et al. (2016) carried out randomized clinical trials to the check the efficacy of topical formulation of cyclosporine loaded in liposomes in mild to moderately stable plaque psoriasis. Results inferred that topical liposomal formulation of cyclosporine, 2.0% w/w, is effective in treatment of limited chronic plaque psoriasis with a satisfactory safety profile, but none of the sites are treated with conventional cyclosporine cream or placebo gel.

Cereda et al. (2016) prepared liposomal butamben gel formulations and evaluated the in vitro cytotoxicity and in vivo analgesic effect and local toxicity of the local anesthetic butamben encapsulated in conventional or elastic liposomes incorporated in gel formulations. The results indicated that both gel formulations of liposomal BTB reduced the cytotoxicity and increased the topical analgesic of BTB, compared to plain BTB gel. The gel formulations presented good rheological properties, and stability-indicating assays detected no differences in physicochemical stability up to 30 days after preparation.

Manca et al. (2016) used combination of phospholipids and argan oil for the development of novel liposomal gel, which improved skin hydration and dermal drug delivery.

It was observed that liposomal gel formulations did not enhance their systemic absorption much, because of poor penetrating ability, but very effective for skin retention of drugs in topical drug delivery (Padamwar and Pokharkar 2006; Seth et al. 2004).

Mitkari et al. (2010) prepared liposomal gel of fluconazole for topical delivery. Results indicated that liposomal formulation had better permeation and skin accumulation than the marketed formulations.

Sutoris et al. (2012) evaluated topical photodynamic therapy of mammary carcinoma using liposomal gel containing hydroxyl-aluminum phthalocyanine and found the liposomal gel to be suitable for carcinoma treatment.

Zhang et al. (2010) reported interior gelation of liposomes that causes changes in physicochemical properties of liposomes, like more rigidity of the lipid bilayers, changed shape, enhanced physical stability, and prolonged release of drug, which affected the performance of liposomes in vivo. The results indicated that gel liposomes showed prolonged residence time in the body as compared to that of conventional liposomes, which increased distribution of drug in various tissues of body.

8.1.2 Modified Liposomal Gels

Transferosomal gels

Transdermal delivery by liposomal gel is not suitable, because of liposomal poor skin permeability, leakage of drug due to breaking and fusion of liposome vesicles (Mourtas et al. 2007; Rakesh et al. 2009). These drawbacks can be overcome by modified liposomal gel formulations.

Transferosome is one of the modified liposomal formulations. They are ultraflexible (ultradeformable) phospholipid supramolecular aggregates, and these can permeate the skin intact. The transferosomes consist of inner aqueous layer that is surrounded with lipid bilayer. The ultraflexible property of transferosome is due to the presence of "edge activators" into the membrane of vesicles. Surfactants, e.g., tween 80, span 80, sodium cholate, and sodium deoxycholate, are generally used as edge activators. Transferosomes rapidly transform their shape in response to external stress and require low energy. Due to their ultraflexible nature of transferosomes, they are found to be good option for transdermal delivery of drug molecules. They can be used as drug carriers for proteins, peptides, and vaccines. Transferosomes are able to permeate through the spaces of stratum corneum layer of skin, which are even smaller than their size and cross the skin easily (Bansal et al. 2012).

The basic components for the formulation of transferosomal vesicles are phospholipids such as soya phosphatidyl choline and egg phosphatidyl choline, edge activator surfactant such as tween 80, sodium cholate for providing flexibility, alcohol like ethanol, methanol as a solvent, optional dye for confocal scanning laser microscopy (CSLM) and other buffering agent like saline phosphate buffer pH 7.4, as a hydrating medium. (Gupta et al. 2012).

These transferosomes are available in the form of suspension, without occlusion. These transferosomes are usually incorporated into gel matrix to convert them into transferosomal gel to enhance their viscosity which helps them to enhance the residence time at the application site transdermal drug delivery systems of various drugs (e.g., fluvastatin sodium, sertraline) using transferosomes incorporated in a gel are formulated, which can control the release of drug through skin, increase the drug bioavailability, and thus decrease the dosing frequency of the drug.

Malakar et al. (2012) formulated transferosomal gel incorporating insulin by reverse phase evaporation mainly for painless drug delivery in the treating insulin-dependent diabetes mellitus. The in vivo studies of optimized gel revealed prolonged hypoglycemic effect in alloxan-induced diabetic rats over 24 h after transdermal application, which proves the efficacy of transferosomal gel for transdermal administration of insulin.

In another study, Marwah et al. (2016) formulated and optimized transferosomes as a transdermal delivery system for insulin, to overwhelm the difficulties related to its subcutaneous delivery. The glucose-lowering study revealed that the prepared gel with chemical penetration enhancer had better glucose-lowering effect as compared to the conventional polymer gel. Therefore, the transferosomal gel may be utilized as a potential substitute to the conventional formulations of insulin with progressive permeation properties for transdermal administration.

Ali et al. (2015) prepared and carried out clinical evaluation of nanotransferosomal gel of papaverine hydrochloride, as an alternative to the patient incompliant painful penile injections. Results revealed that the transferosome-based gel is used as a carrier of papaverine hydrochloride for both diagnosing and treating erectile dysfunction.

Ethosomal gel

Touitou (1998) discovered ethosomes as the modified lipid vesicles containing large concentration of ethanol. Ethosomes can offer better skin permeation and stability than liposomes along with improved entrapment efficiency. The presence of large concentration of ethanol is the main reason for good skin permeation of ethosomes as compared to liposomes. Ethanol is used as a penetration enhancer in ethosomes. It can be penetrated into intercellular lipids, which in turn increase fluidity of lipids and thus decrease the density of lipid layer of cell membrane, which leads to enhanced skin permeability (Touitou et al. 2000). Thus, ethosomes can penetrate easily into the deeper skin layers, where it releases drug in deeper skin layers. The basic components of ethosomes are phospholipids, alcohol such as ethanol and isopropyl alcohol (20–45%), and water. For transdermal drug delivery generally ethosomal gel is used to enhance the stability of formulation and to modify the drug delivery rate as in case of liposomal gel and transferosomal gel.

In a study, thermoreversible nanoethosomal gel of eletriptan hydrobromide for brain-targeted drug delivery via the intranasal route was formulated and evaluated. The results revealed that thermoreversible ethosomal gel could serve as the potential choice for brain targeting through intranasal route and further could subsequently enhance its bioavailability (Shelke et al. 2016).

Wo et al. (2014) concluded that in rabbit ear hypertrophic scar model 5-fluorouracil drug delivery was enhanced by using ethosomal gels as compared to conventional gel formulations. Similarly, ethosomal gel of isotretinoin has been reported with enhanced permeation of the drug through skin (David et al. 2013).

Ahmed et al. (2015) optimized ethosomal systems for enhanced transdermal activity of tramadol. The results demonstrated that encapsulated tramadol increased its biological activity due to the superior skin penetration potential. The preclinical study indicated a significant ($P < 0.05$) extended analgesic effect compared to oral solution using the hot plate test method.

In an investigation, the ethosome-based hydrogel of methoxsalen was formulated for the effective treatment of vitiligo and enhanced topical drug delivery. The ethosomal formulations were prepared according to central composite design. They were characterized for physicochemical characteristics like vesicle shape, size, lamellarity, zeta potential, and drug entrapment. The optimized ethosomal formulations were then incorporated into Carbopol® 934 gel. The ethosomes-based hydrogel formulation revealed enhanced percutaneous permeation of methoxsalen and reduced phototoxicity and erythema problems of formulation (Garg et al. 2016).

Proliposomal gel

Provesicular drug delivery system was developed with the aim to overcome the issues of stability associated with vesicular/colloidal drug delivery. It is made up of water-soluble porous powder, where organic solvent is used to dissolve the drug and to produce free-flowing granular product and used as drug delivery carrier. It can overcome many drawbacks of vesicular dispersions. Examples of provesicular drug delivery system include proliposomes and protransferosomes (Payne et al. 1986; Gupta et al. 2008).

Drug delivery from liposomes has limitations of unstability and poor skin permeation. (Weiner 1998; Yarosh 2001). Furthermore, other problems of sterilization in large-scale manufacturing of liposomes are remained to be solved. To improve the stability of liposomal vesicular system, the proliposomes were introduced. They are free-flowing drug particles which immediately form suspension when placed in contact with water. The stability of liposomes can be enhanced without influencing the intrinsic characteristics due to its solid nature (Deo et al. 1997; Katare et al. 1991). Proliposomes consist of drug, phospholipids, and porous powder that are easy to store and sterilize in a dry state. Furthermore, the size of proliposomes can be controlled and relatively smaller-sized liposomes can be obtained after reconstitution (Yang et al. 1993; Ahn et al. 1995; Wei et al. 2003).

Similarly, protransferosomes can be obtained as free-flowing powders to overcome the limitations of transferosomes.

The proliposomes can be utilized for oral drug delivery and topical/transdermal drug delivery. In case of drug delivery through skin, gels loaded with proliposomes can be used to enhance the skin permeation.

Kurakula et al. (2012) prepared proliposomal gel containing anti-inflammatory agent, a steroid prednisolone intended for topical application. The study concluded prednisolone proliposomal gel released the drug in sustained manner with more anti-inflammatory activity revealing proliposomal drug delivery as an effective potential topical pharmacotherapy in treatment of rheumatoid arthritis.

Jain et al. (2005) carried out a study with aim to formulate and evaluate the performance of new vesicular drug delivery system proliposomes and protransferosomes for transdermal delivery of the contraceptive agent, levonorgestrel. Results indicated better skin permeation potential, sustained-release potential, and better stability of protransferosomes than proliposomal formulation of levonorgestrel.

Jukanti et al. (2011) evaluated the efficiency of proliposomes in transdermal drug delivery of exemestane. Results indicated a significant enhancement in the bioavailability (more than twofold) from proliposome gel as compared to oral suspension. Stability studies revealed that formulations were stable at 4 °C.

Thus, proliposomal gels offered potential and proved to be efficient carrier for sustained and improved transdermal delivery of exemestane.

Pluronic lecithin organogels
Thus, organogels have generated interest in development of dermal/transdermal delivery systems with an aim of improving or retarding permeation of drug through the skin, keeping in mind that for topical drug delivery, excess accumulation of drug in the skin and minimal permeation is required, while for transdermal delivery, the vice versa is required. This gel is an emulsion which consists of aqueous phase —Pluronics (20–30%), lipid phase—lecithin and isopropyl myristate/isopropyl palmitate in equal ratio (20% weight). Pluronic lecithin organogels have been proved odorless, non-irritating to the skin, and allow the fast passage of drug through the skin. These are also effective in moisturizing and revitalizing human keratin-like tissues such as the hair, fingernails, and skin. These gels are composed of reversed polymer as of micelles, which evolve from spherical micelles seen initially, which agglomerate to three-dimensional network in bulk phase and are generally obtained by dissolution of trace amount of water in a non-aqueous solution.

Examples of drugs already incorporated in these gels are nonsteroidal anti-inflammatory drugs (NSAIDS), antiemetics, opioids, hormones, anesthetics, antipsychotics, and calcium channel blockers (Almeida et al. 2012a, b).

8.1.3 Microemulsion-/Nanoemulsion-Based Gels

Microemulsions are optically isotropic drug delivery systems composed of water, oil, surfactant, and cosurfactant having a droplet size less than 200 nm. The microemulsions are beneficial because of their easy preparation, optical clarity, thermodynamic stability, higher diffusion and absorption rates. Furthermore, the ingredients of microemulsion can efficiently overcome the problems of diffusion barrier and thereby penetrate through the stratum corneum layer of the skin. Due to this, it has promising potential in both dermal and transdermal drug delivery. In addition, microemulsions are used for oral drug delivery to augment the bioavailability of poorly soluble drugs (Lala and Awari 2014; Shafiq et al. 2007). But the low viscosity of microemulsion makes them less viable carriers in pharmaceutical industrial applications. To solve these problems, these less viscous microemulsions systems were mixed with gel bases resulting in potential topical drug delivery applications with reduced drug-related toxic effects and erratic drug absorption (Khurana et al. 2013).

Microemulsion-/nanoemulsion-based gel formulations with improved solubility, therapeutic efficacy, skin tolerability, skin targeting, and better patient compliance have been reported (Patel et al. 2011, 2013a).

Microemulsions are formed spontaneously and more thermodynamically stable, whereas nanoemulsions (more stable than common emulsions) are only kinetically stable and formed by mechanical shear. These micro- or nanoemulsions due to their versatility offer a potent way for drug delivery as colloidal drug carriers. The gels can be developed using an appropriate polymer which is capable of providing the desired rheological characteristics to micro/nanoemulsions (Patel et al. 2013b, c).

In a study, Mishra et al. (2016) prepared, optimized, and evaluated Zaltoprofen-loaded microemulsion-based gel for transdermal delivery. Results concluded that microemulsion-based gel can be a desired vehicle for transdermal drug delivery.

Üstündağ et al. (2016) prepared and evaluated the potential use of microemulsion-based hydrogel (MBH) formulations for dermal delivery of benzocaine (BZN). Flurbiprofen microemulsions incorporated in carbopol gel exhibit more efficacious transdermal delivery of flurbiprofen (Naeem et al. 2015).

Jeenger et al. (2016) designed emu oil-based nanoemulsion-based gel of curcumin for anti-inflammatory activity. Arthritic scoring, paw volume, molecular, biochemical, radiological, and histological examinations revealed remarkably improved anti-inflammatory activity of nanoemulsions containing curcumin and emu oil as compared to pure curcumin in nanoemulsion gel.

Harwansh et al. (2016) developed a loaded nanoemulsion that contains catechin-based nanogel for the protection of skin against ultraviolet radiation (UV) rays induced photodamage and to ensure its enhanced bioavailability as well as skin permeability through transdermal route.

Microemulsion-/nanoemulsion-based gels can be utilized for nasal and vaginal drug delivery in addition to topical/transdermal drug delivery vehicles (Bachhav and Paravale 2009).

8.1.4 Nanoparticulate Gels

Lipid nanoparticles include nanostructured lipid carriers and solid lipid nanoparticles. They are much like lipid-based nanoemulsions/microemulsions, but differ in their lipid nature. Lipid nanoemulsions are composed of medium-chain triglyceride, soybean oil, lecithin, and drug that are generally formulated by solvent diffusion technique in aqueous system. The lipid which is of liquid nature in nanoemulsions is replaced by a lipid which is solid at room temperature in SLN. Examples of these solid lipids are high melting glycerides or waxes. The lipid nanoparticles incorporated in gel matrix are found to be good candidates for transdermal delivery.

These combine advantages of micro/nanoemulsions, liposomes, and polymeric nanoparticles. The solid matrix is able to protect incorporated drug against chemical degradation. It can also provide highly modulated drug release profiles. There are various advantages of NLC and SLN, viz., wide application spectrum, the presence of biodegradable physiological lipids that are generally recognized as safe (GRAS).

NLC are comprised of solid lipid matrix and liquid lipid. The presence of liquid lipids in the solid lipid matrix leads to imperfections in crystal lattice of nanoparticles, which leads to improved drug loading and reduced drug leakage during storage. SLN- and NLC-based gels are well suited for use as transdermal/topical delivery systems even on damaged skin as they are based on non-irritating and non-toxic lipids (Bhasker et al. 2009).

8.1.5 Niosomal Gels

Niosomes, another vesicular drug delivery system, also known as nonionic surfactant vesicular system, are used as a drug delivery carrier or reservoir. Niosomes have intrinsic skin penetration enhancing properties due to the presence of surfactant and more stability than other vesicular systems including liposomes, transferosomes, ethosomes. Lecithin-containing vesicular systems get hydrolyzed and oxidized due to the presence of phospholipid, but in niosomes, only surfactant is present, which enhances their stability. They are important as drug delivery carriers and may reduce toxicity of drug and modify drug release behavior as well as bioavailability of the incorporated drugs. They are also known as nonionic surfactant vesicles with lamellar structures of micro- to nanosize, which are resulted from an admixture of nonionic surfactant of the alkyl or dialkyl polyglycerol ether and cholesterol with subsequent hydration in aqueous media. These vesicles gain more attention due to its various advantages such as low cost, chemical stability, and high purity. A variety of surfactants are available for fabrication of niosomal drug delivery. Topically administered niosomes may increase the residence time of drugs in the skin, while decrease the systemic absorption of the drug. Niosomes act as reservoirs for drug, and the change in composition and surface properties can modify the drug release behavior and the affinity to the target site (Pawar et al. 2012; Vyas 2009).

Niosomes are further converted into gel form for topical drug delivery. Niosomal gels as delivery devices have been reported with serratiopeptidase, anti-inflammatory, anticancer, anti-tubercular, anti-leishmanial, hormonal drugs and oral vaccines (Ujwala and Shivkumar 2014; Shirsand et al. 2012; Hao and Li 2011; Di Marzio et al. 2011).

Another modification of niosomes is the proniosomes (dry niosomes) which result into niosomal formulations when hydrated before use. Proniosomes help to decrease aggregation, fusion, and leakage issues of niosomes. These offer a potential topical/transdermal drug delivery vehicle which are hydrated with underlying skin, when applied on the skin (Wen et al. 2014). Proniosomes can be converted into gel form which offers more stability than niosomal gel (Ahmad et al. 2016; Alsarra 2009).

9 Applications of Polymer Gels in Drug Delivery

Polymeric gels are used in various medical technologies, including drug delivery, medical devices, tissue engineering, and stimuli-sensitive actuators. In this chapter, particular emphasis is given on use of polymer gels in drug delivery. The gels are used in pharmaceuticals for drug delivery by oral, parenteral, or topical/transdermal route.

Table 4 Applications of polymer gels for oral drug delivery of pharmaceuticals

Gels for oral drug delivery	
Action site	Type of gel
Oral cavity	• Mucoadhesive gels • Gels as tablet superdisintegrant
Stomach	• pH-sensitive gels • Gastric retention gels • Gastric floating gels • Mucoadhesive gels • Superporous gels
Small intestine	• pH-sensitive gels • Hydrotropic gels
Large intestine	• Enzyme degradable gels • pH-sensitive gels

9.1 Oral Drug Delivery

Gels are used for oral drug delivery of pharmaceuticals in oral fast disintegrating tablets due to their superporous properties, buccoadhesive tablets/patches due to fast swelling and superabsorbent properties, floating gels in gastrointestinal fluid, gastric retentive gels due to mucoadhesive properties and colon targeted gels by the environmental stimuli-based smart gels and in situ gels which respond at particular pH or by the presence of microflora of colon region. This has been depicted in Table 4.

9.2 Parenteral Drug Delivery

Smart gels and in situ gels are used in parenteral formulations. These gels help to control and prolong the release of drug in systemic circulation. For parenteral formulations, hydrogels, viz., in situ cross-linked polymeric systems, in situ polymer precipitation system, temperature-sensitive gelling system, and in situ solidifying organogel are explored to enhance patient compliance and to enhance the dosing intervals without affecting/enhancing the therapeutic efficacy of drug (Bari 2010). An injectable estradiol loaded l-amino acid derivatives organogel was developed to improve the burst release problems and for sustained drug delivery of estradiol (Yang et al. 2012). To enhance patient compliance and therapeutic efficacy, thermosensitive and mucoadhesive vaginal gel of clotrimazole incorporated in β-cyclodextrin was developed for cure of vaginitis (Ravivarapu et al. 2000).

9.3 Ocular Drug Delivery

In situ gels and nanostructured gels are used for ophthalmic drug delivery. Ophthalmic drug delivery has already been used for several drug molecules such as antimicrobials, anti-inflammatory agents, autonomic drugs, etc. Poor bioavailability and low therapeutic efficacy of conventional ophthalmic delivery systems is due to excess of tear fluids and fast drug elimination from eye cavity. To overcome this, in situ gels for ophthalmic drugs have been developed. In situ gels release drug in controlled and prolonged manner due to enhanced pre-corneal contact times compared to other conventional eye drops. A prolonged release of indomethacin was reported to release drug for a period of 8 h (Schoenwald 1990).

9.4 Nasal Drug Delivery Systems

Shelke et al. (2016) formulated and optimized brain-targeted thermoreversible gel of eletriptan hydrobromide through intranasal route. Ethosomes were formulated by soya lecithin and ethanol using ethanol injection technique. Ethosomes were incorporated into thermoreversible and mucoadhesive gel matrix prepared with carbopol 934 and poloxamer 407. Results revealed ethosomal thermoreversible gels as potential alternatives for the brain-targeted drug delivery through intranasal route for improved therapeutic efficacy of drug.

Salunke and Patil (2016) designed mucoadhesive in situ gel containing salbutamol sulfate for nasal drug delivery. The formulations were developed, which can gelate after nasal administration at physiological ionic content. It was concluded that nasal drug delivery by in situ gel containing gellan gum is a reliable approach for salbutamol sulfate. These systems can be appropriate for protein and peptide drug delivery by nasal route (Wu et al. 2007).

9.5 Rectal and Vaginal Drug Delivery

Polymeric in situ gels are used in various drug delivery systems through rectal/vaginal route. Rençber et al. (2016) investigated the mucoadhesive in situ gel of clotrimazole for the treatment of vaginal candidiasis. Results of in vivo distribution studies indicated that polymeric gel formulations retained on the vaginal mucosa for up to 24 h after application.

Masoudi et al. (2016) studied and compared the therapeutic efficacy of the vaginal gel on bacterial vaginosis, containing Berberis vulgaris 5% and *Myrtus communis* L. 2% in metronidazole base with only metronidazole gel.

9.6 Topical/Transdermal Drug Delivery

The nanogels are attractive carriers for transdermal drug delivery applications. Transdermal drug delivery systems have certain advantages such as improving patients compliance, reduced dosing frequency, bypass first-pass metabolism, and easy-to-stop action of drug after removing the system. Transdermal drug delivery is helpful to accumulate sufficient concentration of drug at the targeted area. Nanogels possess high drug loading, biocompatibility, and biodegradability, which are the main features in designing an efficient TDDS (Mavuso et al. 2015).

Topical treatment is main application of dermatologic practice. The delivery of drug by topical route depends on physical chemistry of skin and on the patient way of applying the product. The type of vehicle also affects the efficacy of topical drug. The use of nanogels can provide more effective means for the treatment of patients with skin disease (Feldman 2014). Some of the patents on polymer gels are enlisted in Table 5.

Table 5 Patents on polymer gels

S. No.	Patent No.	Description	Publication date
1	US20140271732A1	Compositions of nitric oxide and/or peptides such as thyrotropin-releasing hormone (trh) and/or gnrh (gonadotropin-releasing hormone) and methods for topical or transdermal delivery and treatment of wounds and/or promoting wound healing	Sept 2014
2	US20140271828A1	Pain reliever gel composition comprised of dextrose, aloe vera concentrate, and some or all of the following ingredients: propylene glycol, caprylic/capric tryglicerides, sodium chloride (or acetic acid), a homeopathic anti-inflammatory extract, dimethyl sulfone [or methylsulfonylmethane (msm)], cetyl myristoleate, and, optionally, a pitcher plant extract	Sept 2014
3	WO2014039185A1	An antiretroviral composition (containing polymeric nanoparticles in thermosensitive gel) that gels upon heating and can be administered prophylactically prior to exposure to a retrovirus following sexual intercourse, and methods of using the same	March 2014
4	US 8394783 B2	Polysaccharide gel formulation having multistage bioactive agent delivery	March 2013
5	US 2010/0068264 A1	Niosome-hydrogel drug delivery system (niosome packed in chitosan polymeric hydrogel network)	March 2010
6	US 8357795 B2	Hyaluronic acid-based gels for lidocaine (hyaluronic acid-based compositions including lidocaine have an enhanced stability and cohesivity)	Jan 2013
7	WO2009028764A1	Combination gel of a complex coacervate and a thermoreversible polysaccharide useful for protein drug delivery	March 2009

10 Future Prospects

Polymer gels can be promising vehicles for protein and peptide drug delivery. Nanogels have proved their potential in drug delivery because of high drug loading, high entrapment efficiency, and modified/controlled release behavior of drugs, which can further be utilized for various protein- and peptide-based drug delivery systems. The process of formulation of smart and nanostructured gels needs to be optimized to maintain stability of biologicals. In addition to drug delivery, polymer gels can be explored for diverse array of applications like medical devices, tissue engineering, and even in food and cosmetic industry.

11 Conclusion

Polymeric gels can be utilized for drug delivery by different routes, viz., oral, parenteral, rectal, vaginal, nasal, ocular and topical and transdermal. Recent advances in development of nanocarrier-based nanostructured gels have enhanced the applications of polymeric gels in drug delivery. The nanostructured gels were reported to be more acceptable than nanocarrier dispersions and conventional gels. In future, the advances in polymeric gels can be utilized to deliver protein- and peptide-based drugs by transdermal route or parenteral route. The bioavailability of drugs through ocular/nasal/vaginal drug delivery system by utilizing nanocarrier-based polymer gelling system can further be enhanced.

References

Ahmad MZ, Mohammed AA, Mokhtar Ibrahim M (2016) Technology overview and drug delivery application of proniosome. Pharm Dev Technol 21:1–10

Ahmed S, Sarim Imam S, Zafar A, Ali A, Aqil M, Gull A (2015) In vitro and preclinical assessment of factorial design based nanoethosomes transgel formulation of an opioid analgesic. Artif Cells Nanomed Biotechnol 20:1–10

Ahn BN, Kim SK, Shim CK (1995) Proliposomes as an intranasal dosage form for the sustained delivery of propranolol. J Control Release 34:203–210

Ali MF, Salem HF, Abdelmohsen HF, Attia SK (2015) Preparation and clinical evaluation of nano-transferosomes for treatment of erectile dysfunction. Drug Des Devel Ther 9:2431–2447

Almeida H, Amaral MH, Lobão P (2012a) Temperature and pH stimuli-responsive polymers and their applications in controlled and selfregulated drug delivery. J Appl Pharm Sci 2(06):1–10

Almeida H, Amaral MH, Lobão P, Lobo JM (2012b) Pluronic® F-127 and Pluronic Lecithin Organogel (PLO): main features and their applications in topical and transdermal administration of drugs. J Pharm Pharm Sci 15(4):592–605

Alsarra IA (2009) Evaluation of proniosomes as an alternative strategy to optimize piroxicam transdermal delivery. J Microencapsul 26(3):272–278

Anand B, Pisal SS, Paradkar AR, Mahadik KR (2001) Applications of organogels in pharmaceuticals. J Sci Ind Res 60:311–318

Bachhav YG, Patravale VB (2009) Microemulsion based vaginal gel of fluconazole: formulation, in vitro and in vivo evaluation. Int J Pharm 365:175–179

Bansal S, Kashyap CP, Aggarwal G, Harikumar SL (2012) A comparative review on vesicular drug delivery system and stability issues. Int J Res Pharm Chem 2(3):704–713

Bari H (2010) A prolonged release parenteral drug delivery system—an overview. Int J Pharm Sci Rev Res 3(1):1–11

Bhardwaj TR, Kanwar M, Lal R (2000) Natural gums and modified natural gums as sustained-release carriers. Drug Dev Ind Pharm 26(10):1025–1038

Bhasha SA, Khalid SA, Duraivel S, Bhowmik D, Kumar KPS (2013) Recent trends in usage of polymers in the formulation of dermatological gels. Indian J Res Pharm Biotech 1(2):161–168

Bhaskar K, Anbu J, Ravichandiran V, Venkateswarlu V, Rao YM (2009) Lipid nanoparticles for transdermal delivery of flurbiprofen: formulation, in vitro, ex vivo and in vivo studies. Lipids Health Dis 8:6

Caló E, Khutoryanskiy VV (2015) Biomedical applications of hydrogels: a review of patents and commercial products. Eur Polymer J 65:252–267

Cereda CM, Guilherme VA, Alkschbirs MI, de Brito Junior RB, Tofoli GR, Franz-Montan M, de Araujo DR, de Paula E (2016) Liposomal butamben gel formulations: toxicity assays and topical anesthesia in an animal model. J Liposome Res 28:1–9

Chaterji S, Kwon K, Park K (2007) Smart polymeric gels: redefining the limits of biomedical devices. Prog Polym Sci 32(8–9):1083–1122

David SRN, Hui MS, Pin CF, Ci FY, Rajabalaya R (2013) Formulation and in vitro evaluation of ethosomes as vesicular carrier for enhanced topical delivery of isotretinoin. Int J Drug Del 5 (1):28–34

Deo MR, Sant VP, Parekh SR, Khopade AJ, Banakar UV (1997) Proliposome-based transdermal delivery of levonorgestrel. J Biomat Appl 12:77–88

Di Marzio L, Marianecci C, Petrone M, Rinaldi F, Carafa M (2011) Novel pH-sensitive non-ionic surfactant vesicles: comparison between Tween 21 and Tween 20. Colloids Surf B Biointerfaces 82:18–24

Dionísio M, Grenha A (2012) Locust bean gum: exploring its potential in biopharmaceutical application. J Pharm Bioallied Sci 4(3):175–185

Dolbow J, Fried E, Ji H (2005) A numerical strategy for investigating the kinetic response of stimulus-responsive hydrogels. Comput Methods Appl Mech Eng 194:4447–4480

Dolbow J, Fried E, Jia HD (2004) Chemically induced swelling of hydrogels. J Mech Phys Solids 52(1):51–84

Dong L, Agarwal AK, Beebe DJ, Jiang HR (2006) Adaptive liquid microlenses activated by stimuli-responsive hydrogels. Nature 442(7102):551–554

Elnaggar YSR, El-Refaie WM, El-Massik MA, Abdallah OY (2014) Lecithin-based nanostructured gels for skin delivery: an update on state of art and recent applications. J Control Release 180:10–24

Feldman SR (2014) Principles of topical treatment: advancement in gel vehicle technology. J Drugs Dermatol 13(4):423–427

Garg BJ, Garg NK, Beg S, Singh B, Katare OP (2016) Nanosized ethosomes-based hydrogel formulations of methoxsalen for enhanced topical delivery against vitiligo: formulation optimization, in vitro evaluation and preclinical assessment. J Drug Target 24(3):233–246

Garg T, Bilandi A, Kapoor B, Kumar S, Joshi R (2011) Organogels: advanced and novel drug delivery system. Int J Drug Res. 2(12):15–21

Goswami S, Naik S (2014) Natural gums and its pharmaceutical application. J Sci Innovative Res 3(1):112–121

Gupta A, Aggarwal G, Singla S, Arora R (2012) Transfersomes: a novel vesicular carrier for enhanced transdermal delivery of sertraline: development, characterization, and performance evaluation. Sci Pharm 80:1061–1080

Gupta V, Barupal AK, Ramteke S (2008) Formulation development and in vitro characterization of proliposomes for topical delivery of aceclofenac. Indian J Pharm Sci 70(6):768–775

Hamidi M, Azadi A, Rafiei P (2009) Hydrogel nanoparticles in drug delivery. Adv Drug Deliv Rev 60(15):1638–1649

Hao YM, Li K (2011) Entrapment and release difference resulting from hydrogen bonding interactions in niosome. Int J Pharm 403:245–253

Harwansh RK, Mukherjee PK, Kar A, Bahadur S, Al-Dhabi NA, Duraipandiyan V (2016) Enhancement of photoprotection potential of catechin loaded nanoemulsion gel against UVA induced oxidative stress. J Photochem Photobiol B 160:318–329

Hoffman AS (2012) Hydrogels for biomedical applications. Adv Drug Deliv Rev 64:18–23

Hong W, Zhaoa X, Zhoua J, Suo Z (2008) A theory of coupled diffusion and large deformation in polymeric gels. J Mech Phys Solids 56:1779–1793

Jadhav KR, Kadam VJ, Pisal SS (2009) Formulation and evaluation of lecithin organogel for topical delivery of fluconazole. Curr Drug Deliv 6(2):174–183

Jain S, Sapre R, Tiwary AK, Jain NK (2005) Proultraflexible lipid vesicles for effective transdermal delivery of levonorgestrel: development, characterization, and performance evaluation. AAPS PharmSciTech 6(3):E513–E522

Jeengar MK, Rompicharla SV, Shrivastava S, Chella N, Shastri NR, Naidu VG, Sistla R (2016) Emu oil based nano-emulgel for topical delivery of curcumin. Int J Pharm 506(1–2):222–236

Jukanti R, Sheela S, Bandari S, Veerareddy PR (2011) Enhanced bioavailability of exemestane via proliposomes based transdermal delivery. J Pharm Sci 100(8):3208–3222. https://doi.org/10.1002/jps.22542 Epub 2011 Mar 14

Kadajji VG, Betageri GV (2011) Water soluble polymers for pharmaceutical applications. Polymers 3:1972–2009

Kashyap N, Kumar N, Kumar M (2005) Hydrogels for pharmaceutical and biomedical applications. Crit Rev Ther Drug Carr Syst 22:107–149

Katare OP, Vyas SP, Dixit VK (1991) Proliposomes of indomethacin for oral administration. J Microencapsul 8:1–7

Khong TT, Aarstad OA, Skjåk-Bræk G, Draget KI, Vårum KM (2013) Gelling concept combining chitosan and alginate—proof of principle. Biomacromolecules 14(8):2765–2771

Khurana S, Jain NK, Bedi PMS (2013) Nanoemulsion based gel for transdermal delivery of meloxicam: physico-chemical, mechanistic investigation. Life Sci 92(6–7):383–392

Kortesuo P, Ahola M, Karlsson S, Kangasniemi I, Yli-Urpo A, Kiesvaara J (2000) Silica xerogel as an implantable carrier for controlled drug delivery-evaluation of drug distribution and tissue effects after implantation. Biomaterials 21(2):193–198

Kumar R, Dogra S, Amarji B, Singh B, Sharma SK, Vinay K, Mahajan R, Katare OP (2016) Efficacy of novel topical liposomal formulation of cyclosporine in mild to moderate stable plaque psoriasis: a randomized clinical trial. JAMA Dermatol [Epub ahead of print]

Kurakula M, Srinivas C, Kasturi N, Diwan PV (2012) Formulation and evaluation of prednisolone proliposomal gel for effective topical pharmacotherapy. Int J Pharm Sci Drug Res. 4(1):35–43

Laftah WA, Hashim S, Ibrahim AN (2011) Polymer hydrogel: a review. Polymer-Plast Technol Eng 50(11):1475–1486

Lala RR, Awari NG (2014) Nanoemulsion-based gel formulations of COX-2 inhibitors for enhanced efficacy in inflammatory conditions. Appl Nanosci 4(2):143–151

Malakar J, Sen SO, Nayak AK, Sen KK (2012) Formulation, optimization and evaluation of transferosomal gel for transdermal insulin delivery. Saudi Pharm J 20(4):355–363

Manca ML, Matricardi P, Cencetti C, Peris JE, Melis V, Carbone C, Escribano E, Zaru M, Fadda AM, Manconi M (2016) Combination of argan oil and phospholipids for the development of an effective liposome-like formulation able to improve skin hydration and allantoin dermal delivery. Int J Pharm 505(1–2):204–211

Marwah H, Garg T, Rath G, Goyal AK (2016) Development of transferosomal gel for trans-dermal delivery of insulin using iodine complex. Drug Deliv 23(5):1636–1644

Masoudi M, Miraj S, Rafieian-Kopaei M (2016) Comparison of the Effects of *Myrtus communis* L, *Berberis vulgaris* and metronidazole vaginal gel alone for the treatment of bacterial vaginosis. J Clin Diagn Res 10(3):QC04-7

Mavuso S, Marimuthu T, Choonara YE, Kumar P, du Toit LC, Pillay V (2015) A review of polymeric colloidal nanogels in transdermal drug delivery. Curr Pharm Des 21(20):2801–2813
Mishra R, Prabhavalkar KS, Bhatt LK (2016) Preparation, optimization, and evaluation of Zaltoprofen-loaded microemulsion and microemulsion-based gel for transdermal delivery. J Liposome Res 19:1–10
Mitkari BV, Korde SA, Mahadik KR, Kokare CR (2010) Formulation and evaluation of topical liposomal gel for fluconazole. Indian J Pharm Edu Res 44(4):324–333
Mourtas S, Fotopoulou S, Duraj S, Sfika V, Tsakiroglou C, Antimisiaris SG (2007) Liposomal drugs dispersed in hydrogels: effect of liposome, drug and gel properties on drug release kinetics. Colloids Surf B: Biointerfaces 55:212–221
Munusamy P, Seleem MN, Alqublan H, Tyler R Jr, Sriranganathan N, Pickrell G (2009) Targeted drug delivery using silica xerogel systems to treat diseases due to intracellular pathogens. Mater Sci Eng C 29(8):2313–2318
Murdan S (2005) Organogels in drug delivery. Expert Opin Drug Deliv 2(3):489–505
Naeem M, Ur Rahman N, Tavares GD, Barbosa SF, Chacra NB, Löbenberg R, Sarfraz MK (2015) Physicochemical, in vitro and in vivo evaluation of flurbiprofen microemulsion. An Acad Bras Cienc 87(3):1823–1831
Omidian H, Rocca JG, Park K (2005) Advances in superporous hydrogels. J Control Release 102:3–12
Omidian H, Rocca JG, Park K (2006) Elastic, superporous hydrogel hybrids of polyacrylamide and sodium alginate. Macromol Biosci 703:710
Padamwar MN, Pokharkar VB (2006) Development of vitamin loaded topical liposomal formulation using factorial design approach: drug deposition and stability. Int J Pharm 320:37–44
Park H, Park K, Kim D (2006) Preparation and swelling behavior of chitosan-based superporous hydrogels for gastric retention application. J Biomed Mater Res 76A:144–150
Patel MR, Patel RB, Parikh JR, Solanki AB, Patel BG (2011) Investigating effect of microemulsion components: in vitro permeation of ketoconazole. Pharm Dev Tech 16:250–258
Patel RB, Patel MR, Bhatt KK, Patel BG (2013a) Formulation consideration and characterization of microemulsion drug delivery system for transnasal administration of Carbamazepine. Bul Fac Pharm Cairo Uni 51:243–253
Patel RB, Patel MR, Bhatt KK, Patel BG (2013b) Paliperidone loaded mucoadhesive microemulsion in treatment of schizophrenia: formulation consideration. J Pharma Inn 8:195–204
Patel RB, Patel MR, Bhatt KK, Patel BG (2013c) Risperidone loaded mucoadhesive microemulsion for intranasal delivery: formulation, development, physicochemical characterization and ex vivo evaluation. J Drug Del Sci Tech 23:261–267
Pawar SD, Pawar RG, Kodag PP, Waghmare AS (2012) Niosome: an unique drug delivery system. Int J Biol Pharm Allied Sci 3:406–416
Payne NI, Timmins P, Ambrose CV, Ward MD, Ridgway F (1986) Proliposomes: a novel solution to an old problem. J Pharm Sci 75(4):325–329
Quintanar-Guerrero D, Ganem-Quintanar A, Nava-Arzaluz MG, Piñón-Segundo E (2009) Silica xerogels as pharmaceutical drug carriers. Expert Opin Drug Deliv 6(5):485–498
Rakesh P, Hardik H, Ashok H (2009) Formulation and evaluation of Carbopol gel containingliposomes of ketoconazole. Int J Drug Deliv 1:42–45
Ravivarapu HB, Moyer KL, Dunn RL (2000) Sustained activity and release of leuprolide acetate from an in situ forming polymeric implant. AAPS Pharm Sci Tech 1(1):1–8
Reddy PC, Chaitanya KSC, Rao YM (2011) A review on bioadhesive buccal drug delivery systems: current status of formulation and evaluation methods. Daru 19(6):385–403
Rençber S, Karavana SY, Şenyiğit ZA, Eraç B, Limoncu MH, Baloğlu E (2016) Mucoadhesive in situ gel formulation for vaginal delivery of clotrimazole: formulation, preparation, and in vitro/in vivo evaluation. Pharm Dev Technol 7:1–11
Sabale V, Vora S (2012) Formulation and evaluation of microemulsion-based hydrogel for topical delivery. Int J Pharm Investig 2(3):140–149

Salunke SR, Patil SB (2016) Ion activated in situ gel of gellan gum containing salbutamol sulphate for nasal administration. Int J Biol Macromol 87:41–47

Schoenwald RD (1990) Ocular drug delivery: pharmacokinetic considerations. Clin Pharmacokinet 18:255–269

Seth AK, Misra A, Umrigar D (2004) Topical liposomal gel of idoxuridine for the treatment of herpes simplex: pharmaceutical and clinical implications. Pharm Dev Technol 9:277–289

Shafiq S, Faiyaz S, Sushma T, Ahmad FJ, Khar RK, Ali M (2007) Design and development of oral oil in water ramipril nanoemulsion formulation: in vitro and in vivo evaluation. J Biomed Nanotech 3:28–44

Shelke S, Shahi S, Jadhav K, Dhamecha D, Tiwari R, Patil H (2016) Thermoreversible nanoethosomal gel for the intranasal delivery of Eletriptan hydrobromide. J Mater Sci Mater Med 27(6):103

Shirsand SB, Para MS, Nagendrakumar D, Kanani KM, Keerthy D (2012) Formulation and evaluation of Ketoconazole niosomal gel drug delivery system. Int J Pharm Investig 2(4):201–207

Singh S, Vardhan H, Kotla NG, Maddiboyina B, Sharma D, Webster TJ (2016) The role of surfactants in the formulation of elastic liposomal gels containing a synthetic opioid analgesic. Int J Nanomed 11:1475–1482

Stamatialis DF, Papenburg BJ, Gironés M, Saiful S, Bettahalli SNM, Schmitmeier S, Wessling M (2008) Medical applications of membranes: drug delivery, artificial organs and tissue engineering. J Membr Sci 308(1–2):1–34

Sutoris K, Vetvicka D, Horak L, Benes J, Nekvasil M, Jezek P, Zadinova M, Pouckova P (2012) Evaluation of topical photodynamic therapy of mammary carcinoma with an experimental gel containing liposomal hydroxyl-aluminium phthalocyanine. Anticancer Res 32(9):3769–3774

Tian W, Schulze S, Brandl M, Winter G (2010) Vesicular phospholipid gel-based depot formulations for pharmaceutical proteins: development and in vitro evaluation. J Control Release 142:319–325

Thakur VK, Thakur MK (2014a) Recent trends in hydrogels based on psyllium polysaccharide: a review. J Clean Prod 82:1–15

Thakur VK, Thakur MK (2014b) Recent advances in graft copolymerization and applications of Chitosan: a review. ACS Sustain Chem Eng 2(12):2637–2652

Thakur VK, Thakur MK (2015) Recent advances in green hydrogels from lignin: a review. Int J Biol Macromol 72:834–847

Touitou E, Dayan N, Bergelson L, Godin B, Eliaz M (2000) Ethosomes-novel vesicular carriers for enhanced delivery: characterization and skin penetration properties. J Control Release 65:403–418

Touitou E (1998) Composition of applying active substance to or through the skin. US patent, 5,716,638

Ujwala AS and Shivkumar SK (2014) Serratiopeptidase niosomal gel with potential in topical delivery. J Pharm 1–9. Article ID 382959

Ulker Z, Erkey C (2014) An emerging platform for drug delivery: aerogel based systems. J Control Release 177:51–63

Üstündağ Okur N, Çağlar EŞ, Arpa MD, Karasulu HY (2016) Preparation and evaluation of novel microemulsion-based hydrogels for dermal delivery of benzocaine. Pharm Dev Technol 7:1–11

Venugopalarao G, Gowtham MS, Sarada NC (2013) Formulation evaluation and stability studies of hydrogel tablets containing Cefditoren Pivoxil. J Pharm Res 7(3):230–234

Vintiloiu A, Leroux JC (2008) Organogels and their use in drug delivery—a review. J Control Release 125(3):179–192

Vyas SP (2009) Theory and practice in novel drug delivery system. CBS Publishers and Distributors, New Delhi, pp 284–298

Wei NN, Lu BI (2003) Preparation, morphology and in vitro release of chitosan coated liposomes of fluorouracil for colon targeting. Acta Pharmacol Sin 38:53–56

Weiner N (1998) Targeted follicular delivery of macromolecules via liposomes. Int J Pharm 162:29–38

Wen MM, Farid RM, Kassem AA (2014) Nano-proniosomes enhancing the transdermal delivery of mefenamic acid. J Liposome Res 24(4):280–289

Wo Y, Zhang Z, Zhang Y, Zhang Z, Wang K, Mao X, Su W, Li K, Cui D, Chen J (2014) Enhanced in vivo delivery of 5-fluorouracil by ethosomal gels in rabbit ear hypertrophic scar model. Int J Mol Sci 15(12):22786–22800

Wu J, Wei W, Wang LY, Su ZG, Ma G (2007) A thermosensitive hydrogel based on quaternized chitosan and poly (ethylene glycol) for nasal delivery system. Biomaterials 28:2220–2232

Xuan XY, Cheng YL, Acosta E (2012) Lecithin-linked microemulsion gelatin gels for extended drug delivery. Pharmaceutics 4:104–129

Yang Y, Xu L, Gao Y, Wang Q, Che X, Li S (2012) Improved initial burst of estradiol organogel as long-term in situ drug delivery implant: formulation, in vitro and in vivo characterization. Drug Dev Ind Pharm 38(5):550–556

Yang ZJ, Hino T, Kawashima Y (1993) Studies on the size of rehydrated new liposomes from scutellaria proliposomes. J China Pharma Univ 24:161–164

Yarosh DB (2001) Liposomes in investigative dermatology. Photoimmunol Photomed 17:203–212

Zhang B, Lu Y, Chen J, Wu W (2010) Effects of interior gelation on pharmacokinetics and biodistribution of liposomes encapsulating an anti-cancer drug cytarabine. J Biomed Nanotechnol 6(6):704–709

Chapter 11
Emerging Trends of Organogels in Drug Chemistry

Selvaraj Mohana Roopan and Duraipandi Devipriya

Abstract Organogels are one of the major constituents in class of gels. They show three-dimensional and cross-linked network in organic liquid phase. Organogels are non-glassy thermoplastic solid materials and are non-crystalline and viscoelastic in nature. Despite of the liquid composition, some differences are seen in the morphological appearance of solids and the rheological behavior. These organogel systems have specific molecular requirements like gelation, depending on the physical and fiber interactions. Organogels are grown rapidly with more stability than other gels and also have features like moisture intensive, economic. Also, organogels are having lower hydration polymeric or low molecular weight of organogelators compared to other polymers. In general, organogels are thermodynamically stable in nature and utilized mainly for drug delivery of bioactive agents in scarce toxicology. Over the decades, organogels are having tremendous applications in various fields like pharmaceuticals, cosmetics, art conservation, and food. This chapter comprises of recent research work on organogels, its preparation, properties, characteristic parameters and various applications in different industries.

Keywords Organogel polymer · Gel preparations · Gel application

1 Introduction

A gel is defined as a semisolid material having polar and nonpolar form, together with an immobilized external solvent within the spaces available in a three-dimensional network as a structure (Vintiloiu and Leroux 2008). According to the bonds present in the gelator network, it can be divided into two types in Fig. 1.

S. M. Roopan (✉) · D. Devipriya
Chemistry of Heterocycles & Natural Product Research Laboratory, Department of Chemistry, School of Advances Sciences, Vellore Institute of Technology, Vellore 632014, Tamilnadu, India
e-mail: mohanaroopan.s@gmail.com; mohanaroopan.s@vit.ac.in

© Springer Nature Singapore Pte Ltd. 2018

V. K. Thakur et al. (eds.), *Polymer Gels*, Gels Horizons: From Science to Smart Materials, https://doi.org/10.1007/978-981-10-6080-9_11

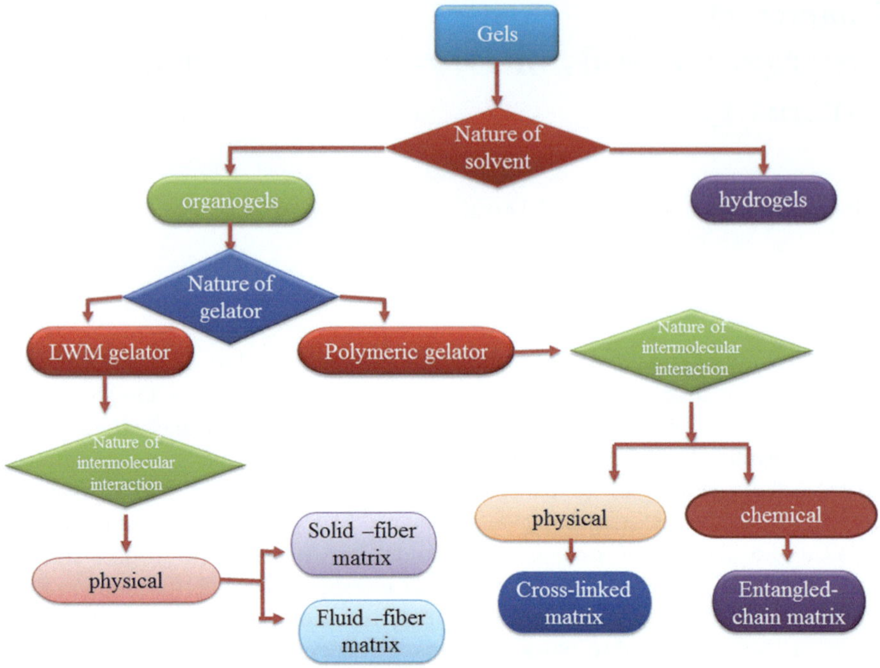

Fig. 1 Types of gels

- Physical gels,
- Chemical gels.

1.1 Physical Gels

Physical gels are arranged by weaker physical forces of attraction like as van der Waals interaction and also having hydrogen interaction. Physical gels are thermoreversible, and it has reversible cross-link polymer-containing system. It was obtained from cooled solutions, the conformational transition of the gelatin chain links. In the physical gels, the micellar crystallites are present and these crystallites are found in helix formation. In microphase separation, these gels are found to be composed of hydrogen or ionic bonds.

Example: gelatin.

1.2 Chemical Gels

Chemical gels are arranged by means of covalent bonds. Chemical gels are opposite to the physical gels, having cross-linking of polymers, induced by certain reagents. In this, reagents were added to aqueous solution. It is also called as permanent gel, which is due to covalent links. Here, permanent network is created in the high-temperature state.

Example: oils in solution, beef gelatin, alkaline, in buffer

Polar hydrogels are composed like as water in a three-dimensional polymeric network (Vintiloiu and Leroux 2008). It can be used for studying various ranges of drugs. The organogel has bicontinuous systems, made by gelators. Here, gelators may not contain water molecules in their three-dimensional polymeric network structures.

Organogels are not having semisolid nature. It is made of macromolecules containing twisted matted stands (Lim et al. 2008). The crystalline amorphous solids are bound together by strong type of van der Waals forces. Organogels, at minimum hydration, possess drug-dissolving polymers which are transported through the cross-link chains. Thereby, polymer gels having cross-link system exhibit increased hydrophobicity nature. It also minimizes the diffusion rate of drug. Organogels having gelators with concentration less than 15% may undergo physical or chemical interaction. It also forms self-assembled structures of the gelators. Chemical interactions of gelators form self-assembled fibrous structures which give rise to a three-dimensional network structure. This inhibits the movement of peripheral apolar phase. Examples of gelators contain sterols, lecithin, cholesteryl anthraquinone derivatives, and sorbitan monostearate. Organogel substances are fabricated in different shapes like rods, disks, films, and microparticles. It depends on the gelator applications and administration sites. Organogels having cross-link polymer system and thermoreversible property show potential usage in drug delivery system (Pal et al. 2007). The system basically utilizes polymers which undergo transformation from solution to gel like consistency, due to change in their physicochemical properties.

Thermoreversible property is one of the best properties for drug delivery system. Based on this property, organogels have been used for drug delivery system in human body (Pal et al. 2006a, b). Organogels being thermostable also develop fibrous structures at lower energy state. Gel-to-solution conversion is carried out at room temperature, with an external involved in it; the organogels decompose to normal solution state from a three-dimensional structure. In addition to temperature sensitivity, it is also sensitive against moisture which paved way in developing controlled delivery systems. Some of the examples using gellable organic solvents consist of aliphatic and aromatic hydrocarbons, alcohols, dimethyl sulfoxide, silicone oil, and vegetable oils (Pal et al. 2008). Organogels are opposite to hydrogels, in which the gelator is normally a polymer, and most of the organogelators are moderately small molecules termed as low molecular weight organogelators. Most of Organogel-based formulations have been used as control for bioactive agents, in various routes administration (Pal et al. 2008, 2009).

1.3 Advantages

Organogel does not form semisolids at normal conditions. It is also moisture sensitive. It can be used to improve skin penetration and transport of the molecules. It has organic character and also resistance against microbial contamination (Toro-Vazquez et al. 2007). Organogels can reduce the diffusion rate of drug, due to its dissolving nature and easily get transported between the chains. It stores for long time due to the structural integrity of organogels (Wright and Marangoni 2006). It is simple and easy to handle for various purposes. Most of the organogels are made of biocompatible, biodegradable, and non-immunogenic materials which make it safe for long-term use (Plourde et al. 2005; Garner et al. 1998).

1.4 Limitations

Organogels are less stable in temperature, so it exists sometimes in gel form and sometime as solution nature. If any impurities are present, then no gelling will occur. The manufacturing process is very expensive. It should be stored in proper conditions like temperature, pressure (Suzuki et al. 2007).

2 Organogelators

In the formation of organogels, one of the major roles in designing part is that of the organogelators (Fig. 2). There are two types of organogelators present based on the ability of formation of hydrogen bonding. The examples of those forming hydrogen bonds include amino acids, amide, urea moieties, and carbohydrates (Suzuki and Hanabusa 2010), whereas those which do not form the hydrogen bonding include anthracene, anthraquinone, and steroid-based molecules. There are various types of organogelators present due to their application. Gels are comparatively short-chain n-alkenes such as hexadecane and other organic liquids (Suzuki et al. 2003). Some of examples included in organogelators are substituted fatty acids (ex 12-hydroxyoctadecanoic acid), 1,3:2,4-di-O-benzylidene-D sorbitol (D-DBS), sorbitan monostearate, a nonionic surfactant, steroids and their derivatives, anthryl derivatives (ex 2,3-bis-n-decycloxyanthracene, macrocyclic gelators (ex calixarenes), ALS compounds (an aromatic moiety attached to a steroidal group by a linker

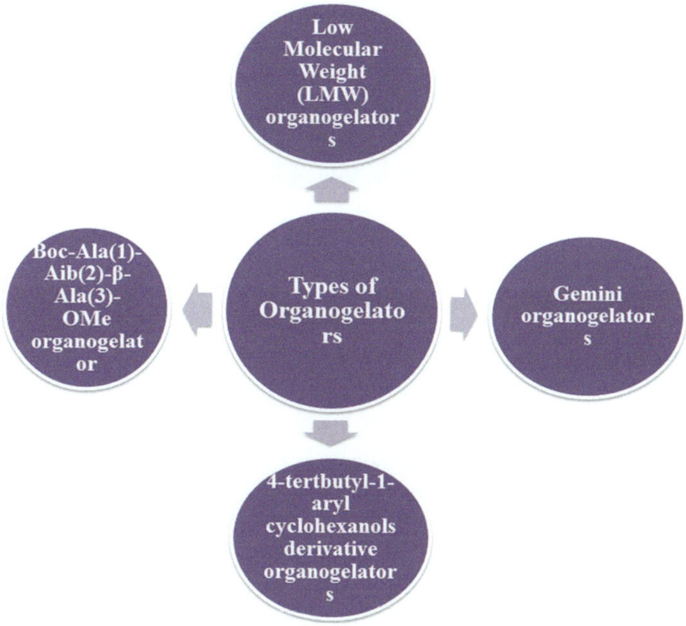

Fig. 2 Types of organogelators

segment), cyclo(dipeptide)s, biurea compounds, bisamide, bolaform amides derived from amino acids, n-alkyl perfluoroalkanamides, carbohydrate derivatives, perfluoroalkanes (Malik et al. 2002).

2.1 Types of Organogelators

- Polymer organogelators,
- Gemini organogelators,
- 4-tertbutyl-1-arylcyclohexanol derivative organogelators,
- Boc-Ala(1)-Aib(2)-β-Ala(3)-OMe organogelator,
- Low molecular weight (LMW) organogelators.

2.2 Polymer Organogelators

Organogelators mostly used are polymeric structures. L-lysine derivatives are a common example of polymeric organogelators. Also to be mentioned are the conventional polymers like polyethylene glycol, polycarbonate, polyester, and poly-alkaline (Toshiyuki et al. 2003). In this type, the organogelators include organogelation even at very low concentration. One of the most important factors controlling polymer properties is the forces between polymer chains due to gelation properties (Abdallah et al. 2000). The gels are prepared from polymer organogelators. This type of gelators is having lower gel-sol transition temperature and also higher gel strength compared to other lower molecular weight organogelators (Esch and Feringa 2000).

2.3 Gemini Organogelators

Gemini organogelators, based on L-lysine derivatives, were synthesized by (Suzuki et al. 2003). Gemini is a Latin word, meaning twin. This type of organogelators has two L-lysine derivatives joined by alkaline spacer chains with different chain lengths. A variety of polar compounds were immobilized by using bis (Nε-lauroyl-L-lysine ethyl ester) oxalyl amide organogelator (Shchipunov and Shumilina 1995). Oxalyl amide derivatives comprising numerous numbers of alkyl ester group also showed relatively good organogelation property like decyl, dodecyl, hexyl, 2-ethyl-1-hexyl, and 3,5,5-trimethylhexyl.

2.4 4-Tertbutyl-1-Arylcyclohexanol Derivative Organogelators

4-tertbutyl-1-arylcyclohexanol derivative organogelators are grouped among aryl-cyclohexanol derivatives. It is mainly used for designing thermoreversible organogels. In this type, organogelators have solid-state nature at room temperature, partially soluble in apolar solvents like cyclohexane, carbon tetrachloride, and benzene. Formation of these organogels happens only when the phenyl group is in the axial position; otherwise, if it is present in equatorial position, the organogel formation will be hindered. The organogels belonging to this category will be transparent. The organogels belonging to this category will be transparent or turbid and also are dependent on the apolar solvent type (Kumar and Katare 2005).

2.5 Low Molecular Weight Organogelators

The Low molecular weight compounds, like fatty acids, n-alkynes containing gelators called as low molecular weight organogelators (Kantaria et al. 1999). It is soluble in apolar solvent even at low concentration like <2%. In these types, organogelators, depending on physical intermolecular interactions, produce solid–fiber matrix or fluid–fiber matrix. The solid–fiber matrix can be designed if the combination of the organogelators (heated) and bounds of organogelators were ventilated at lower temperature (Nasseria et al. 2003). This results in structures similar to fiber which undergo physical relations, to form a gelled structure. It aligns as bundle like structure (Upadhyay et al. 2007). Once water is added, tubular micelles will gain their function which interact with both polar and apolar solvents which result in the formation of gelled structure known as organogels. Combination of this solid–fiber organogels has some mechanical applications associated with organogels of fluid–fiber matrix which can interact easily with fluid–fiber matrix organogels (Díaz et al. 2008).

2.6 Structure of Organogel and Mechanism of Organogelling Techniques

In organogelling or gelation of the lecithin solution, polar solvents are a major factor to induce the gelling in organic solvents (Dasgupta et al. 2009). The composite shifts from spherical form to cylindrical form were identified by monolayer of amphiphile. Some of the polar solvent stuffs incorporated into lecithin micelles which were spherical in morphology with cross-sectional interpretation of polar region were synthesized (Guenet 2006).

2.7 Gelators Homemade

More gelators in the form of fibers, tubules, platelets, rods, wormlike chains, ribbons, and fan-like structure have been reported. Cumulative thickness ranges from a fraction of a nanometer to microns. The gelator molecules self-arrange themselves depending on the nature of gelators. In these types, the gelator component has substituted groups due to the forces of interactions involved in gelator self-assembly. For example, ionic or nonionic surfactant sorbitan monostearate is thought to assemble into bilayers, combined with tubules. Another example of fibrillar aggregates using 12-hydroxyoctadecanoic acid has extensive axial hydrogen bonding and dipolar interaction. The number of forces is included in gelator molecules like dipole–dipole interactions, hydrogen bonding, π-stacking, London dispersion forces, electron transfer, ionic interactions, and solvophobic effects (Avramiotis et al. 2007). This type of interaction depends upon the chemical structures of the gelators. The gelators are not arranged in a neat manner but if it is arranged well, it is due to the force of interaction. Gelation technique plays a fundamental role in liquid phase of gels. The liquid phase in turn determines the solubility or insolubility balance (Chen et al. 2007). The gelators dissolve, or are distributed, even at higher temperature, but it comes easily when the cooling is started. Molecular shape of the solvent molecule can also have a profound effect of gelation.

3 Organogel Preparation

Various organogels are prepared by using conventional method. For the formation of organic compound, liquid material from organic solution and gelators were mixed which resulted in dispersion of organogels, and further, it was cooled for the formation of gel structure (Scartazzini and Luisi 1988). The gelators are dissolved under heating condition. The gelators convert liquid phase into solid phase at lower temperature. In the gelator–gelator interactions, it aggregates itself for the formation of fibers, rods, and tubules. Organogels are having thermoreversible property, because of that they were held with non-covalent forces (Murdan et al. 1999a, b, c). The gels dissolve in organic liquid, and when the cooling is started, the hot sol phase results in gelation (Fig. 3).

Basically, there are three methods involved in the preparation of organogels which are listed and briefly described below:

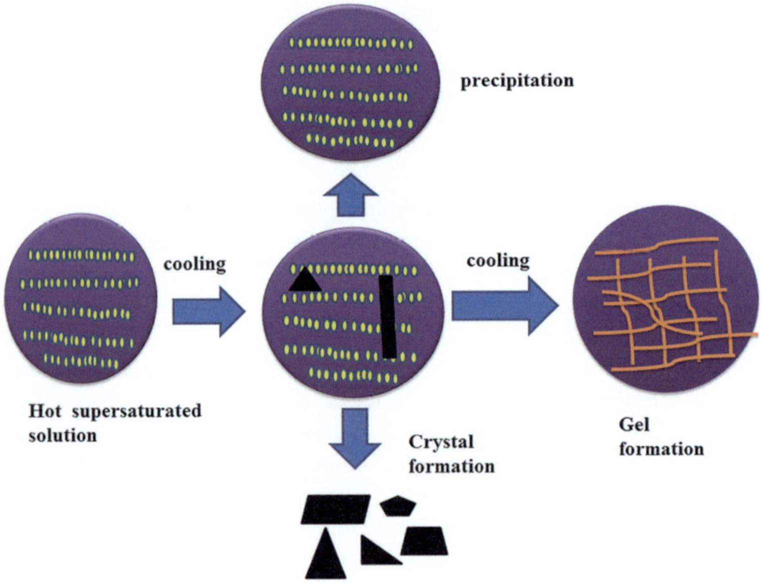

Fig. 3 Preparation of organogels

- Fluid-filled fiber method,
- Solid–fiber method,
- Hydration method.

3.1 Fluid-Filled Fiber Method

It is one of the well-known methods to prepare organogels, and this method involves dissolving surfactant mixture in apolar solvent for the formation of reverse micelles. Further, addition of water leads to the formation of tubular reverse micelles. The elongated tubular reverse micelle also gets the 3D network formation from side to side entangled (Terech and Weiss 1997) (Fig. 4).

3.2 Solid–Fiber Mechanism

In this method, we prepared organogels by using apolar solvent and solid organogelator. Apolar solvent and solid organogelators were heated, and then, apolar solution of organogelators was formed. Once RT has attained, organogelators precipitate out as a fiber forming a 3D network structure, which paralyzes apolar solvent. From that solid–fiber mechanism method, one can achieve a fine degree of subdivision of the particle and gelatinous character of those particles (Engelkamp et al. 1999).

3.3 Hydration Method

Hydration method is one of the efficient methods for synthesis of various organic materials, which results in the preparation of organogel dispersion. Various substances were utilized for gel formation, namely hydroxyl propyl cellulose, propyl gallate, and propylene glycol which results in enhancement of gel formation (Fages 2005).

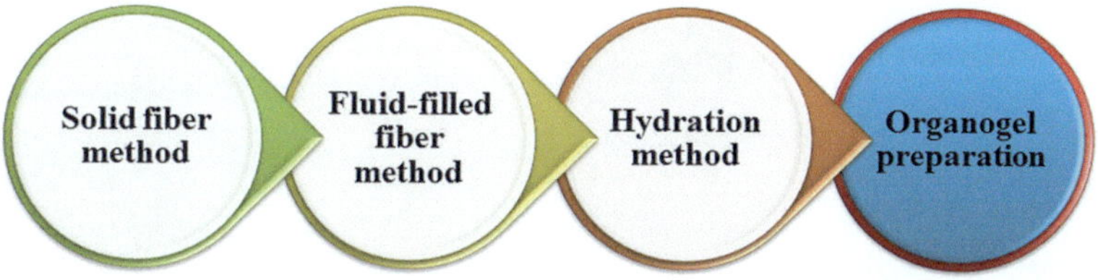

Fig. 4 Organogels preparation methods

4 Properties of Organogels

Organogels have various physicochemical properties which are mentioned here (Fig. 5)

- Viscoelasticity,
- Non-birefringence,
- Thermoreversibility,
- Biocompatibility,
- Thermostability,
- Optical property,
- Chiral nature.

4.1 Viscoelasticity

Viscoelasticity means it deals with materials having properties of viscosity and elasticity. The organogels exhibit Maxwell's model of viscoelasticity property. They have the three-dimensional structures, which arise as a result of the physical interactions in the gelator molecules. The organogels are solid at lower temperature and also have an elastic property. When there is an increase in shear stress, the interactions become weak and the organogel begins to flow (Motulsky et al. 2005).

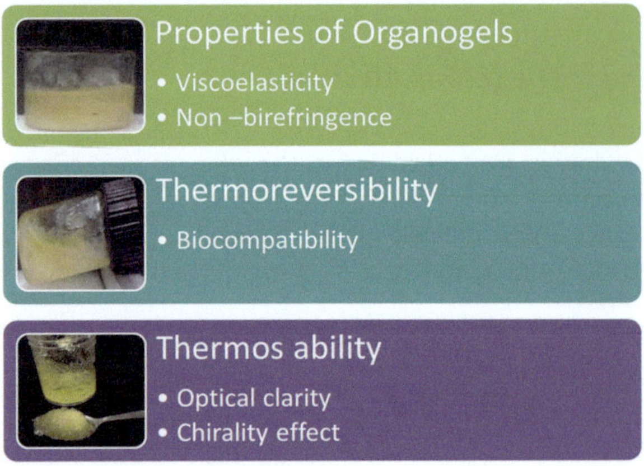

Fig. 5 Properties of organogels

4.2 Non-birefringence

Commonly, organogels appear as a dark matrix in polarized light. It does not permit the polarized light, to travel through the matrix because of having well isotropic in nature. So it is called as a non-birefringence material (Schurtenberger et al. 1990).

4.3 Thermoreversibility

Thermoreversibility means when an organogel is heated up above the critical temperature, it fails its solid nature and gets converted into liquid nature. This shows the disturbance in physical nature of the gelators in their thermal interaction. On the contrary, as the thermal energy loses, it changes into their original mode, like a solid (Scartazzini and Luisi 1988).

4.4 Biocompatibility

In older days, organogels were established using non-biocompatible organogels. In recent years, organogels have been used in most of biomedical applications, so now research is going on with various biocompatible constituents (Belgamwar et al. 2008).

4.5 Thermostability

Organogels are one of the essential gels in nature. The immovability of organogelators can be determined by their capacity to self-assemble under the adapted environment and to convert into an organogel. During the formation, gelators have very less thermostability which is caused by the decrease in the total free energy system. According to this property of organogels, it is mainly used as delivery vehicle for bioactive agents. It is also useful for cosmetic applications (Murdan 2005).

4.6 Optical Property

In accordance with the composition of the organogels, they are transparent or opaque in nature. For example, lecithin organogels are transparent in nature and sorbitan monostearate organogels are opaque in nature (Xenexlabs 2010).

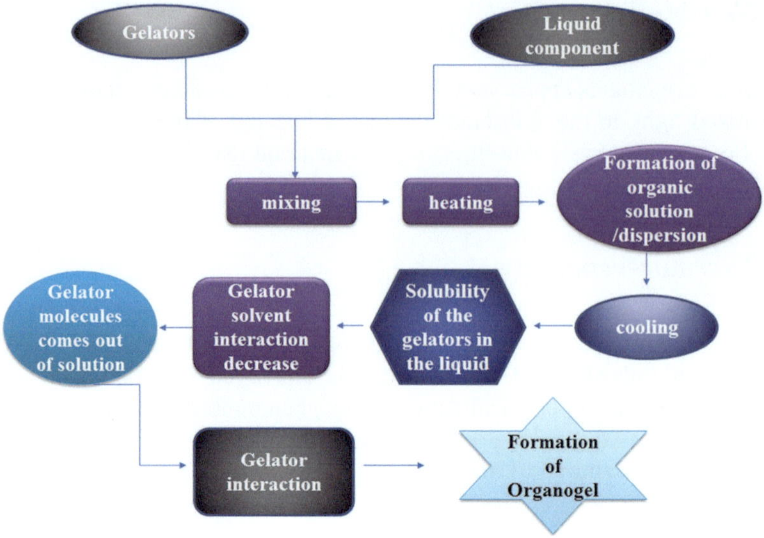

Fig. 6 Formation of organogels

4.7 Chiral Nature

Chirality is associated with the formation of the self-assembled solid–fiber network, which is the reason for thermoreversibility of organogels. In case of LMW gelators, the presence of chirality will affect both the growth and the stability of the solid–fiber networks (Lim et al. 2006). Good solid–fiber gelator has a chiral center, but it does not affect fluid–fiber gelators. The gelators having chiral center help in the formation of a compact molecular packing, which helps to maintain thermodynamic and kinetic stability of the organogels. For example, crown ether phthalocyanine organogels are known to be an excellent chiral organogels (Sawant and Liu 2002) (Fig. 6).

5 Factors Affecting Oranogels

There are many factors which should be considered for the preparation of organogels. The various parameters that can affect the preparation and stability of organogels are listed (Kang et al. 2005) (Fig. 7).

- Organic solvents,
- Phase transition temperature,
- Salt addition,
- Temperature,

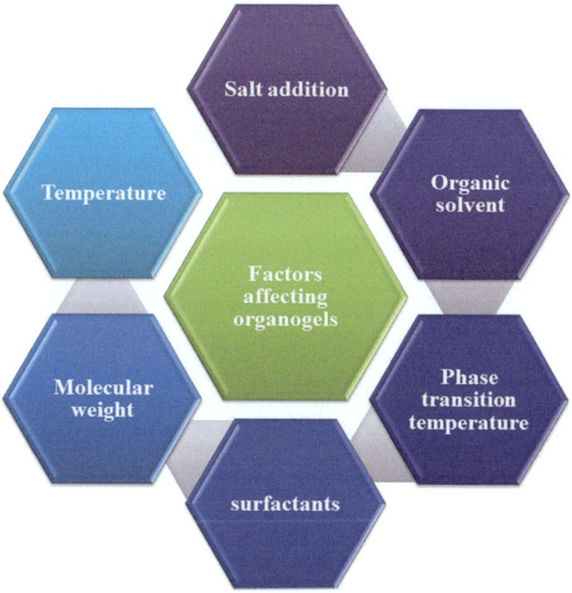

Fig. 7 Factors affecting organogels

- Molecular weight,
- Surfactants,
- Physicochemical properties.

5.1 Organic Solvents

Organic solvents are divided into two solvent systems such as polar solvent and non-aqueous solvent (Haering and Luisi 1986).

Polar solvent—in lecithin micelles, the polarity of the solvent is affected due to increase in the cross-sectional area of the lecithin polar region.

Non-aqueous solvent—various non-aqueous solvents could replace for water bacterial cellulose hydrogel, for example, dimethyl ether and polyethylene glycol (Zhao et al. 2006).

5.2 Phase Transition Temperature

Phase transition temperature gives an insight to the microstructure nature to form cross-linked network. For example, PTT gels consist of homogenous microstructures (Santos et al. 2008).

5.3 Salt Addition

Salting out means attraction of water hydration of polymer in the form of salt *via* intermolecular secondary bond (Liu et al. 2007).

5.4 Temperature

The formation of the polymer in its reaction mechanism is determined by effect of temperature and its scheme of relations with the intermediate. If the temperature decreases, in that gel part, the degree of hydration is decreased, and the gelation starts (Murdan et al. 1999a, b, c). Gel is formed using the chemical cross-linking. It cannot be changed even if temperature changes.

5.5 Molecular Weight

These are called as low molecular weight polymers. And the polymers have high viscosity to form gel structure (Willimann and Luisi 1992).

5.6 Surfactants

The characteristics of gel formation could be modified via varying the percentage and meditation of the ingredients. Polyoxyethylene functions as a surfactant (Aboofazeli et al. 2002).

5.7 Physicochemical Properties

Polymers containing charged groups are useful for the mucoadhesion. The polycations are particularly ideal for polycarboxylates. The mobility of polymers may increase the swell contact with air moisture due to the mucoadhesives. The bond formation occurs due to interface and exposing more sites. Molecular weight of polymers might change due to the predicament and the dealings with mucins (Hoffman et al. 2002).

6 Common Types of Organogels

There are various organogels available due their property, advantages, and applications. These are listed below (Fig. 8):

- Lecithin organogels,
- Premium lecithin organogel (PrLOs),
- Pluronic lecithin organogels (PLOs),
- Limonene GPl/PG organogels,
- Gelatin-stabilized microemulsion-based organogel (MBG)
- Polyethylene organogels,
- Eudragit organogels,
- Fatty acid-derived sorbitan organogels,
- Sorbitan monostearate organogels.

6.1 Lecithin Organogels

Lecithin is one of the phospholipids, which is extracted out from plants and animal tissues apart from egg yolk. Scartazzaini and Luisi discovered organogels designed with lecithin in 1988 (Willis-Goulet et al. 2003). Advanced research on lecithin-based organogels is being going on. The production of lecithin derived via gels since natural ordinary sources, which can be converted into the gelled structures, and these structures have been developed due to the presence of unsaturated

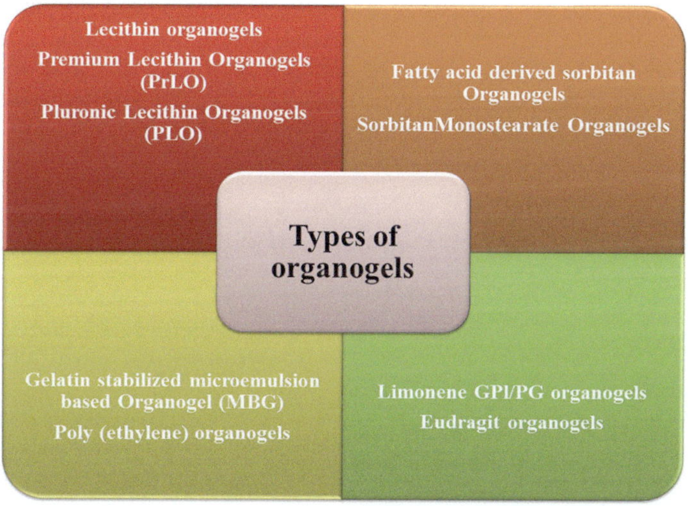

Fig. 8 Types of organogels

chemistry. Lecithin organogels developed in the synthetic way failed to develop organogels. In the preparation of organogels, lecithin preparation plays an important role. And so the lecithin derived from plant source must be pure which otherwise can affect the formation of organogels (Mayer et al. 2010). Lecithin had failed in initiating the gelling process, as it contains less than 95% of phosphatidyl content. Lecithin-based organogels have good thermodynamically stable nature (Hadidi et al. 2009) and also various properties like viscoelastic, biocompatible, thermoreversible, transparent, and non-errant. Lecithin organogels have isotropic structure; it involves moreover in the solubilization of numerous guest molecules within its arrangement. Lecithin organogels have great potential, and it is used for some controlled manner reactions. In lecithin organogels, statically the molar percentage of water to lecithin varies between 1 and 12 (Sinha et al. 2009). In the formation of lecithin-based organogels, it might attribute to the mortification of fluid–fiber reverse micellar tubular structures. In conclusion, lecithin organogels are having three different components: an apolar phase, a polar phase, and a surfactant (Murdan et al. 1999a, b, c).

6.2 Premium Lecithin Organogels (PrLOs)

Premium lecithin organogels (PrLOs) are one of the types of lecithin organogel. One of the main advantages of this PrLO is that it has higher thermostability and also has non-greasy and non-tacky nature (Murdan et al. 1999a, b, c). PrLO has main application in which utilized as cosmetics. This type of gels avoids skin irritation and local skin intolerance reactions due to the absence of pluronic derivative, in the PrLO premixed gels. Using PrLO gels as carriers for drug delivery shows that these gels have helped in achieving good bioavailability in the tissues and also improved the penetration of bioactive agents (Upadhyay et al. 2007). PrLO gels are used for the intradermal drug delivery. It is also magnificently used to accommodate various bioactive agents, for example, ketoprofen, progesterone, ibuprofen, diclofenac (Uchegbu and Vyas 1998).

6.3 Pluronic Lecithin Organogels (PLOs)

Another type of lecithin organogel is called as a pluronic lecithin organogels (PLOs). PLO has isopropyl myristate or isopropyl palmitate, water, and pluronic F127. Pluronic F127 is also called as Poloxamer 407 (Plourde et al. 2005a, b). Sorbic acid is one of the main preservative used for the PLO organogels. It appears as a yellow colored, odorless, opaque gel, which is quickly absorbed by our skin. PLO gel similar to the lecithin organogels consists of entangled tubular reverse micelle structure and formation of temporal three-dimensional structures. PLO gels have 22% of apolar phase, which is observed as microemulsion-based gel. In general, PLO is thermostable, viscoelastic, and biocompatible in nature. PLO organogels produce minimal skin irritation (Lawrence and Rees 2000). PLO organogels are used to transport for hydrophobic and hydrophilic molecules in topical and transdermal parts (Murdan et al. 2005).

6.4 Limonene GP1/PG Organogel

Limonene, one of the terpene compounds, is an excellent penetration enhancer. Due to the enhancing property, limonene organogels are used for transdermal formulation of various bioactive agents to improve their penetration character (Bhatnagar and Vyas 1994). This also helps in increasing the bioavailability of bioactive agent inside the dermal tissue. One of the most important biocompatible organogels is limonene merged within dibutyllauroylglutamide (GP1) in propylene glycol (PG) (Dreher et al. 1995). GP1 is an organogelator, which is characterized under amino acid-type gelators. This type of organogelators undergoes general intermolecular hydrogen bonding because of amide group in its structure and also has hydrophobic interactions due to the alkyl chains (Dreher et al. 1996). GP1/PG is an amino acid-type organogelator, which can form solid–fiber-based matrix. The preparation of GP1/PG organogels is done by mixing of the suitable amount of GP1, limonene and propylene glycol followed by incubation of the same at 120 °C (Nastruzzi 1994). This mixture when cooled forms a white gel. The limonene in GP1/PG organogels changed the rheological properties, and no change in chemical stability was found (Bonina et al. 1995). In addition to limonene, numerous other biocompatible terpene-based penetration enhancers have been successfully incorporated, for example, linalool, farnesol, cineole. The presence of penetraction enhancers in the organogels resulted in the enlargement of the rate permeation of the bioactive agents (Fujii et al. 1996).

6.5 Gelatin Stabilized Microemulsion-Based Organogel (MBG)

Gelatin is a protein, used as structuring agent in various food planning techniques. It has surplus aqueous phase. This type of organogels has gelled structure at 35–40 °C temperature (Dreher et al. 1997). According to the above observations, gelatin is one of the aqueous phases shielded with reverse micelles; it is the formation of water-in-oil microemulsion. This technique is mainly used for the phase separation of the aqueous gel, and the system will be cloudy in nature. Adding gelatin to water-in-microemulsion method gives rise to transparent gels. Microemulsion technique is one of the important factors in forming gelatin-stabilized organogels, due to thermostable nature and its easiness to prepare (Kantaria et al. 2003).

Water-in-oil microemulsion technique is utilized in grounding of MBG which consists of isopropyl myristate, AOT, Tween 85, and water. The MBG drugs have been reported for the use in topical and transdermal organized delivery conveyance for bioactive agents (Ferry 1980).

6.6 Polyethylene Organogel

Polyethylene organogels are commonly colorless gel. It is formed when lower molecular weight polyethylene is liquefied in mineral oil at less than 130 °C and then continuously shocked and cooled (Terech 1997). Polyethylene organogels are extensively used for ointment bases. These types of organogels can form gelled structure, used for the physical interactions of solid–fibers formed (Esch et al. 1999).

6.7 Eudragit Organogels

This type of organogels is different from the other types. Eudragit organogels are mixture like R and S forms (Zoumpanioti et al. 2003). This type of organogels used for drug delivery show high gel rigidity and stability at lower concentration. Examples for Eudragit organogels are polyhydric alcohols such as liquid polyethylene glycol, glycerol, and propylene glycol (Bastiat et al. 2010).

6.8 Fatty Acid-Derived Sorbitan Organogels

Fatty acid-derived sorbitan organogels are having sorbitan monostearate and sorbitan monopalmitate. These organogelators are hydrophobic nonionic molecules (Kantaria et al. 1999). It also has surface active properties and incapacitates several solvents, for example, isopropyl myristate and vegetable oils. Fatty acid-derived sorbitan organogelators heated with the solution forms solid–fiber matrix cooled down in apolar solvent. The formation of gel continuously forms toroidal reverse micelles at lower temperature (Moniruzzaman et al. 2009). The formation of toroidal reverse micelles rearranges itself to form rod-shaped tubules, and these undergo physical interaction to forms a three-dimensional network structure. The gels thus formed are having unique properties like opaque, thermoreversible, and thermostable even at the room temperature. Organogels using fatty acid gelators are made by dissolving gelators at high temperature in water-in-oil emulsion following a decrease in the temperature of emulsion (Carretti et al. 2005). As the temperature decreases, the gelator solubility decreases which entangles to form a gel structure.

6.9 Sorbitan Monostearate Organogels

These are also another type of sorbitan monostearate organogels having sorbitan monopalmitate also. This type of organogel is soluble in various numbers of organic solvents at the lower concentration. It is prepared like as heating with water bath at 60 °C and cooling of the resulting formation of suspension. This type of organogel is used for delivery of hydrophilic vaccines and sorbitan monostearate (Scartazzini and Luisi 1988).

6.10 L-Alanine Derivative Organogels

The preparation of N-lauroyl-L-alanine methyl ester gels is carried out by using soybean oil and triglycerides. This formation appears in gel state at room temperature. It has been used as a sustained release implant and also used for the delivery of rivastigmine and leuprolide drug (Schurtenberger et al. 1990).

7 Application of Organogels

Organogels have lot of applications in various fields, but one of the main among them is drug delivery. According to that the drug having following ingredients through various routes like as shown in Fig. 9.

- Oral drug delivery,
- Topical drug delivery,
- Parenteral delivery,
- Transdermal drug delivery.

7.1 Oral Drug Delivery

Applications of organogels in drug delivery system have been reported. In the report, cyclosporine A and 12-hydroxystearic acid showed good activity for the drug delivery system (Luisi et al. 1990). These two compounds used for the improvement of organogels have soybean oil as an apolar phase. The discharge studies showed increase in the organogelator concentration into the organogel, and release rate of the organogels was decreased. Organogel was used as a controlled delivery vehicle for invivo rat studies.

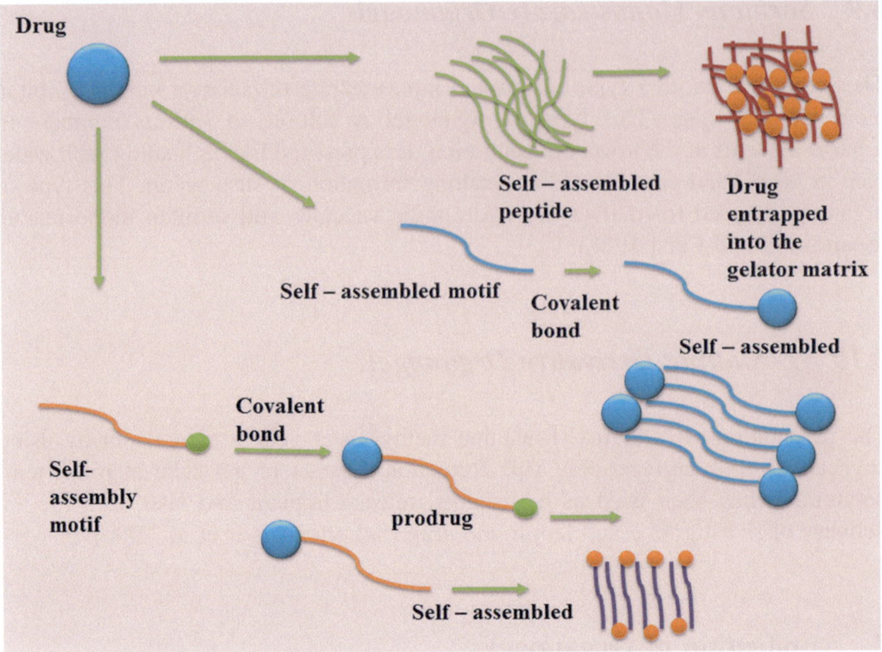

Fig. 9 Drug release function of organogels

7.2 *Topical Drug Delivery*

Lecithin-based organogels are used as the medium for transdermal delivery systems. It is also used for aromatic amines, amino acids, and peptides (Angelico et al. 1998). Lecithin organogels have long-standing biocompatibility and lower irritability potential, due to histological studies. Organogels have various applications in therapeutic agents, like different chemical and physicochemical backgrounds (Vintiloiu and Leroux 2008). It is mainly used for the muscle relaxants, antiemetic, analgesics, steroid hormones, and cardiovascular agents incorporated within these gels for getting good results. It is also used for cosmetic and personal care markets. In the topical drug delivery, the drug product has normal turnover, using rapid clearance and suspension of dosage forms. Also other properties like good tolerability, protection from of conjunctival adhesion and establishment of a protective film over the cornea (Wendel 1995). In the treatment of glaucoma, methazolamide was found to be an ineffective ophthalmic solution, so was assimilated within carbomer and poloxamer gels, to be effective for the treatment (Dreher et al. 1997).

7.3 Parenteral Delivery

The sorbitan monostearate organogels have a very short half-life at the injection site. Because of this, distribution of water molecules in that gelled structure occurs which result in successive distraction of the gelled structure (Hadgraft 1999). The same group also reported organogels having continuous delivery of model antigen and radiolabeled bovine serum albumin after intramuscular administration of the same in mice. The reports identified that L-alanine-based organogels for in situ forming delivery of macromolecular bioactive agents (Willimann and Luisi 1991). In the presence of the hydrophilic solvent, organogels like L-alanine derivatives, N-stearoyl 1-alanine, and methyl esters were used for immobilizing vegetable and synthetic oil. The formation of gel-to-sol transition, of the 1-alanine-based organogels due to the concentration of gelator and presence of solvent nature. These types of organogels are thermoreversible in nature (Shchipunov and Shumilina 1995). According to the experimental results, most of the organogels injected in the rat release bioactive agents subcutaneously. The time taken was 14–25 days with subsequent degradation of the gelled structure (Capitani et al. 1996).

7.4 Transdermal Drug Delivery

One of the main important uses of organogel system is transdermal transport of different therapeutic compounds (Shchipunov and Schmiedel 1996). The organogels enhance the solubility of drugs. Various drugs like nifedipine, clonidine, scopolamine, and broxaterol were dissolved less in IPP solution, but it was noted to increase in lecithin–IPP solution (Aliotta et al. 2000). Also, organogels will be increasing the lifetime of drugs, for example, nicardipine, one of the calcium channel blocker, but has shorter life period for low dose (Angelico et al. 2004).

7.5 Some Other Application of Organogels

Organogels are having important applications in areas of bioadhesives, cosmetic, dosage forms, gelatins gels, and microbiological media (Schurtenberger et al. 1993). According to the bioadhesives of pharmaceutical compounds for organogel were mucoadhesives, which suggests the medium adhesion is mucous itself. Most of the alternative routes of administration (buccal, ophthalmic, nasal, vaginal, etc.) provide bioadhesives in the presence of mucosal tissue. Gels are engaged in variety of products which include shampoo, fragrance products, dentifrices, and skin and hair care preparation. Glycerogelatins are frequently used as a basis for medicated plastiles (Shchipunov et al. 1999). They are used in the formulation of suppositories. Gelatin gels are used in the microbial media and also are used for diffusion of antibiotics, vitamins, enzymes, and antiseptics in the microbiological assay through the culture media.

8 Characterization of Organogels

The formation of the organogels is characterized by various parameters like physiochemical properties using spectroscopic techniques, namely NMR and FTIR. Rheological behavior of organogels like viscoelasticity, swelling, water content is analyzed by using near-infrared spectroscopy. To identify phase transition temperature, microscopy and high-sensitivity differential scanning calorimeter are used. Gelation kinetics is identified using inverse method and turbidimetry method and then in vitro drug release using Franz diffusion cell. Safety and skin compatibility studies using irritation, structural features like molecular and hydrogen bonding using NMR and FTIR spectroscopy.

9 Conclusion

In this chapter, we highlighted the organogel preparation and its properties along with the usage of organogels as drug carriers. It has longer self-life, can be easily prepared, and has thermoreversible nature of organogel-based formulation. The capacity of the organogels to provide accommodations for both hydrophilic and hydrophobic compounds has expanded the future scope of organogels in innumerable delivery systems. These organogels will take over the pharmaceuticals and nutraceuticals as they can replace the conventional dosing systems, once the biocompatibility profile of the organogels is known or available.

References

Abdallah DJ, Sirchio S, Weiss R (2000) Hexatriacontane organogels. The first determination of the conformation and molecular packing of a low-molecular-mass organogelator in its gelled state. Langmuir 16(20):7558–7561

Aboofazeli R, Zia H, Needham TE (2002) Transdermal delivery of Nicardipine: an approach to in vitro permeation enhancement. Drug Delivery 9(4):239–247

Aliotta F, Vasi C, Lechner RE, Ruffle B (2000) Evidence of percolative phenomena in a lecithin-based gel. Physica B: Condens Matter 276–278:347–348

Angelico R, Ceglie A, Colafemmina G, Lopez F, Murgia S, Olsson U, Palazzo G (2004) Biocompatible Lecithin organogels: structure and phase equilibria. Langmuir 21(1):140–148

Angelico R, Palazzo G, Colafemmina G (1998) Water diffusion and headgroup mobility in polymer-like reverse micelles: evidence of a sphere-to-rod-to-sphere transition. J Phys Chem B 102(16):2883–2889

Avramiotis S, Avramiotis S, Papadimitriou V, Hatzara E, Bekiari V, Lianos P, Xenakis A (2007) Lecithin organogels used as bioactive compounds carriers. A microdomain properties investigation. Langmuir 23(8):4438–4447

Bastiat G, Plourde F, Motulsky A, Furtos A, Dumont Y, Quirion R, Fuhrmann G, Leroux JC (2010) Tyrosine-based rivastigmine-loaded organogels in the treatment of Alzheimer's disease. Biomaterials 31(23):6031–6038

Belgamwar V, Almeida H, Amaral MH, Lobão P, Lobo JM (2008) Pluronic lecithin organogel. Asian J Pharm 2(3):134–138

Bhatnagar S, Vyas S (1994) Organogel-based system for transdermal delivery of propranolol. J Microencapsul 11(4):431–438

Bonina FP, Montenegroa L, Scrofania N, Espositob E, Cortesib R, Menegattib E, Nastruzz C (1995) Effects of phospholipid based formulations on in vitro and in vivo percutaneous absorption of methyl nicotinate. J Controlled Release 34(1):53–63

Capitani D, Segre AL, Dreher F, Walde P, Luisi PL (1996) Multinuclear NMR investigation of phosphatidylcholine organogels. J Phys Chem 100(37):15211–15217

Carretti E, Dei L, Weiss R (2005) Soft matter and art conservation. Rheoreversible gels and beyond. Soft Matter 4(2):17–22

Chen Z, Prof F, Yang H, Yi T, Huang C (2007) A thermostable and long-term-stable ionic-liquid-based gel electrolyte for efficient dye-sensitized solar cells. ChemPhysChem 8(9):1293–1297

Dasgupta D, Srinivasan S, Rochas C, Ajayaghosh A, Guenet J (2009) Hybrid thermoreversible gels from covalent polymers and organogels. Langmuir 25(15):8593–8598

Díaz D, Tellado J, Velázquez D, Ravelo A (2008) Polymer thermoreversible gels from organogelators enabled by ['click'] chemistry. Tetrahedron Lett 49(8):1340–1343

Dreher F, Walde P, Luisi PL, Elsner P (1995) Human skin irritation of a soybean lecithin microemulsion gel and of liposomes. Proc Controlled Release Soc 22:640–641

Dreher F, Walde P, Luisi PL, Elsner P (1996) Human skin irritation studies of a lecithin microemulsion gel and of lecithin liposomes. Skin Pharmacol 9(2):124–129

Dreher F, Waldea P, Waltherb P, Wehrlib E (1997) Interaction of a lecithin microemulsion gel with human stratum corneum and its effect on transdermal transport. J Controlled Release 45(2):131–140

Engelkamp H, Middelbeek S, Nolte RJ (1999) Self-assembly of disk-shaped molecules to coiled-coil aggregates with tunable helicity. Science 284(5415):785–788

Esch J, Feringa B (2000) New functional materials based on self-assembling organogels: from serendipity towards design13. Angew Chem Int Ed 39(13):2263–2266

Esch J, Schoonbeek F, Loos MD, Veen EM, Kellogg RM, Feringa BL (1999) Low molecular weight gelators for organic solvents. Supramol Sci: Where it is and Where it is Going 6(3): 233–259

Fages F (2005) Low molecular mass gelators: design, self-assembly, function topics in current chemistry. 256(2):7–5. Germany: Springer Berlin Heidelberg

Ferry JD (1980) Viscoelastic properties of polymers. 3rd edn. Wiley, New York

Fujii M, Shiozawa K, Henmi T, Yamanouchi S, Suzuki H, Yamashita N, Matsumot M (1996) Skin permeation of indomethacin from gel formed by fatty-acid ester and phospholipid. Int J Pharm 137(1):117–124

Garner C, Terech P, Allegraud J, Mistrot B, Nguyen P, Geyer A, Rivera D (1998) Thermoreversible gelation of organic liquids by arylcyclohexanol derivatives: synthesis and characterisation of the gels. J Chem Soc, Faraday Trans 94:7–9

Guenet J-M (2006) Microfibrillar networks: polymer thermoreversible gels versus organogels. Macromol Symp 241(1):45–50

Hadgraft J (1999) Passive enhancement strategies in topical and transdermal drug delivery. Int J Pharm 184(1):1–6

Hadidi N, Nazari N, Aboofazeli R (2009) Formulation and optimization of microemulsion-based organogels containing propranolol hydrochloride using experimental design methods. DARU 17(3)

Haering G, Luisi PL (1986) Hydrocarbon gels from water-in-oil microemulsions. J Phys Chem 90 (22):5892–5895
Hoffman SB, Yoder AR, Trepanier LA (2002) Bioavailability of transdermal methimazole in a pluronic lecithin organogel (PLO) in healthy cats. J Vet Pharmacol Ther 25(3):189–193
Kang L, Liu XY, Sawant PD, Ho PC, Chan YW, Chan SY (2005) SMGA gels for the skin permeation of haloperidol. J Controlled Release 106(1–2):88–98
Kantaria S, Rees GD, Lawrence MJ (2003) Formulation of electrically conducting microemulsion-based organogels. Int J Pharm 250(1):65–83
Kantaria S, Rees G, Lawrence M (1999) Gelatin-stabilised microemulsion-based organogels: rheology and application in iontophoretic transdermal drug delivery. J Controlled Release 60 (2–3):355–365
Kumar R, Katare P (2005) Lecithin organogels as a potential phospholipid-structured system for topical drug delivery: a review. AAPS Pharm Sci Tech 6(2):E298–E310
Lawrence MJ, Rees GD (2000) Microemulsion-based media as novel drug delivery systems. Adv Drug Deliv Rev 45(1):89–121
Lim PFC, Lim PF, Liu XY, Kang L, Ho PC, Chan YW, Chan SY (2006) Limonene GP1/PG organogel as a vehicle in transdermal delivery of haloperidol. Int J Pharm 311(1–2):157–164
Lim P, Liu XY, Kang L, Ho PC, Chan YW, Chan SY (2008) Physicochemical effects of terpenes on organogel for transdermal drug delivery. Int J Pharm 358(1–2):102–107
Liu H, Wang Y, Han F, Yao H, Li S (2007) Gelatin-stabilised microemulsion-based organogels facilitates percutaneous penetration of Cyclosporin A and dermal pharmacokinetics. J Pharm Sci 96(11):3000–3009
Luisi PL, Scartazzini R, Haering G, Schurtenberger P (1990) Organogels from water-in-oil microemulsions. Colloid Polym Sci 268(4):356–374
Malik S, Maji S, Banerjee A, Nandi A (2002) A synthetic tripeptide as organogelator: elucidation of gelation mechanism. J Chem Soc Perkin Trans 2:1177–1186
Mayer J, Wagner R, Taeymans O (2010) Advanced diagnostic approaches and current management of thyroid pathologies in Guinea pigs. Vet Clin North Am: Exotic Anim Pract 13(3):509–523
Moniruzzaman M, Sahin A, Winey K (2009) Improved mechanical strength and electrical conductivity of organogels containing carbon nanotubes. Carbon 47(3):645–650
Motulsky A, Lafleur M, Couffin-Hoarau AC, Hoarau D, Boury F, Benoit JP, Leroux JC (2005) Characterization and biocompatibility of organogels based on L-alanine for parenteral drug delivery implants. Biomaterials 26(31):6242–6253
Murdan S (2005) A review of pluronic lecithin organogel as a topical and transdermal drug delivery system. Hospital Pharmacist 12(7):267–270
Murdan S, Andrýsek T, Son D (2005) Novel gels and their dispersions—oral drug delivery systems for ciclosporin. Int J Pharm 300(1–2):113–124
Murdan S, Gregoriadis G, Florence AT (1996) Non-ionic surfactant based organogels incorporating niosomes. STP Pharm Sci 6(1):44–48
Murdan S, Gregoriadis G, Florence AT (1999a) Interaction of a nonionic surfactant based organogel with aqueous media. Int J Pharm 180(2):211–214
Murdan S, Gregoriadis G, Florence A (1999b) Sorbitan monostearate/polysorbate 20 organogels containing niosomes: a delivery vehicle for antigens. Eur J Pharm Sci 8(3):177–185
Murdan S, Gregoriadis G, Florence A (1999c) Novel sorbitan monostearate organogels. J Pharm Sci 88(6):608–614
Murdan S, van den Bergh B, Gregoriadis G, Florence AT (1999d) Water-in-sorbitan monostearate organogels (water-in-oil gels). J Pharm Sci 88(6):615–619
Nasseria A, Aboofazelib R, Zia H, Needhama T (2003) Lecithin—stabilized microemulsion—based organogels for topical application of Ketorolac Tromethamine. II. In vitro release study. Iran J Pharm Res 2:117–123
Nastruzzi C (1994) antitumor activity of (Trans) dermally delivered aromatic tetra-amidines. J Controlled Release 29(1–2):53–62

Pal K, Banthia A, Majumdar D (2006a) Preparation of novel pH-sensitive hydrogels of carboxymethyl cellulose acrylates: a comparative study. Mater Manuf Processes 21(8):877–882
Pal K, Banthia A, Majumdar D (2006b) Polyvinyl alcohol-gelatin patches of salicylic acid: preparation, characterization and drug release studies. J Biomater Appl 0885328206056312
Pal K, Banthia A, Majumdar D (2007) Biomedical evaluation of polyvinyl alcohol–gelatin esterified hydrogel for wound dressing. J Mater Sci: Mater Med 18(9):1889–1894
Pal K, Banthia A, Majumdar D (2008) Effect of heat treatment of starch on the properties of the starch hydrogels. Mater Lett 62(2):215–218
Pal K, Banthia A, Majumdar D (2009) Polymeric hydrogels: characterization and biomedical applications. Des Monomers Polym 12:197–220
Plourde F, Motulsky A, Couffin-Hoarau A, Hoarau D, Ong H, Leroux J (2005) First report on the efficacy of L-alanine-based in situ-forming implants for the long-term parenteral delivery of drugs. J Controlled Release 108(2–3):433–441
Santos P, Watkinson AC, Hadgraft J, Lane ME (2008) Application of microemulsions in dermal and transdermal drug delivery. J Pharmacol Biophys Res 21(5):246–259
Sawant PD, Liu X (2002) Formation and novel thermomechanical processing of biocompatible soft materials. Chem Mater 14(9):3793–3798
Scartazzini R, Luisi R (1988) Organogels from lecithins. J Phys Chem 92(3):829–833
Schurtenberger P, Peng Q, Leser ME, Luisi PL (1993) Structure and phase behavior of lecithin-based microemulsions: a study of the chain length dependence. J Colloid Interface Sci 156(1):43–51
Schurtenberger P, Scartazzini R, Magid JL, Leser ME, Luisi PL (1990) Structural and dynamic properties of polymer-like reverse micelles. J Phys Chem 94(9):3695–3701
Shchipunov YA, Dürrschmidt T, Hoffmann H (1999) Electrorheological effects in lecithin organogels with water and glycerol. J Colloid Interface Sci 212(2):390–401
Shchipunov YA, Shumilina EV (1995) Lecithin bridging by hydrogen bonds in the organogel. Mater Sci Eng, C 3(1):43–50
Shchipunov YA, Schmiedel P (1996) Phase behavior of lecithin at the oil/water interface. Langmuir 12(26):6443–6445
Sinha V, Kumar R, Singh G (2009) Ketorolac tromethamine formulations: an overview. Expert Opin Drug Deliv 6(9):961–975
Suzuki M, Hanabusa K (2010) Polymer organogelators that make supramolecular organogels through physical cross-linking and self-assembly. Chem Soc Rev 39:455–463
Suzuki M, Nigawara T, Yumoto M, Kimura M, Shirai H, Hanabusa K (2003) L-lysine based gemini organogelators: their organogelation properties and thermally stable organogels. Org Biomol Chem 1(22):4124–4131
Suzuki M, Setoguchi C, Shirai H, Hanabusa K (2007) Organogelation by polymer organogelators with a L-Lysine derivative: formation of a three-dimensional network consisting of supramolecular and conventional polymers. Chem Eur J 13(29):8193–8200
Terech P (1997) Low-molecular weight organogelators. Spec Surf 3(2):208–268
Terech P, Weiss R (1997) Low molecular mass gelators of organic liquids and the properties of their gels. Chem Rev 97(8):3133–3160
Toro-Vazquez J, Morales-Rueda JA, Dibildox-Alvarado E, Charo-Alonso M, Alonzo-Macias M, González-Chávez M (2007) Thermal and textural properties of organogels developed by Candelilla wax in safflower oil. J Am Oil Chem Soc 84(11):989–1000
Toshiyuki S, Daisuke O, Kenji H (2003) Viscoelastic behavior of organogels. Riron Oyo Rikigaku Koenkai Koen Ronbunshu 52:477–478
Uchegbu I, Vyas S (1998) Non-ionic surfactant based vesicles (niosomes) in drug delivery. Int J Pharm 172(1–2):33–70
Upadhyay K, Tiwari C, Khopade AJ, Bohidar HB, Jain SK (2007) Sorbitan ester organogels for transdermal delivery of Sumatriptan. Drug Dev Ind Pharm 33(6):617–625
Vintiloiu A, Leroux JC (2008) Organogels and their use in drug delivery—a review. J Controlled Release 125:179–192
Wendel AKO (1995) Encyclopedia of chemical technology 15:192

Willimann HL, Luisi PL (1991) Lecithin organogels as matrix for the transdermal transport of drugs. Biochem Biophys Res Commun 177(3):897–900
Willimann H, Luisi PL (1992) Lecithin organogel as matrix for transdermal transport of drugs. J Pharm Sci 81(9):871–874
Willis-Goulet H, Schmidt BA, Nicklin CF, Marsella R, Kunkle A, Tebbett I (2003) Comparison of serum dexamethasone concentrations in cats after oral or transdermal administration using pluronic lecithin organogel (PLO): a pilot study. Vet Dermatol 14(2):83–89
Wright A, Marangoni A (2006) Formation, structure, and rheological properties of ricinelaidic acid-vegetable oil organogels. J Am Oil Chem Soc 83(6):497–503
Xenexlabs. PLO Gel Transderma (2010) [cited 2010 April 25]. Available from: http://www.xenexlabs.com/catalogue.php?cid=4&pid=516
Zhao XY, Quan C, Li-Qiang Z, Gao-Yong Z (2006) Rheological properties and microstructures of gelatin-containing microemulsion-based organogels. Colloids Surf, A 281(1–3):67–73
Zoumpanioti M, Stamatis H, Xenakis A (2003) Microemulsion-based organogels as matrices for lipase immobilization. Biotechnol Adv 28(3):395–406

Chapter 12
Design and Application of Injectable Gels in Tissue Engineering and Drug Delivery

B. A. Aderibigbe

Abstract Injectable gels offer several advantages such as minimally invasive, targeted delivery, regenerate damaged tissues and organs, support cell proliferation and differentiation, and facilitate tissue growth. They are prepared from carbon-based biomaterials such as carbon nanotubes and graphene oxide, natural and synthetic polymers. In the design of injectable gels, the properties of the material used influence their application. Injectable gels have been investigated for delivery of bioactive agents such as anticancer, antimalarials, antihyperglycemic, antiviral, anti-inflammatory, antimicrobial. They are also used for wound dressing and regenerative medicine such as bone regeneration. This chapter will focus on the design and therapeutic efficacy of the currently developed injectable gels.

Keywords Injectable gels · Tissue engineering · Drug delivery
Regenerative medicine

1 Introduction

Injectable hydrogels are potential biomedical devices for drug delivery and tissue engineering applications. They contain high water content similar to tissue, their physical properties can be modified or tailored to suit specific biomedical application, they encapsulate cells homogeneously, exhibit minimal invasive delivery, and do not require surgical procedures for administration, reduce pain, healing time, scarring, and there is reduced risk of infection (Tan and Marra 2010; Nguyen and Lee 2010; Patenaude et al. 2014). On administration by injection, the gels undergo a solution-to-gelation transition in situ as a result of their response to external stimuli (Tan and Marra 2010).

They are prepared from carbon-based materials, synthetic and natural polymers. Natural polymers are useful due to several advantages they offer and their impacts

B. A. Aderibigbe (✉)
Department of Chemistry, University of Fort Hare, Alice Campus, Alice, South Africa
e-mail: blessingaderibigbe@gmail.com

© Springer Nature Singapore Pte Ltd. 2018

V. K. Thakur et al. (eds.), *Polymer Gels*, Gels Horizons: From Science to Smart Materials, https://doi.org/10.1007/978-981-10-6080-9_12

on the gels such as biocompatibility, non-toxic nature, affordability, biodegradability, and hydrophilicity. The natural polymers are classified as plant-based or animal-based. However, the natural polymers used are limited due to their inability to control their batch-to-batch variability, poor mechanical properties, and processing limitations (Malafaya et al. 2007). Incorporation of synthetic polymers and carbon-based materials onto gel networks enhances their mechanical properties.

Injectable gels are prepared by several methods such as physical self-assembly, ionic interaction, thermal gelation, photo-polymerization, and by chemical cross-linking using cross-linkers (Tan and Marra 2010). The biomedical application of injectable hydrogels is for targeted, controlled, and sustained delivery of bioactive agents such as antimalarial, anticancer, antiviral, anti-inflammatory drugs, for tissue repair and regeneration, and for the management of fecal incontinence (Fig. 1). Their biomedical application is dependent on their design, materials used for their preparation, and physicochemical properties. This chapter will elaborate on the design and therapeutic efficacy reports of recently developed injectable gels.

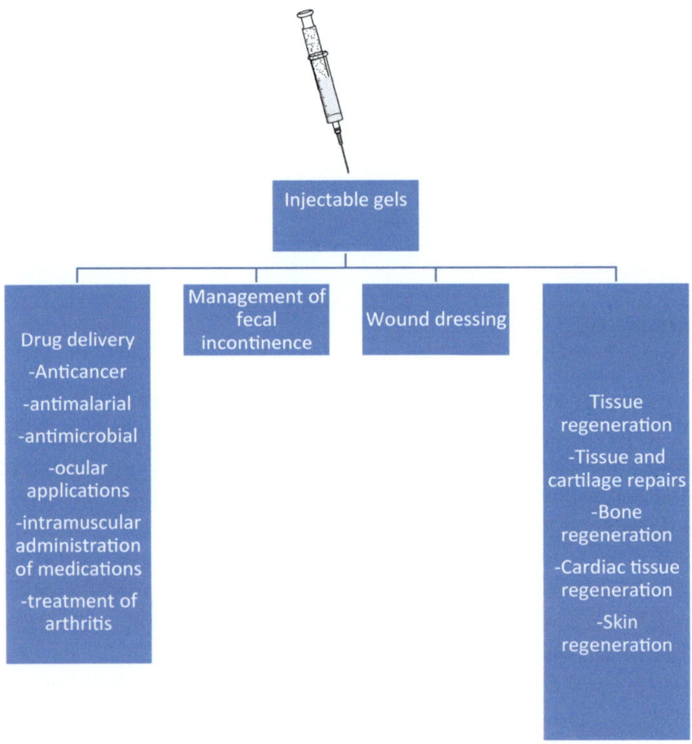

Fig. 1 Biomedical applications of injectable gels

1.1 Preparation of Injectable Gels

They are prepared by physical and chemical cross-linking using cross-linkers.

1.1.1 Physical Cross-Linking

Physically cross-linked hydrogels are prepared by several methods such as (Patenaude et al. 2014; Ebara et al. 2014) ionic interactions; hydrogen bonding interactions; stereo-complexation, hydrophobic interactions; and supramolecular chemistry. They are prepared at physiological pH and room temperature (Ebara et al. 2014). However, the physically cross-linked hydrogels suffer from some shortcomings such as difficulty in controlling the degradation and gelling time; pore size of the network; and chemical functionalization (Patenaude et al. 2014; Ebara et al. 2014). The aforementioned factors affect the flexibility of the hydrogel, and the strength of the gel is dependent on the properties of the constituent gelators. The physical cross-linked gels are suitable for short-term administration (Patenaude et al. 2014).

1.1.2 Chemical Cross-Linking

Chemical cross-linking methods involve the use of polymerization; radiation; small-molecule cross-linking; polymer-polymer cross-linking (Patenaude et al. 2014; Ebara et al. 2014; Thakur and Thakur 2014a, b, 2015). Photo-irradiation, organic solvents, and cross-linking agents are used to prepare chemical cross-linked hydrogels (Nguyen and Lee 2010). The aforementioned factors limit their applications, and as such, the physically cross-linked methods are preferred.

2 Application of Injectable Gels in Wound Dressing/Skin Regeneration

Wound healing occurs in four phases (Satteson and Molnar 2015), namely hemostasis; inflammation, proliferation; and finally, remodeling (Fig. 2). The first phase known as the hemostasis occurs immediately after a wound is inflicted on the skin. During this phase, there is formation of fibrin and degranulation of platelets. In the inflammation phase, immune system gets rid of bacteria and damaged tissue from the wound. At the proliferative phase, angiogenesis, re-epithelialization, fibroblast proliferation, and migration occur which is then followed by the last phase, remodeling. The last phase can occur for a period of 6–12 months (Satteson and Molnar 2015). An ideal wound dressing is expected to have the following properties: non-toxic, not penetrable by bacteria, non-allergenic, allows gaseous exchange, comfortable, affordable, can be easily removed without causing more harm to the wound, maintain a high humidity, good thermal insulation with long shelf life (Jones et al. 2006). The presently used wound dressing can be classified as low adherent dressings, semi-permeable films, hydrocolloids, hydrogels, foams, composites, alginates, etc. (Jones et al. 2006; Synthetic wound dressing). Hydrogels are suitable for wound dressing because they provide water molecules to the surface of the wound resulting in a moist environment around the wound. They also absorb wound exudate and allow gaseous exchange. However, they have a limitation because some of them can only be used in combination with a secondary dressing to avoid bacterial permeability and for the management of sloughy or necrotic wounds (Jones et al. 2006).

Balakrishnan et al. (2005) developed wound dressing in situ hydrogels from alginate, gelatin, and borax. The hydrogels were biodegradable and non-toxic with a fluid uptake of 90%, thereby preventing accumulation of wound exudates. The hydrogel provided a moist environment over the wound. In vivo study performed on rat model over a period of 2 weeks revealed the formation of new epithelium with no significant side effects. The borax was believed to act as an antiseptic (Balakrishnan et al. 2005). Tran et al. (2011) prepared in situ gel from rutin and tyramine-conjugated chitosan derivatives for dermal wound repair. Rutin enhanced extracellular matrix production and accumulation during wound healing. In vivo wound studies on rat dorsal wounds over a period of 14 days indicated that the gels enhanced the rate of wound healing when compared to the commercially available wound dressing known as Duoderm. The gels enhanced the formation of neo-epithelium and thicker granulation (Tran et al 2011). Kim et al. (2016) developed thermosensitive injectable hydrogel composed of soluble extracellular matrix isolated from human adipose tissue and methylcellulose. At low temperatures, the hydrogel exhibited a viscous state, and at a high temperature, a gel-like

Fig. 2 Phases of wound healing

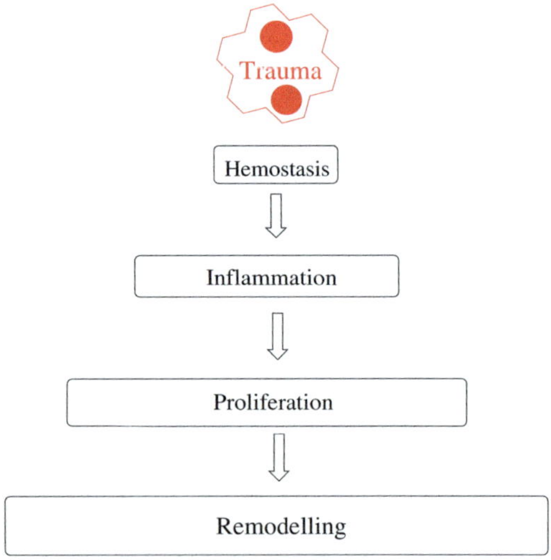

elastic state was observed. An injection of the formulation of the gel to the dorsal area of a female rat, Sprague–Dawley, with a full-thickness circle cutaneous wound 169 mm^2 in area resulted in rapid healing by re-epithelialization and neovascularization. Scar formation was minimum suggesting they are promising injectable devices for tissue regeneration and stem cell delivery (Kim et al 2016). Cerqueira et al. (2014) prepared spongy-like hydrogels from gellan gum–hyaluronic acid and human adipose stem cells to recapitulate skin matrix and promote enhanced neoskin vascularization. In vivo analysis was performed by transplantation into mice full-thickness excisional wounds. The application of the hydrogels resulted in the formation of dense granulation tissue. The hydrogel was degradable with a progressive wound closure, re-epithelialization, and matrix remodeling. The hydrogel exhibited a superior neovascularization which is attributed to the human adipose stem cells (Cerqueira et al. 2014). Li et al. (2015) developed in situ gel for dermal traumas from antimicrobial peptide (AP-57) and poly(L-lactic acid)-Pluronic L35-poly(L-lactic acid) (PLLA-L35-PLLA). The hydrogels exhibited reduced cytotoxicity and antioxidant activity. In vivo wound healing analysis on dermal defect model of rats over a period of 14 days revealed an enhanced rate of wound healing (Li et al. 2015). Du et al. (2012) developed multifunctional in situ forming hydrogel to relieve pain, stop bleeding, inhibit inflammation with enhanced healing. It was prepared from a mixture of poloxamers 407, 188 loaded-chitosan and loaded with aminocaproic acid, povidone iodine, and lidocaine. The gel did not show cytotoxicity effects in fibroblasts. It lowered the average bleeding and hemostasis time. The hydrogel also exhibited excellent antibacterial activity suggesting they are promising formulation for wound healing in emergency situations (Du et al. 2012).

Griffin et al. (2015) developed injectable hydrogels for tissue regeneration. In vitro studies showed that cells incorporated onto the gels resulted in cell proliferation forming an extensive three-dimensional network over a period of 48 h. In vivo studies further revealed that the gels enhanced cell migration resulting in regeneration of cutaneous-tissue and tissue-structure formation at a rapid rate (Griffin et al. 2015). Nguyen et al. prepared polyvinyl alcohol–silver nanoparticles–chitosan injectable-based hydrogel for wound dressing. The hydrogels were flexible with good biodegradable and swelling rate (Nguyen et al. 2015). Del Gaudio et al. prepared hydrogels containing Annexin A1 *N*-terminal derived peptide. The hydrogels exhibited sustained release. In vivo wound healing studies on mice dorsal wounds revealed that the gels quicken the wound healing process suggesting their potential application in dermal wound healing (Del Gaudio et al. 2015). Gaharwar et al. (2014) developed shear-thinning nanocomposite hydrogels composed of silicate nanoplatelets and gelatin. These hydrogels stopped bleeding and formed clot-gel systems that were stable. They were found to be potential wound dressing materials for out-of-hospital and emergency conditions (Gaharwar et al. 2014). Zohdi et al. prepared cross-linked honey-based hydrogel for wound dressing. In vivo studies on burn wounds in rats showed that the hydrogel enhanced wound closure and re-epithelialization (2012). Lao et al. developed new recombinant human epidermal growth factor (rhEGF) containing hydrogels for wound healing. In vivo analysis was performed on fifty-six wounded diabetic rats; the results suggested that the hydrogels enhanced wound healing and are potential dressings for diabetic wounds (Lao et al. 2012).

3 Application of Injectable Gels for Bone Regeneration

The integrity of bone is threatened by factors such as aging, injury, disease. In the clinical treatment of bone defects, autografts and allografts are used which present serious limitations such as they require invasive surgery, infections, lack of histocompatibility, retarded vascular penetration, reduced rate of bone formation, increased bone resorption, slow graft incorporation and the allograft exhibit delayed union than the autograft (Kalfas 2001; Friedlaender et al. 1984; Habal 1992; Zdeblick and Ducker 1991). To overcome the aforementioned limitations associated with autografts and allografts, the design of biocompatible scaffolds that support cell migration, division, and proliferation with noninvasive treatment such as injectable hydrogels has been found to be a pressing need. Several researchers have reported the design of injectable gels for bone regeneration.

Although injectable hydrogels are useful for bone regeneration, they are also limited due to their poor mechanical properties. The addition of nanofillers has been found to improve their mechanical properties. Yasmeen et al. (2014) developed thermosensitive injectable composite hydrogels by cross-linking chitosan with chitosan-grafted carbon nanotubes and chitosan–hydroxyapatite complexes resulting in composites with homogeneous interaction between the materials and enhanced properties. The composite gels exhibited rapid sol–gel transition at body temperature, enhanced bone bonding, with good mechanical properties suggesting that they are potential scaffolds for bone regeneration. Shavandi et al. (2016) evaluated squid pen chitosan-based injectable gels as potential scaffolds for bone tissue engineering applications. The gels were prepared from glycerol phosphate, crab or squid pen chitosan and selected concentrations of calcium phosphate compounds. The gels were formed rapidly over a period of 3 min at body temperature. The gels exhibited high water uptake ability, enhanced cytocompatibility, and biocompatibility with good mechanical properties. These attributes of the gels showed that they are useful for bone regeneration (Shavandi et al. 2016). Sang et al. developed thermally sensitive hydrogel via copolymerization of *N*-isopropylacrylamide, single-walled carbon nanotubes, and polyethylene glycol diacrylate–dodecylamine–1-(2-aminoethyl) piperazine (Sang et al. 2016). The hydrogels promoted the growth of SH-SY5Y cells, nerve tissue regeneration, and scar tissue formation in spinal cord injury model. These observations confirmed that the hydrogels are potential scaffolds for spinal cord regeneration and neuron network reconstruction (Sang et al. 2016). Liu et al. (2014) developed injectable system from poly(L-lactide-co-ε-caprolactone) nanoyarns suspended in type I collagen hydrogel by electrospinning method. The nanoyarns exhibited enhanced mechanical properties, and in vitro study on human mesenchymal stem cells revealed that the nanoyarns enhanced cell proliferation. They were found to be potential scaffolds for bone regeneration (Liu et al. 2014). Yan et al. developed injectable hydrogels for bone regeneration. Thermosensitive triblock copolymers prepared were loaded with simvastatin (Yan et al. 2015). The osteogenic property of the hydrogels was tested in vitro on MC-3T3 E1. The hydrogels enhanced osteoblastic differentiation and cell proliferation. In vivo analysis on rat model with bone defect displayed enhanced new bone formation, suggesting that they are potential therapeutics for bone healing (Yan et al. 2015). Ren et al. (2015) grafted poly(*N*-isopropylacrylamide) onto gelatin via atom transfer radical polymerization. The gel exhibited sol-to-gel transformation at physiological temperature. In vivo studies on bone defect regeneration in a cranial model indicated that the hydrogels were biocompatible with enhanced bone regeneration. The hydrogel exhibited targeted delivery of bone mesenchymal stem cells resulting in enhanced bone regeneration (Ren et al. 2015). Fang et al. (2014) prepared injectable gel graft from transglutaminase cross-linked gelatin loaded with bone morphogenetic protein-2. Injection of the gel formulation into the induced cranial defect site resulted in good bone regeneration and involved minimal invasive surgical procedures. The findings suggested that they are potential systems suitable for the treatment of irregular-shaped bone defects and controlled release of therapeutic agents (Fang et al. 2014). Dorati et al. (2012) developed injectable bone graft substitute from biomaterials of natural origin, chitosan. They were physically stable over a long period of time, biocompatible, biodegradable. They were suitable for tissue regeneration in dental and orthopedic surgery (Dorati et al. 2012). Yu et al. (2013) developed injectable bioactive cement for complex bone defect prepared from calcium phosphate cement and bioactive glass. The injectable material exhibited enhanced mechanical properties, and they were degradable and bioactive. They also promoted cell proliferation, attachment, and differentiation with enhanced bone formation (Yu et al. 2013). Zhang et al. (2016) prepared injectable macroporous calcium phosphate cements by syringe-foaming using hydrophilic polymeric solution as foaming agent. In vivo study was performed on the distal femoral sites of rabbits. The formulation exhibited good mechanical properties and supported bone regeneration (Zhang et al. 2016). Kaigler et al. (2013) reported the application of injectable hydrogels to deliver vascular endothelial growth factor. In vivo analysis was performed on defects made on rat calvariae. The injectable gels enhanced early angiogenesis and bone regeneration and exhibited controlled release mechanism of angiogenic growth factors (Kaigler et al. 2013).

4 Application of Injectable Gels in Tendon/Cartilage Repairs

Tendons are structures that connect muscle with bone, and they are found in every joint. They are very useful for movement, and as such, tendon injury can affect movement [Knee treatment (Internet)]. The natural healing of tendons is slow.

On the other hand, articular cartilage is a tissue that covers the condyle surfaces where bones meet and it provides friction between the bones and aids movement [Knee treatment (Internet)]. When the cartilage is damaged, it becomes rough, causing pain and affecting mobility. Articular cartilage lesion is a major unsolved clinical problem because it is characterized by poor intrinsic repair capacity (Orth et al. 2014). The presently available treatment option cannot restore the integrity of articular cartilage resulting in the tissue inability to withstand mechanical stresses during physical activities of the patient (Orth et al. 2014). Hoemann et al. (2005) developed cytocompatible chitosan-based injectable gels that bond to the cartilage and bone. The system was loaded with primary articular chondrocytes in order to enhance cartilage formation in vitro and in vivo. In vivo analysis performed by subcutaneous injection of the gels in nude mice revealed the formation of subcutaneous dorsal implants. The gel was retained 1 week after injection (2005). Chang et al. (2014) invented an injectable gel with minimal invasive for in situ repair and regeneration of an injured ligament or tendon in a mammalian. The gels were able to deliver growth factors, therapeutic agents, and cells into the tendon or ligament injury. The injectable gel solution upon administration polymerized to form an in situ hydrogel where it promoted repairs and regeneration to the tendon (Chang et al. 2014). Qi et al. (2013) prepared injectable hydrogel from chitosan and polyvinyl alcohol that were thermosensitive and loaded with gene-modified bone marrow mesenchymal stromal cells. In vivo studies on rabbits with defective articular cartilage revealed regeneration of cartilage (Qi et al. 2013). Farnebo et al. (2014) prepared biocompatible hydrogel loaded with extracellular matrix from human tendons. In vivo application of the gel on rat model by percutaneous administration at tendon injury revealed that the gel solution polymerized resulting in the formation of porous gel. The gel was found to be a potential option for the treatment of acute tendon injuries (Farnebo et al. 2014). Crowe et al. (2016) developed a gel from fresh human cadaveric flexor tendon. A comparison study of the hydrogel and its reconstituted lyophilized form was performed. In vitro evaluation was performed by seeding rat adipose-derived stem cells in the gel formulation, followed by assessment of the cell viability over a period of 7 days. The reconstituted lyophilized hydrogel formulation enhanced platelet-rich plasma activation and cell proliferation than the non-reconstituted lyophilized form. After prolonged storage, lyophilized hydrogels exhibited superior properties with wider clinical application (Crowe et al. 2016). De France et al. developed in situ-gelling nanocomposite hydrogels from poly(oligoethylene glycol methacrylate). Cellulose nanocrystals were incorporating onto the hydrogels (De France et al. 2016). They exhibited improved mechanical strength with potential biomedical applications for stiffer tissues (De France et al. 2016). Oliviera et al. loaded autologous cells onto gellan gum-based hydrogels. The ability of the hydrogels to regenerate rabbit articular cartilage defects over a period of eight weeks was evaluated (Oliviera et al. 2010). Gellan gum was encapsulated with chondrogenic pre-differentiated rabbit adipose stem cells. The hydrogels were able to treat articular cartilage defects (Oliviera et al. 2010). Miljkovic et al. (2009) prepared injectable gels from gelatin, poly(ethylene glycol), and genipin. The injectable hydrogel was combined with cyanoacrylate-based surgical sealant, and in vivo studies were evaluated in osteochondral knee defect rats over a period of 14 days. The combination was found to be biocompatible and safe for maintaining the hydrogel in the defect (Milijkovic et al. 2009). D'Este et al. (2016) developed thermoresponsive brush copolymer from hyaluronan for cartilage repair. The brush copolymer was obtained by grafting poly (*N*-isopropylacrylamide) to hyaluronan. In vivo studies on osteochondral defect rabbits revealed that they were not toxic (D'Este et al. 2016). Cakmak et al. (2013) reported injectable hydrogels prepared from commercially available fibrin sealant. In vivo studies on immunocompetent New Zealand white rabbits revealed the formation of mature and immature cartilage. Around the newly formed cartilage, side effects were significant such as abscess formation, inflammatory reactions, and reaction to foreign body (Cakmak et al. 2013). The cartilage formation was not influenced by the source of the cartilage, cultured chondrocyte, or concentrations of the commercially available fibrin glue (Cakmak et al. 2013). Lee et al. developed injectable composite gels composed of collagen, hyaluronic acid, and fibrinogen. It was loaded with synovium-derived mesenchymal stem cells (Lee et al. 2012a, b). In vivo study on rabbit model showed that the composite gel regenerated and repaired osteochondral defects in the knee (Lee et al. 2012a, b). Pereira et al. (2009) reported injectable hydrogels prepared from carrageenan, fibrin, hyaluronic acid and seeded with chondrogenic cells. The injectable hydrogel seeded with human articular chondrocytes regenerated and repaired a lesion produced experimentally in bovine articular cartilage (Pereira et al. 2009). Funayama et al. (2008) prepared injectable type II collagen gel scaffold containing type II collagen obtained from nasal septa of a sheep and polyethylene glycol embedded with chondrocytes. In vivo studies were performed on New Zealand white rabbits by injecting the gel into the cartilage defects made on the patellar groove of the femur. The gel bonded to the cartilage resulted in cartilage regeneration with excellent chondrocyte morphology (Funayama et al. 2008). Passaretti et al. (2004) cultured chondrocytes encapsulated with fibrin polymers having ability to produce new cartilage in vivo. The polymer encapsulated with chondrocytes was injected subcutaneously on nude mice resulting in the formation of cartilaginous nodules. Cultured cells alone that were not encapsulated onto a polymer did not produce cartilage (Passaretti et al. 2004). Igarashi et al. (2012) prepared in situ forming gel for the delivery of bone marrow stromal cells. In vivo studies performed on osteochondral defects in canine model revealed the formation of matured hyaline-like cartilage repair of osteochondral defects in a canine model (Igarashi et al. 2012). Sa-Lima et al. (2010) prepared chitosan–starch, chitosan-β-glycerophosphate, and chitosan-β-glycerophosphate-1% starch thermosensitive hydrogels. The degradation profile of the hydrogels was enhanced by the addition of starch. The hydrogels were also found to biodegradable and enhance chondrogenic differentiation of adipose-derived stromal cells suggesting their potential application for cartilage-engineered regeneration via minimal invasive techniques (Sa-Lima et al. 2010). Sancho-Tella et al. (2015) synthesized poly(ethyl acrylate-co-hydroxyethyl acrylate) copolymer seeded with allogenic chondrocytes. In vivo studies performed on chondral defect and subchondral bone injury showed that neotissue growth appeared 1 week after implantation, followed by cartilaginous tissue after 3 months. Native chondrocytes also proliferated at the edge of the defects. The injectable scaffolds enhanced cartilaginous tissue growth (Sancho-Tella et al. 2015). Chen et al. (2016) prepared enzymatically cross-linked injectable hydrogel system containing carboxymethyl pullulan-tyramine and chondroitin sulfate-tyramine conjugates. Porcine auricular chondrocytes were encapsulated onto the hydrogel system. The system exhibited good cytocompatibility with enhanced cell proliferation and cartilaginous extracellular matrix deposit facilitating chondrogenesis (Chen et al. 2016).

5 Application of Injectable Gels for Drug Delivery

Injectable gels have been found to be useful for the delivery of several bioactive agents such as antimalarials, anticancer, antihyperglycemic, antiviral agents. The application of injectable hydrogels offers several advantages such as reduced risk of infection and scar formation, patient compliance, affordability, biocompatibility, biodegradability and is useful for targeted delivery (Hou et al. 2004; Hoare and Kohane 2008). There are several research reports on the application of injectable gels for delivery of bioactive agents.

5.1 Injectable Gels for Delivery of Antimalarials

Malaria is a chronic and life-threatening disease, and it varies in manifestation and epidemiology in different parts of the world (Bloland 2001). These variations are attributed to the different species of malaria parasites, their susceptibility to the available antimalarials, environmental conditions, and the level of immunity of the exposed human populations (Bloland 2001). Over half of the world's population are at risk of the disease, especially children, pregnant women, people living with HIV/AIDS, and low-immune travelers from malaria-free areas (WHO 2016). However, malaria can be treated and prevented. There are five parasite species that cause malaria in human beings: *P. falciparum*, *P. vivax*, *P. ovale*, *P. malariae*, and *P. knowlesi.* Most deaths are caused by *P. falciparum* and *P. vivax* (WHO 2005). The currently used antimalarials suffer from severe pharmacological limitations such as drug resistance and toxicity resulting in treatment failure. Poor patient compliance to duration of dosing regimen of antimalarials is another factor that contributes to treatment failure. Injectable hydrogels are potential delivery systems that can deliver antimalarials with enhanced therapeutic efficacy.

Nishi and Jayakrishnan (2007) developed gum arabic-based injectable hydrogels containing primaquine, an 8-aminoquinoline. In vitro release of primaquine at physiological pH and temperature was dependent on the density of cross-linking and drug payload. The biodegradability of the gels was also dependent on the cross-linking density. In vitro analysis showed that the gel was useful for sustained drug delivery of antimalarials (Nishi and Jayakrishnan 2007). Fawaz et al. (2004) developed rectal quinine-based pediatric gels for the treatment of acute uncomplicated malaria. In vivo studies on rabbits showed that the gels exhibited sustained release behavior and no adverse effects such as damage on the rectal mucosa of the rabbit were visible. The gels displayed a similar mean residence time (i.e., 8.87 ± 0.96 and 8.73 ± 0.95 h) which was higher than the rectal solution suggesting their sustained release behavior in the rabbit (Fawaz et al. 2004). Gaudin et al. (2008) also prepared similar rectal-based gels for the delivery of artesunate. The rate of artesunate was slow, suggesting sustained release profile.

5.2 *Injectable Gels for Delivery of Anti-diabetic Drugs*

Diabetes is a chronic condition that is progressive, and as such, glycemic control is very important. It results in damage of body nerves, blood vessels classified as microvascular, macrovascular, and neuropathic complications (Type 2 Diabetes Mellitus, Internet). The global prevalence of diabetes was reported to be 9% among adults aged 18 years and above in 2014 (WHO 2014). The currently used anti-diabetic drugs are expensive and suffer from severe pharmacological limitations such as weight gain, decreased effectiveness after a prolonged period of usage, increased risk of cardiovascular conditions, gastrointestinal problems, peripheral edema, liver damage (Schramm et al. 2008; Del et al. 2007; Bell 2002). Due to the aforementioned limitations, there is a pressing need to develop drug delivery systems which can reduce the side effects associated with anti-diabetic drugs.

Gu et al. (2013) prepared injectable and acid-degradable polymeric network for self-regulated delivery of insulin. In vivo studies by subcutaneous administration of the gel on type 1 diabetic mice improved glucose control in the diabetic mice. A single injection of the system was effective in stabilizing the blood glucose level for 10 days (Gu et al. 2013). Chou et al. (2016) prepared an injectable gel from carboxymethyl-hexanoyl chitosan, integrated lysozyme, and loaded with insulin. The injectable was designed for controlled insulin release for long-term management of diabetes. The amount of lysozyme in the gel formulations influenced the degradation and rate of release of insulin. In vivo work performed on a diabetic mouse model revealed that a single injection of the gel maintained the blood glucose level for 10 days (Chou et al. 2016). Anand et al. (2012) prepared gels loaded with insulin glargine and evaluated the biological effects of drug release from the gels in type 2 diabetic Zucker diabetic fatty rats. The injectable gels were prepared from polylactic-co-glycolic acid using plasticizers that are biodegradable. The administration of the gel formulation loaded with insulin glargine subcutaneously to rats decreased the blood glucose levels of the rats below 200 mg/dL within 2 h which was stable for up to 6 h. The action of the gels continued over a period of ten days. A further decrease in the level of blood glucose was significant in the gels containing zinc sulfate. These gels were found to be promising materials for the treatment of type 2 diabetes (Anand et al. 2012).

Peng et al. (2013) designed chitosan-based thermosensitive hydrogels loaded with poly(3-hydroxybutyrate-co-3-hydroxyhexanoate) containing insulin phospholipid complex. The system was designed for controlled and long-term sustained delivery of insulin. In vitro release studies revealed that only 19.11% of total insulin was released over a period of 1 month. In vivo analysis on diabetic rats suggested that the hypoglycemic effect after a single subcutaneous injection lasted more than five days. The pharmacological availability insulin was enhanced significantly by the gels suggesting their great potential long-term delivery of antidiabetic agents (Peng et al. 2013). Wu et al. (2012) developed glucose-responsive injectable gels loaded with insulin for self-regulated delivery. It was prepared from poly(*N*-isopropylacrylamide), dextran, and poly(3-acrylamidophenylboronic acid). In vitro studies on L-929 mouse fibroblast cells confirmed good biocompatibility of the gels. The composition of dextran and the concentration of glucose in the gels influenced the rate of release of insulin from the gel. In vivo experiments performed on diabetic rats further revealed that the gels decreased the blood glucose levels of the rats. The blood sugar level was stable when compared to the free insulin. The difference in hypoglycemic effect of the nanogels loaded with insulin and the free insulin after administration on the rats was not significant (Wu et al. 2012).

Huynh et al. (2008) prepared poly(β-amino ester)-based sensitive injectable hydrogel for controlled delivery of insulin. The hydrogel was pH- and temperature-sensitive because of the presence of biodegradable triblock copolymer of poly(ethylene glycol)-poly(ε-caprolactone). Insulin release from the gel was controlled by the degradation of the copolymer. Sol–gel phase transition behavior of the copolymer from a sol to a gel phase was dependent on the temperature and pH. The release mechanism of insulin from the gel was chemically and diffusion-controlled. In vivo pharmacokinetic release studies of insulin from the gel on male Sprague–Dawley rats revealed that the insulin level was maintained for two weeks. The insulin level dropped after two weeks which is attributed to chemically and diffusion-controlled release (Huynh et al. 2008). Khodaverdi et al. (2015) prepared supramolecular injectable hydrogel from poly(caprolactone)–poly(ethylene glycol)–poly(caprolactone) and γ-cyclodextrin by ring-opening polymerization method. In vitro insulin release from the hydrogel was sustained with over 80% of insulin released over a period of three weeks (Khodaverdi et al. 2015). Hyun et al. (2014) prepared thermoresponsive copolymers from methoxy poly(ethylene glycol) and hydrophobic biodegradable polyesters. The hydrogel revealed enhanced drug release in vitro and in vivo (Hyun et al. 2014).

5.3 Injectable Gels for Delivery of Anticancer Drugs

Cancer is the biggest cause of mortality worldwide. Increasing urbanization is believed to add to the environmental and dietary factors conducive to carcinogenesis. The rapid spread of HIV infection is also contributing to the global burden of cancer. The currently used anticancer drugs are limited by toxic side effects and drug resistance. And due to these limitations, the therapeutic effect of anticancer drugs is enhanced by the development of bioactive agents that regulate the molecular mechanisms and pathways linked with tumor progression and by incorporating anticancer drugs onto drug delivery systems. The latter have been reported to be more effective with a reduced exposure of normal tissues to the toxic effects of the drug (Vicent et al. 2005). There are several reports on the application of gels for delivery of anticancer drugs.

Smith et al. (1995) encapsulated selected chemotherapeutic agents (cisplatin, doxorubicin, fluorouracil, floxuridine, vinblastine, mitomycin, or paclitaxel) onto protein-based injectable gels and evaluated their effects on human pancreatic adenocarcinoma cell lines growth, BxPC-3. In vivo evaluation by intratumoral administration of the gels on nude mice bearing by BxPC-3 xenografts over a period of 28 days indicated a decrease in the size of the tumors by 72–79%. The injectable gel when administered intratumorally improved tumor response to treatment with less systemic toxicity (Smith et al. 1995). Nagahama et al. (2015) developed injectable gel composed of copolymer micelles, poly(D,L-lactide-co-glycolide)-b-poly(ethylene glycol)-b-poly(D,L-lactide-co-glycolide), clay nanodisk, and doxorubicin. The gel exhibited long-term sustained release and controlled release mechanism of doxorubicin. In vivo investigation using a single injection of the gel on mice with human xenograft tumors revealed long-term antitumor effects of the gels (Nagahama et al. 2015). Singh et al. (2013) prepared in situ injectable gels containing cytarabine, an anticancer agent. The formulation was prepared from carbopol 934 NF and sodium alginate. They exhibited sustained release mechanism for 24 h and were stable with no signs of irritation (Singh et al. 2013). Yu et al. (1995) developed injectable hydrogels containing fluorouracil and epinephrine for enhanced antitumor effects. In vitro evaluation of the drug-loaded gels on mouse tumor model revealed good antitumor effects. The release of the drugs from the gel was sustained with over 50% released after 4 h and 80% released in 24 h. In vivo evaluation by intratumoral administration of the gel on RIF-1 mouse tumors further confirmed the good antitumor effects of the drugs loaded to the gels (Yu et al. 1995). Shen et al. (2012, 2016) evaluated doxorubicin-loaded zein in situ gels. In vivo anticancer activity of the gels was investigated using BALB/c nude tumor-bearing mice. Efficient accumulation of the drug in the tumor with low drug concentration in blood and normal organs was observed. This resulted in effective inhibition of tumor growth and few side effects. In vitro release of doxorubicin from the gel was sustained for up to 7 days suggesting they are potential devices for the treatment of colorectal cancer (Shen et al. 2012, 2016). Monette et al. (2016) prepared injectable chitosan-based thermogels using sodium hydrogen carbonate and phosphate buffer as gelling agents. The gel exhibited excellent mechanical properties and cytocompatibility. The gels were encapsulated with $CD8^+$ T lymphocytes and cells encapsulated onto the formulation retain their anticancer functions, and this system can be further developed to complement the existing immunotherapies (Monette et al. 2016).

Ye and Hu (2016) prepared acetylated-β-cyclodextrin-based nanoparticles injectable gels encapsulated with camptothecin, an anticancer drug. The drug release behavior of the gel was dependent on the pH. The injectable gel exhibited short gelation time with a reversible sol–gel transition in alkaline pH. In vitro drug release profile of the gel suggested that the gels exhibited controlled drug release behavior (Ye and Hu 2016). Hayashi et al. (2016) developed smart injectable gels with reversible sol–gel transition in tumors generating heat in response to an alternating magnetic field resulting in simultaneous release of the encapsulated anticancer drug. The system was synthesized from Fe_3O_4 nanoparticles, alginate, and cysteine and was useful for magnetic resonance imaging. In vivo experiments revealed that the sol–gel transition of the gel in the tumor hindered the tumor growth. The gel was found to be very useful for suppressing tumor growth and magnetic resonance imaging (Hayashi et al. 2016). Gupta et al. prepared L-alanine-based small-molecule hydrogelator. Selected anticancer drugs were loaded onto the hydrogel (Gupta et al. 2016). The highest drug entrapment was observed in gel loaded with 5-fluorouracil, while small polar anticancer drugs such as cisplatin and carboplatin loading efficiency were poor. The gel was entrapped with hydrophobic drugs, such as tamoxifen and docetaxel with a high drug-loading efficiency. The in vitro drug release of the drugs from the gels was slow and sustained suggesting they are suitable for cancer therapeutic applications (Gupta et al 2016).

Singh et al. (2014) reported L-alanine low molecular weight-based hydrogelators injectable gel loaded with doxorubicin. Subcutaneous injection of the gel loaded with doxorubicin resulted in significant decrease in tumor than by intravenous injection of doxorubicin. The gel loaded with doxorubicin was very effective at the early stage of tumor progression (Singh et al. 2014). Lee et al. prepared injectable gels from hydrophobically modified chitosan and liposomes via hydrophobic interactions (Lee et al. 2012a, b). The gels remained at the target site for an extended period of time. The gels also displayed sustained release profile over a period of one week (Lee et al. 2012a, b). Lopez-Noriega et al. (2014) developed injectable thermoresponsive chitosan-based hydrogel containing doxorubicin-loaded thermosensitive liposomes. The drug release from the gels was dependent on the temperature suggesting that they are promising systems for local chemotherapy (Lopez-Noriega et al. 2014). Mamaghani et al. (2015) prepared dual thermo-pH-responsive injectable hydrogel loaded with doxorubicin hydrochloride. The gel loaded with doxorubicin exhibited significant cytotoxicity effects against MCF7 cells than the free drug because of the sustained release mechanism of the drug from the gel (Mamaghani et al. 2015). Cho et al. (2012) prepared biodegradable poly(organophosphazene)-based injectable hydrogels loaded with hydrophobic silibinin. The gel exhibited sol–gel transition dependent on temperature, and the rate of degradation was high at pH 6.8 than pH 7.4. Silibinin release mechanism from the gel was sustained. In vivo work was performed, whereby the hydrogels were injected intratumorally to HT-29 xenografted mice model. The gel formulation exhibited enhanced antitumor effect than the free drug (Cho et al. 2012). Ci et al. (2014) reported injectable formulation prepared from poly(D,L-lactide-co-glycolide)-b-poly(ethylene glycol)-b-poly(D,L-lactideco-glycolide) and containing irinotecan, an antitumor drug. Subcutaneous injection of the formulation on nude mice with xenografted SW620 human colon tumors revealed a significant regressed with mild side effects (Ci et al. 2014). Wu et al. (2016) prepared injectable gel formulations by Schiff base reaction for the co-delivery of metformin and 5-fluorouracil. In vitro degradation and drug release studies showed that both metformin and 5-fluorouracil were released from the gels in a controlled manner which was pH-dependent. In vivo analysis on incubated mouse colon adenocarcinoma cells demonstrated the synergistic inhibitory effects of the gels loaded with both drugs on the cell cycle process and cell proliferation in colon cancer cells. In vivo evaluation on tumors of BALB/c mice inoculated with C26 cells using a single subcutaneous injection of the hydrogel containing both drugs displayed enhanced therapeutic effects resulting from a combination of p53-mediated G1 arrest and apoptosis in C26 cells. These findings suggested the potential therapeutic applications of the gels in the treatment of colon cancer (Wu et al. 2016).

5.4 Injectable Hydrogels for Delivery of Antimicrobial Agents

Antimicrobial agents are used for the treatment of different types of infections such as fungal and bacterial infections. Fungal infections are difficult to treat. However, the currently used antimicrobial agents suffer from severe toxicity and drug resistance. In order to overcome the drug resistance, several drug delivery systems are presently used to deliver them with enhanced therapeutic effects. Hudson et al. (2010) developed dextran-based injectable local antifungal formulation loaded with amphotericin B for direct administration into sites of fungal infection. The gel retained antifungal efficacy against *C. albicans* over a period of three weeks (Hudson et al. 2010).

Navath et al. prepared biodegradable hydrogel loaded with amoxicillin for sustained delivery to cervicovaginal region (Navath et al. 2011) from 4-poly(amidoamine) [G4-$(NH_2)_{64}$] dendrimer and polyethylene glycol. In vivo evaluation in pregnant guinea pig model revealed that the gels exhibited antibacterial activity which is attributed to the sustained release of amoxicillin from the gels. The residence time of the gel loaded with amoxicillin was 72 h. The formulation was retained in maternal tissues and did not cross the fetal membranes into the fetus. The gel is a therapeutic option for intracervically delivery of antimicrobial drugs in pregnant woman for the treatment of genital infections (Navath et al. 2011). Giano et al. (2014) prepared syringe-injectable bioadhesive hydrogels from polydextran aldehyde and branched polyethylenimine. These hydrogels were effective against Gram-negative and Gram-positive bacteria. A murine infection model was used to demonstrate its antibacterial efficacy, and it was effective against *Streptococcus pyogenes* with minimal inflammatory effects (Giano et al. 2014). Mateescu et al. (2015) developed injectable gel containing antimicrobial peptide to protect implant surfaces from infection. The gels adhere strongly to implant surfaces and gingiva; they also exhibited potent antimicrobial activities against *P. gingivalis*, with reduced virulence. The gels were biocompatible, non-toxic, and were found to be suitable for prevention and management of peri-implant diseases (Mateescu et al. 2015). In another report, Salick et al. developed peptide-based, injectable hydrogel effective against methicillin-*resistant Staphylococcus aureus* (MRSA) (Salick et al. 2009). They were applied as coatings to inhibit MRSA infection on contaminated surface by killing MRSA on contact. Ji et al. (2010) prepared chitosan-based injectable gels containing chlorhexidine for periodontal treatment. In vitro drug release studies showed that chlorhexidine was released over 18 h in artificial saliva pH with excellent antibacterial inhibitory effects (Ji et al. 2010). Kaushik et al. (2015) developed an antibiotic encapsulated biomimetic nanomatrix gel for enhanced natural healing process in root canal infections. The antibiotics selected for the studies were ciprofloxacin, metronidazole, and minocycline. The injectable self-assembled biomimetic nanomatrix gel loaded with antibiotics was effective against Enterococcus faecalis and Treponema denticola and suitable for the root canal disinfection (Kaushik et al. 2015).

5.5 *Injectable Gels for Ocular Applications*

Several delivery systems are used for the administration of drugs to the eyes. They are administered subcutaneously, intravenously, orally, and by topical application. However, some of these administrations suffer from some limitations such as very low doses of the drug get to the target tissue of the eye; failure to cross the barriers, *e.g.,* cornea and blood–retina barrier, and rapid degradation of the protein or peptide (Lavik et al. 2011). Chen et al. (2011) developed in situ chitosan-based thermosensitive gel-based formulations for ocular drug delivery system. The sol–gel transition of the formulation increased with temperature. The sol-to-gel phase transition of the formulations as influenced by the concentrations of chitosan and the model drug, levocetirizine dihydrochloride. In vivo studies were performed on New Zealand albino rabbits, and the gel exhibited rapid initial drug release, followed by a sustained release mechanism. The formulations enhanced the cornea penetration of the drug with prolonged ocular residence time when compared with the free drug. The gels also exhibited effective anti-allergic conjunctivitis effects than the free drug. These results showed that the gels are promising ocular drug delivery system (Chen et al. 2011). Xi et al. (2014) prepared biodegradable thermosensitive injectable gel from poly(trimethylene carbonate)$_{15}$-F127-poly(trimethylene carbonate)$_{15}$ for ocular glaucoma filtration surgery in vivo and in vitro. The drug of choice mitomycin was loaded onto this gel, and the gels exhibited excellent sol–gel transition temperature at 37 °C. In vitro studies showed that the drug release from the gels was sustained for more than two weeks. In vivo evaluation of the gels in a rabbit glaucoma filtration surgery model revealed scar tissue formation, new collagen deposition, reduced myofibroblast generation with no significant side effects (Xi et al. 2014). Famili et al. (2014) developed injectable ocular drug delivery system from poly(*N*-isopropylacrylamide) coupled to a poly(serinol hexamethylene urea) backbone. The release kinetics of the poorly soluble corticosteroid triamcinolone acetonide from the gels was sustained over a period of one year. In vivo studies indicated that the gels were well tolerated in the rats after intravitreal injection (Famili et al. 2014). Rauck et al. (2014) studied in vivo release profile of bevacizumab from poly(ethylene glycol)-poly-(serinol hexamethylene urethane)-based injectable thermoresponsive gel administered intravitreally on New Zealand white rabbit eyes. The injectable gels were well tolerated in vivo with minimal cell death in vitro with no evidence of inflammation. The drug release was sustained for over a period of 9 weeks, and the drug concentration was maintained when compared to the free drug (Rauck et al. 2014). Xu et al. (2013) developed polysaccharide-based injectable hydrogels encapsulated with avastin for the treatment of age-related proliferative diabetic retinopathy and macular degeneration. The content of oxidized alginate in hydrogels influenced the rate of degradation of the hydrogels. The release of the drug from the gel was sustained over a period of 3 days (Xu et al. 2013). Manda et al. (2012) developed and evaluated in situ gel-forming ophthalmic drug delivery system containing moxifloxacin hydrochloride. The gels exhibited sustained release of drug for 10 h. In vivo work performed on male albino rabbit showed that the formulations were non-irritating with no significant ocular damage to the cornea, iris, or conjunctiva (Manda et al. 2012). Turturro et al. (2011) prepared poly(*N*-isopropylacrylamide)-based injectable drug delivery for delivery of drugs to the retina. The results obtained suggested that the injection of the gel resulted in a small transient effect on retinal function without long-term effects (Turturro et al. 2011). Horvat et al. (2015) designed a new formulation in situ gelling, mucoadhesive drug delivery system from thiolated poly(aspartic acid) for ocular applications. The release of sodium diclofenac from the gel was sustained over a period of 24 h (Horvat et al. 2015). Elhayek et al. (2014) evaluated the release mechanism of bevacizumab from intravitreal gel-based injectables. In vivo evaluation of the gels showed that the release profiles were sustained over a period of 4 weeks. No retinal abnormalities were observed (Elhayek et al. 2014).

5.6 Injectable Gels for Treatment of Arthritis

Arthritis is a rheumatic condition in which the symptoms are pain, stiffness, and swelling in the joints. There are different types of arthritis namely rheumatoid arthritis, osteoarthritis. It can result in death, disability, hospitalization, and some may require joint replacement (Nicoles 2015). In order to treat arthritis effectively, few researchers have studied the application of injectable gels. There are few recent research reports on the application of injectable gels for the treatment of arthritis.

Venkatesh et al. (2013) developed injectable gel formulations loaded with methotrexate for the treatment of rheumatoid arthritis. The gels were thermosensitive, thermoresponsive, and the drug release from in situ gels was sustained for 96–120 h. The release profile was greatly influenced by the nature and concentration of polymers used (Venkatesh et al. 2013). Shinde et al. (2016) developed lipid-based smart gel loaded with methotrexate for the treatment of rheumatic diseases. The gel exhibited a drug entrapment efficiency of 69.53% and was thermosensitive. It exhibited sustained release mechanism with a release of over 92.41% of the drug in 108 h. In vivo experiment proved that the gel significantly decreased joint swelling in rat model over a period of 28 days. The result confirmed that the formulation is a potential therapeutic for rheumatic conditions (Shinde et al. 2016). Jun et al. (2011) developed polysaccharide-based injectable gels containing cationic macromolecules and recombinant human IL-1Ra (anakinra) for the treatment of osteoarthritis. The gels exhibited sustained release pattern with minimal initial bursts (Jun et al. 2011). Miao et al. (2011) developed injectable thermosensitive poly(ε-caprolactone)-poly(ethylene glycol)-poly(ε-caprolactone)-based gel for intra-articular drug delivery. The gels reduced the clearance rate of methotrexate. The maximum plasma concentrations of methotrexate in rats injected were reduced when compared to the free drug. These results suggest that the gels are potential intra-articular delivery platform for treating arthritis diseases (Miao et al. 2011). Wei et al. (2013) developed injectable in situ gel from β-cyclodextrin-modified poly(ε-caprolactone)-poly(ethylene glycol)-poly(ε-caprolactone) copolymer. Indomethacin was encapsulated onto the gels. The drug-loaded gel exhibited controlled release mechanism. In vivo work on carrageenan-induced acute and complete Freund's adjuvant-induced adjuvant arthritis models confirmed the sustained therapeutic efficacy of indomethacin when encapsulated onto injectable gels for enhanced anti-inflammatory effect (Wei et al 2013). Lapasin et al. (2015) developed injectable gels for viscosupplementation. Viscosupplementation, an intra-articular symptomatic treatment of mild osteoarthritis, was prepared from cross-linked high molecular weight hyaluronan. The gels were stable, injectable, and exhibited good properties suggesting that they are suitable for viscosupplementation (Lapasin et al. 2015).

6 Injectable Gels for Cardiac Tissue Regeneration

Heart diseases are among the major causes of death worldwide. Poor regenerative ability of the cardiac tissue and the inability of cardiac tissues to tolerate ischemia limit the successful treatment of cardiac injuries (Hasan et al. 2015). Presently, the treatment of cardiac damage is by heart transplantation, surgery therapies, medications, and ventricular assist devices (Hasan et al. 2015). Heart transplant is the effective treatment and is limited by the severe shortage of organ donors. Due to the aforementioned factor, there is a need to develop injectable gels for cardiac tissue repairs. There are few reports on the application of injectable gels for cardiac tissue repairs.

Pape et al. prepared injectable gel formulation for the treatment the myocardium. The formulation was a long flexible catheter that was noninvasive and specific for the heart (Pape et al. 2015). The hydrogel's ability to switch from gel to a solution state was dependent on the environmental triggers. The gels behaved as Newtonian fluid at basic pH making it easily injectable and as a gel at physiological pH (Pape et al. 2015). Plotkin et al. (2014) developed injectable hydrogels from Tetronic-fibrinogen and PEG-fibrinogen conjugates with enhanced matrix stiffness. In vivo work on infarcts created on the left anterior descending coronary artery in a rodent model indicated increased wall thickness, arterial density, and viable cardiac tissue. Significant improvement of heart function and neovascularization was observed (Plotkin et al. 2014).

Ifkovits et al. (2010) evaluated hyaluronic acid-based injectable hydrogels for the treatment of myocardial infarction. Different formulations were prepared with varied concentration of hyaluronic acid. In vivo work was performed on clinically ovine myocardial infarction model. Injection of the hydrogels 30 min after infarction resulted in the maintenance of tissue thickness. The tissue thickness increased when hydrogel with high concentration of hyaluronic acid was used (Ifkovits et al. 2010). Wu et al. (2011) developed polyester-based temperature-sensitive injectable scaffolds for cardiac repair and stabilizing a myocardial infarct. Vascular endothelial growth factor was conjugated to the hydrogels. In vivo studies on coronary ligation in rats revealed that the gel improved the ventricular function of the rat (Wu et al. 2011). Wang et al. (2014) also developed chitosan-based injectable hydrogels loaded with adipose-derived stem cells. In vivo studies on infarcted rat hearts revealed that the hydrogels hindered adverse matrix remodeling, enhanced angiogenesis, and maintained heart function suggesting it is a potential scaffold for myocardium regeneration (Wang et al. 2014). Paul et al. developed a low-modulus methacrylated gelatin injectable from polyethylenimine functionalized GO nanosheets for myocardial therapy. It was loaded with vascular endothelial growth factor-165 pro-angiogenic gene (Paul et al. 2014). Intramyocardially administration of the gel in the peri-infarct regions of rat model with acute myocardial infarction revealed a significant increase in myocardial capillary density and reduced scar area in the infarcted hearts. The results suggested that the hydrogel is a potential system for ischemic heart diseases (Paul et al. 2014). Navaei et al. (2016) synthesized temperature-responsive poly(*N*-isopropylacrylamide)-gelatin-based injectable hydrogel for cardiac tissue engineering. In vitro evaluation showed that the gel exhibited good viscoelastic behavior, high water content and accommodated cardiac cells and cardiomyocytes. The encapsulated cells displayed a 90% survival 7 days after encapsulation (Navaei et al. 2016). Wan et al. (2014) prepared dextran-poly(e-caprolactone)-2-hydroxylethylmethacrylate-poly(*N*-isopropylacrylamide)-based hydrogel with cardio-protective effects. The administration of the gel in vivo resulted in post-infarct ventricular remodeling. The injectable containing short-hairpin RNA of angiotensin-converting enzyme was injected on rat after myocardial infarction. The hydrogel increased the duration of gene expression in vivo. Decreased ACE expression and infarct size, inhibited cell apoptosis, and improved cardiac function were significant 1 month after myocardial infarction in rats (Wan et al. 2014). In a study by Das et al., a protein, syndecan-4, was encapsulated to injectable gels for the treatment of ischemia in the diabetic diseased mice. The delivery of the syndecan-4 via injectable gels resulted in significant blood perfusion, vessel density, and tubule formation in the ischemic hind limbs of the diseased mice (Das et al. 2016). These findings suggested that injectable gels are potential delivery systems for treatment of ischemia (Das et al. 2016).

7 Application of Injectable Gels for Management of Fecal Incontinence

Fecal incontinence also known as anal incontinence is the loss of control of the bowels resulting in involuntarily loss of feces from the rectum (AmeriHealth Caritas 2015). It is a symptom of many underlying causes. It impacts negatively on daily activities and quality of life (AmeriHealth Caritas 2015; Bharucha 2015). Reports have shown that fecal incontinence can be treated with injectable gels. NASHA gels known as Solesta™ is prepared from hyaluronic acid, patented, and developed by Q-Med AB, Uppsala, Sweden. It is used for the treatment of fecal incontinence (NASHA™ 2008). One milliliter of the injectable gel consists of dextranomer microspheres (50 mg/mL) and stabilized sodium hyaluronate (15 mg/mL) in a phosphate-buffered 0.9% sodium chloride solution. It is administered by injection into the deep submucosal layer in the proximal part of the high-pressure zone of the anal canal (NASHA™ 2008; APhA DrugInfoLine [Internet]).

Graf et al. (2011) investigated the efficacy of injectable hydrogels for anal canal. Injection of dextranomer in stabilized hyaluronic acid (NASH-Dx) was used for the studies. In the randomized clinical trial, patients aged between 18 and 75 years were administered with either sham or transanal submucosal injections. There was a 50% reduction in the number of episodes of incontinence in patients administered the transanal submucosal injections than the patients administered sham treatment. Anal injection of gel formulation is an effective treatment for fecal incontinence (Graf et al. 2011). Dodi et al. (2010) also evaluated the efficacy of NASHA-Dx gel as a treatment for fecal incontinence. A 50% reduction in the number of fecal incontinence episodes occurred over a period of 12 months in patients administered the formulation (Dodi et al. 2010).

8 Other Injectable Gels for Intramuscular Administration

Administration of medications to the muscles is performed by use of injections. The medication is absorbed quickly into the bloodstream because the muscle has more blood supply. Several medications are administered intramuscularly such as vaccine, antibiotics, corticosteroids, and hormonal therapies (Shatsky 2009). There are some reports on the design of injectable gels for intramuscular administration of medications with enhanced therapeutic effects.

Jabarian et al. (2013) developed poloxamer-based injectable gels for intramuscular administration of octreotide. Octreotide is used to treat hormonal disorder that results from too much growth hormone. The formulation containing octreotide exhibited higher serum half-life, and its biological activity was retained when compared to the free octreotide. The release mechanism of octreotide from formulation was sustained, and the animal studies further confirmed the sustained release behavior of the gel (Jabarian et al. 2013).

Ahmed et al. (2016) developed poly(D,L-lactide-co-glycolide) and polyethylene glycol-based injectable gel for delivery of atorvastatin. It was prepared from poly(D, L-lactide-co-glycolide) and polyethylene glycol. Atorvastatin is a cholesterol-lowering medication. In vivo study of the gel loaded with atorvastatin by intramuscular administration on male New Zealand white rabbits revealed insignificant initial drug burst. This finding confirmed that the prolonged drug release mechanism is a potential therapeutics for patients with chronic long-term therapy (Ahmed et al. 2016).

Zhang et al. (2015) developed poloxamer-based temperature-sensitive injectable hydrogel for the delivery of hydrophilic macromolecules by intramuscular administration. The release kinetics of the gel was sustained and useful for intramuscular administration of hydrophilic therapeutics (Zhang et al. 2015).

Li et al. (2016) evaluated the pharmacokinetic release of enrofloxacin from injectable gel. Enrofloxacin, an antibiotic, is used for the treatment of bacterial infections in animals. Pharmacokinetic parameters obtained from in vivo studies on dogs administered with the gel intramuscularly showed that loading the drug to gels reduced the drug dosing frequency in long-term treatment. The enrofloxacin loaded injectable in situ gel has potential application in veterinary clinic (Li et al. 2016).

Park et al. (2016) developed thermoreversible gel for parathyroid autotransplantation from polyethyleneglycol-polyalanine-co-phenylalanine. In vivo studies were performed on Sprague–Dawley rats with excised parathyroid glands. In the rats administered with the thermoreversible gel, the parathyroid hormones level was restored. The injectable gel was found to be a good delivery system for intramuscular parathyroid autotransplantation (Park et al. 2016).

9 Conclusion and Future Perspective

Significant research and efforts have been made toward injectable gels for drug delivery and tissue engineering. Injectable gels offer several advantages such as minimal invasive delivery, do not require surgical procedures for administration, reduce pain, healing time, scarring, and reduced risk of infection. Their biomedical application depends on their design, types of materials used for their preparation, and their physicochemical properties. They have been prepared from a wide range of polymers and carbon-based materials such as graphene oxide, synthetic and natural polymers for enhanced therapeutic effects. Presently, most of the gels developed have not reached clinical trials. However, very few are presently used in clinics. There is a pressing need for these novel injectable gels to reach clinical trials. Based on the in vitro and in vivo results obtained so far, there is no doubt that in the next few years, the development of injectable gels will lead to a safer, better, and safer drug delivery and tissue engineering systems for biomedical applications.

Acknowledgements The financial assistance of Medical Research Council (Self-Initiated Research) and National Research Foundation, South Africa towards this research are hereby acknowledged. The views and opinions expressed in this manuscript are those of the author and not of MRC or NRF.

References

Ahmed TA, Alharby YA, El-Helw ARM, Hosny KM, El-Say KM (2016) Depot injectable atorvastatin biodegradable in situ gel: development, optimization, in vitro, and in vivo evaluation. Drug Des Devel Ther 10:405–415

AmeriHealth Caritas Pennsylvania Clinical Policy Title: Injectable bulking agents for fecal incontinence Clinical policy number: 08.02.04 Effective Date: October 1, 2015

Anand O, Almoazen H, Mehrotra N, Johnson J, Shukla A (2012) Controlled release of modified insulin glargine from novel biodegradable injectable gels. AAPS PharmSciTech 13:313–322

Balakrishnan B, Mohanty M, Umashankar PR, Jayakrishnan A (2005) Evaluation of an in situ forming hydrogel wound dressing based on oxidized alginate and gelatin. Biomaterials 26:6335–6342

Bell DSH (2002) Current status of diabetes treatment. South Med J, vol 95. http://www.medscape.com/viewarticle/426918_6. Accessed 24 Dec 2015

Bharucha AE, Dunivan G, Goode PS, Hamilton FA (2015) Epidemiology, pathophysiology, and classification of fecal incontinence: state of the science summary for the National Institute of Diabetes and Digestive and Kidney Diseases (NIDDK) workshop. Am J Gastroenterol 110:127–136

Bloland PB (2001). Drug resistance in malaria. WHO, Geneva

Cakmak O, Babakurban ST, Akkuzu HG, Bilgi S, Ovalı E, Kongur M, Altintas H, Yilmaz B, Bilezikçi B, Celik ZY, Yakicier MC, Sahin FI (2013) Injectable tissue-engineered cartilage using commercially available fibrin glue. Laryngoscope 123:2986–2992

Cerqueira MT, da Silva LP, Santos TC, Pirraco RP, Correlo VM, Reis RL, Marques AP (2014) Gellan gum-hyaluronic acid spongy-like hydrogels and cells from adipose tissue synergize promoting neoskin vascularization. ACS Appl Mater Interfaces 6:19668–19679

Chang J, Woon C, Pham H, Farnebo S (2014) Injectable composition for in-situ repair and regeneration of an injured ligament or tendon and methods of use. WO 2014144215 A1. 18 Sept 18 2014
Chen X, Li X, Zhou Y, Wang X, Zhang Y, Fan Y, Huang Y, Liu Y (2011) Chitosan-based thermosensitive hydrogel as a promising ocular drug delivery system: preparation, characterization, and in vivo evaluation. J Biomater Appl 27:391–402
Chen F, Yu S, Liu B, Ni Y, Yu C, Su Y, Zhu X, Yu X, Zhou Y, Yan D (2016) An injectable enzymatically crosslinked carboxymethylated pullulan/chondroitin sulfate hydrogel for cartilage tissue engineering. Sci Rep 6(20014):1–12
Cho J-K, Park JW, Song S-C (2012) Injectable and biodegradable poly(organophosphazene) gel containing silibinin: its physicochemical properties and anticancer activity. J Pharm Sci 101:2382–2391
Chou HS, Larsson M, Hsiao MH, Chen YC, Röding M, Nydén M, Liu DM (2016) Injectable insulin-lysozyme-loaded nanogels with enzymatically-controlled degradation and release for basal insulin treatment: In vitro characterization and in vivo observation. J Control Release 224:33–42
Ci T, Chen L, Yu L, Ding J (2014) Tumor regression achieved by encapsulating a moderately soluble drug into a polymeric thermogel. Sci Rep 4:5473, 13 p
Crowe CS, Chattopadhyay A, McGoldrick R, Chiou G, Pham H, Chang J (2016) Characteristics of reconstituted lyophilized tendon hydrogel. Plast Reconstr Surg 137:843–851
D'Este M, Sprecher CM, Milz S, Nehrbass D, Dresing I, Zeiter S, Alini M, Eglin D (2016) Evaluation of an injectable thermoresponsive hyaluronan hydrogel in a rabbit osteochondral defect model. J Biomed Mater Res Part A 104:1469–1478
Das S, Monteforte AJ, Singh G, Majid M, Sherman MB, Dunn AK, Baker AB (2016) Syndecan-4 enhances therapeutic angiogenesis after hind limb ischemia in mice with type 2 diabetes. Adv Healthc Mater 5:1008–1013
De France KJ, Chan KJ, Cranston ED, Hoare T (2016) Enhanced mechanical properties in cellulose nanocrystal-poly(oligoethylene glycol methacrylate) injectable nanocomposite hydrogels through control of physical and chemical cross-linking. Biomacromolecules 17:649–660
Del PS, Bianchi C, Marchetti P (2007) Beta cell function and anti-diabetic pharmacotherapy. Diabetes Metab Res Rev 23:518–527
Del Gaudio P, De Cicco F, Aquino RP, Picerno P, Russo P, Dal Piaz F, Bizzarro V, Belvedere R, Parente L, Petrella A (2015) Evaluation of in situ injectable hydrogels as controlled release device for ANXA1 derived peptide in wound healing. Carbohydr Polym 115:629–635
Dodi G, Jongen J, de la Portilla F, Raval M, Altomare DF, Lehur PA (2010) An open-label, noncomparative, multicenter study to evaluate efficacy and safety of NASHA/Dx gel as a bulking agent for the treatment of fecal incontinence. Gastroenterol Res Pract 2010:467136
Dorati R, Genta I, Conti B, Klöss H, Martin K (2012) An injectable in situ forming composite gel to guide bone regeneration: design and development of technology platform. Eur Cells Mater 24:14 p
Du L, Tong L, Jin Y, Jia J, Liu Y, Su C, Yu S, Li X (2012) A multifunctional in situ—forming hydrogel for wound healing. Wound Rep Reg 20:904–910
Ebara M, Kotsuchibashi Y, Narain R, Idota N, Kim Y-J, Hoffman JM, Uto K, Aoyagi T (2014) Smart Biomaterials (Chapter 2). National Institute for Materials Science, Japan. Published by Springer Japan
Elhayek RF, Sawhney A, Jarrett P, Guedez S, Rosales C (2014) Sustained release of bevacizumab from hydrogel depots for intravitreal injections. Investig Ophthalmol Vis Sci 55:5264
Famili A, Kahook MY, Park D (2014) A combined micelle and poly(serinol hexamethylene urea)-co-poly(nisopropylacrylamide) reverse thermal gel as an injectable ocular drug delivery system. Macromol Biosci 14:1719–1729
Fang J, Yang Z, Tan S, Tayag C, Nimni ME, Urata M, Han B (2014) Injectable gel graft for bone defect repair. Regen Med 9:41–51

Farnebo S, Woon CYL, Schmitt T, Joubert LM, Kim M, Pham H, Chang J (2014) Design and characterization of an injectable tendon hydrogel: a novel scaffold for guided tissue regeneration in the musculoskeletal system. Tissue Eng Part A 20:1550–1561

Fawaz F, Koffi A, Guyot M, Millet P (2004) Comparative in vitro–in vivo study of two quinine rectal gel formulations. Int J Pharm 280:151–162

Friedlaender GE, Strong DM, Sell KW (1984) Studies on the antigenicity of bone. II. Donor-specific anti-HLA antibodies in human recipients of freeze-dried allografts. J Bone Joint Surg Am 66:107–112

Funayama A, Niki Y, Matsumoto H, Maeno S, Yatabe T, Morioka H, Yanagimoto S, Taguchi T, Tanaka J, Toyama Y (2008) Repair of full-thickness articular cartilage defects using injectable type II collagen gel embedded with cultured chondrocytes in a rabbit model. J Orthop Sci 13:225–232

Gaharwar AK, Avery RK, Assmann A, Paul A, McKinley GH, Khademhosseini A, Olsen BD (2014) Shear-thinning nanocomposite hydrogels for the treatment of hemorrhage. ACS Nano 8:9833–9842

Gaudin K, Barbaud A, Boyer C, Langlois M-H, Lagueny A-M, Dubost J-P, Millet P, Fawaz F (2008) In vitro release and stability of an artesunate rectal gel suitable for pediatric use. Int J Pharm 353:7 p

Giano MC, Ibrahim Z, Medina SH, Sarhane KA, Christensen JM, Yamada Y, Brandacher G, Schneider JP (2014) Injectable bioadhesive hydrogels with innate antibacterial properties. Nat Commun. https://doi.org/10.1038/ncomms5095

Graf W, Mellgren A, Matzel KE, Hull T, Johansson C, Bernstein M (2011) NASHA Dx study group. Efficacy of dextranomer in stabilised hyaluronic acid for treatment of faecal incontinence: a randomised, sham-controlled trial. Lancet 377:997–1003

Griffin DR, Weaver WM, Scumpia PO, Carlo DD, Segura T (2015) Accelerated wound healing by injectable microporous gel scaffolds assembled from annealed building blocks. Nat Mater 14:737–744

Gu Z, Aimetti AA, Wang Q, Dang TT, Zhang Y, Veiseh O, Cheng H, Langer RS, Anderson DG (2013) Injectable Nano-network for glucose-mediated insulin delivery. ACS Nano 7:4194–4201

Gupta S, Singh M, Amarendar Reddy M, Yavvari PS, Srivastava A, Bajaj A (2016) Interactions governing the entrapment of anticancer drugs by low-molecular-weight hydrogelator for drug delivery applications. RSC Adv 6:19751–19757

Habal MB (1992) Different forms of bone grafts. In: Habal MB, Reddi AH (eds) Bone grafts and bone substitutes. WB Saunders, Philadelphia, pp 6–8

Hasan A, Khattab A, Islam MA, Hweij KA, Zeitouny J, Waters R, Sayegh M, Hossain M, Paul A (2015) Injectable hydrogels for cardiac tissue repair after myocardial infarction. Adv Sci 2:18 p

Hayashi K, Sakamoto W, Yogo T (2016) Smart ferrofluid with quick gel transformation in tumors for MRI-guided local magnetic thermochemotherapy. Adv Funct Mater 26:1708–1718

Hoare TR, Kohane DS (2008) Hydrogels in drug delivery: progress and challenges. Polymer 49:1993–2007

Hoemann CD, Sun J, Legare A, McKee MD, Buschmann MD (2005) Tissue engineering of cartilage using an injectable and adhesive chitosan-based cell-delivery vehicle. Osteoarthr cartilage 13:318–329

Horvát G, Gyarmati B, Berkó S, Szabó-Révész P, Szilágyi BÁ, Szilágyi A, Soós J, Sandri G, Bonferoni MC, Rossi S, Ferrari F, Caramella C, Csányi E, Budai-Szűcs M (2015) Thiolated poly(aspartic acid) as potential in situ gelling, ocular mucoadhesive drug delivery system. Eur J Pharm Sci 67:1–11

Hou Q, De Bank PA, Shakesheff KM (2004) Injectable scaffolds for tissue regeneration. J Mater Chem 14:1915–1923

Hudson SP, Langer R, Fink GR, Kohane DS (2010) Injectable in situ cross-linking hydrogels for local antifungal therapy. Biomaterials 31:1444–1452

Huynh DP, Nguyen MK, Pi BS, Kim MS, Chae SY, Lee KC, Kim SB, Kim SW, Lee DS (2008) Functionalized injectable hydrogels for controlled insulin delivery. Biomaterials 29:2527–2534

Hyun H, Park SH, Kwon DY, Khang G, Lee HB, Kim MS (2014) Thermo-responsive injectable MPEG-polyester diblock copolymers for sustained drug release. Polymers 6:2670–2683

Ifkovits JL, Tous E, Minakawa M, Morita M, Robb JD, Koomalsingh KJ, Gorman JH, Gorman RC, Burdick JA (2010) Injectable hydrogel properties influence infarct expansion and extent of postinfarction left ventricular remodeling in an ovine model. Proc Natl Acad Sci U S A 107:11507–11512

Igarashi T, Iwasaki N, Kawamura D, Kasahara Y, Tsukuda Y, Ohzawa N, Ito M, Izumisawa Y, Minami A (2012) Repair of articular cartilage defects with a novel injectable in situ forming material in a canine model. J Biomed Mater Res Part A 100A:180–187

Jabarian LE, Rouini MR, Atyabi F, Foroumadi A, Nassiri SM, Dinarvand R (2013) In vitro and in vivo evaluation of an in situ gel forming system for the delivery of PEGylated octreotide. Eur J Pharm Sci 48:87–96

Ji QX, Zhao QS, Deng J, Lu R (2010) A novel injectable chlorhexidine thermosensitive hydrogel for periodontal application: preparation, antibacterial activity and toxicity evaluation. J Mater Sci Mater Med 21:2435–2442

Jones V, Grey JE, Harding KG (2006) ABC of wound healing wound dressings. BMJ 332:777–780

Jun J-B, Kim JK, Kim T-H, Na Y-I, Choi CH, Kim Y-H (2011) Inhibition of the IL-1β-induced expression of matrix metalloproteinases by controlled release of IL-1 receptor antagonist using injectable and thermo-reversible gels in human osteoarthritis chondrocytes. J Rheum Dis 18:85–93

Kaigler D, Silva EA, Mooney DJ (2013) Guided bone regeneration using injectable vascular endothelial growth factor delivery gel. J Periodontol 84:230–238

Kalfas IH (2001) Principles of bone healing. Neurosurg Focus 10:E1. http://www.medscape.com/viewarticle/405699_7

Kaushik SN, Scoffield J, Andukuri A, Alexander GC, Walker T, Kim S, Choi SC, Brott BC, Eleazer PD, Lee J-Y, Wu H, Childers NK, Jun H-W, Park J-H, Cheon K (2015) Evaluation of ciprofloxacin and metronidazole encapsulated biomimetic nanomatrix gel on Enterococcus faecalis and Treponema denticola. Biomater Res 19:10 p

Khodaverdi E, Heidari Z, Tabassi SAS, Tafaghodi M, Alibolandi M, Sadat F, Tekie M, Khameneh B, Hadizadeh F (2015) Injectable supramolecular hydrogel from insulin-loaded triblock PCL-PEG-PCL copolymer and γ-cyclodextrin with sustained-release property. AAPS PharmSciTech 16:140–149

Kim EJ, Choi JS, Kim JS, Choi YC, Cho YW (2016) Injectable and thermosensitive soluble extracellular matrix and methylcellulose hydrogels for stem cell delivery in skin wounds. Biomacromolecules 17:4–11

Lao G, Yan L, Yang C, Zhang L, Zhang S, Zhou Y (2012) Controlled release of epidermal growth factor from hydrogels accelerates wound healing in diabetic rats. J Am Podiat Med Assoc 102:89–98

Lapasin R, Segatti F, Mercuri D, Conti GD, Spagnul C, Fusi S (2015) Rheological studies dedicated to the development of a novel injectable polymeric blend for visco supplementation treatment. Chem Biochem Eng Q 29:511–518

Lavik E, Kuehn MH, Kwon YH (2011) Novel drug delivery systems for glaucoma. Eye 25:578–586

Lee JC, Lee SY, Min HJ, Han SA, Jang J, Lee S, Seong SC, Lee MC (2012a) Synovium-derived mesenchymal stem cells encapsulated in a novel injectable gel can repair osteochondral defects in a rabbit model. Tissue Eng Part A 18:2173–2186

Lee J-H, Oh H, Baxa U, Raghavan SR, Blumenthal R (2012b) Biopolymer-connected liposome networks as injectable biomaterials capable of sustained local drug delivery. Biomacromolecules 13:3388–3394

Li X, Fan R, Tong A, Yang M, Deng J, Zhou L, Zhang X, Guo G (2015) In situ gel-forming AP-57 peptide delivery system for cutaneous wound healing. Int J Pharm 495:560–571

Li HM, Tian J, Zhang ZR, Luo XQ, Yu ZG (2016) Pharmacokinetics studies of enrofloxacin injectable in situ forming gel in dogs. J Vet Pharmacol Ther 39:144–148

Liu W, Zhan J, Su Y, Wu T, Ramakrishna S, Liao S, Mo X (2014) Injectable hydrogel incorporating with nanoyarn for bone regeneration. J Biomater Sci Polym Ed 25:168–180

López-Noriega A, Hastings CL, Ozbakir B, O'Donnell KE, O'Brien FJ, Storm G, Hennink WE, Duffy GP, Ruiz-Hernández E (2014) Hyperthermia-induced drug delivery from thermosensitive liposomes encapsulated in an injectable hydrogel for local chemotherapy. Adv Healthc Mater 3:854–859

Malafaya PB, Silva GA, Reis RL (2007) Natural–origin polymers as carriers and scaffolds for biomolecules and cell delivery in tissue engineering applications. Adv Drug Deliv Rev 59:207–233

Mamaghani PY, Kaffashi B, Salehi R, Davaran S (2015) Synthesis, characterization, and viscoelastic behavior of thermothickening poly(*N*-isopropylacrylamide-methacrylicacide-vinylpyrrolidone) nanogels as an injectable biocompatible drug carrier. Int J Polym Mater Polym Biomater 64:55–63

Manda S, Thimmasetty MKJM, Prabhushankar GL, Geetha MS (2012) Formulation and evaluation of an in situ gel-forming ophthalmic formulation of moxifloxacin hydrochloride. Int J Pharm Investig 2:78–82

Mateescu M, Baixe S, Garnier T, Jierry L, Ball V, Haikel Y, Metz-Boutigue MH, Nardin M, Schaaf P, Etienne O, Lavalle P (2015) Antibacterial peptide-based gel for prevention of medical implanted-device infection. PLoS ONE 10:e0145143

Miao B, Song C, Ma G (2011) Injectable thermosensitive hydrogels for intra-articular delivery of methotrexate. J Appl Polym Sci 122:2139–2145

Miljkovic ND, Lin YC, Cherubino M, Minteer D, Marra KG (2009) A novel injectable hydrogel in combination with a surgical sealant in a rat knee osteochondral defect model. Knee Surg Sports Traumatol Arthrosc 17:1326–1331

Mohd Zohdi R, Abu Bakar Zakaria Z, Yusof N, Mohamed Mustapha N, Abdullah MNH (2012) Gelam (Melaleuca spp.) honey-based hydrogel as burn wound dressing. Evid Based Complement Alternat Med, 7 p

Monette A, Ceccaldi C, Assaad E, Lerouge S, Lapointe R (2016) Chitosan thermogels for local expansion and delivery of tumor-specific T lymphocytes towards enhanced cancer immunotherapies. Biomaterials 75:237–249

Nagahama K, Kawano D, Oyama N, Takemoto A, Kumano T, Kawakami J (2015) Self-assembling polymer micelle/clay nanodisk/doxorubicin hybrid injectable gels for safe and efficient focal treatment of cancer. Biomacromolecules 16:880–889

NASHA™ (2008) The monograph. Bengt Ågerup, Q-Med AB, Uppsala, Sweden

Navaei A, Truong D, Heffernan J, Cutts J, Brafman D, Sirianni RW, Vernon B, Nikkhah M (2016) PNIPAAm-based biohybrid injectable hydrogel for cardiac tissue engineering. Acta Biomater 32:10–23

Navath RS, Menjoge AR, Dai H, Romero R, Kannan S, Kannan RM (2011) Injectable PAMAM dendrimer-PEG hydrogels for the treatment of genital infections: formulation, in-vitro and in-vivo evaluation. Mol Pharm 8:1209–1223

Nguyen MK, Lee DS (2010) Injectable biodegradable hydrogels. Macromol Biosci 10:563–579

Nguyen X-T, Toi VV, Nguyen T-H (2015) Development of a new injectable PVA–Ag NPs/chitosan hydrogel for wound dressing application. In: 5th International conference on biomedical engineering in Vietnam, vol 46, pp 321–324

Nishi KK, Jayakrishnan A (2007) Self-gelling primaquine-gum arabic conjugate: an injectable controlled delivery system for primaquine. Biomacromolecules 8:84–90

Oliveira JT, Gardel LS, Rada T, Martins L, Gomes ME, Reis RL (2010) Injectable gellan gum hydrogels with autologous cells for the treatment of rabbit articular cartilage defects. J Orthop Res 28:1193–1199

Orth P, Rey-Rico A, Venkatesan JK, Madry H, Cucchiarini M (2014) Current perspectives in stem cell research for knee cartilage repair. Stem Cells Cloning 7:1–17

Pape ACH, Bakker MH, Tseng CCS, Bastings MMC, Koudstaal S, Agostoni P, Chamuleau SAJ, Dankers PYW (2015) An injectable and drug-loaded supramolecular hydrogel for local catheter injection into the pig heart. J Vis Exp 100:e52450. https://doi.org/10.3791/52450

Park HS, Jung SY, Kim HY, Ko DY, Chung SM, Jeong B, Kim HS (2016) Feasibility of injectable thermoreversible gels for use in intramuscular injection of parathyroid autotransplantation. Eur Arch Otorhinolaryngol, pp 1–8. https://doi.org/10.1007/s00405-016-3990-9

Passaretti D, Silverman RP, Huang W, Kirchhoff CH, Ashiku S, Randolph MA, Yaremchuk MJ (2004) Cultured chondrocytes produce injectable tissue-engineered cartilage in hydrogel polymer. Tissue Eng 7:805–815

Patenaude M, Smeets NMB, Hoare T (2014) Designing injectable, covalently cross-linked hydrogels for biomedical applications. Macromol Rapid Commun 35:598–617

Paul A, Hasan A, Kindi HA, Gaharwar AK, Rao VTS, Nikkhah M, Shin SR, Krafft D, Dokmeci MR, Shum-Tim D, Khademhosseini A (2014) Injectable graphene oxide/hydrogel-based angiogenic gene delivery system for vasculogenesis and cardiac repair. ACS Nano 8:8050–8062

Peng Q, Sun X, Gong T, Wu CY, Zhang T, Tan J, Zhang ZR (2013) Injectable and biodegradable thermosensitive hydrogels loaded with PHBHHx nanoparticles for the sustained and controlled release of insulin. Acta Biomater 9:5063–5069

Pereira RC, Scaranari M, Castagnola P, Grandizio M, Azevedo HS, Reis RL, Cancedda R, Gentili C (2009) Novel injectable gel (system) as a vehicle for human articular chondrocytes in cartilage tissue regeneration. J Tissue Eng Regen Med 3:97–106

Plotkin M, Vaibavi SR, Rufaihah AJ, Nithya V, Wang J, Shachaf Y, Kofidis T, Seliktar D (2014) The effect of matrix stiffness of injectable hydrogels on the preservation of cardiac function after a heart attack. Biomaterials 35:1429–1438

Qi B-W, Yu A-X, Zhu S-B, Zhou M, Wu G (2013) Chitosan/poly(vinyl alcohol) hydrogel combined with Ad-hTGF-β1 transfected mesenchymal stem cells to repair rabbit articular cartilage defects. Exp Biol Med (Maywood) 238:23–30

Rauck BM, Friberg TR, Mendez CAM, Park D, Shah V, Bilonick RA, Wang Y (2014) Biocompatible reverse thermal gel sustains the release of intravitreal bevacizumab in vivo. Invest Ophthalmol Vis Sci 55:469–476

Ren Z, Wang Y, Ma S, Duan S, Yang X, Gao P, Zhang X, Cai Q (2015) Effective bone regeneration using thermosensitive poly(*N*-isopropylacrylamide) grafted gelatin as injectable carrier for bone mesenchymal stem cells. ACS Appl Mater Interfaces 7:19006–19015

Salick DA, Pochan DJ, Schneider JP (2009) Design of an injectable *b*-hairpin peptide hydrogel that kills methicillin-resistant staphylococcus aureus. Adv Mater 21:4120–4123

Sá-Lima H, Caridade SG, Mano JF, Reis RL (2010) Stimuli-responsive chitosan-starch injectable hydrogels combined with encapsulated adipose-derived stromal cells for articular cartilage regeneration. Soft Matter 6:5184–5195

Sancho-Tello M, Forriol F, Gastaldi P, Ruiz-Saurí A, de Llano JJM, Novella-Maestre E, Antolinos-Turpín CM, Gómez-Tejedor JA, Ribelles JJG, Carda C (2015) Time evolution of in vivo articular cartilage repair induced by bone marrow stimulation and scaffold implantation in rabbits. Int J Artif Organs 38:210–223

Sang L, Liu Y, Hua W, Xu K, Wang G, Zhong W, Wang L, Xu S, Xing MMQ, Qiu X (2016) Thermally sensitive conductive hydrogel using amphiphilic crosslinker self-assembled carbon nanotube to enhance neurite outgrowth and promote spinal cord regeneration. RSC Adv 6:26341–26351

Satteson ES, Molnar JA (2015). In: Elston DM (ed) Materials for wound closure. http://emedicine.medscape.com/article/1127693-overview. 13 Feb 2015. Accessed 10 Mar 2016

Schramm TK, Gislason GH, Hospet R, Rasmussen S, Jørgensen CH, Folke F, Hansen ML, Fosbøl EL, Bretler D-M, Køber L, Vaag A, Torp-Pedersen C (2008) Differences in risk of cardiovascular death according to type of oral glucose-lowering therapy in patients with diabetes: a nationwide study. Circulation 118:S1117

Shatsky M (2009) Evidence for the use of intramuscular injections in outpatient practice. Am Fam Physician 79:297–300

Shavandi A, Bekhit AEDA, Sun Z, Ali MA (2016) Injectable gel from squid pen chitosan for bone tissue engineering Applications. J Sol Gel Sci Technol 77:675–687

Shen N, Hu J, Zhang L, Zhang L, Sun Y, Xie Y, Wu S, Liu L, Gao Z (2012) Doxorubicin-loaded zein in situ gel for interstitial chemotherapy of colorectal cancer. Acta Pharm Sin B 2:610–614
Shen N, Hu J, Zhang L, Zhang L, Sun Y, Xie Y, Wu S, Liu L, Gao Z (2016) Doxorubicin-loaded zein in situ gel for interstitial chemotherapy of colorectal cancer. Biomaterials 75:237–249
Shinde CG, Pramod kumar TM, Venkatesh MP, Rajesh KS, Srivastava A, Osmani RAM, Sonawane YH (2016) Intra-articular delivery of a methotrexate loaded nanostructured lipid carrier based smart gel for effective treatment of rheumatic diseases. RSC Adv 6:12913–12924
Singh K, Kaur H, Kumar SLH (2013) Design and development of sustained release injectable in situ gel of cytarabine. Pharmacophore 4:252–267
Singh M, Kundu S, Amarendar Reddy M, Sreekanth V, Motiani RK, Sengupta S, Srivastava A, Bajaj A (2014) Injectable small molecule hydrogel as a potential nanocarrier for localized and sustained in vivo delivery of doxorubicin. Nanoscale 6:12849–12855
Smith JP, Stock E, Orenberg EK, Yu NY, Kanekal S, Brown DM (1995) Intratumoral chemotherapy with a sustained-release drug delivery system inhibits growth of human pancreatic cancer xenografts. Anticancer Drugs 6:717–726
Tan H, Marra KG (2010) Injectable, biodegradable hydrogels for tissue engineering applications. Materials 3:1746–1767
Thakur VK, Thakur MK (2014a) Recent trends in hydrogels based on psyllium polysaccharide: a review. J Clean Prod 82:1–15
Thakur VK, Thakur MK (2014b) Recent advances in graft copolymerization and applications of chitosan: a review. ACS Sustain Chem Eng 2(12):2637–2652
Thakur VK, Thakur MK (2015) Recent advances in green hydrogels from lignin: a review. Int J Biol Macromol 72:834–847
Tran NQ, Joung YK, Lih E, Park KD (2011) In situ forming and rutin-releasing chitosan hydrogels as injectable dressings for dermal wound healing. Biomacromolecules 12:2872–2880
Turturro SB, Guthrie MJ, Appel AA, Drapala PW, Brey EM, Pérez-Luna VH, Mieler WF, Kang-Mieler JJ (2011) The effects of cross-linked thermo-responsive PNIPAAm-based hydrogel injection on retinal function. Biomaterials 32:3620–3626
Venkatesh MP, Anis S, Kumar P (2013) Design and development of an injectable in methotrexate for the treatment of rheumatoid arthritis. J Drug Deliv Sci Technol 23:445–453
Vicent MJ, Greco F, Nicholson RI, Paul A, Griffiths PC, Duncan R (2005) Polymer therapeutics designed for a combination therapy of hormone-dependent cancer. Angew Chem Int Ed 44:4061–4066
Wan WG, Jiang XJ, Li XY, Zhang C, Yi X, Ren S, Zhang XZ (2014) Enhanced cardioprotective effects mediated by plasmid containing the short-hairpin RNA of angiotensin converting enzyme with a biodegradable hydrogel after myocardial infarction. J Biomed Mater Res A 102:3452–3458
Wang H, Shi J, Wang Y, Yin Y, Wang L, Liu J, Liu Z, Duan C, Zhu P, Wang C (2014) Promotion of cardiac differentiation of brown adipose derived stem cells by chitosan hydrogel for repair after myocardial infarction. Biomaterials 35:3986–3998
Wei X, Lv X, Zhao Q, Qiu L (2013) Thermosensitive β-cyclodextrin modified poly(ε-caprolactone)-poly(ethylene glycol)-poly(ε-caprolactone) micelles prolong the anti-inflammatory effect of indomethacin following local injection. Acta Biomater 9:6953–6963
WHO (2014) Global status report on non-communicable diseases 2014. WHO, Geneva
WHO Malaria Factsheet Updated January 2016 (2016) http://www.who.int/mediacentre/factsheets/fs094/en/. Accessed 18 Feb 2016
World Health Organization (2005) Malaria control in complex emergencies. WHO, Switzerland
Wu J, Zeng F, Huang XP, Chung JC, Konecny F, Weisel RD, Li RK (2011) Infarct stabilization and cardiac repair with a VEGF-conjugated, injectable hydrogel. Biomaterials 32:579–586
Wu Z, Zhang X, Guo H, Li C, Yu D (2012) An injectable and glucose-sensitive nanogel for controlled insulin release. J Mater Chem 22:22788–22796
Wu X, He C, Wu Y, Chen X (2016) Synergistic therapeutic effects of Schiff's base cross-linked injectable hydrogels for local co-delivery of metformin and 5-fluorouracil in a mouse colon carcinoma model. Biomaterials 75:148–162

Xi L, Wang T, Zhao F, Zheng Q, Li X, Luo J, Liu J, Quan D, Ge J (2014) Evaluation of an injectable thermosensitive hydrogel as drug delivery implant for ocular glaucoma surgery. PLoS ONE 9:e100632
Xu X, Weng Y, Xu L, Chen H (2013) Sustained release of Avastin® from polysaccharides cross-linked hydrogels for ocular drug delivery. Int J Biol Macromol 60:272–276
Yan Q, Xiao L-Q, Tan L, Sun W, Wu T, Chen L-W, Mei Y, Shi B (2015) Controlled release of simvastatin-loaded thermo-sensitive PLGA-PEG-PLGA hydrogel for bone tissue regeneration: in vitro and in vivo characteristics. J Biomed Mater Res Part A 103A:3580–3589
Yasmeen S, Lo ML, Bajracharya S, Roldo M (2014) Injectable scaffolds for bone regeneration. Langmuir 30:12977–12985
Ye Y, Hu X (2016) A pH-Sensitive injectable nanoparticle composite hydrogel for anticancer drug delivery. J Nanomater, Article ID 9816461, 8 p
Yu NY, Orenberg EK, Luck EE, Brown DM (1995) Antitumor effect of intratumoral administration of fluorouracil/epinephrine injectable gel in C3H mice. Cancer Chemother Pharmacol 36:27–34
Yu L, Li Y, Zhao K, Tang Y, Chang Z, Chen J, Zang Y, Wu J, Kong L, Liu S, Lei W, Wu Z (2013) A novel injectable calcium phosphate cement-bioactive glass composite for bone regeneration. PLoS ONE 8:e62570
Zdeblick TA, Ducker TB (1991) The use of freeze-dried allograft bone for anterior cervical fusions. Spine 16:726–729
Zhang K, Shi X, Lin X, Yao C, Shen L, Feng Y (2015) Poloxamer-based in situ hydrogels for controlled delivery of hydrophilic macromolecules after intramuscular injection in rats. Drug Deliv 22:375–382
Zhang J, Liu W, Gauthier O, Sourice S, Pilet P, Rethore G, Khairoun K, Bouler J-M, Tancret F, Weiss P (2016) A simple and effective approach to prepare injectable macroporous calcium phosphate cement for bone repair: syringe-foaming using a viscous hydrophilic polymeric solution. Acta Biomater 31:326–338

References (Internet Source)

APhA DrugInfoLine. http://www.aphameeting.pharmacist.com. Accessed 5 May 2016
Knee treatment. http://www.premier-ortho.com/patient-education/tendon-and-cartilage-repair/. Accessed 1 May 2016
Nicoles H (9 Dec 2015). Arthritis. http://www.medicalnewstoday.com/articles/7621.php. Accessed 12 May 2016
Synthetic wound dressing. http://www.dermnetnz.org/procedures/dressings.html. Accessed 10 Mar 2016
Type 2 Diabetes Mellitus. http://emedicine.medscape.com/article/117853-overview. Accessed 24 Dec 2015

Chapter 13
Thermo-intelligent Injectable Implants: Intricate Mechanisms and Therapeutic Applications

Pradeep Kumar, Yahya E. Choonara and Viness Pillay

Abstract The importance, influence and effectiveness of thermo-intelligent systems are evident from the fact that thermo-induced bioactive and biointeractive therapy features among the only two stimuli-responsive systems undergoing clinical trials. This chapter provides a detailed mechanistic account of the most widely used thermosensitive polymers such as chitosan-β-glycerophosphate, poly(ethylene glycol)-g-chitosan, methylcellulose, Pluronics, amphiphilic poly(ethylene glycol)/polyhydroxy acids block copolymers and poly(*N*-isopropylacrylamide). The inherent thermogelling mechanisms governing these specialized injectable implants include, but not limited to, attractive and repulsive interchain interactions, formation of dehydrated hydrophobic architectures, steep decline in the mobility of the polymeric side chains, creation of reversible crosslinking loci, creation of highly stabilized hydrophobically crosslinked networks, lattice restructuring, decrease in critical micelle concentration, advent of frictional forces between the copolymer micelles, entropy change from gauche state to antistate, increase in the volume fraction of micelles and chain collapse before aggregation followed by polymer scattering. Additionally, the chapter evaluates the behaviour and integration of injectable thermogelling systems towards the fabrication and development of therapeutic paradigms. Furthermore, the combinatorial stigmergy of multiplatform multi-responsive conjugated systems is proposed as an alternative and futuristic approach.

Keywords Thermoresponsive · Thermosensitive · Thermogels
Hydrogels · Injectable · Thermogelation mechanism · Drug delivery
Polymers · Implant

P. Kumar · Y. E. Choonara · V. Pillay (✉)
Wits Advanced Drug Delivery Platform Research Unit,
Department of Pharmacy and Pharmacology, Faculty of Health Sciences,
School of Therapeutic Sciences, University of the Witwatersrand,
Johannesburg, South Africa
e-mail: viness.pillay@wits.ac.za

© Springer Nature Singapore Pte Ltd. 2018

V. K. Thakur et al. (eds.), *Polymer Gels*, Gels Horizons: From Science to Smart Materials, https://doi.org/10.1007/978-981-10-6080-9_13

1 Introduction

The phrase "*As You like It*" appropriately explains the expediency provided by the thermosensitive, thermoresponsive or thermoreversible polymeric/biomaterial systems via their fabrication and functional flexibility in terms of (i) the design, (ii) the in situ phase transition and related temperatures and (iii) the passive targeting thereby increasing the efficacy and decreasing the toxicity of the bioactives (Bikram and West 2008). As discussed later in the chapter, by varying the (a) polymer/biomaterial composition; (b) molecular weight of the polymeric system; (c) polymer concentration; (d) type of crosslinking mechanism; (e) lower critical solution temperature; (f) length of hydrophilic and hydrophobic chains; and (g) the method of fabrication—a "thermo-intelligent drug delivery system" of choice can be customized and designed. With the potential of being stimulated both internally and externally, these versatile systems are perfect candidates for thermo-induced bioactive and biointeractive therapy such as ThermoDox—doxorubicin-loaded temperature-sensitive liposomes for breast cancer (phase II clinical trial) and hepatocellular carcinoma (phase III clinical trial) interventions and NanoTherm—iron oxide-loaded magnetic-sensitive fluid system for the treatment of glioblastoma (Mura et al. 2013). Although there are several robust reports available describing the intricate temperature-induced gelation of injectable thermogels, the scientific community is widely divided over the definite "thermogelation mechanism" governing these specialized environmentally responsive polymeric archetypes. This chapter provides the first ever exhaustive and comparative account of various thermo-induced and thermoreversible gelation mechanisms converging into their biomedical applicability.

2 Mechanism(s) of Thermo-induced and Thermoreversible Gelation

The thermogelling systems can be broadly divided into, but not limited to, natural (chitosan/glycerophosphate), synthetic (poly-*n*-isopropylacrylamide), semisynthetic (chitosan-poly(ethylene glycol); methylcellulose) and conjugated (poly(ethylene oxide)-poly(propylene oxide)-poly (ethylene oxide); poly(ethylene glycol)-poly-hydroxy acids) polymers and biomaterials. These specialized systems follow unique physico-mechanical sol–gel and gel–sol transition mechanisms inherent to their physico-chemical, material and/or crosslinker composition as detailed below under representative example(s) from each class.

2.1 Chitosan-β-Glycerophosphate

Chenite and co-workers, 2000, described for the first time, the thermo-induced gelation of a chitosan solution by the incorporation a β-glycerophosphate (βGP) solution—an approach devoid of covalent crosslinkers, organic solvents or detergents (Chenite et al. 2000). The βGP acts as a weakly basic salt capable of increasing the pH of the acidic chitosan solution to 7.0–7.4 forming an "endothermically gelling chitosan solution" allowing the controlled formation of a hydrogel at 37 °C while still maintaining the solution phase with no precipitation and gelation at 10 °C (Chenite et al. 2000, 2001; Ruel-Gariepy et al. 2000). The initial investigations revealed the involvement of interconnected (1) electrostatic repulsion within chitosan fragments; (2) electrostatic attraction between cationic-charged amine groups ($NH3^{+}$) of chitosan and anionic phosphate moiety of βGP; (3) selective hydrophobic attraction and hydrogen bonding among chitosan fragments; and (4) the water-restructuring capability of the glycerol moieties (Chenite et al. 2001). A right balance among these attractive and repulsive interchain interactions is required for the temperature-induced gelation which may translate into (1) reduced chitosan polarity leading to an increase in matrix hydrophobicity; (2) dehydration of the chitosan chain because of water-restructuring function of glycerol and hence bringing the chitosan fragments together; and (3) proton transfer from chitosan to βGP causing a reduction in charge density and chitosan-βGP attraction creating opportunities for interchain hydrophobic attraction combined with hydrogen bonding (Ruel-Gariepy et al. 2000; Chenite et al. 2001).

In a further study by Cho and co-workers, 2005, the researchers described the decrease in intrachain hydrogen bonding interactions as a reason for thermo-induced gelation in chitosan-βGP complex solution (Cho et al. 2005). Briefly, with an increase in temperature, the enclosed hydrated structures surrounding the chitosan fragments undergo enhanced vibrational and rotational energetic transition leading to a disorientation and removal of dipolar water molecules oriented in the vicinity of chitosan chains. The associating dehydrated hydrophobic architectures hence change the inherent molecular conformation of chitosan from a compact and coil-locked physical gel junctions (higher intramolecular H-bonding) at low temperatures to an unfolding 3D structure capable of creating contacts between junction zones causing desirable gelation (reduced intramolecular H-bonding and increased intermolecular H-bonding) at physiological temperatures (Cho et al. 2005; Kim et al. 2010).

2.2 Poly(Ethylene Glycol)-g-Chitosan (PEG-g-Chitosan)

Analogous to chitosan-βGP thermogelling system, poly(ethylene glycol)-g-chitosan or simply PEG-g-chitosan is one of the chitosan-based physical hydrogel obtained

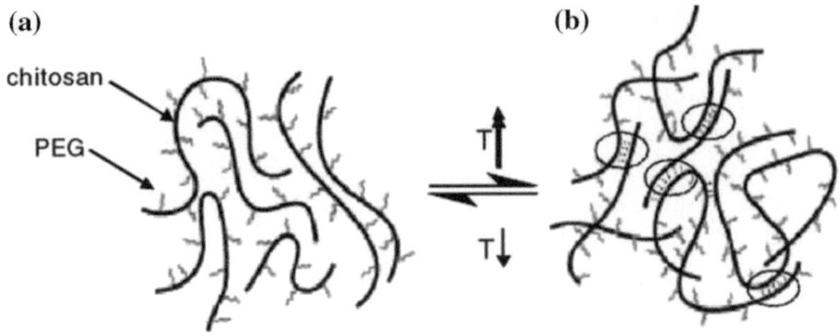

Fig. 1 A schematic representation of thermoreversible hydrogel formulation of aqueous PEG-g-chitosan solution (Bhattarai et al. 2005a). Reproduced with permission from Elsevier B.V. Ltd. © 2005

without any crosslinking agent. Although, there are more than twenty different methods of PEG-g-chitosan synthesis reported in ≈200 research studies, only few reports mentioned PEG-g-chitosan as a thermosensitive polymer forming thermogels at physiological temperatures (Casettari et al. 2012). In two closely related studies, Bhattarai and co-workers, 2005, discussed the sol–gel–sol transition of the PEG-g-chitosan copolymers as depicted in Fig. 1 (Bhattarai et al. 2005a, b).

Concisely, at low temperatures and with PEG grafting content >36%, the hydrophilic component of PEG remain in close vicinity of water molecules making the copolymer aqueous soluble. With an increase in temperature, the hydrophobic interactions dominated among the polymers chains leading to a steep decline in the mobility of the PEG side chains and thence causing the association of chitosan fragments forming a long-range stiff hydrogel network. As there are no external factors involved, such as salt or electrostatic interactions, the possibility and feasibility of forming a thermogel depend on an optimal balance between the hydrophilic/hydrophobic content of the PEG-g-chitosan (PEG grafting $\geq 45\%$ and $\leq 55\%$) (Bhattarai et al. 2005a, b). Ganji and Abdekhodaie (2008) reviewed the thermogelation of a chitosan/PEG copolymer as a function of temperature-induced decreased solubility of the ethylene glycol subgroups present in chitosan polysaccharide as well as the substituted PEG chains. The conformational conversion of ethylene glycol chains to a less polar form lead to the precipitation of the component polymers from the aqueous solutions causing a phase separation and hence hydrogel formation (Ganji and Abdekhodaie 2008).

2.3 *Methylcellulose*

Methylcellulose (MC), a cellulose derivative appearing on the United States Food and Drug Administration's Generally Regarded AS Safe (US FDA GRAS) list, is a hydrophobically modified cellulose derivative obtained by heterogeneously and partially substituting the –OH moieties at C2, C3 and C6 positions of the anhydroglucose repeat units (AGU) with—CH_3 moieties forming a hydrophilic polymer with substantial aqueous solubility, viscosity and thermoreversibility at intermediate substitution levels (Hirren et al. 1996). Although the thermosensitivity of MC was discovered way back in 1936, the scientific community is yet to reach a common conclusion over the mechanism of MC thermogelation (Arvidson et al. 2013). Apart from the very usual and initial thermo-induced dehydration-based gelation mechanism, several related and differing MC gelation mechanisms have been proposed by various researchers as listed below:

i. trimethyl glucose units exist as crystalline sequences forming "reversible crosslinking loci" forming a gel network supported by hydrogen bonds, hydrophobic bonds and dipolar interactions (Kato et al. 1978);
ii. unsubstituted methylcellulose chains pack into aggregated "bundles" via hydrophobic clustering (at low temperatures) → methyl groups are exposed to aqueous environment via the opening of bundles leading to increase in volume (at moderate temperatures) → methyl side chains lose constituent water forming a thermal hysteresis in the form of "highly stabilized hydrophobically crosslinked networks" (at higher temperatures) (Haque and Morris 1993; Fig. 2);
iii. the MC thermogelation is a two-stage process, wherein (1) hydrophobic associations between the methyl groups lead to the formation of a cluster followed by (2) the polymer–polymer and polymer–water hydrogen bonding inequity (lattice restructuring) leading to phase separation-induced gelation at high temperatures (Kobayashi et al. 1999; Bodvik et al. 2010); and
iv. formation of fibrillar structures as a consequence of "aggregation of polymer chains preferentially orientating parallel to each other to minimize water–hydrophobic group interactions" (Tanaka and Ishida 1995).

2.4 *Poly(Ethylene Oxide)-Block-Poly(Propylene Oxide)-Block-Poly(Ethylene Oxide) (PEO-PPO-PEO) Copolymers (Pluronics or Poloxamers)*

In line with the difficulties and controversies associated with other thermoresponsive polymers, (PEO-PPO-PEO) copolymers have also attracted disagreements related to the mechanism of thermogelation as explained below:

i. an inherent transformation of the intrinsic micellar properties culminating from micellar aggregation and symmetry as evidenced from decreased critical micelle concentration with an increase in temperature (Rassing and Attwood 1983);

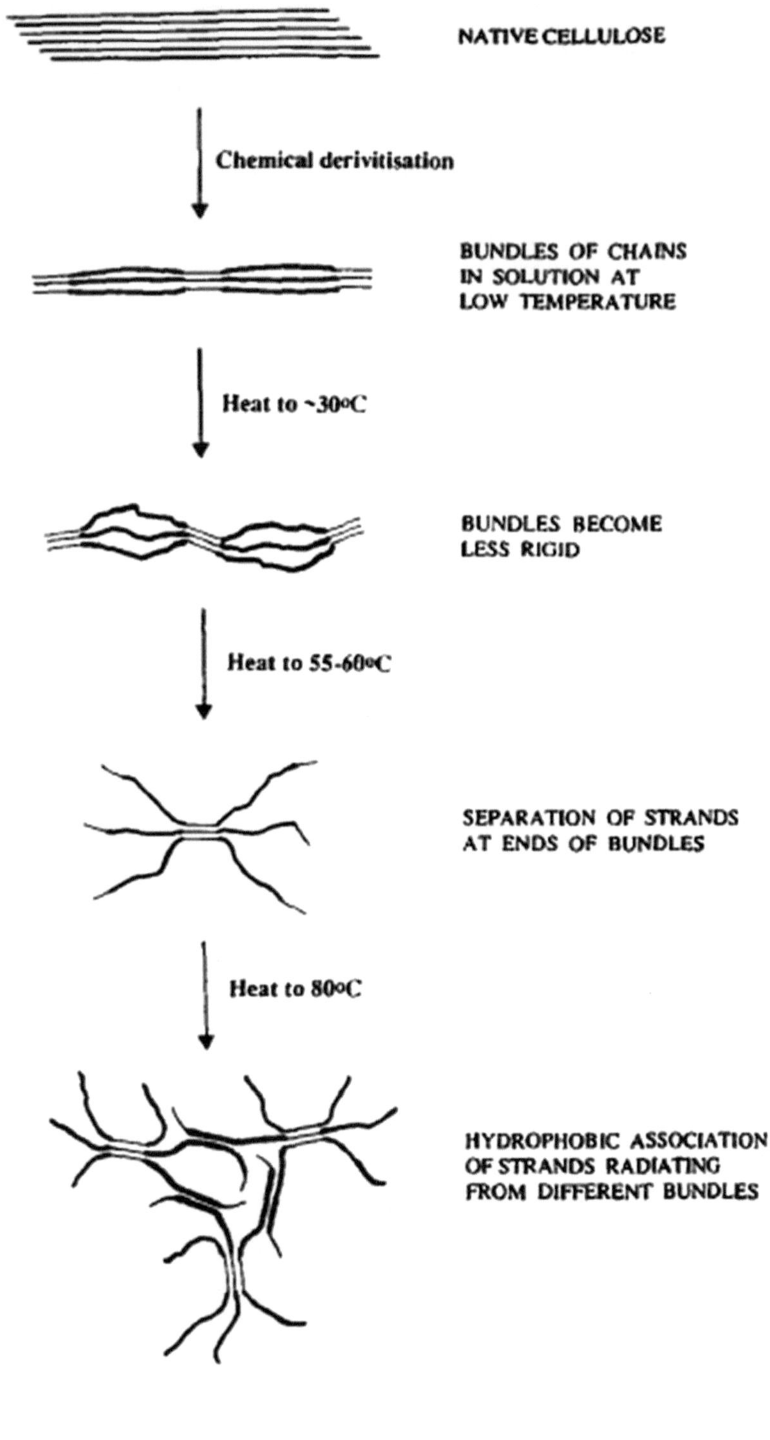
NATIVE CELLULOSE
Chemical derivitisation
BUNDLES OF CHAINS IN SOLUTION AT LOW TEMPERATURE
Heat to ~30°C
BUNDLES BECOME LESS RIGID
Heat to 55-60°C
SEPARATION OF STRANDS AT ENDS OF BUNDLES
Heat to 80°C
HYDROPHOBIC ASSOCIATION OF STRANDS RADIATING FROM DIFFERENT BUNDLES

◂**Fig. 2** Schematic illustration of the postulated structures and processes involved in the thermogelation of methocel. Faint lines denote unsubstituted or sparingly substituted chain segments; bold lines denote regions of dense substitution (Haque and Morris 1993). Reproduced with permission from Elsevier B.V. Ltd. © 1993

ii. the dehydration of the PPO components resulted in the advent of frictional forces between the copolymer micelles which further ensued with an increase in viscosity of the system forming a gel phase (Rassing 1984);
iii. an entropy change from the polar (gauche) state to nonpolar (anti) state for (O–C–C–O) groups caused by local reordering of water molecules around the core eventually driving the sol-to-gel transition (Vadnere et al. 1984); and
iv. an equilibrium exists between the copolymer micelles and the copolymer unimers above the critical micelle concentration (CMC) (at low temperature) which shifts towards micellar formation and hence an increase in volume fraction of micelles as compared to unimers (at high temperature) triggering a sol-to-gel transformation referred to as hard-sphere crystallization (Jeong et al. 2012).

2.5 Amphiphilic Poly(Ethylene Glycol)/Polyhydroxy Acids Block Copolymers

A series of PEG/polyhydroxy acids, such as poly(ethylene glycol)-polycaprolactone-poly(ethylene glycol) (PEG-PCL-PEG) and polycaprolactone-poly(ethylene glycol)-polycaprolactone-poly(ethylene glycol) (PCL-PEG-PCL) block copolymers, have been extensively studied owing to their stimuli responsive, biodegradability and biocompatibility. Figure 3 depicts a detailed account of thermogelling mechanism inherent to PCL-PEG-PCL copolymer system. At low temperature, small micelles flow freely in the aqueous solution (fully sol state) → increase in micelle size with slight increase in the temperature (favourably sol state) → rapid increase in micelle size near sol-to-gel transition temperature (favourably gel state) → aggregation and packing between micelles (denser gel formation) (Gong et al. 2009).

*2.6 Poly(*N*-Isopropylacrylamide) (PNIPAAm) Thermogels*

PNIPAAm thermogels are chemically crosslinked three-dimensional hydrogels synthesized via free radical initiation of *N*-isopropylacrylamide in either organic solutions or aqueous media. The thermogelling mechanism of PNIPAAm consists of various factors such as free energies of mixing, elasticity, osmotic effects and specific interactions for the hydrogel gel in the constituent aqueous media.

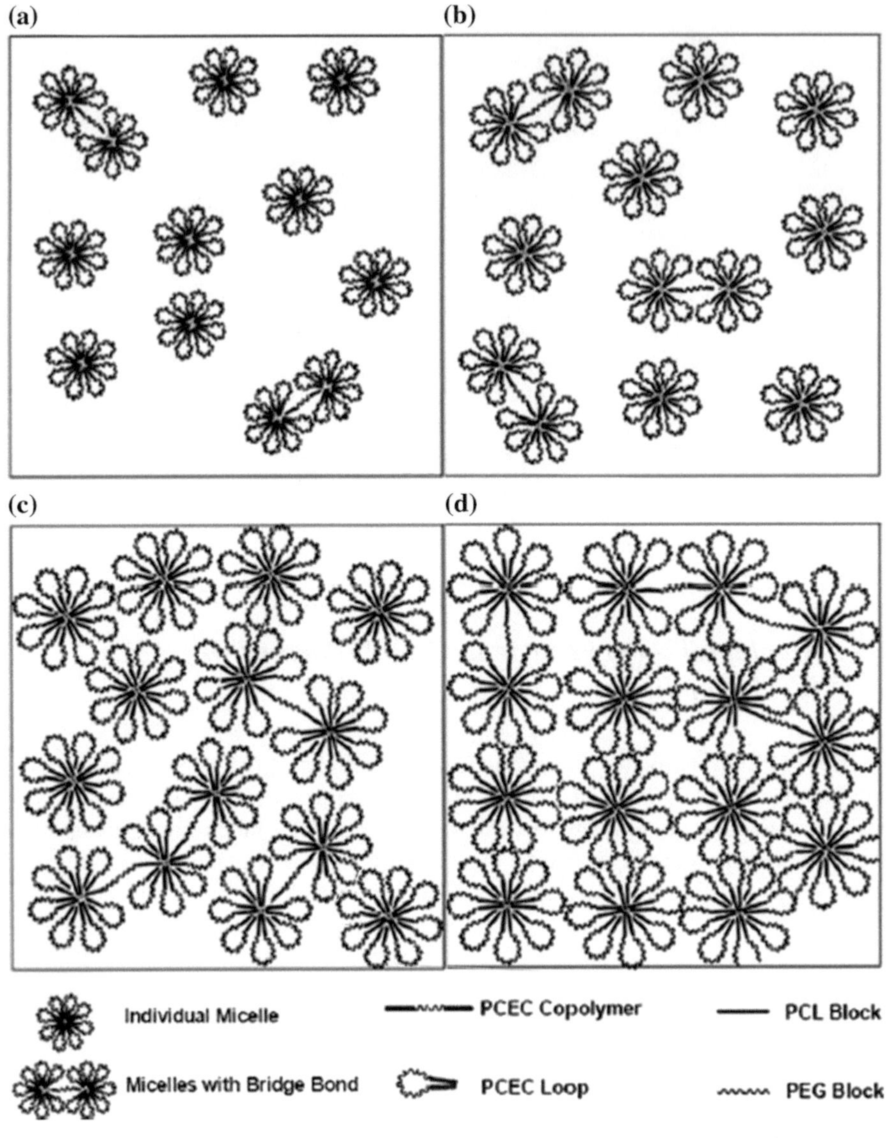

Fig. 3 A schematic diagram of the micellar aggregation mechanism for PCEC copolymer (PCL-PEG-PCL) aqueous solution as temperature increase. **a** At low temperature, micelles with small size flow freely in the aqueous solution. **b** The micelle size increases slightly as temperature increases. **c** When the hydrogel concentration is above the critical gelling concentration (CGC), further increasing the temperature to around the sol–gel transition temperature causes the micelle size to increase rapidly, resulting in sol–gel transition. **d** The aggregation and packing between micelles increase with increasing temperature to form a denser gel (Gong et al. 2009). Reproduced with permission from Elsevier B.V. Ltd. © 2009

Conclusively, below the sol-to-gel transition temperature, PNIPAAm exists as isolate and flexible coils extended across the aqueous medium. At and after reaching the sol-to-gel transition temperature, these extended isolated chains collapse before being aggregated followed by polymer scattering and finally the formation of a low polarity polymer network suspended in a relatively water-rich phase (Schild 1992).

3 Thermo-intelligent Drug Delivery Systems (TiDDS)

The substantial difference between the storage temperatures of drug delivery systems (refrigerated ≈ 2–8 °C; room temperature ≈ 20 °C) and the physiological temperature of human body (normal temperature ≈ 37 °C; fever up to ≈ 43 °C; inflammation > 37 °C; tumour cell temperature ≈ 39 °C) forms the basis and motivation for designing intelligent thermosensitive drug delivery systems. Another important aspect of designing TiDDS is their ability to form bioactive-loaded hydrogel solid implants capable of releasing the bioactives over extended periods depending on the intrinsic or extrinsic temperature stimuli. TiDDS can be broadly divided into single platform (1) hydrogel-based sol-to-gel systems; and (2) conjugated polymers-derived particulate systems, as discussed henceforth in detail.

3.1 Hydrogel-Based Sol-to-Gel TiDDS

The polysaccharide hydrogel-based systems such as chitosan/PEG, chitosan/βGP, and methylcellulose as well as acrylate copolymer-based systems such as poly(*N*-isopropyl acrylamide) have been employed to deliver various bioactives either via direct incorporation of bioactives onto the hydrogels or by incorporating bioactive-loaded polymeric particles into the said hydrogels.

Chitosan/βGP hydrogels have been extensively explored for the delivery of various anti-cancer drugs such as ellagic acid and docetaxel with a localized treatment potential for solid tumours in particular. The effect of ionic drugs on the performance of chitosan/βGP hydrogels is due to the fact that these hydrogels operate via a hydrophobic mechanism rather than a more common ionic crosslinking as observed in composite films and an excess of ionic charges can thus influence the porous structure of the polymer matrix via complexation (Kim et al. 2010). However, loading neutral poorly water-soluble drug docetaxel into chitosan/βGP hydrogels provided with a controlled delivery of the drug for 3 weeks when administered intratumorally. Interestingly, apart from prolonging the mean retention time (↑MRT), the thermosensitivity of the delivery system increased the concentration of the drug in the tumour (↑AUC_{tumor}) along with a very low distribution index in the normal tissues (↓DI_{normal}) thereby confirming the localized confinement of TiDDS with increased efficiency and efficacy (Li et al. 2014).

Aliaghaie and co-workers, 2012, while carrying out a mechanistic rheological study to elucidate the mechanism of thermogelation of chitosan/βGP hydrogels, concluded that the molecular weight of chitosan and meglumine antimoniate (MGA)-loaded microparticle incorporation altered the gelation temperature of TiDDS providing a prolonged release of MGA with significantly reduced burst release of the drug perfect for intralesional administration of MGA (Aliaghaie et al. 2012). In line with the above study, Peng and co-workers, 2013, incorporated insulin phospholipid complex-loaded (poly(3-hydroxybutyrate-co-3-hydroxyhexanoate)) (PHBHHx) nanoparticles in chitosan/βGP hydrogel and concluded that ≈ 19% of total insulin was released from the nanoparticle-loaded hydrogel (NP-CS/GP). The thermosensitive characteristic of the chitosan/βGP hydrogel provided a porous implant architecture with regulated degradation profile capable of acting as a barrier for the insulin release delineating the customized physiological function of insulin (Fig. 4) (Peng et al. 2013). Another TiDDS based on chitosan, PEG-g-chitosan, has shown potential towards injectable implantable drug delivery of proteins and cyclosporine A producing effective plasma concentrations at par with chitosan/βGP hydrogels for more than five weeks in vivo. The thermogelling capability of the PEG-g-chitosan led to the formation of a robust implant exhibiting no burst release (Jiang et al. 2014). From the above discussion, it can be concluded that the release of bioactives—hydrophobic drugs, ionic drugs and peptides—can be controlled and sustained via custom-designed TiDDS applying variable approaches ranging from molecular weight of chitosan through βGP concentration to nano/microparticle incorporation.

Rationally and practically, the hydrogel-based TiDDS are emerging as the approach of choice among the drug delivery fraternity, especially concerning

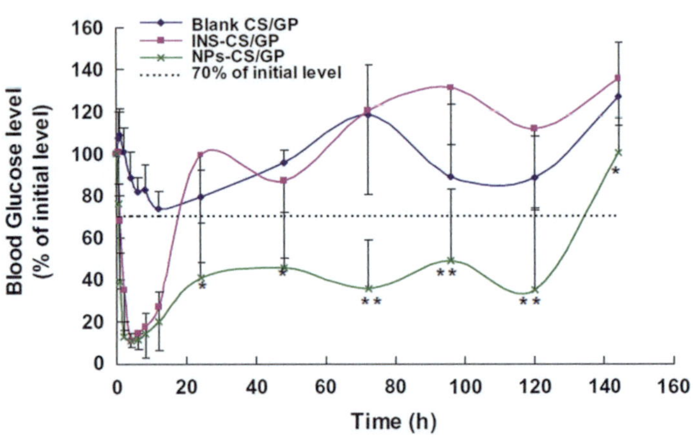

Fig. 4 Blood glucose level–time curve after s.c. injection of blank CS/GP (filled diamond), INS-CS/GP at a dose of 4 IU kg^{-1} (filled square) and NP-CS/GP at a dose of 6 IU kg^{-1} (×) to male diabetic SD rats (Peng et al. 2013). Reproduced with permission from Elsevier B.V. Ltd. © 2013

peptide and protein delivery, as the operating temperatures for the fabrication of the hydrogel system range from 4 to 20 °C—a temperature range where the peptides and proteins retain their stability. Additionally, the use of buffered aqueous-based systems adds to the advantage of using TiDDS for biological therapeutics. Akash and co-workers, 2012, employed Pluronic F127 (PF127)-based TiDDS for the delayed and prolonged delivery of interleukin-1 receptor antagonist (IL-1Ra) for the potential treatment of Type-2 diabetes mellitus via prolongation of the hypoglycemic effect and the inhibition of IL-1β-stimulated IL-6 (Akash et al. 2012). In a recent study, Payyappilly and co-workers, 2014, compared the mechanical effectiveness of using PEG-PCL-PEG hydrogels as compared to PF127 hydrogels and concluded that PEG-PCL-PEG hydrogels are capable of providing thermoresponsive pulsatile release of insulin similar to PF127 hydrogels while maintaining the first-order Fickian diffusion and the preserving the secondary structure of insulin (Payyappilly et al. 2014).

Although majority of the TiDDS research is focused on injectable and implantable drug delivery applications, their applicability in bioactive delivery to mucosal cavities cannot be ignored. In particular, the thermogelling systems offer great potential towards cavities, such as vaginal and nasal cavities, with a possibility of leakage of the delivered content under the influence of gravity. Once applied, the thermogelling systems can rapidly cover and take the shape of mucosal lining, thereby forming a robust long-lasting localized hydrogel application (Li et al. 2012). Apart from providing localized delivery of bioactives; TiDDS have demonstrate capability of acting as a satellite platform for delivering bioactives to remote site such as a "nasal platform" to deliver and target curcumin to "brain tissue". A PF127- and Poloxamer 188-based nasal hydrogel developed by Chen and co-workers, 2013, exhibited shorter gelation time, longer mucociliary transport time and prolonged curcumin retention in the nasal cavity up to 14 days after single application, thereby providing enhanced brain targeting of curcumin to the cerebrum, cerebellum, hippocampus and olfactory bulb as compared to i.v. administration (Chen et al. 2013).

Highlighting the further functionality of the TiDDS towards cavities; thermogelling systems provide innovative strategies for the "processing and templating of bioactive-loaded polymeric neural architectures" capable of bridging the neuronal gap inherent to traumatic spinal cord injuries. In addition to providing support and microenvironment for the proliferation of neural cells, meticulous combination of methylcellulose with hyaluronic acid or agarose can be employed as an intrathecally injectable scaffold for the incorporation and delivery of various bioactives (such as nimodipine and erythropoietin) to an injured spinal cord, thereby covering the five R's of spinal cord injury intervention—restriction, repair, regeneration, restoration and reorganization—alone or in combination with nanoparticulate strategies, as reviewed elsewhere by the authors (Kumar et al. 2014, 2015).

Multiplatform polymeric blend systems form a subclass of hydrogel-based TiDDS, wherein two thermosensitive polymers are mixed together to form a hydrogel–micelle system such as low molecular weight methylcellulose-based thermoreversible gel/Pluronic micelle combination system (a gel/micelle combination) with a capability to encapsulating a hydrophobic drug (e.g. docetaxel) in the micelle system and the hydrogel capable of providing a carrier system for localized drug delivery (e.g. to tumour sites) and hence overcoming the drawbacks of both the gel and the micelle system individually (Kim et al. 2012).

3.2 Conjugated Polymers-Derived Particulate TiDDS

Thermosensitive copolymer micelles provide a unique platform for the delivery of bioactives with specific advantages including (1) the formation of a core–shell architecture to enhance the solubility of hydrophobic drugs architecture; (2) modification of related pharmacokinetics; (3) extended circulation and stay in the plasma; and (4) site-specific accumulation and release of the drugs (Yang et al. 2014). In an exemplary study, Nakayama and co-workers, 2006, described the synthesis and performance of a range of amphiphilic thermoresponsive micelles with identical hydrophilic polymer chain (poly(*N*-isopropylacrylamide-co-*N*,*N*-dimethylacrylamide)) and varying hydrophobic fragments ((poly(D,L-lactide), poly(ε-caprolactone) or poly(D,L-lactide-co-ε-caprolactone)) leading to controlled and customized thermoresponsive release of doxorubicin as well as thermoresponsive modulation of the hydrophobicity of the micelle core (Fig. 5). The researchers observed no enhancement of the drug release when the temperature was increased from 37 °C (below the lower critical solution temperature (LCST)) to 42.5 °C (above LCST) for P(IPAAmco-DMAAm)-b-PCL, while P(IPAAm-co-DMAAm)-b-poly(LA-co-CL) micelles showed a significant increase in doxorubicin release rate when temperature was increased above the LCST (Nakayama et al. 2006).

In line with the above study, Yang and co-workers, 2014, studied the thermo-induced micellization behaviour of thermosensitive mPEG-b-PA-g-PNIPAM comb block copolymer consisting of methoxy poly(ethylene glycol) (mPEG) blocks and hydrophobic polyacrylate (PA) backbones with thermosensitive poly(*N*-isopropylacrylamide) (PNIPAM) graft chains. The researchers concluded that the block copolymers were capable of continuous CPT release with minimal burst release and the CPT release was accelerated once the environmental temperature was raised above LCST (40–44.5 °C) confirming the thermo-induced controlled drug delivery (Yang et al. 2014). Table 1 presents a brief account of various conjugated polymers-derived particulate systems used for biomedical applications. The method of fabrication of the micellar system further dictates the performance of TiDDS. Wang and co-workers, 2012, designed and functionalized poly(e-caprolactone) (PCL)-poly(ethylene glycol)-poly(e-caprolactone) amphiphilic copolymeric micelles employing two different methods as described below and in Fig. 6:

Formulation 1: the copolymer solution was first mixed with paclitaxel at room temperature, and then the gel was allowed to form at 37 °C; and

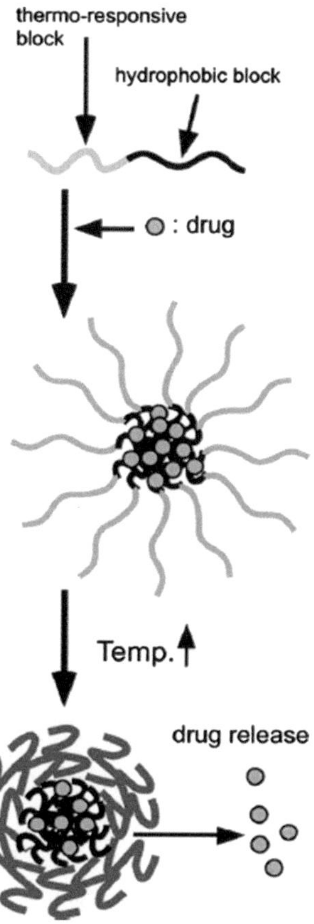

Fig. 5 Mechanism of thermoresponsive polymeric micelle for selective drug release when the temperature rises. The formation of a thermoresponsive polymeric micelle results from the assembly of block copolymers composed of a thermoresponsive block and a hydrophobic block. Drug molecules are incorporated into the hydrophobic inner core, and the thermoresponsive character is possessed by the outer shell of the polymeric micelles. When the temperature is below the phase transition temperature of the thermoresponsive block, micelles are formed with the hydrated outer shell and the hydrophobic inner core. When the temperature exceeds the phase transition temperature, the outer shell shrinks and becomes hydrophobic. Intermicellar aggregates may form on the basis of both a micelle concentration and the strength of the shrunken outer shell layer's hydrophobic interactions. A rise in temperature may enhance the drug release (Nakayama et al. 2006). Reproduced with permission from Elsevier B.V. Ltd. © 2006

Table 1 Representative examples of recent conjugate polymeric particulate TiDDS

Thermogelling system	Drug incorporated	Responsive temperature	Salient drug delivery features	Ref.
Thiolated Pluronic decorated magnetic nanoparticles	Doxorubicin	37 °C	Doxorubicin was incorporated to the magnetic nanoparticles via thermoresponsive manners; higher the amount of Pluronic moieties → higher the drug-loading efficiency	Park et al. (2012)
Superhydrophobically modified photo-crosslinked dextran-methacrylated/poly(*N*-isopropylacrylamide) hydrogels particles	Insulin; Bovine serum albumin (BSA)	37 °C	The particles exhibited temperature-sensitive swelling, porosity and protein release rate, with the responsiveness tunable by the dextran-MA/PNIPAAm weight ratio	Lima et al. (2011)
PEGylated-PNIPAm microgels	BSA	38–45 °C	The molecular weight and concentration of PEG chains controlled the morphology of the microgels and influenced the temperature response and biocompatibility	Trongsatitkul and Budhlall (2013)
Poly(*N*-isopropylacrylamide-co-*N*,*N*-dimethylacrylamide-b-lactide) and poly(*N*-isopropylacrylamide-co-*N*,*N*-dimethylacrylamide-b-e-caprolactone) diblock copolymers micelles	Adriamycin	40–42 °C	↑ intracellular uptake of adriamycin above 40 °C (×4 than commercial formulation); ↑ cytotoxicity; ↑ cell surviving at 42 °C due to the intermicellar aggregation; thermoresponsive and pH-responsive drug release	Li et al. (2011)
Self-crosslinked poly(*N*-isopropylacrylamide-co-methacrylic acid) semi-hollow spheres	Doxorubicin (DOX)	43 °C	DOX conjugated to the carboxylic groups on the core; DOX release controlled by temperature; shell thickness; and disulphide bonds degradation by glutathione.	Qian and Wu (2013)
Poly(ε-caprolactone)-poly(ethylene glycol)-poly(ε-caprolactone) triblock copolymer nanomicelles	Paclitaxel (PTX)	37 °C	High drug-loading content and efficiency; core–shell morphology; longer systemic circulation time; slower plasma-elimination rate; ↑ inhibition of EMT6 breast tumour growth; ↓ toxicity of PTX	Zhang et al. (2012)
β-cyclodextrin-modified poly (ε-caprolactone)-poly(ethylene glycol)-poly (ε-caprolactone) micelles	Indomethacin (IND)	37–40 °C	↑ Drug encapsulating capacity by introducing β-CD to PCEC copolymer; ↑ anti-inflammatory effect of IND in rats; sustained therapeutic efficacy	Wei et al. (2013)

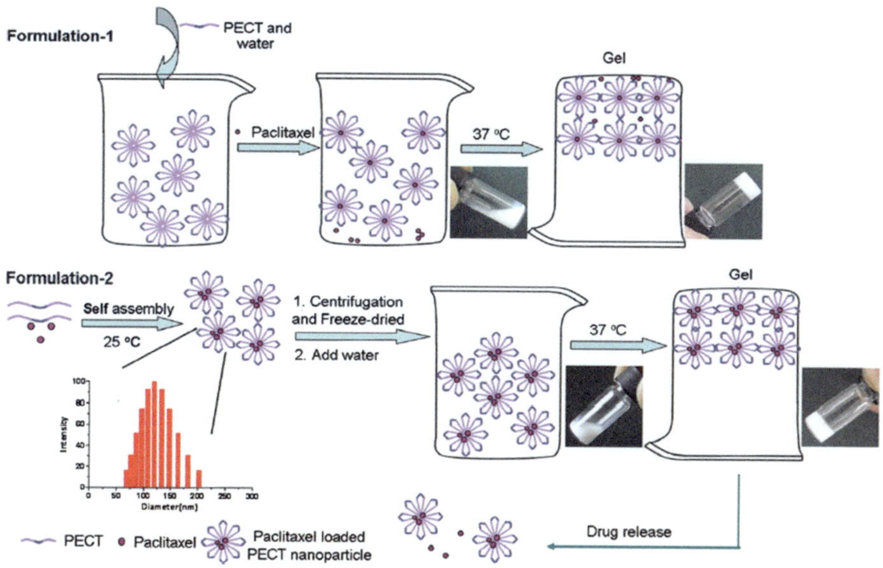

Fig. 6 A schematic diagram illustrating the preparation of the different types of paclitaxel formulation and paclitaxel release from the formulations (Wang et al. 2012). Reproduced with permission from Elsevier B.V. Ltd. © 2012

Formulation 2: first paclitaxel-containing nanoparticles were prepared followed by lyophilisation to form a powder. The final Formulation 2 was then reconstituted by dissolving the above nanoparticle powder in water and the gel was then allowed to form at 37 °C (Wang et al. 2012).

4 Conclusion and Future Perspective

This chapter evaluates the compositional and conformational changes occurring within the complex thermo-influenced polymeric architectures being applied as "injectable thermogel implants" for various biomedical applications. These single platform thermo-intelligent drug delivery systems can be divided into composite hydrogel-based sol-to-gel systems and conjugated polymers-derived particulate systems, wherein their design and composition is determined by the therapeutic application, site of injection or implantation and the physico-chemical properties of the incorporated bioactive (50). Considering the aforementioned thermogelling mechanisms, it can be concluded that the physico-chemical and physico-mechanical changes occurring in situ are very complex as well as abrupt making it very challenging for the researchers to devise precise mechanism governing the performance of the thermogels. To overcome this “thermogel mystery”, we hereby propose three classical approaches that might assist future researchers to elucidate thermoresponsive mechanisms:

(1) *Molecular modelling*: It would be of considerable interest to computationally analyse the polymeric architectures employing molecular mechanics and dynamics simulations in vacuum and solvent conditions for the interpretation and understanding of the mechanisms that control the formation, design, complexation, interaction and stereospecificity inherent in injectable thermogels.
(2) *Addition of a third component to generate mechanistic insights*: Incorporation of a third component to the thermogelling system (thermoresponsive polymer + aqueous medium) may provide a concentrate on dependent interaction and complexation profile that can be studied through physico-chemical and physico-mechanical characterization of the complex network.
(3) *Real-time monitoring of the gelation using a marker*: It is well conceived that during thermogelation, a certain amount of aqueous medium is released or absorbed during an increase or and decrease in environmental temperature, respectively. The movement of these solvent molecules through the network system can be traced employing specialized marker molecules capable of being monitored in real time, thereby providing important insights into the release of bioactives from these specialized archetypes.

5 Executive Summary

- The existing research and review reports w.r.t. the gelation mechanism of various thermosensitive systems lack a common consensus, and a better theoretical–experimental protocol is urgently due for the same.
- Multiplatform gel/micelle combination has the potential of being TiDDS of choice as both the “hydrophobic drug encapsulation” as well as the “localized carrier” requirements can be met using these microgel/hydrogel systems.
- The drug loading in and release pattern from injectable thermogels can be effectively and efficiently controlled and customized by varying the formulation and environmental factors.
- Thermo-intelligent drug delivery systems are emerging as the approach of choice for peptide and protein delivery as the low operating temperatures for the fabrication of the hydrogel systems can help retain and sustain the stability of these bioactives.

Financial and Competing Interests Disclosure The authors have no relevant affiliations or financial involvement with any organization or entity with a financial interest in or financial conflict with the subject matter or materials discussed in the manuscript. This includes employment, consultancies, honoraria, stock ownership or options, expert testimony, grants or patents received or pending or royalties. No writing assistance was utilized in the production of this manuscript.

References

Akash MSH, Rehman K, Li N, Gao JQ, Sun H, Chen S (2012) Sustained delivery of IL-1Ra from pluronic F127-based thermosensitive gel prolongs its therapeutic potentials. Pharm Res 29:3475–3485

Aliaghaie M, Mirzadeh H, Dashtimoghadam E, Taranejoo S (2012) Investigation of gelation mechanism of an injectable hydrogel based on chitosan by rheological measurements for a drug delivery application. Soft Matter 8:7128–7137

Arvidson SA, Lott JR, McAllister JW, Zhang J, Bates FS, Lodge TP (2013) Interplay of phase separation and thermoreversible gelation in aqueous methylcellulose solutions. Macromolecules 46:300–309

Bhattarai N, Matsen FA, Zhang M (2005a) PEG-grafted chitosan as an injectable thermoreversible hydrogel. Macromol Biosci 5:107–111

Bhattarai N, Ramay HR, Gunn J, Matsen FA, Zhang M (2005b) PEG-grafted chitosan as an injectable thermosensitive hydrogel for sustained protein release. J Control Rel 103:609–624

Bikram M, West JL (2008) Thermo-responsive systems for controlled drug delivery. Expert Opin Drug Deliv 5(10):1077–1091

Bodvik R, Dedinaite A, Karlson L, Bergstrom M, Baverback P, Pedersen JS, Edward K, Karlsson G, Varga I, Claesson PM (2010) Aggregation and network formation of aqueous methylcellulose and hydroxypropylmethylcellulose solutions. Colloid Surf A: Physicochem Eng Aspects 354:162–171

Casettari L, Vllasaliu D, Castagnino E, Stolnik S, Howdle S, Illum L (2012) PEGylated chitosan derivatives: synthesis, characterizations and pharmaceutical applications. Prog Polym Sci 37:659–685

Chen X, Zhi F, Jia X, Zhang X, Ambardekar R, Meng Z, Paradkar AR, Hu Y, Yang Y (2013) Enhanced brain targeting of curcumin by intranasal administration of a thermosensitive poloxamer hydrogel. J Pharm Pharmacol 65:807–816

Chenite A, Buschmann M, Wang D, Chaput C, Kandani N (2001) Rheological characterization of thermogelling chitosan/glycerol-phosphate solutions. Carbohydr Polym 46:39–47

Chenite A, Chaput C, Wang D, Combes C, Buschmann MD, Hoemann CD, Leroux JC, Atkinson BL, Binette F, Selmani A (2000) Novel injectable neutral solutions of chitosan form biodegradable gels in situ. Biomaterials 21:2155–2161

Cho J, Heuzey MC, Begin A, Carreau PJ (2005) Physical gelation of chitosan in the presence of β-glycerophosphate: the effect of temperature. Biomacromolecules 6:3267–3275

Ganji F, Abdekhodaie MJ (2008) Synthesis and characterization of a new thermosensitive chitosan–PEG diblock copolymer. Carbohydr Polym 74:435–441

Gong CY, Shi S, Wu L, Yang B, Qi XR, Guo G, Gu YC, Zhao X, Wei YQ, Qian ZY (2009) Biodegradable in situ gel-forming controlled drug delivery system based on thermosensitive PCL–PEG–PCL hydrogel. Part 2: sol–gel–sol transition and drug delivery behaviour. Acta Biomater 5:3358–3370

Haque A, Morris ER (1993) Thermogelation of methylcellulose. Part I: molecular structures and processes. Carbohydr Polym 22:161–173

Hirren M, Desbrières J, Rinaudo M (1996) Physical properties of methylcelluloses in relation with the conditions for cellulose modification. Carbohydr Polym 31:243–252

Jeong B, Kim SW, Bae YH (2012) Thermosensitive sol–gel reversible hydrogels. Adv Drug Deliv Rev 64:154–162

Jiang G, Sun J, Ding F (2014) PEG-g-chitosan thermosensitive hydrogel for implant drug delivery: cytotoxicity, in vivo degradation and drug release. J Biomater Sci Polym Ed 25(3):241–256

Kato T, Yokoyama M, Takahashi A (1978) Melting temperatures of thermally reversible gels IV. Methyl cellulose-water gels. Colloid Polym Sci 256:15–21

Kim S, Nishimoto SK, Bumgardner JD, Haggard WO, Gaber MW, Yang Y (2010) A chitosan/β-glycerophosphate thermo-sensitive gel for the delivery of ellagic acid for the treatment of brain cancer. Biomaterials 31:4157–4166

Kim JK, Won YW, Lim KS, Kim YH (2012) Low-molecular-weight methylcellulose-based thermo-reversible gel/pluronic micelle combination system for local and sustained docetaxel delivery. Pharm Res 29:525–534

Kobayashi K, Huang C, Lodge TP (1999) Thermoreversible gelation of aqueous methylcellulose solutions. Macromolecules 32:7070–7077

Kumar P, Choonara YE, Modi G, Naidoo D, Pillay V (2014) Nanoparticulate strategies for the 5 R's of traumatic spinal cord injury intervention: restriction, repair, regeneration, restoration and reorganization. Nanomedicine 9(2):331–348

Kumar P, Choonara YE, Modi G, Naidoo D, Pillay V (2015) Multifunctional therapeutic delivery strategies for effective neuro-regeneration following traumatic spinal cord injury. Curr Pharm Des 21:1517–1528

Li W, Li J, Gao J, Li B, Xia Y, Meng Y, Yu Y, Chen H, Dai J, Wang H, Guo Y (2011) The fine-tuning of thermosensitive and degradable polymer micelles for enhancing intracellular uptake and drug release in tumors. Biomaterials 32:3832–3844

Li N, Yu M, Deng L, Yang J, Dong A (2012) Thermosensitive hydrogel of hydrophobically-modified methylcellulose for intravaginal drug delivery. J Mater Sci: Mater Med 23:1913–1919

Li C, Ren S, Dai Y, Tian F, Wang X, Zhou S, Deng S, Liu Q, Zhao J, Chen X (2014) Efficacy, pharmacokinetics, and biodistribution of thermosensitive chitosan/β-glycerophosphate hydrogel loaded with docetaxel. AAPS Pharm Sci Tech 15(2):417–424

Lima AC, Song W, Blanco-Fernandez B, Alvarez-Lorenzo C, Mano JF (2011) Synthesis of temperature-responsive dextran-ma/pnipaam particles for controlled drug delivery using superhydrophobic surfaces. Pharm Res 28:1294–1305

Mura S, Nicolas J, Couvreur P (2013) Stimuli-responsive nanocarriers for drug delivery. Nat Mater 12:991–1003

Nakayama M, Okano T, Miyazaki T, Kohori F, Sakai K, Yokoyama M (2006) Molecular design of biodegradable polymeric micelles for temperature-responsive drug release. J Control Rel 115:46–56

Park S, Kim HS, Kim WJ, Yoo HS (2012) Pluronic@Fe_3O_4 nanoparticles with robust incorporation of doxorubicin by thermo-responsiveness. Int J Pharm 424:107–114

Payyappilly S, Dhara S, Chattopadhyay S (2014) Thermoresponsive biodegradable PEG-PCL-PEG based injectable hydrogel for pulsatile insulin delivery. J Biomed Mater Res Part A 102(5):1500–1509

Peng Q, Sun X, Gong T, Wu CY, Zhang T, Tan J, Zhang ZR (2013) Injectable and biodegradable thermosensitive hydrogels loaded with PHBHHx nanoparticles for the sustained and controlled release of insulin. Acta Biomater 9:5063–5069

Qian J, Wu F (2013) Thermosensitive PNIPAM semi-hollow spheres for controlled drug release. J Mater Chem B 1:3464–3469

Rassing J, Attwood D (1983) Ultrasound velocity and light scattering studies on the polyoxyethylene–polyoxypropylene copolymer Pluronic F127 in aqueous solution. Int J Pharm 13:47–55

Rassing J, Mckenna W, Bandyopadhyay S, Eyring EM (1984) Ultrasonic and 13C-NMR studies on gel formation in aqueous solution of ABA block copolymeric Pluronic F-127. J Mol Liq 27:165–178

Ruel-Gariepy E, Chenite A, Chaput C, Guirguis S, Leroux J (2000) Characterization of thermosensitive chitosan gels for the sustained delivery of drugs. Int J Pharm 203:89–98

Schild HG (1992) Poly(N-isopropylacrylamide): experiment, theory and application. Prog Polym Sci 17:163–249

Tanaka F, Ishida M (1995) Thermoreversible gelation of hydrated polymers. J Chem Soc Faraday Trans 91(16):2663–2670

Trongsatitkul T, Budhlall BM (2013) Temperature dependence of serum protein adsorption in PEGylated PNIPAm microgels. Colloid Surf B: Biointerfaces 103:244–252

Vadnere M, Amidon G, Lindenbaum S, Haslam JL (1984) Thermodynamic studies on the gel–sol transition of some Pluronic polyol. Int J Pharm 22:207–218

Wang W, Deng L, Liu S, Li X, Zhao X, Hu R, Zhang J, Han H, Dong A (2012) Adjustable degradation and drug release of a thermosensitive hydrogel based on a pendant cyclic ether modified poly(ε-caprolactone) and poly(ethylene glycol)co-polymer. Acta Biomater 8: 3963–3973

Wei X, Lv X, Zhao Q, Qiu L (2013) Thermosensitive b-cyclodextrin modified poly (ε-caprolactone)-poly(ethylene glycol)-poly(ε-caprolactone) micelles prolong the anti-inflammatory effect of indomethacin following local injection. Acta Biomater 9: 6953–6963

Yang XL, Luo YL, Xu F, Chen YS (2014) Thermosensitive mPEG-b-PA-g-PNIPAM comb block copolymer micelles: effect of hydrophilic chain length and camptothecin release behaviour. Pharm Res 31:291–304

Zhang L, He Y, Ma G, Song C, Sun H (2012) Paclitaxel-loaded polymeric micelles based on poly (ε-caprolactone)-poly(ethylene glycol)-poly(ε-caprolactone) triblock copolymers: in vitro and in vivo evaluation. Nanomed: Nanotechnol Biol Med 8:925–934

Chapter 14
Gelled Microparticles/Beads of Sterculia Gum and Tamarind Gum for Sustained Drug Release

Amit Kumar Nayak, M. Saquib Hasnain and Dilipkumar Pal

Abstract In recent years, a variety of biocompatible natural polysaccharides are extensively investigated as pharmaceutical excipients in designing various pharmaceutical formulations. Among these, sterculia gum and tamarind gum have emerged as two popular biopolymer candidates, which have found their applications in designing of various sustained drug releasing dosage forms. Presently, both sterculia gum and tamarind gum have been employed to design oral multiple unit sustained release systems such as microparticles, beads. The present chapter deals with a helpful and comprehensive discussion on already reported different microparticles/beads made of sterculia gum and tamarind gum, which have shown a promise in sustained drug releasing capacity over a longer period after oral administration. This chapter will be helpful in applying appropriate strategies or achieving desired controlled sustained drug release profiles from microparticles/beads made of sterculia gum and tamarind gum.

Keywords Sterculia gum · Tamarind gum · Sustained drug release Microparticles · Beads

A. K. Nayak (✉)
Department of Pharmaceutics, Seemanta Institute of Pharmaceutical Sciences, Mayurbhanj 757086, Odisha, India
e-mail: amitkrnayak@yahoo.co.in

M. Saquib Hasnain
Department of Pharmacy, Shri Venkateshwara University, NH-24, Rajabpur, Gajraula, Amroha 244236, UP, India

D. Pal
Department of Pharmaceutical Sciences, Guru Ghasidas Vishwavidyalaya, Koni, Bilaspur 495009, Chhattisgarh, India

© Springer Nature Singapore Pte Ltd. 2018

V. K. Thakur et al. (eds.), *Polymer Gels*, Gels Horizons: From Science to Smart Materials, https://doi.org/10.1007/978-981-10-6080-9_14

1 Introduction

Recently, the majority of the population is in the favor of the consumption of naturally derived materials and their utilizations as excipient in food and biomedical applications (Pal and Mitra 2010; Pal et al. 2012; Nayak and Pal 2012). Thus, the importance of the naturally derived excipients is also increasing day by day because these are mostly nontoxic, biologically biodegradable, and readily available in nature in comparison with various synthetic excipients (Pal et al. 2010; Avachat et al. 2011; Nayak et al. 2010, 2012, 2015). Moreover, the extraction methodologies of these naturally derived excipients are very simple and economic (Nayak and Pal 2015, 2016a, b). Currently, natural polysaccharides are considered as one of the popular naturally derived polymeric groups for the use in various biomedical applications including drug delivery (Hasnain et al. 2010). The biodegradation capability of the natural polysaccharides into various physiological metabolites makes these useful biopolymeric candidates for the formulations of different kinds of drug delivery (Nayak 2016). In general, natural polysaccharides are of complex and branched structures with different monosaccharide residues connected to each other via the O-glycosidic linkages (Kaur et al. 2012a, b). Natural polysaccharides possess several important physicochemical characteristics such as aqueous solubility, high swelling ability in aqueous environment, good stability to pH alterations (Nayak 2016). These are extracted mainly from plants, animals, algals, microbials, and biotechnology (Prajapati et al. 2013). In general, plant polysacchariades are extracted from leaves, pods, fruits, seeds, rhizomes, roots, exudates (Nayak and Pal 2015). In recent years, different plant polysaccharides have been extracted as well as investigated for their potential and diverse applications as drug delivery excipients such as gelling agents, suspending agents, emulsifying agents, granulating agents, binders, mucoadhesive agents, disintegrants, enteric resistants, matrix-forming agents, release retardants, coating agents, film-forming agents (Prajapati et al. 2013). Thus, in recent years, various plant polysaccharides extracted from different plant sources are being employed as useful drug delivery excipients in various drug delivery dosage forms such as emulsions, suspensions, gels, tablets, capsules, transdermal patches, buccal films (Avachat et al. 2011; Prajapati et al. 2013). These plant polysaccharides are also being used to formulate various oral sustained drug releasing systems to achieve target site specificity, minimize fluctuations of drug concentrations within the therapeutic ranges, minimize risk of side effects and dose dumping, reduce dosing frequency, and enhance bioavailability and patient compliances (Nayak and Pal 2015).

Numerous oral sustained drug releasing dosage forms have been already designed and investigated by various drug delivery researchers and formulators (Nayak 2016). In the designing of these oral sustained releasing dosage forms, both single unit systems (such as tablets, capsules) and multiple unit systems (such as pellets, spheroids, beads, microparticles) have been developed and generally employed for the medications of patients (Kulkarni et al. 2005; Malakar et al. 2012). The multiple unit sustained drug releasing systems for oral administration displayed few potential benefits over the single unit systems, as these are capable of mixing with the gastric juices and distributing over a larger gastrointestinal tract (GIT). This facilitates fewer chances of impairing functions because of the failure of a few multiple units and comparatively more predictable sustained drug releasing kinetics (Guru et al. 2013). In addition, these multiple unit systems are able to avoid the vagaries of gastric emptying, to prevent the exposure of higher drug concentrations, to reduce the chances of localized mucosal tissue damages, and to decrease the occurrences of dose dumping (Malakar and Nayak 2012; Guru et al. 2013; Nayak et al. 2013a, b, c, d). At present, the design of various multiple unit sustained drug releasing systems for oral use, such as nanoparticles, microparticles, beads, spheroids, pellets, is recognized as a popular research area in the pharmaceutical drug delivery. Recently, the utilization of various plant polysaccharides to design and formulate different orally used gelled multiple unit systems for sustained drug releasing has been the significant focus of the pharmaceutical drug delivery research. The present chapter deals with a comprehensive and helpful discussion on the already reported different gelled multiple unit sustained drug releasing systems made of two important plant polysaccharides, namely sterculia gum and tamarind gum for oral use.

2 Sterculia Gum: Source, Chemical Composition, Properties, and Use

2.1 Source

Sterculia gum (commonly known as karaya gum) is a medicinally important plant-derived polysaccharide (Leung 1980). It is extracted from exudate of the tree, *Sterculia urens* (family: sterculiaceae) (Cerf et al. 1990). *S. urens* tree is a small- to medium-sized tree with a pale-colored trunk. It is indigenous to Burma and India. The crude gum is conventionally collected through peeling and/or cutting the bark of the tree (Leung 1980).

2.2 Chemical Composition

In general, sterculia gum is a partially acetylated polysaccharide (Cerf et al. 1990). The chemical structure of sterculia gum is shown in Fig. 1. Sterculia gum is composed of three dissimilar polysaccharide chains (Cerf et al. 1990; Bera et al. 2015b) in which one polysaccharide chain (i.e., 50% of the total polysaccharide in the sterculia gum) comprises the repeating units of four galacturonic acid residues, which contains β-D-galactose branches and L-rhamnose residues at the reducing end of the unit. The second polysaccharide chain of sterculia gum (i.e., 17% of the total polysaccharide in the sterculia gum) covers an oligorhamnan, containing D-galacturonic acid branch

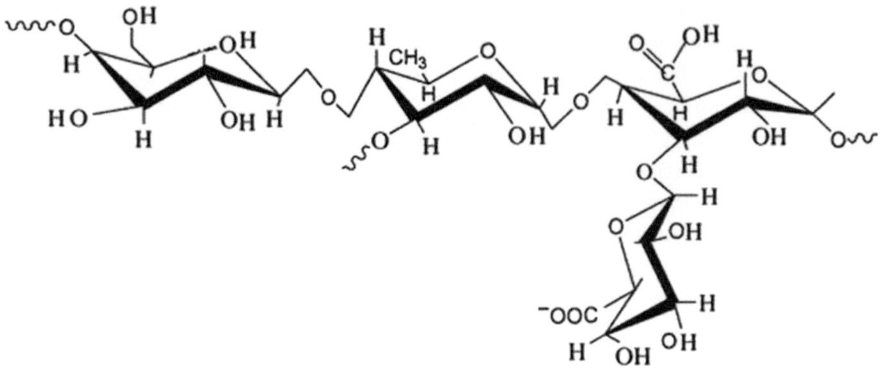

Fig. 1 Chemical structure of sterculia gum

residues and occasionally interrupted by D-galactose residues. The third polysaccharide chain of sterculia gum (i.e., 33% of the total polysaccharide in the sterculia gum) possesses D-glucuronic acid residues, which contain approximately 15–30% rhamnose, 13–26% galactose, and around 40% uronic acid residues (Kulkarni et al. 2014; Nayak and Pal 2016a, b).

2.3 *Properties and Uses*

Sterculia gum displays a number of unique characteristics such as excellent viscosity, higher swelling ability in aqueous medium, water retention capability, better acidic stability (Gauthami and Bhat 1992; Singh and Sharma 2011). In USA, it is graded as 'Generally Recognized as Safe' ('GRAS') material (Anderson 1989). Sterculia gum is described as nontoxic, nonallergic, and nonteratogenic in several studies (Gauthami and Bhat 1992; Singh and Sharma 2011; Singh et al. 2011). It also possesses antimicrobial capability (Gauthami and Bhat 1992). It is also capable of reducing cholesterol to enhance the glucose metabolism without hindering the mineral balances (Behall et al. 1987).

Sterculia gum has been investigated for its utilizations in the treatment of various diseases such as ulcers (Zide and Bevin 1980), diarrhea (Huttel 1983), chronic colonic disorders (Guerre and Neuman 1979), and irritable bowel syndrome (Capron et al. 1981). It is also reported as balk laxative (Meier et al. 1990). No side effect of the consumption of sterculia gum has been reported till date (Singh et al. 2011). It is employed as stabilizer, emulsifier, and thickener in various food products (Anderson and Wang 1994). Due to their similar physical characteristics, it is also used as adulterate gum tragacanth (Weiping 2000).

In the previous literature, gelation of sterculia gum through the influence of polyvalent metal cations such as 'egg-box model' has been reported (Singh et al. 2010). This occurrence could be because of the arrangement of intermolecular junction zones comprising ionotropic interaction between homogalacturonic segments of the sterculia gum chains and the polyvalent metal cations. In addition, the ionotropic interaction of sterculia gum occurs between two or more galacturonic acid residues on the different chains or glucuronic in branched or nonregular chain segments of the sterculia gum structure (Siva et al. 2003).

Since the past few decades, sterculia gum is distinguished as a potential biodegradable as well as biocompatible polysaccharide, which is employed as excipient in the drug delivery applications (Munday and Philip 2000; Sreenivasa et al. 2000; Park and Munday 2004; Deshmukh et al. 2009). However, a lot of effort on the physical as well as chemical modifications of sterculia gum has been carried out to design hydrogel materials by different research groups for controlling the hydration rate and swelling to tailor the drug releasing profiles over a prolonged period. The modifications of sterculia gum, such as polymer blendings (Singh et al. 2010; Guru et al. 2013), cross-linkings (Singh and Vashishtha 2008; Singh and Pal 2012), interpenetrated polymer network (IPN) formations (Kulkarni et al. 2014; Bera et al. 2015a), polyelectrolyte complex (PEC) formations (Lankalapalli and Kolapalli 2012), and polymer graft formations (Singh and Sharma 2008; Singh et al. 2011), were already explored and researched to design and develop different sterculia gum-based hydrogel systems for the better delivery of numerous drugs.

3 Sterculia Gum-Based Gelled Microparticles/Beads for Sustained Drug Release

3.1 Floating and Nonfloating Sterculia Gum-Alginate Beads for Sustained Pantoprazole Release

During the past few years, polymer blends of sterculia gum and sodium alginate have been widely investigated for the development of a range of hydrogel-based systems in drug delivery applications (Singh et al. 2010; Singh and Chauhan 2011; Guru et al. 2013). Sodium alginate (sodium salt of alginic acid) is an anionic natural heteropolysaccharide, which is biodegradable, nontoxic, and extensively employed as biopolymeric excipient in several dosage forms (Sriamornsak and Sungthongjeen 2007; Goh et al. 2012; Nayak et al. 2013b; Jana et al. 2015). Through ionotropic gelation, it is capable of forming hydrogel-based microparticles/beads by the influence of trivalent and divalent metal cations such as Al^{3+}, Ca^{2+}, Zn^{2+}, Ba^{2+} (Nayak et al. 2012). The encapsulation of numerous drugs and therapeutic molecules has been investigated in the ionotropically gelled alginate hydrogel-based microparticles/beads (Yoo et al. 2006; Al-Kassas et al. 2007; Nayak et al. 2011; Pal and Nayak 2011; Banerjee et al. 2013). However, some drawbacks such as lower encapsulation efficiency and premature release of drugs have been associated with these ionotropically gelled alginate hydrogel-based microparticles/beads (Nayak et al. 2012; Sinha et al. 2015a, b).

To overcome these drawbacks, a number of researches have been carried out to develop these ionotropically gelled alginate-based microparticles/beads as controlled drug releasing matrices using polymer blends of sodium alginate with the second natural biocompatible polymers (Nayak et al. 2010, 2013c; Malakar et al. 2013; Nayak and Pal 2015; Sinha et al. 2015a).

Only some hydrogel beads made of sterculia gum-alginate blends have been studied for oral delivery of drugs (Singh et al. 2010; Singh and Chauhan 2011; Guru et al. 2013; Kulkarni et al. 2014). Trivalent and divalent metal cations such as Al^{3+}, Ca^{2+}, Zn^{2+}, and Ba^{2+} were used as ionotropic cross-linkers in preparing most of hydrogel microparticles/beads made of sterculia gum-alginate blends (Singh et al. 2010; Singh and Chauhan 2011; Guru et al. 2013; Bera et al. 2015b). In fact, both the sterculia gum and sodium alginate structures comprise anionic groups (i.e., –COOH groups). Both the gums display the typical properties of electrostatic ionotropic interaction by the influence of polyvalent metal cations (Singh et al. 2010). Thus, an ionotropic interaction occurs between positively charged polyvalent metal cations and –COOH groups of both these gums (i.e., sodium alginate and sterculia gum), when these polymeric blends appear in contact with polyvalent metal cations. The polyvalent metal ions compete with the monovalent Na^+ ions present at the anionic sites of the sterculia gum-sodium alginate polymeric blends and replace it. These polyvalent metal ions as ionotropic cross-linkers accommodate in the interstices of polyuronate chains of sterculia gum-sodium alginate containing close ion pair interactions with the $-COO^-$ anions and adequate coordination by means of electronegative oxygen atoms (Rees 1981). In addition, the hydrogen bonding occurs between the polysaccharide chains present in the sterculia gum-sodium alginate polymeric blends (Singh et al. 2010). The possible interactions between divalent metal cations (M^{2+}), sterculia gum, and sodium alginate are shown in Fig. 2.

Singh et al. (2010) investigated pantoprazole (an antiulcer drug) encapsulated sterculia gum-alginate beads as well as floating sterculia gum-alginate beads using $CaCl_2$ as an ionotropic cross-linking agent. The rational use of sterculia gum-alginate blends was on the basis of pharmacological significance associated with the antiulcer activities of both the polymers, alginate and sterculia gum. Following the preliminary assessment, sodium alginate and sterculia gum polymeric blends containing 2% w/v sodium alginate and 1.25% w/v sterculia gum with 0.1 M $CaCl_2$ aqueous ionotropic cross-linking solution were considered to prepare the optimized formulations of sterculia gum-alginate beads for sustained pantoprazole release and the effervescent technology was utilized to impart floatation in these beads for sustained pantoprazole release. These pantoprazole-loaded sterculia gum-alginate floating beads were formulated by adding 2% w/v $CaCO_3$ in the polymeric blend solutions containing 1.25% w/v sterculia gum and 2.00% w/v sodium alginate. These solutions of sterculia gum-sodium alginate polymeric blend were added drop-wise to the cross-linking solutions containing 0.1 M $CaCl_2$ and 10% v/v acetic acid. In case of pantoprazole-loaded sterculia gum-alginate floating beads, ionotropic gelation was occurred internally and externally. Ionotropic gelation of sterculia gum-alginate blends was occurred upon contact with an acidic

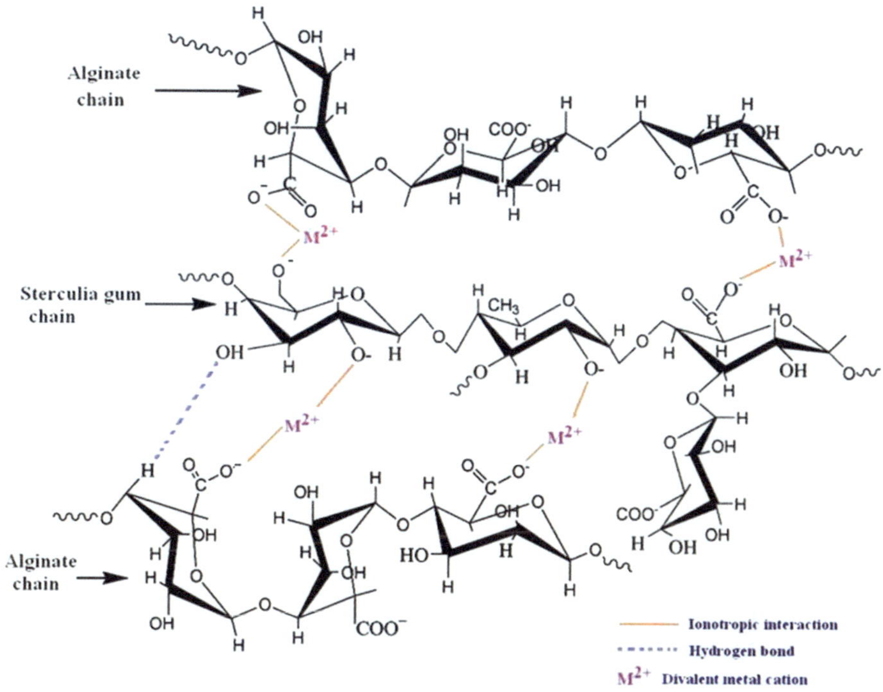

Fig. 2 Possible interactions between divalent metal cations (M^{2+}), sterculia gum, and sodium alginate

medium by the influence of divalent Ca^{2+}-ions to facilitate gel barrier at the bead surface. The $CaCO_3$ effervesced with liberating CO_2 gas and Ca^{2+} cations. The liberated CO_2 gas was trapped into the gel network to facilitate floatation to these floating beads. Subsequently, the divalent Ca^{2+}-ions ionotropically interacted with –COOH groups of sterculia gum and sodium alginate polymeric blend to produce ionotropically gelled three-dimensional gel network so as to control further diffusion of CO_2 gas. Since the concentration of sodium alginate increased in the polymeric blend solutions, spherically shaped Ca^{2+}-ion-induced ionotropically gelled sterculia gum-alginate beads containing pantoprazole. This phenomenon can be attributed that mean diameters of these gelled beads of sterculia gum-alginate blends enlarged because of increasing microviscosity of the polymeric blends with the increase in sodium alginate concentrations. Nevertheless, concentrations of sterculia gum in the polymeric blends solutions and concentrations of cross-linker ($CaCl_2$) did not affect the bead shape as well as bead diameter. The mean diameters of the optimized Ca^{2+}-ion-induced ionotropically gelled sterculia gum-alginate beads containing pantoprazole were measured 1.35 ± 0.30 mm (Singh et al. 2010). The increments in mean bead diameters by means of the incorporation of $CaCO_3$ into the bead formulations might be because of the interaction of $CaCO_3$ with acetic acid contained in the ionotropic cross-linking solutions; CO_2 gas was generated.

The optimized Ca^{2+}-ion-induced ionotropically gelled sterculia gum-alginate nonfloating beads containing pantoprazole possessed rough bead surface by the photomicrographs in scanning electron microscopy (SEM) analyses, while comparatively smooth surface was shown in photomicrographs of optimized Ca^{2+}-ion-induced ionotropically gelled sterculia gum-alginate floating beads containing pantoprazole formulated through the addition of $CaCO_3$ in these bead formulations. The formation of these sterculia gum-alginate floating beads might be contributed through the divalent Ca^{2+}-ions generated from $CaCO_3$. This occurrence may be accountable for the smooth surface of these sterculia gum-alginate floating beads. This occurrence can be explained by the event of burst effect of CO_2 gas liberated before the bead matrix walls get sufficiently hardened.

Characterization of optimized ionotropically gelled sterculia gum-alginate floating and nonfloating beads containing pantoprazole was carried out through electron dispersion X-ray (EDX) analyses, which indicated the presence of hydrogen, oxygen, and carbon as chief elements of both the polysaccharides contained in the polymeric blends. Therefore, these elemental peaks supported the fact of Ca^{2+}-ion-induced cross-linked ionotropically gelation of these sterculia gum-alginate beads. Fourier transform infrared (FTIR) analyses of both the floating and nonfloating beads containing pantoprazole displayed different significant and characteristic peaks of alginate and sterculia gum.

In in vitro swelling study of these sterculia gum-alginate floating and nonfloating beads containing pantoprazole, irregular trends of swelling of these sterculia gum-alginate beads in distilled water were observed up to 5 h. However, a declined swelling pattern of these beads was observed after attainment of the swelling equilibrium after 24 h. The swelling pattern was also found increased with the increasing concentration of sterculia gum. In case of these sterculia gum-alginate beads prepared with 1.25% w/v sterculia gum, maximum swelling pattern was noticed. The influence of increasing $CaCl_2$ concentration exhibited a decline in their swelling pattern. Even the maximum swelling pattern was noticed in case of the beads formulated with 0.1 M $CaCl_2$ as a cross-linker. The influence of pH of the swelling medium on the swelling pattern of these beads was also assessed. From these results, it was found that the swelling behavior of these beads in pH 7.4 was found to be enlarged up to 1 h of swelling and, subsequently, dissolution of these beads observed. Also, the swelling pattern of these beads was observed greater when compared to the swelling pattern in pH 7.4 with that in pH 2.2 and distilled water as the swelling mediums. Comparable patterns of swelling behavior of pantoprazole-loaded optimized ionotropically gelled sterculia gum-alginate floating beads in the swelling mediums of different pH such as distilled water, pH 7.4, and pH 2.2 were seen. Majority of these sterculia gum-alginate floating beads containing pantoprazole demonstrated their buoyancy over a prolonged time.

The drug loadings in these pantoprazole-loaded optimized ionotropically gelled sterculia gum-alginate floating and nonfloating beads were calculated as 66.10 and 83.90%, respectively. The sustained drug releasing behavior of these pantoprazole-loaded floating and nonfloating beads was observed over a period of 24 h in pH 2.2 and distilled water. These sustained drug releasing ionotropically gelled sterculia gum-alginate beads were followed Fickian diffusion mechanism of drug release (Singh et al. 2010).

The same research group, Singh and Chauhan (2011), has also explored the formulation of similar kinds of ionotropically gelled sterculia gum-alginate blends floating beads containing pantoprazole employing $BaCl_2$ as ionotropic cross-linker. The drug loadings in divalent Ba^{2+}-ion cross-linked optimized sterculia gum-alginate floating and nonfloating beads containing pantoprazole were calculated as 61.60 and 67.90%, respectively. Hence, comparatively greater drug loading was found in the Ca^{2+}-ion cross-linked sterculia gum-alginate beads containing pantoprazole formulated in the earlier study by the similar research group than that of Ba^{2+}-ion cross-linked beads.

The swelling pattern and drug release pattern of Ba^{2+}-ion cross-linked ionotropically gelled sterculia gum-alginate floating beads containing pantoprazole were found much greater in pH 7.4 than distilled water and pH 2.2. In comparison with the swelling patterns of sterculia gum-alginate floating and nonfloating beads containing pantoprazole, lesser swelling was reported in case of nonfloating beads than the floating beads composed of ionotropically gelled sterculia gum-alginate matrices. However, Ba^{2-} ion cross-linked sterculia gum-alginate beads demonstrated comparatively more stability than that of the Ca^{2+}-ion cross-linked beads in pH 7.4. Actually, bonding to alginate by these two divalent cations, Ba^{2+} and Ca^{2+}, is likely to take place in a planner two-dimensional mode inside these ionotropically gelled alginate-based beads. However, Ca^{2+}-ions (1.14 Å) possess the comparatively smaller radius than that of Ba^{2+}-ions (1.74 Å) (Singh and Chauhan 2011). Thus, it is thought to fill a larger space between the cross-linking chains of the ionic polymers, which creates a tight arrangement with a smaller void. Therefore, the exchange of the divalent bigger Ba^{2+}-ions in these ionotropically gelled sterculia gum-alginate beads with the monovalent Na^{+}-ions and also their elimination in the form of insoluble barium phosphate are hindered, which may result in the comparatively lower water uptake and higher stability by these beads. A sustained drug releasing pattern from these Ba^{2+}-ion cross-linked ionotropically gelled sterculia gum-alginate floating beads containing pantoprazole over 24 h was found in all tested mediums (pH 2.2 and distilled water), and the Fickian diffusion mechanism of drug release from these beads was found to be followed.

3.2 Gastroretentive Sterculia Gum-Alginate Buoyant Beads (Oil-Entrapped) for Sustained Aceclofenac Release

For the gastroretentive delivery of aceclofenac (a nonsteroidal anti-inflammatory drug), Guru et al. (2013) have formulated and optimized oil (liquid paraffin, relative density = 0.84 g/cm^3)-entrapped buoyant beads made of sterculia gum-sodium alginate blends through ionotropic emulsion-gelation method using an ionotropic cross-linker, $CaCl_2$. In the formulation optimization of these sterculia gum-alginate buoyant beads containing aceclofenac, a computer-aided optimization technique based on the 3^2 (2 factors and 3 levels) factorial design was employed, where the influences of different process variables such as polymer-to-drug ratio and sodium alginate-to-sterculia gum ratio on drug entrapment efficiency and drug release at 7 h of these emulsion-gelled sterculia gum-alginate buoyant beads of aceclofenac were investigated. By the numerical optimization, the optimized formula of sterculia gum-alginate buoyant beads of aceclofenac was obtained, where the polymer-to-drug ratio was considered as 4.99 and the sodium alginate-to-sterculia gum ratio was considered 2.17. The drug entrapment efficiency of oil-entrapped optimized sterculia gum-alginate buoyant beads of aceclofenac was measured as $83.73 \pm 0.81\%$, while the drug entrapment efficiencies of all these beads were in between 63.28 ± 0.55 and $90.92 \pm 2.34\%$. With the decrease in sodium alginate-to-sterculia gum ratio and increase in polymer-to-drug ratio in the bead formula, the drug entrapment efficiency of these beads was found to be increased. In fact, the enhancement in viscosity of the solutions of polymer blends through increasing the addition of sterculia gum quantity might enhance the drug entrapment efficiencies in these buoyant beads of aceclofenac by preventing the drug leaching to the cross-linking solutions during preparation. Increasing polymer-to-drug ratio in these emulsion-gelled sterculia gum-alginate buoyant beads might create entanglement of elevated amounts of drug inside the intricate Ca^{2+}-ion cross-linked sterculia gum-alginate gel network, which may help to increase drug entrapment efficiency in these beads.

The average sizes of these emulsion-gelled sterculia gum-alginate buoyant beads containing aceclofenac were found in the range of 1.32 ± 0.04 to $1.72 \pm 0.12\%$ mm, while the average size of the optimized buoyant beads containing aceclofenac was 1.62 ± 0.08 mm. All these sterculia gum-alginate buoyant beads containing aceclofenac comprised of densities lesser than the density of simulated gastric fluid (pH 1.2) impart the buoyancy. Within 6 min after being placed in the simulated gastric fluid (pH 1.2), these buoyant beads were floated well and confirmed in vitro buoyancy over 7 h. The entrapped oil material (liquid paraffin) is a low-density mineral oil, which was accountable for imparting the good buoyancy of these emulsion-gelled sterculia gum-alginate buoyant beads of aceclofenac.

The surface morphology of the optimized oil-entrapped emulsion-gelled sterculia gum-alginate buoyant beads of aceclofenac was analyzed by performing SEM analyses. A rough and corrugated surface morphology with the presence of small channels or pores was seen in the SEM photograph of these buoyant beads of aceclofenac (Fig. 3). Also, an orange-peel-like appearance with irregular corrugations was observed in the SEM photograph. The drug excipient compatibility within these optimized oil-entrapped emulsion-gelled sterculia gum-alginate buoyant beads of aceclofenac was analyzed by FTIR spectroscopy analyses. The results of the FTIR analyses suggested the compatibility of the encapsulated drug (i.e., aceclofenac) with polymers employed as polymeric blends (i.e., sterculia gum and sodium alginate). The state of the encapsulated drug with these optimized buoyant beads of aceclofenac was compared with the pure drug by the XRD analyses. XRD results recommended the crystalline nature of aceclofenac in the pure drug, and it was also observed that the crystalline nature of encapsulated aceclofenac was decreased

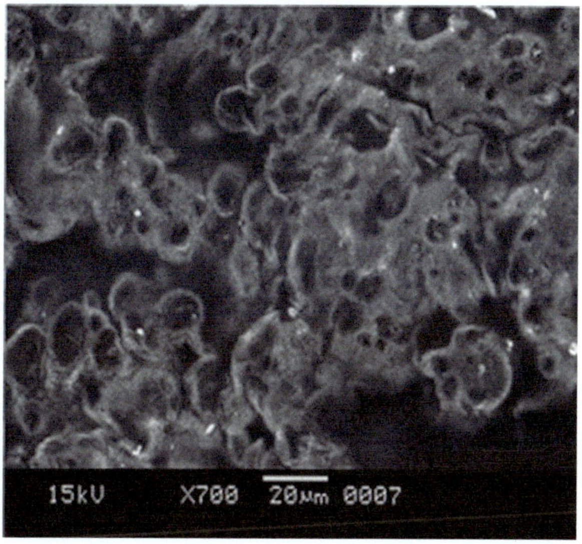

Fig. 3 SEM photograph of oil-entrapped emulsion-gelled sterculia gum-alginate buoyant beads of aceclofenac at magnification of ×700 (Guru et al. 2013). Copyright © 2012 with permission from Elsevier B.V.

significantly in these oil-entrapped emulsion-gelled sterculia gum-alginate buoyant beads of aceclofenac. This change in the state of the encapsulated drug could be because of the influences of the polymers used and/or the formulation process.

In vitro drug release from various oil-entrapped emulsion-gelled sterculia gum-alginate buoyant beads of aceclofenac was investigated in simulated gastric fluid (pH 1.2) as dissolution medium. The results of in vitro drug releases from these buoyant beads clearly indicated a sustained drug releasing pattern over a period of 7 h (Fig. 4). As the polymer-to-drug ratio was increased, an enhancement of the aceclofenac release from these sterculia gum-alginate buoyant beads was

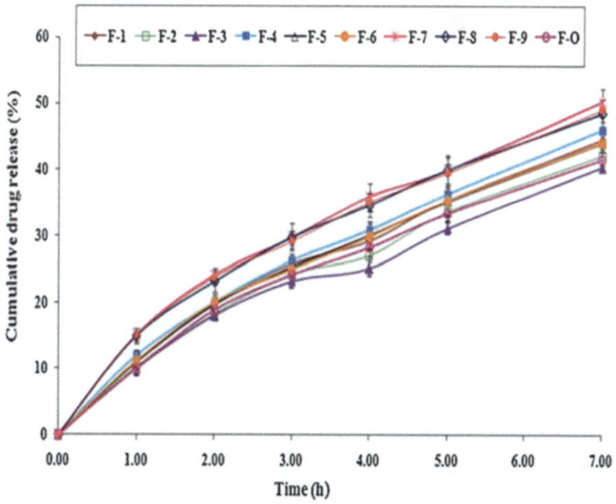

Fig. 4 In vitro drug releasing pattern of various oil-entrapped emulsion-gelled sterculia gum-alginate buoyant beads of aceclofenac in simulated gastric fluid (pH 1.2) (Guru et al. 2013). Copyright © 2012 with permission from Elsevier B.V.

reported. In addition, the comparatively sustained drug releasing pattern was evidenced as the oil entrapment in these buoyant beads was increased, which can be illustrated by the fact that the majority of the drug amounts remained as dispersed as well as saturated in the oil pockets to form drug-oil dispersed matrices in these emulsion-gelled sterculia gum-alginate buoyant beads. Actually, the drug release from oil-entrapped emulsion-gelled sterculia gum-alginate buoyant beads of aceclofenac may take place in two steps: That is, the drug molecules may diffuse out of the oil pockets into the emulsion-gelled sterculia gum-alginate matrix and, also, it may diffuse out of the emulsion-gelled sterculia gum-alginate matrix into the release medium. The in vitro drug releasing pattern of these buoyant beads of aceclofenac was observed to sustain over 7 h. The Koresmeyer-Peppas model kinetics of drug releasing was identified, when curve fitting of the drug releases was analyzed by various kinetic models. The in vitro drug releasing pattern of these buoyant beads of aceclofenac was found to follow non-Fickian drug releasing mechanism (anomalous transport), indicating both diffusion and swelling controlled drug release. Therefore, these oil-entrapped emulsion-gelled sterculia gum-alginate buoyant beads of aceclofenac were capable of possessing higher drug entrapment, excellent buoyancy with a minimum lag time, and prolonged sustained drug releasing pattern. And these capabilities are considered as effective and important parameters for gastroretentive drug delivery. Additionally, for such a formulation loaded with a NSAID-like aceclofenac, the antiulcer activity of sterculia gum and alginate might be useful, which often can produce ulceration as a result of prolonged use of NSAIDs.

3.3 Gastroretentive Alginate-Sterculia Gum Gel-Coated (Oil-Entrapped) Alginate Buoyant Beads for Sustained Risperidone Release

Bera et al. (2015b), in a research, have formulated olive oil-entrapped alginate beads containing risperidone (an atypical antipsychotic drug) coated with Ca^{2+}-ion cross-linked alginate-sterculia gum gel for the gastroretentive risperidone release in sustained manner over a longer period. They have formulated olive oil-entrapped emulsion-gelled sterculia gum-alginate buoyant beads of risperidone. To impart a combination mechanism of floatation and mucoadhesion in the GIT, they employed further coating composed of Ca^{2+}-ion cross-linked alginate-sterculia gum gel. In 1.0% w/w sodium alginate-sterculia gum (1:1) aqueous dispersion, optimized olive oil-entrapped alginate beads of risperidone were placed for coating and, subsequently, transferred to 5% w/v $CaCl_2$ solution for 10 min to make these rigid. After that, these alginate-sterculia gum gel-coated buoyant beads of risperidone were washed using distilled water and placed overnight at room temperature to dry these. Drug entrapment efficiency of these alginate-sterculia gum gel-coated buoyant beads of risperidone was found 81.63 ± 1.54%, while mean diameter and density were measured as 2.49 ± 0.12 mm and 0.66 ± 0.15 g/cm^3, respectively.

These uncoated optimized olive oil-entrapped alginate buoyant beads of risperidone and alginate-sterculia gum gel-coated buoyant beads of risperidone were of spherically shaped as shown in the SEM photographs (Fig. 5a, b). The SEM photographs of the uncoated beads of risperidone displayed a bead surface morphology with wrinkles and/or pores resembling an orange-peel-like appearance. This surface morphology might be formed because of the migration of the water molecules during drying as well as subsequent shrinkage of the polymeric gel network. Even drug crystals were not present on the uncoated bead surface, which indicated the entrapped drug was in finely dispersed state in the oil-entrapped emulsion-gelled alginate matrix. SEM analyses also demonstrated smooth surface of the alginate-sterculia gum gel-coated buoyant beads as compared to uncoated beads (Fig. 5a, b). The coated membrane composed of alginate-sterculia gum gel onto uncoated optimized olive oil-entrapped alginate buoyant beads of risperidone might decrease the number of cracks and/or pores on the outer bead surface. The cross

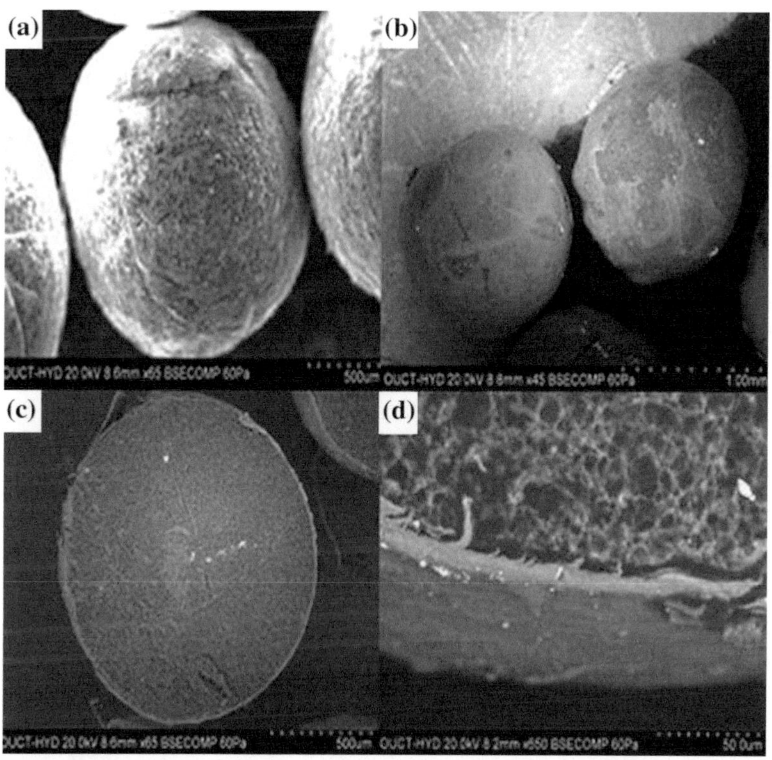

Fig. 5 SEM images of the surface of the uncoated optimized olive oil-entrapped alginate buoyant beads of risperidone (**a**), alginate-sterculia gum gel-coated olive oil-entrapped alginate buoyant beads of risperidone (**b**), cross-sectional view of the alginate-sterculia gum gel-coated buoyant beads with less magnification of = × 65 (**c**), and high magnification of = × 650 (**d**) (Bera et al. 2015b). Copyright © 2014 with permission from Elsevier Ltd.

section of the coated beads in SEM photograph showed a sponge-like structure, in which oil was entrapped (Fig. 5c, d). Additionally, a thin-film membrane deposited onto the surface of the oil-entrapped beads as coating was shown by this.

The uncoated optimized olive oil-entrapped alginate buoyant beads of risperidone and alginate-sterculia gum gel-coated optimized beads of risperidone were characterized by FTIR spectroscopy and P-XRD analyses. The results of FTIR analyses confirmed that drug excipients' compatibility and state of the encapsulated drug in the alginate-sterculia gum gel-coated optimized beads of risperidone. The P-XRD patterns indicated the state of the encapsulated drug within the alginate-sterculia gum gel-coated optimized beads of risperidone, which suggested very less crystalline state of risperidone as compared to that of pure form. The reduction of the crystalline peaks of risperidone in P-XRD patterns suggested that polymers and encapsulated drug interacted strongly at their molecular level, probably drug formed solid solution in occurrence of polymers and oil.

The in vitro floating pattern of these uncoated optimized olive oil-entrapped alginate buoyant beads of risperidone and alginate-sterculia gum gel-coated optimized buoyant beads of risperidone was also investigated. All these beads were found floated within 6 min. From the in vitro floating results, it was also obvious that the percentage buoyancy of the uncoated oil-entrapped alginate buoyant beads found depended on the concentration of polymers, where higher concentration of polymers in the bead formulations reduced the percentage buoyancy of these beads. The alginate-sterculia gum gel-coated optimized buoyant beads of risperidone showed greater percentage buoyancy than that of the uncoated buoyant beads of risperidone (Fig. 6). The enhanced buoyant capability of the alginate-sterculia gum gel-coated optimized buoyant beads of risperidone might be because of their lower density and the occurrence of an air compartment in the middle of polymeric membrane of alginate-sterculia gum gel and alginate core beads.

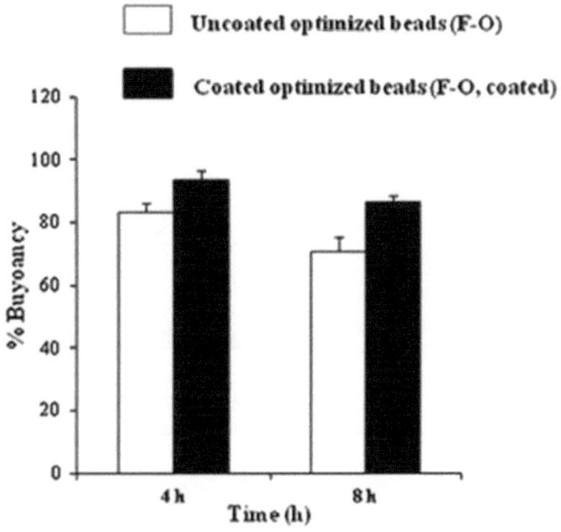

Fig. 6 Percentage buoyancy of uncoated and alginate-sterculia gum gel-coated optimized buoyant alginate beads of risperidone in simulated gastric fluid (pH 1.2). Results are presented as mean ± SD; SD denoted by error bars (Bera et al. 2015b). Copyright © 2014 with permission from Elsevier Ltd.

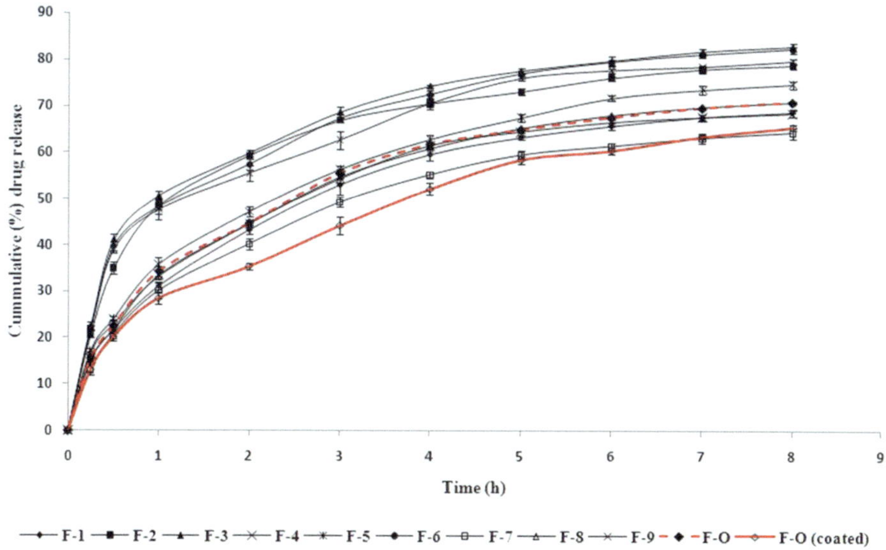

Fig. 7 In vitro drug release profiles of uncoated and alginate-sterculia gum gel-coated optimized buoyant alginate beads of risperidone in simulated gastric fluid (pH 1.2). Results are presented as mean ± SD; SD denoted by error bars (Bera et al. 2015b). Copyright © 2014 with permission from Elsevier Ltd.

The in vitro dissolution of uncoated and alginate-sterculia gum gel-coated optimized buoyant alginate beads of risperidone in simulated gastric fluid (pH 1.2) was tested. It was found to maintain sustained in vitro drug releasing pattern over 8 h in case of coated beads, which was also comparatively slower than the uncoated beads (Fig. 7). The alginate-sterculia gum gel coating membrane might act as a drug release retarding barrier from the coated buoyant beads. Furthermore, the surface pores of the uncoated buoyant beads could be blocked by the alginate-sterculia gum gel coating membrane, which might direct to decreased water penetration and lowering of drug releasing rate. These alginate-sterculia gum gel-coated optimized buoyant alginate beads of risperidone were followed by the Koresmeyer-Peppas model kinetics and the Fickian diffusion mechanism over 8 h.

The swelling index of the uncoated and alginate-sterculia gum gel-coated optimized buoyant alginate beads of risperidone was investigated in simulated gastric fluid (pH 1.2). For both the kinds of buoyant beads poor swelling patterns were observed (Fig. 8). Poor swelling pattern of uncoated olive oil-entrapped alginate buoyant beads of risperidone might be due to the fact that –COOH groups of the polymer formed through decross-linking by the extraction of Ca^{2+}-ions by the Cl^{-}-ions of the acidic swelling medium, which might remain unionized in the acidic swelling medium and could not bring any charge repulsion to enhance the gel porosity of the bead matrices. The lower swelling was observed in case of alginate-sterculia gum gel-coated optimized olive oil-entrapped buoyant alginate

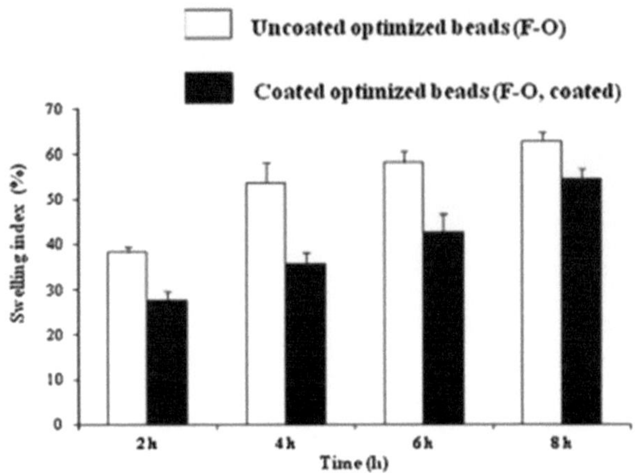

Fig. 8 Swelling patterns of uncoated and alginate-sterculia gum gel-coated optimized buoyant alginate beads of risperidone in simulated gastric fluid (pH 1.2). Results are presented as mean ± SD; SD denoted by error bars (Bera et al. 2015b). Copyright © 2014 with permission from Elsevier Ltd.

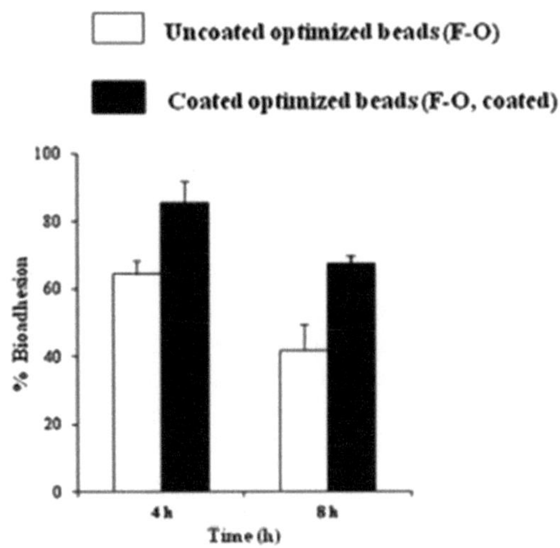

Fig. 9 Bioadhesion results (ex vivo wash off using goat stomach mucosal membrane) of uncoated and alginate-sterculia gum gel-coated optimized buoyant alginate beads of risperidone in simulated gastric fluid (pH 1.2) (Bera et al. 2015b). Copyright © 2014 with permission from Elsevier Ltd.

beads of risperidone, which could possibly be because of the slower water-imbibition into the bead matrices.

Ex vivo wash off evaluation of uncoated and alginate-sterculia gum gel-coated optimized buoyant alginate beads of risperidone was performed using goat stomach mucosal membrane in simulated gastric fluid (pH 1.2). The results of ex vivo wash off evaluation showed that the ex vivo wash off uncoated alginate buoyant beads of risperidone was comparatively faster than the alginate-sterculia gum gel-coated optimized buoyant alginate beads of risperidone (Fig. 9).

3.4 Interpenetrating Polymeric Network (IPN) Alginate-Sterculia Gum Microparticles for Sustained Release of Repaglinide

In recent years, various IPN systems have been investigated in the polymeric research more and more to develop various advanced polymeric materials for various biomedical applications including drug delivery (Lee et al. 2000; Babu et al. 2007; Rokhade et al. 2007; Angadi et al. 2010; Kumar and Singh 2010). Actually, IPN systems are made of the polymeric blends comprising two or more polymers (Kulkarni et al. 2011; Kaity et al. 2013). These IPN systems are formed as at least one polymeric networking is cross-linked and/or synthesized independently in instantaneous presence of other polymers (Banerjee et al. 2010, 2012). According to International Union of Pure and Applied Chemistry (IUPAC), an IPN system is described as polymeric blend systems consisting of two or more polymeric networks that are at least to some extent interlocked with the molecular level but cannot be disconnected if the chemical bonds of polymers are not split (Jenkins et al. 1996). The mixtures of two or more preformed polymeric network systems are not considered as IPNs (Nayak and Pal 2015; Pal and Nayak 2016). Recently, various IPN systems that composed of various natural polymers were designed and evaluated for the use in sustained release of different drugs (Jana et al. 2013a, b; Das et al. 2014; Kulkarni et al. 2012, 2014).

In a recent study by Kulkarni et al. (2014), IPN-based microparticles composed of sterculia gum-sodium alginate blends were formulated by means of sequential ionotropic gelation and emulsion covalent cross-linking technique. In the formulations of these IPN microparticles, $AlCl_3$, $CaCl_2$, and $BaCl_2$ were used as ionotropic cross-linking agents and glutaraldehyde as covalent cross-linking agent. Repaglinide (an antidiabetic agent) was encapsulated in these sterculia gum-alginate IPN microparticles. Kulkarni et al. (2014) have elucidated that the exchange of Na^+-ions took place with the ionotropic counterions Ca^{2+}-, Ba^{2+}- and Al^{3+}-ions at the $–COO^-$-sites of sodium alginate structure, while a subsequent strands might be connected to these cations by the development of connections in which the ionotropic counterions might be connected to 2 or 3 strands of sodium alginate structure together. The further treatment with covalent cross-linking agent (i.e., glutaraldehyde), an acetal structure, had been formed in ionotropically cross-linked sterculia gum-alginate microparticles between the –OH groups of sodium alginate-sterculia gum and –CHO groups of glutaraldehyde to structure an IPN of sterculia gum alginate. The possible interactions between sterculia gum and sodium alginate to form an IPN structure to prepare sterculia gum-alginate IPN microparticles are shown in Fig. 10.

The formulated sterculia gum-alginate IPN microparticles of repaglinide exhibited drug encapsulation efficiency of 81.10–91.70%. A reduction of drug encapsulation efficiency in these IPN microparticles was observed with the reduction of the concentration of sodium alginate within the polymeric blends. The drug encapsulations of the sterculia gum-alginate microparticles of repaglinide formulated via Al^{3+}-ions

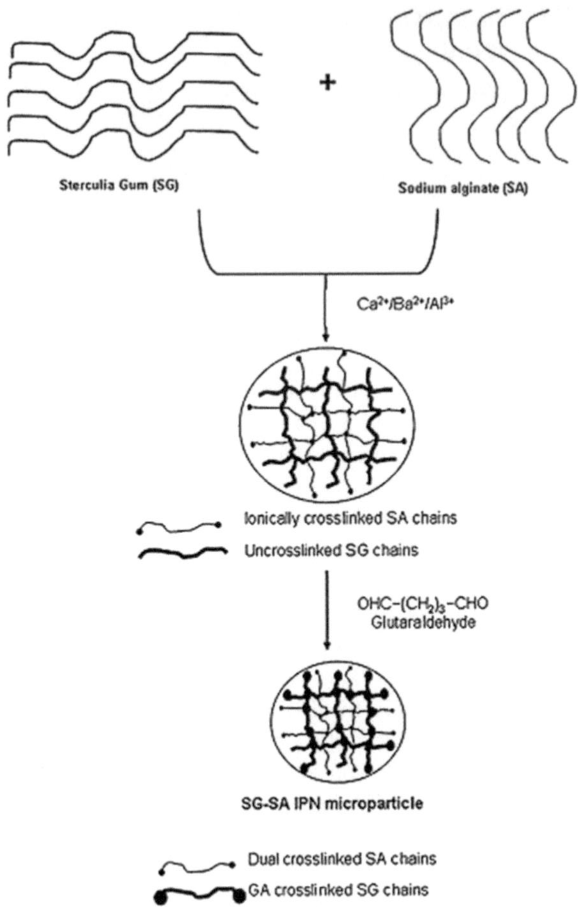

Fig. 10 Possible interactions between sterculia gum and sodium alginate to form an IPN structure to prepare sterculia gum-alginate IPN microparticles (Kulkarni et al. 2014). Copyright © 2014 with permission from Elsevier B.V.

induced cross-linking were observed as the highest. The order of drug encapsulations of IPN microparticles formulated via different ionotropic counterion-induced cross-linking was observed as: Ca^{2+}-ions < Ba^{2+}-ions < Al^{3+}-ions. During formulation, in the Ca^{2+}-ion-induced cross-linked microparticles of repaglinide, the polymeric network of the microparticles might loose-fitting with the larger pores that could cause the leakage of encapsulated drug amounts into the external medium from the polymeric matrix, which may cause lower drug encapsulation. However, in the Al^{3+}-ion-induced cross-linked microparticles of repaglinide, the polymeric network of the microparticles might be comparatively rigid because of faster rate of cross-linking with the occurrence of additional positive charge as each Al^{3+}-ion is capable of combining with additional $-COO^{-}$-ions of the alginate structure (Banerjee et al. 2013). The sterculia gum-alginate IPN microparticles of repaglinide formulated by means of dual cross-linking (i.e., sequential ionotropic gelation and emulsion covalent cross-linking) exhibited higher drug encapsulation. This comparatively higher drug encapsulation in the sterculia gum-alginate IPN microparticles can be described by the development of comparatively stiffer polymeric matrix that probably could reduce the leakage of encapsulated drug from the sterculia gum-alginate IPN matrix. The average sizes of these IPN microparticles of repaglinide were calculated within the range of 19.75–61.52 μm. The average size of these IPN microparticles of repaglinide was increased with the increment of the concentration of sodium alginate in the polymer blends, and it was also found that the microparticle size was also observed to be dependent on the use of cross-linking agents' type. The sequence of producing smaller sterculia gum-alginate microparticles of repaglinide was as follows: Al^{3+}-ions produced smaller microparticles of repaglinide than that of Ba^{2+}-ions, which in turn produced comparatively smaller microparticles of repaglinide than that of Ca^{2+}-ions. During cross-linking of sterculia gum-sodium alginate polymer blends, the polymeric network in these IPN microparticles of repaglinide might have undergone the syneresis that might produce comparatively smaller microparticles at the higher cross-linking densities. When the comparison between ionotropic gelation cross-linked and dual cross-linked sterculia gum-alginate microparticle of repaglinide was made, comparatively smaller microparticles were noticed in case of dual cross-linked microparticle of repaglinide. This might occurrence might be because of rapid shrinkage of polymeric matrix through the covalent cross-linking between the sterculia gum and alginate chains to form rigid IPN network.

Surface morphology of these sterculia gum-alginate IPN microparticles of repaglinide was visualized by SEM analyses, and microphotograph suggested smooth surface of the IPN microparticles. FTIR spectroscopy suggested the formation of IPN structure between two natural polymers, sterculia gum, and alginate, in these IPN microparticles of repaglinide. Thermogravimetric studies also supported the formation of IPN. Both DSC and XRD analyses indicated the occurrence of uniform and amorphous dispersion of the encapsulated repaglinide in these IPN microparticles of repaglinide IPN-containing sterculia gum-alginate.

The in vitro drug releasing from these sterculia gum-alginate IPN microparticles of repaglinide was assessed in acidic medium (0.1 N HCl, pH 1.2) for initial 2 h and, then, in alkaline medium (phosphate buffer, pH 1.2) for the next hours of the study. These IPN microparticles of repaglinide exhibited sustained drug releasing pattern over a period of 24 h (Fig. 11). It was found that Al^{3+}-ions cross-linked microparticles of repaglinide showed comparatively slower drug release as compared to Ba^{2+}-ions cross-linked microparticles of repaglinide, while Ba^{2+}-ions cross-linked microparticles of repaglinide showed comparatively slower than that of Ca^{2+}-ion cross-linked microparticles of repaglinide. The further treatment of ionotropically gelled cross-linked microparticles by covalent cross-linking agent (i.e., glutaraldehyde) produced slower drug releasing pattern of these sterculia gum-alginate IPN microparticles of repaglinide. As the concentration of sodium alginate was increased in the polymer blends used, the slower drug releasing pattern of these sterculia gum-alginate IPN microparticles of repaglinide was observed. The ionotropically gelled cross-linked microparticles discharged encapsulated repaglinide rapidly in the drug releasing medium, while the dual cross-linked IPN microparticles of repaglinide exhibited a comparatively sustained in vitro drug

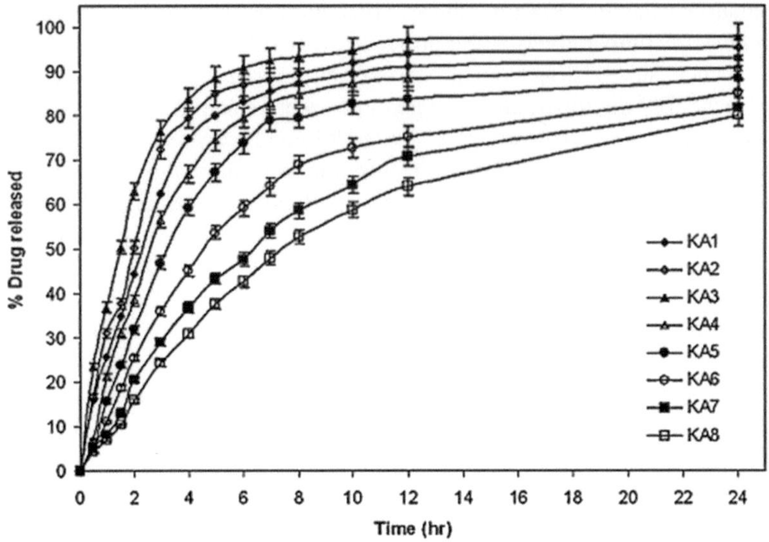

Fig. 11 In vitro drug releasing pattern of sterculia gum-alginate IPN microparticles of repaglinide [key: KA1, KA2, and KA3 were Ca^{2+}-ions cross-linked sterculia gum-alginate microparticles of repaglinide; KA4 was Ba^{2+}-ions cross-linked sterculia gum-alginate microparticles of repaglinide; KA5 was Al^{3+}-ions cross-linked sterculia gum-alginate microparticles of repaglinide; KA6 was dual cross-linked sterculia gum-alginate IPN microparticles of repaglinide (sequential ionotropic gelation by Ca^{2+}-ions and emulsion covalent cross-linking by glutaraldehyde); KA7 was dual cross-linked sterculia gum-alginate IPN microparticles of repaglinide (sequential ionotropic gelation by Ba^{2+}-ions and emulsion covalent cross-linking by glutaraldehyde); KA8 was dual cross-linked sterculia gum-alginate IPN microparticles of repaglinide (sequential ionotropic gelation by Al^{3+}-ions and emulsion covalent cross-linking by glutaraldehyde) (Kulkarni et al. 2014). Copyright © 2014 with permission from Elsevier B.V.

releasing pattern over a longer period than that of the ionotropically gelled cross-linked microparticles of repaglinide. The non-Fickian mechanism of drug release was measured for the dual cross-linked sterculia gum-alginate IPN microparticles of repaglinide.

On streptozotocin-induced diabetic rats, the in vivo antidiabetic activity of KA8 dual cross-linked sterculia gum-alginate IPN microparticles of repaglinide (sequential ionotropic gelation by Al^{3+}-ions and emulsion covalent cross-linking by glutaraldehyde) was assessed. The in vivo antidiabetic activity revealed that the pristine repaglinide a sudden decrease in the blood glucose level in the streptozotocin-induced diabetic rats up to 3 h and after that, it was found recovered. The reduction (%) of glucose level in the diabetic rats was found slower, when treated with sterculia gum-alginate IPN microparticles of repaglinide as compared to that of the pristine repaglinide within 3 h of treatment, and it was found increased over 24 h (Fig. 12). Therefore, the formulated KA8 dual cross-linked sterculia gum-alginate IPN microparticles of repaglinide may be useful for the effective treatment of noninsulin-dependent diabetes mellitus (NIDDM).

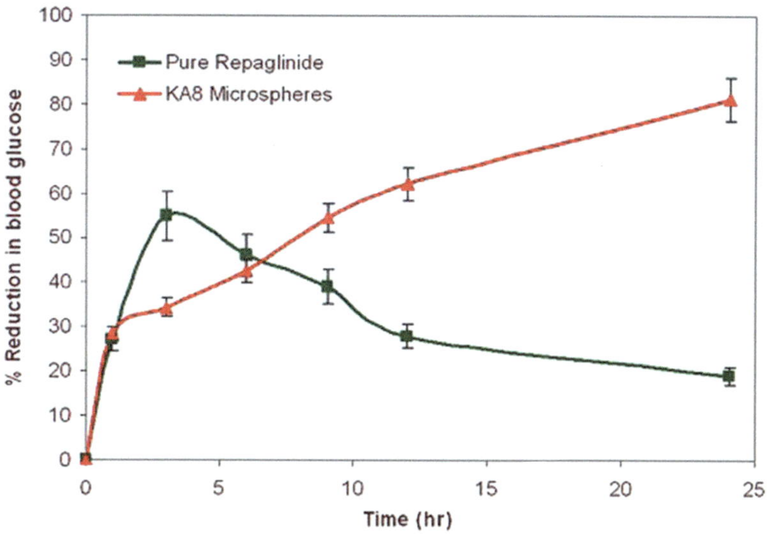

Fig. 12 Reduction (%) of blood glucose in streptozotocin-induced diabetic rats treated with pristine repaglinide and KA8 microparticles [dual cross-linked sterculia gum-alginate IPN microparticles of repaglinide (sequential ionotropic gelation by Al^{3+}-ions and emulsion covalent cross-linking by glutaraldehyde)] (Kulkarni et al. 2014). Copyright © 2014 with permission from Elsevier B.V.

3.5 *Gastroretentive Pectinate-Sterculia Gum IPN Beads for Sustained Ziprasidone HCl Delivery*

Recently, extensive research endeavors have been made for designing various oral drug delivery systems made of pectin because of its biodegradability and biocompatibility (Sriamornsak et al. 2007, 2010; Das et al. 2010). It is a plant-derived natural polysaccharide consisting of a linearly connected α-(1-4)-D-galacturonic acid, which is regularly interspersed with α-(1-2)-linked α-L-rhamnopyranose residues (Munarin et al. 2012). In general, low methoxy pectin has the ionotropic cross-linking ability to prepare gel microparticles/beads by the influence of various divalent metal cations (such as Ca^{2+}, Zn^{2+}) (Nayak and Pal 2014). These ionotropically cross-linked pectinate gel microparticles/beads has been intensively exploited as carrier matrices for the delivery of different drugs (Nayak et al. 2013c, 2014b). These pectinate gel microparticles/beads are suffering from poor drug encapsulation and premature drug releasing because of the high aqueous solubility and swelling of ionotropically cross-linked pectinate gel matrices (Nayak et al. 2014a, b). To overcome these limitations, several modifications of ionotropically cross-linked pectinate gel microparticles/beads have been examined for the better drug delivery (Saravanan et al. 2002; Desai 2005; Atyabi et al. 2005; Wong and Nurjaya 2008; Chakraborty et al. 2010; Oliveira et al. 2010; Saravanan and Rao 2010; Sharma and Ahuja 2011; Soares et al. 2013; Prezotti et al. 2014). Recently, blendings of low methoxy pectin and other biocompatible polymers have been employed as an effectual approach to limit the above-discussed shortcomings of the ionotropically cross-linked pectinate gel matrices (Nayak et al. 2013c, 2014a; Nayak and Pal 2014).

On the basis of the polymer-blendings approach, Bera et al. (2015a) have developed dual cross-linked pectinate-sterculia gum IPN mucoadhesive-floating beads for gastroretentive ziprasidone HCl (an atypical antipsychotic drug) delivery over a longer period. The ziprasidone HCl-containing IPN beads of low methoxy pectin-sterculia gum blends were prepared through simultaneous ionotropic gelation by zinc acetate and covalent cross-linking by glutaraldehyde. For the formulation optimization of pectinate-sterculia gum IPN mucoadhesive-floating beads for gastroretentive ziprasidone HCl, a computer-aided optimization technique based on the 3^2 (2 factors and 3 levels) factorial design was employed, where the effects of low methoxy pectin and sterculia gum amounts on the drug encapsulation and drug release were investigated. Strong noncovalent associations such as hydrogen bonding and hydrophobic interactions between polymeric chains of the pectinate gel network could be induced by the influence of divalent Zn^{2+}-ions due to its lower coordination number (Das et al. 2010). In addition, the intermolecular zones such as 'egg-box model' could be formed by the ionotropically gelation interactions among the homogalacturonic residues, galacturonic acid residues, and/or glucuronic acid residues of sterculia gum by the Zn^{2+}-ions (Singh et al. 2010). The hydrogen bonding as well as electrostatic interactions between both sterculia gum chains and pectinate chains might have played an important role in the formation of IPN gel matrix. However, the formation of IPN gel matrix was led by the formation of acetal linkage with the reaction of –OH groups of the polysaccharides (sterculia gum and pectin) and –CHO groups of glutaraldehyde. The drug encapsulation efficiency of 87.98 ± 1.15% and mean bead diameter of 2.17 ± 0.13 mm were measured for the optimized pectinate-sterculia gum IPN beads of ziprasidone HCl.

The surface morphology of optimized pectinate-sterculia gum IPN beads of ziprasidone HCl was analyzed by SEM analyses. Spherically shaped beads were seen in the SEM photographs with characteristic cracks and/or large wrinkles on the bead surface resembling an orange-peel like appearance (Fig. 13a, b). In the SEM photographs, drug crystals were not observed on the bead surface of these IPN beads of ziprasidone HCl, which indicated that the drug (i.e., ziprasidone HCl) was present in the finely dispersed state within the IPN matrix. Additionally on these IPN bead surface, polymeric debris was observed, which may be owing to the bead preparation methodology (i.e., simultaneous gel bead formation and formation of the polymeric matrix) (Malakar et al. 2013). FTIR analyses suggested the formation of pectinate-sterculia gum IPN structure along with the drug excipient compatibility of these beads, while DSC and XRD analyses suggested the physical state of the encapsulated ziprasidone HCl in the pectinate-sterculia gum IPN beads of ziprasidone HCl.

The pectinate-sterculia gum IPN beads of ziprasidone HCl exhibited in vitro sustained drug releasing behavior over a period of 8 h in the simulated gastric fluid (pH 1.2) (Fig. 14). These kinds of delayed drug releasing might possibly because of

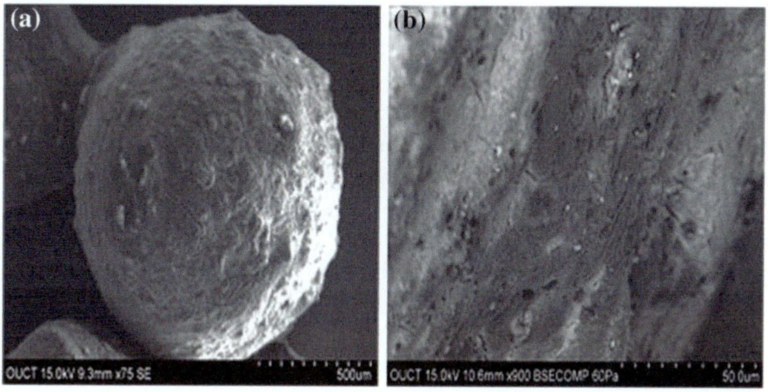

Fig. 13 SEM photographs of optimized pectinate-sterculia gum IPN beads of ziprasidone HCl [×75 (**a**) and ×900 (**b**)] (Bera et al. 2015a). Copyright © 2015 with permission from Elsevier Ltd.

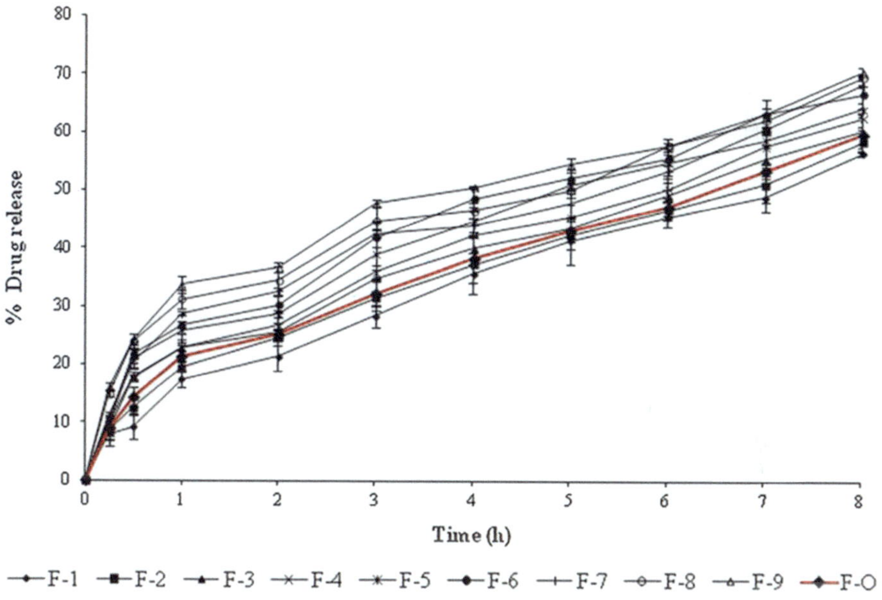

Fig. 14 In vitro drug releasing patterns of pectinate-sterculia gum IPN beads of ziprasidone HCl in simulated gastric fluid (pH 1.2) (Bera et al. 2015a). Copyright © 2015 with permission from Elsevier Ltd.

decreased free volume spaces of glutaraldehydetreated pectinate-sterculia gum IPN matrices that could restrict the diffusion of drug from that (Kulkarni et al. 2014). The increase in the polymer contents in these IPN bead formula displayed slower rate of drug releasing. The IPN beads containing higher contents of hydrophilic polymers might help to attach more tightly to the water molecules. This could result in the formation of viscous gel, and viscous gel could block the pores on the bead surface, which might produce sustained drug releasing pattern. The in vitro drug releasing behavior of most of these IPN beads of ziprasidone HCl obeyed the Higuchi kinetics with the non-Fickian (anomalous) transport mechanism.

The in vitro swelling patterns of all these pectinate-sterculia gum IPN beads of ziprasidone HCl in the acidic medium (simulated gastric fluid, pH 1.2) revealed a smaller degree of gradual increases of in vitro swelling up to 6 h by the beads. The reduced in vitro swelling pattern of these IPN beads of ziprasidone HCl can be characterized by the fact that the $-COO^-$ groups of pectin backbone formed by the decross-linking through the withdrawal of Zn^{2+}-ions by the Cl^--ions present in the acidic swelling medium used of pH 1.2, which were remained unionized and unable to persuade the charge repulsion to improve the gel porosity of IPN bead matrix.

The in vitro floatation of all these pectinate-sterculia gum IPN beads of ziprasidone HCl was studied in the acidic medium (simulated gastric fluid, pH 1.2). These IPN beads of ziprasidone HCl displayed excellent buoyancy (%) along with the lower floating lag time, which was measured <2 min. The buoyancy (%) of these floating IPN beads was measured as 63% at 8 h of floatation study in the simulated gastric fluid, pH 1.2 (Fig. 15). Even the buoyancy (%) of these floating IPN beads was found to be correlated well with the density of these beads. Meanwhile, the average density of these pectinate-sterculia gum IPN beads of ziprasidone HCl was measured as lower than 1.004 g/cm^3, which is regarded as the density of the gastric fluid. For this reason, these newly developed pectinate-sterculia gum IPN beads exhibited good floatation. The measured buoyancy (%) as well as floating lag times of these IPN beads was found to be decreased

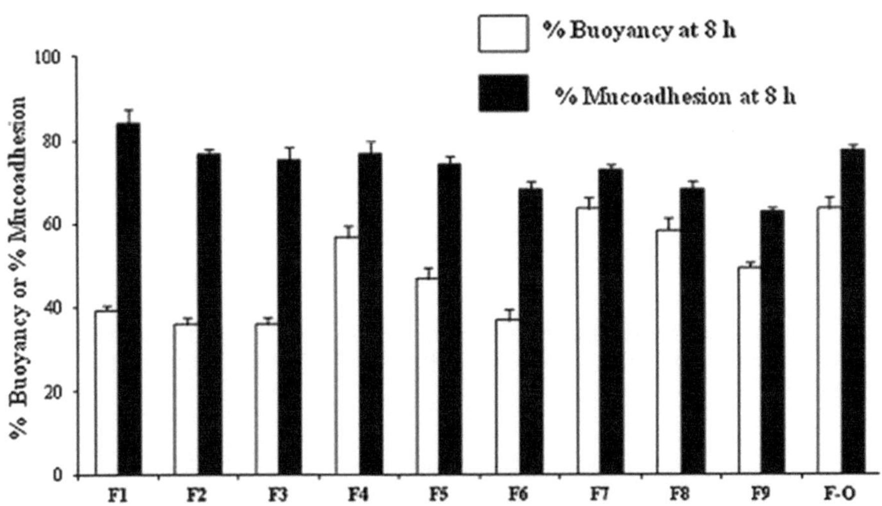

Fig. 15 In vitro buoyancy (%) (**a**) and ex vivo mucoadhesion (**b**) of the pectinate-sterculia gum IPN beads of ziprasidone HCl at 8 h in simulated gastric fluid (pH 1.2) (Bera et al. 2015a). Copyright © 2015 with permission from Elsevier Ltd.

as the concentrations of polymers in the polymer blends (i.e., polymer blends of low methoxy pectin and sterculia gum) were increased. This occurrence in the flotation results could probably because of their lower density as well as greater expansion of polymer matrix in the aqueous media. The good buoyancy behavior of these pectinate-sterculia gum IPN beads of ziprasidone HCl can also be attributed to the swelling as well as expansion of the matrix volume of the polymeric bead structure in the aqueous media studied, which also led to the further decreasing of the bead density (Guru et al. 2013).

The ex vivo mucoadhesivity potential of these pectinate-sterculia gum IPN beads of ziprasidone HCl was evaluated using goat gastric mucosal tissue as biocoadhesive mucosal membrane in the simulated gastric fluid (pH 1.2). In this study, the IPN beads (%) adhered to the goat stomach mucosal membrane were measured within the range, 84.00 ± 3.36% and 62.67 ± 1.15% after 8 h (Fig. 15).

The –OH groups of the hydrophilic polymers employed to prepare IPN bead matrix in this work (i.e., low methoxy pectin and sterculia gum) might attach to the mucosal tissue through various kinds of interactions such as hydrogen bonding, van der Waal's forces, and ionic interactions to impart the good mucoadhesion (Pliszczak et al. 2012). From the results, it was also observed that the IPN beads containing higher contents of hydrophilic polymers exhibited comparative slower ex vivo wash off than those IPN beads containing lower contents of hydrophilic polymers. The slower wash off for the IPN beads containing higher contents of hydrophilic polymers might be occurred through higher rate of unionization of the –COOH moieties of sterculia gum and pectinate in the acidic medium of pH 1.2, which resulted in improved adhesive property by the tested matrix.

4 Tamarind Gum: Source, Chemical Composition, Properties and Use

4.1 Source

Tamarind gumis an important plant-derived polysaccharide, which is extracted from the seeds of the tree *Tamarindus indica*, family Leguminosae (commonly known as 'Indian date,' '*imli*' in Hindi, '*Tentul*' in Bengali) (Khanna et al. 1987; Joseph et al. 2012). *T. indica* tree is a large evergreen tree usually found in India, Southeast Asia, and some parts of the Africa. The mature tamarind seeds contain 20–30% of seed coat (testa), 70–75% of kernel (seed endosperm), and 67.10 g/Kg of crude fibers along with higher amounts of carbohydrate sugars (Gupta et al. 2010; Nayak 2016). The mature tamarind seeds contain natural polysaccharide, which present as cell-wall storage material (Rao et al. 1946; Khanna et al. 1987). Extraction of tamarind gum is done from the kernel powder of tamarind seeds by using several established extraction techniques (Rao et al. 1946). The mature tamarind seeds are processed into the kernel powder through the process of tamarind seed collection, removal of seed coat, separation of seed kernel, milling, grinding, and then sieving

(Avachat et al. 2011). The extraction process of tamarind gum was first reported by Rao et al. (1946). On laboratory scales, Rao and Srivastava (1973) and Nandi (1975) further modified the extraction process of tamarind gum. In general, tamarind gum is extracted through the conventional chemical method, where kernel powder of tamarind seeds is soaked by boiled water and, then, extracted mucilage is filtered. The filtered mucilage content is added to the equal amount of ethyl alcohol or acetone to precipitate tamarind gum. The precipitation of tamarind gum is separated, which is then dried and sieved (Nayak and Pal 2011).

4.2 Chemical Composition

According to the chemical structure, tamarind gum is a branched structured polysaccharide. Chemically, tamarind gum contains (1 → 4)-β-D-glucan backbone substituted with the side chains of α-D-xylopyranose and *β*-D-galactopyranosyl (1 → 2)-α-D-xylopyranose-linked (1 → 6) glucose residues, where glucose, xylose, and galactose units are present in the ratio 2.80:2.25:1.00 (Fig. 16) (Kaur et al. 2012a, b). Therefore, tamarind gum is a galactoxyloglucan. In the chemical structure of tamarind gum, approximately 80% of the glucose residues are substituted with the 1–6 linked xylose residues and, also, partly substituted with the p-1-2 galactose residues (Lang et al. 1992; Nayak 2016).

4.3 Properties and Uses

Tamarind gum is a neutral and nonionic polysaccharide (Khanna et al. 1987). It is a aqueous soluble hydrophilic and gel-forming material (Gupta et al. 2010). In the

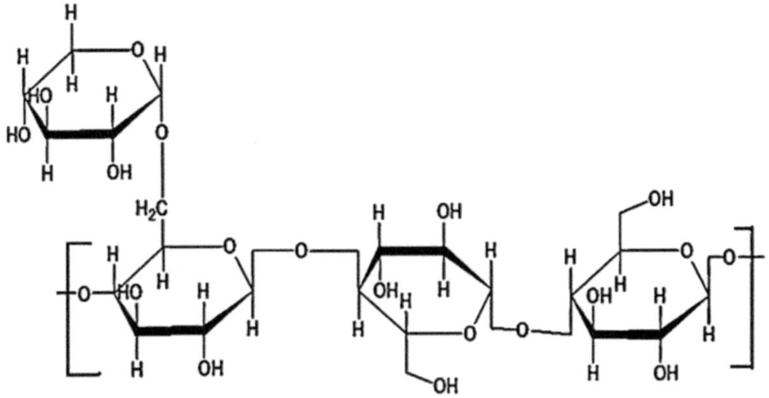

Fig. 16 Chemical structure of tamarind gum

aqueous solutions of tamarind gum, the higher substitution degree of glucan chain in its molecular structure generates stiffer conformation with the larger volume occupancy. Within the dispersion of tamarind gum in the aqueous solvents, it generally exhibits self-aggregation, which comprises the lateral assemblies of single polysaccharide, which can be suitably elucidated by Kuhn's model (Gupta et al. 2010; Joseph et al. 2012). The aqueous solution of tamarind gum shows a typical non-Newtonian rheology and pseudoplastic nature (Joseph et al. 2012). Tamarind gum is insoluble in various commonly used organic solvents such as ethanol, methanol, ether, and acetone (Nayak 2016). It is reported as stable in the medium of acidic pH (Kaur et al. 2012a). In various research works, tamarind gum has also been described for its biomucoadhesivity and mucomimetic character (Kaur et al. 2012a, b). This gum is also reported as nontoxic, noncarcinogenic, nonirritant, biocompatible, and hemostatic (Avachat et al. 2011; Joseph et al. 2012). In the reported literature, tamarind gum has shown anti-inflammatory, antidiabetic, and hepatoprotective potentials (Samal and Dangi 2014). It has a unique property of film-forming; even these films possess good mechanical parameters such as higher degree of flexibility as well as tensile strength (Gupta et al. 2010; Manchanda et al. 2014).

Tamarind gum has already been investigated and employed as potential and useful pharmaceutical excipients in the formulations of various drug deliveries. It has been already studied as suspending agent, emulsifier, thickener, gelling agent, film former, binder, disintegrating agent, drug release-modifier, matrix former, mucoadhesive agent, etc., in various kinds of pharmaceutical formulations (such as suspensions, emulsions, gels, tablets, nanoparticles, microparticles, beads, transdermal, and buccal patches). (Khanna et al. 1987; Kulkarni et al. 2005; Chanda et al. 2008; Gupta et al. 2010; Avachat et al. 2011; Joseph et al. 2012; Parajapati et al. 2013; Nayak 2016). Tamarind gum is also employed in the development of oral (Deveswaran et al. 2009; Chandramouli et al. 2012; Pal and Nayak 2012; Nayak et al. 2013d), buccal (Avachat et al. 2013), colon (Mishra and Khandare 2011), nasal (Datta and Bandyopadhyay 2006), and ocular (Gheraldi et al. 2000) drug delivery systems. Recently, because of its biomucoadhesive capacity, it is widely used in the formulations of several kinds of mucoadhesive drug deliveries, such as oral, ocular, nasal, transdermal, buccal (Gheraldi et al. 2000; Datta and Bandyopadhyay 2006; Pal and Nayak 2012; Avachat et al. 2013; Nayak et al. 2014a).

5 Tamarind Gum-Based Gelled Microparticles/Beads for Sustained Drug Release

5.1 Tamarind Gum-Alginate Mucoadhesive Gelled Beads for Sustained Metformin HCl Release

Mucoadhesive gelled beads for sustained release of metformin HCl (an oral antidiabetic drug) were developed using sodium alginate-tamarind gum polymer blends and $CaCl_2$ as ionotropic cross-linking agent (Nayak and Pal 2013; Nayak et al. 2016). For the formulation optimization of these ionotropically gelled tamarind gum-alginate mucoadhesive beads containing metformin HCl, a computer-aided optimization technique based on the 3^2 (2 factors and 3 levels) factorial design was employed, where the effects of sodium alginate-to-tamarind gum ratio (polymer-blends ratio) and concentration of $CaCl_2$ on drug encapsulation efficiency and in vitro drug releases at 2 and 10 h were analyzed by response surface methodology (Nayak and Pal 2013). From the results of the statistical optimization performed, an increment in the drug encapsulation efficiency of these tamarind gum-alginate beads of metformin HCl was measured with the reduction of sodium alginate-to-tamarind gum ratio and increase of $CaCl_2$ concentration. With the increment of both sodium alginate-to-tamarind gum ratio and concentration of $CaCl_2$, a decrease in metformin HCl release from these tamarind gum-alginate beads was detected at 2 h (acidic pH), while a decrease in metformin HCl release from these beads at 10 h was observed with the reduction of sodium alginate-to-tamarind gum ratio and concentration of $CaCl_2$. The drug encapsulation efficiencies of these tamarind gum-alginate beads of metformin HCl were found in the range of 71.86 ± 1.88 to 94.86 ± 3.92%, while the bead sizes (average diameters) were measured in the range of 1.05 ± 0.03 to 1.52 ± 0.09 mm (Nayak and Pal 2013). The drug encapsulation efficiency of the optimized tamarind gum-alginate beads of metformin HCl was found 94.86 ± 3.92%, while the size of these beads was found 1.24 ± 0.07 mm.

The SEM photograph of ionotropically gelled tamarind gum-alginate beads containing metformin HCl was studied for bead surface morphology analyses, which revealed very rough and corrugated bead surface with characteristic wrinkles and polymeric derbies (Fig. 17) (Nayak et al. 2016). FTIR spectroscopy was performed to reveal any interaction between the encapsulated metformin HCl and the sodium alginate-tamarind gum polymer blends utilized in these ionotropically

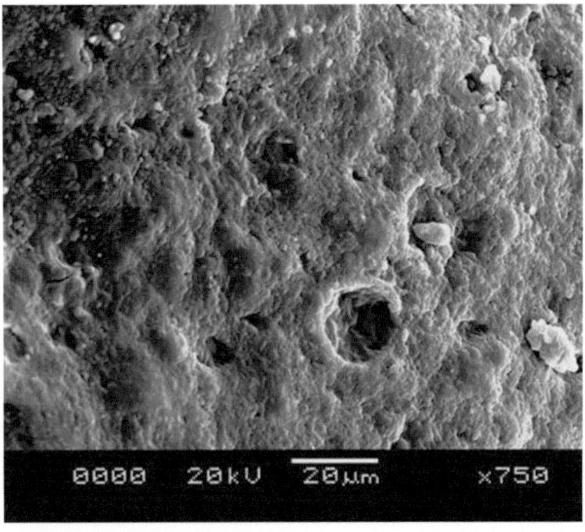

Fig. 17 SEM photograph of optimized ionotropically gelled tamarind gum-alginate beads containing metformin HCl (Nayak et al. 2016). Copyright © 2015 with permission from Elsevier B.V.

gelled beads formation. From the results of the FTIR analyses, no interaction was found between the drug and polymer blends.

The in vitro drug releasing from various ionotropically gelled tamarind gum-alginate beads of metformin HCl was tested in acidic gastric pH (1.2) for initial 2 h and then in alkaline intestinal pH (7.4) for next 8 h. After 2 h of initial drug release study, the drug release was measured slower (<17%) and, after that, it was found comparatively faster. The overall drug release from these beads of metformin HCl displayed a sustained drug releasing pattern over 10 h. A decreasing drug release was observed after 2 h of drug releasing study in case of the ionotropically gelled tamarind gum-alginate beads made of higher ratio of sodium alginate than the tamarind gum because of the shrinkage of alginate at acidic pH. The major amount of initial drug releasing after 2 h could probably be because of the presence of the surface adhered drug particles. A comparatively faster drug release from these tamarind gum-alginate beads was seen in alkaline pH (7.4) because of the higher swelling rate of these beads in alkaline medium. The drug releasing from these ionotropically gelled beads was found higher as sodium alginate-to-tamarind gum ratio was decreased and Ca^{2+} concentration was increased during formulation. The optimized ionotropically gelled tamarind gum-alginate beads containing metformin HCl showed 15.18 ± 0.86% and 69.48 ± 2.52% of metformin HCl releases after 2 and 10 h, respectively (Fig. 18) (Nayak et al. 2016). This drug release from these beads followed a zero-order kinetics (controlled drug release) pattern with case-II transport mechanism.

The swelling index of optimized ionotropically gelled tamarind gum-alginate bead containing metformin was initially found as lesser in acidic gastric pH (1.2) (initial 2 h of dissolution) as compared with that of in alkaline intestinal pH (7.4), demonstrating a pH-sensitive swelling performance (Fig. 19) (Nayak et al. 2016). Maximum swelling of these beads was observed in between 2 and 3 h in alkaline pH, and after which, erosion and dissolution of the swollen beads took place. The equilibrium swelling of optimized ionotropically gelled tamarind gum-alginate beads containing metformin in both acidic gastric pH (1.2) and alkaline intestinal

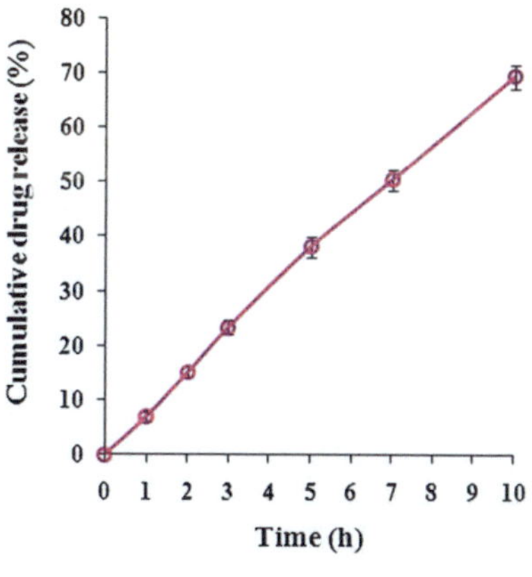

Fig. 18 In vitro drug releasing pattern of optimized ionotropically gelled tamarind gum-alginate beads containing metformin HCl (Nayak et al. 2016). Copyright © 2015 with permission from Elsevier B.V.

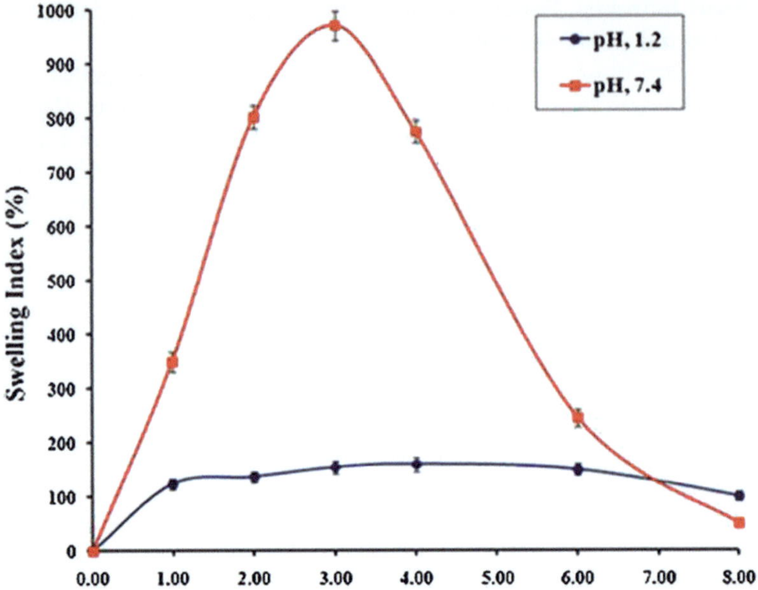

Fig. 19 Swelling index of optimized ionotropically gelled tamarind gum-alginate beads containing metformin HCl in acidic gastric pH (1.2) as compared with that of in alkaline intestinal pH (7.4) (Nayak et al. 2016). Copyright © 2015 with permission from Elsevier B.V.

Fig. 20 Equilibrium swelling of optimized ionotropically gelled tamarind gum-alginate beads containing metformin HCl in acidic gastric pH (1.2) as compared with that of in alkaline intestinal pH (7.4) (Nayak et al. 2016). Copyright © 2015 with permission from Elsevier B.V.

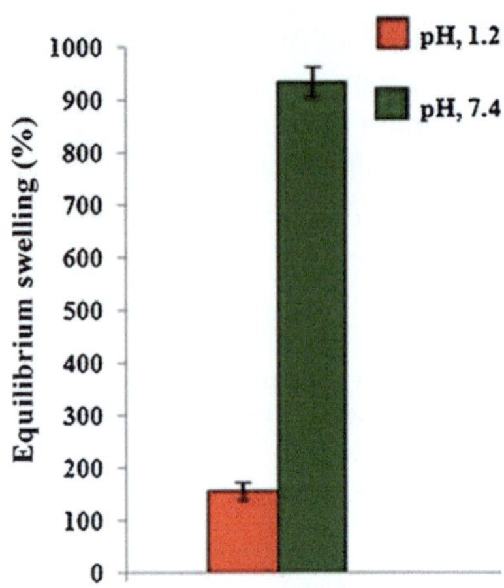

pH (7.4) exhibited higher degree of swelling of these ionotropically gelled beads in pH 7.4 as compared to that in pH 1.2 (Fig. 20) (Nayak et al. 2016).

Ex vivo wash off characteristics of these optimized ionotropically gelled tamarind gum-alginate beads containing metformin HCl in acidic gastric pH (1.2) and in alkaline intestinal pH (7.4) was studied using goat intestinal mucosa as the biomucoadhesive membrane. Faster ex vivo wash off of these beads was measured in the intestinal pH as compared to that in gastric pH (Nayak and Pal 2013). The results of ex vivo wash off suggested that these optimized tamarind gum-alginate beads of metformin HCl possessed good biomucoadhesion with the goat intestinal mucosa. In the in vivo study using alloxan-induced diabetic albino rats, a significant hypoglycemic effect over a prolonged period after oral administration of ionotropically gelled optimized tamarind gum-alginate mucoadhesive beads containing metformin HCl was measured (Nayak and Pal 2013).

5.2 Tamarind Gum-Alginate Gelled Beads for Sustained Diclofenac Sodium Release

In another study, diclofenac sodium was loaded within the ionotropically gelled beads made of tamarind gum-alginate blends (Nayak and Pal 2011). The preparation of these beads of diclofenac sodium was similar to the previous one (i.e., ionotropically gelled tamarind gum-alginate mucoadhesive beads of metformin HCl) as the ionotropic gelation method using $CaCl_2$ solution as cross-linking agent was employed. The gelled tamarind gum-alginate beads of diclofenac sodium were instantaneously formed when the dispersion mixtures containing drug and polymer blends were dropped into the solutions of $CaCl_2$. These ionotropically gelled beads were formed because of the electrostatic interaction between negatively charged $-COO^-$ groups of the alginate backbone and positively charged Ca^{2+}-ions present in the cross-linking solutions. A computer-aided optimization technique based on the 3^2 factorial design was employed for the formulation optimization of these ionotropically gelled tamarind gum-alginate beads containing diclofenac sodium, where the effects of sodium alginate-to-tamarind gum ratio (polymer blends ratio) and concentration of $CaCl_2$ on drug encapsulation efficiency and in vitro drug release at 10 h were analyzed by response surface methodology. From the results obtained through the analyses of the statistical optimization performed, the increase in drug encapsulation efficiency and the decrease in drug release at 10 h were identified as sodium alginate-to-tamarind gum ratio was decreased and concentration of $CaCl_2$ was increased. Using numerical optimization process, the optimized formula for these beads of diclofenac sodium was selected as sodium alginate-to-tamarind gum ratio of 1 and $CaCl_2$ concentration of 12%. The optimized tamarind gum-alginate beads of diclofenac sodium possessed the drug encapsulation efficiency of $97.32 \pm 4.03\%$, average bead size of 0.71 ± 0.03 mm, and in vitro drug release at 10 h of $69.08 \pm 2.36\%$.

The bead surface morphology of ionotropically gelled optimized tamarind gum-alginate beads of diclofenac sodium was visualized by the SEM photograph of the bead surface, which displayed a dense morphology with rough surfaces and absence of cracks (Fig. 21). The encapsulated drug excipient compatibility with these formulated optimized beads of diclofenac sodium was analyzed by FTIR and nuclear magnetic resonance (^1H NMR) studies. Both these instrumental studies

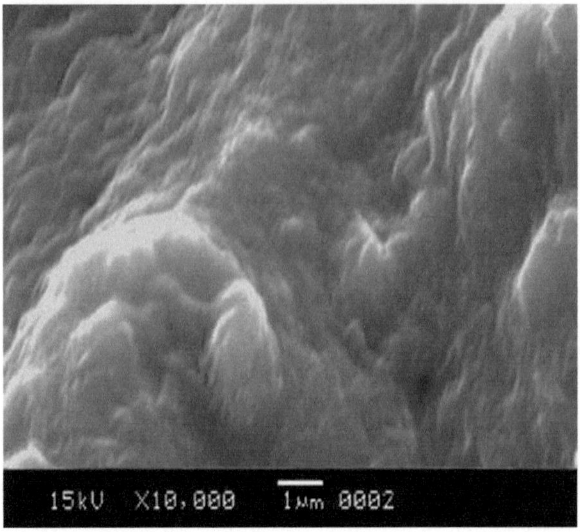

Fig. 21 SEM photographs of optimized tamarind gum-alginate beads of diclofenac sodium through ionotropic gelation method (Nayak and Pal 2011). Copyright © 2011 with permission from Elsevier Ltd.

suggested the compatibility of the diclofenac sodium with tamarind gum-alginate blends used to formulate these beads through ionotropic gelation method.

The in vitro drug releasing behavior of various ionotropically gelled tamarind gum-alginate beads of diclofenac sodium was estimated in the acidic medium (0.1 N HCl, pH 1.2) for initial 2 h and then in intestinal medium (phosphate buffer, pH 7.4) for next hours of the study. The results displayed a slower and sustained drug releasing behavior of these tamarind gum-alginate beads over a period of 10 h (Fig. 22). During the initial period of in vitro drug releasing, very negligible amounts of diclofenac sodium (<7%) because of the poor solubility of alginate in lower acidic drug releasing medium as it is pH sensitive. The percentage in vitro drug releasing after 10 h from ionotropically gelled tamarind gum-alginate beads was measured within the range 69.08 + 2.36 to 96.07 ± 3.54%. It was observed as higher with the decreasing sodium alginate to tamarind gum and increasing concentration of $CaCl_2$ as formulation parameters. The percentage in vitro drug releasing after 10 h from these beads formulated with the higher degree of cross-linking was observed to be comparatively sustained than that of the lower degree of cross-linking by the Ca^{2+}-ions. A controlled release (zero-order kinetics) drug releasing pattern with case-II transport mechanism controlled by swelling and relaxation of drug releasing matrix was seen in this study. The swelling index of ionotropically gelled tamarind gum-alginate beads of diclofenac sodium was found lesser in the acidic medium (0.1 N HCl, pH 1.2) than that in intestinal medium (phosphate buffer, pH 7.4), initially because of the shrinkage of alginate at the acidic pH. Maximum swelling of these ionotropically gelled was noticed within 2–3 h in phosphate buffer. After that, erosion of the swelled beads and then dissolution occurred subsequently.

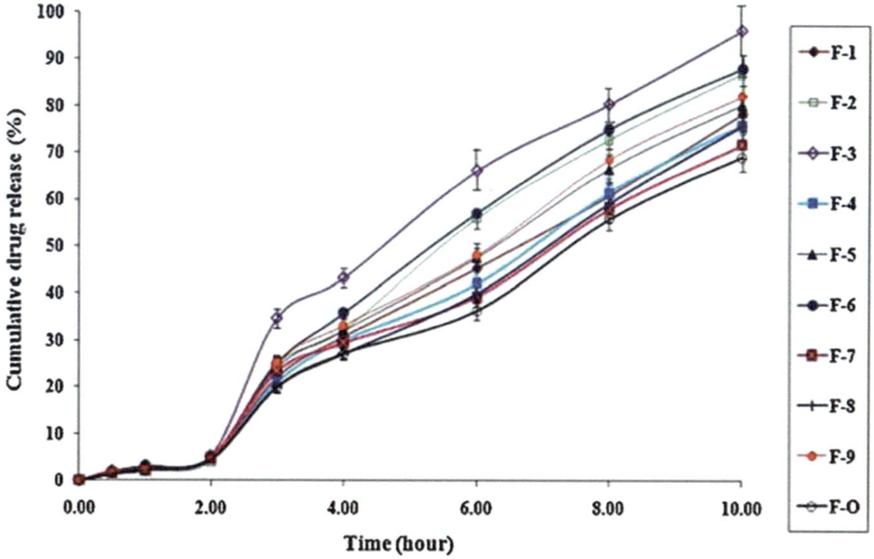

Fig. 22 In vitro drug releasing behavior of various ionotropically gelled tamarind gum-alginate beads of diclofenac sodium (Nayak and Pal 2011). Copyright © 2011 with permission from Elsevier Ltd.

5.3 Gastroretentive Tamarind Gum-Alginate Buoyant Beads (Oil-Entrapped) for Sustained Diclofenac Sodium Release

For the use in gastroretentive drug delivery, tamarind gum-alginate buoyant beads were developed on the basis of low-density oil entrapment technique and ionotropic emulsion-gelation technique (Nayak et al. 2013d). In the buoyant beads' formulation, tamarind gum-sodium alginate as polymer blends and $CaCl_2$ as ionotropic cross-linker were employed, while groundnut oil as low-density oil was entrapped to attain buoyancy for longer period. Because of the electrostatic interaction in between positively charged Ca^{2+}-ions and negatively charged sodium alginate, emulsion-gelled spherical tamarind gum-alginate beads of diclofenac sodium were generated instantly, when the emulsions containing tamarind gum, sodium alginate, groundnut oil, and diclofenac sodium were added into the cross-linking solutions containing $CaCl_2$. By applying a computer-assisted optimization process based on the 2^3-factorial design and response surface methodology, formulation development of these groundnut oil-entrapped tamarind gum-alginate of diclofenac sodium was carried out, where the influences of three independent variables (factors), i.e., polymer-to-drug ratio, oil-to-water ratio, and sodium alginate-to-tamarind gum ratio on the dependable variables (responses), i.e., drug entrapment efficiency, density of the beads, and cumulative drug release after 8 h were analyzed. The drug entrapment efficiency of optimized beads containing diclofenac sodium was found 82.48 ± 2.34%, while density was 0.88 ± 0.07 g/cm^3.

These groundnut oil-entrapped tamarind gum-alginate beads of diclofenac sodium were characterized by SEM, FTIR, and ^{1}H NMR. SEM analyses showed occurrence of rough surface of these beads with small pores, channels, and corrugations. Both FTIR and ^{1}H NMR analyses confirmed the compatibility of the diclofenac sodium with tamarind gum and sodium alginate employed to formulate the groundnut oil-entrapped tamarind gum-alginate beads of diclofenac sodium.

The results of in vitro floatation in simulated gastric fluid (pH 1.2) of these beads showed fine buoyancy over 8 h with floating lag time less than 10 min because of their low density. The entrapped low-density groundnut oil in these tamarind gum-alginate beads was accountable for the buoyancy. Actually, entrapped low-density oils can lesser the density of these beads. The in vitro drug releasing behaviors of these groundnut oil-entrapped tamarind gum-alginate buoyant beads of diclofenac sodium were analyzed in acidic stomach pH (using simulated gastric fluid, pH 1.2). The cumulative drug released from these buoyant beads of diclofenac sodium in 8 h was within the range from 33.00 ± 1.52 to 44.98 ± 1.22%. All groundnut oil-entrapped tamarind gum-alginate buoyant beads of diclofenac sodium showed sustained release of diclofenac sodium over a period of 8 h. The release of diclofenac sodium from these buoyant beads was found to follow controlled release of drug with super-case-II transport mechanism.

By using carrageenan-induced rat paw edema model, anti-inflammatory effect was evaluated for the optimized oil-entrapped tamarind gum-alginate buoyant beads of diclofenac sodium. For all time interval points, subsequent to oral administration of pure diclofenac sodium and optimized buoyant beads of diclofenac sodium, the mean increase in paw volumes was found significantly ($p < 0.05$) lower than that of the aqueous solution of sodium carboxymethyl cellulose (0.5% w/v). Again, after oral administration of optimized buoyant beads of diclofenac sodium, the mean paw volume was found significantly ($p < 0.05$) lower as compared to that of the pure diclofenac sodium after 2 h of oral administration. The oral administration of pure diclofenac sodium showed comparatively rapid paw edema inhibition than that of the optimized buoyant beads of diclofenac sodium tested indicating a rapid anti-inflammatory action up to the third hour of the treatment. After that, this was found to be decreased. The optimized oil-entrapped tamarind gum-alginate buoyant beads of diclofenac sodium demonstrated a slower paw edema inhibition as well as maintained increasing paw edema inhibition over a period of 8 h after oral administration indicating a sustained action of diclofenac sodium by the oral administration of the optimized beads.

5.4 *Tamarind Gum-Alginate IPN Microbeads Containing Diltiazem-Indion 254® Complex*

In a research, Kulkarni et al. (2014) have developed tamarind gum-alginate blends of IPN microbeads loaded with the controlled release of diltiazem HCl (a water-soluble drug used in hypertension) by combined ionotropic gelation/covalent cross-linking. In this research work, Kulkarni et al. (2014) have prepared a complex of diltiazem HCl for its greater solubility. They have used Indion 254® (a cation exchange resin) to prepare the diltiazem-Indion 254® complex, and the prepared diltiazem-Indion 254® was encapsulated within the tamarind gum-alginate IPN microbeads. These tamarind gum-alginate IPN microbeads encapsulated with diltiazem-Indion 254® showed drug entrapment efficiency within the range of 78.15–92.15%. Increasing drug entrapment efficiency in these IPN beads was observed with the decreasing concentration of $CaCl_2$ in the cross-linking solutions. This can be attributed by the fact that more Ca^{2+}-ions might be diffused into the diltiazem-Indion 254® complex, when higher concentration of $CaCl_2$ in the cross-linking solutions was employed. Accordingly, a greater quantity of encapsulated drug might be displaced from the diltiazem-Indion 254® complex. Thus, the free drug could be diffused out of the IPN microbeads, which might be resulted in decreased drug entrapment. The average sizes of these tamarind gum-alginate IPN microbeads were found within the range of 986–1257 mm. As compared to the only ionotropically cross-linked beads, the size of the combined ionotropic gelation/covalent cross-linked IPN microbeads was measured smaller and this might be because of rapid shrinking of the polymeric network after glutaraldehyde treatment. Again, with the increase in the concentration of glutaraldehyde, the sizes of these IPN microbeads were found decreased and this might be because of the development of stiffer polymeric -matrix at the higher densities of cross-linking agents during preparation.

These tamarind gum-alginate IPN microbeads containing diltiazem-Indion 254® complex were characterized by SEM, FTIR, XRD, TGA, and DSC analyses. SEM analyses demonstrated a spherically shaped structure of these IPN microbeads with very rough surface morphology. The formation of the IPN structure between alginate and tamarind gum within the formulated IPN microbeads of diltiazem-Indion 254® complex was confirmed by FTIR and TGA analyses. DSC and XRD analyses suggested that diltiazem was uniformly dispersed in an amorphous state in the tamarind gum-alginate IPN matrix.

In the in vitro drug release analyses, tamarind gum-alginate IPN microbeads formulated with diltiazem-Indion 254® complex showed drug release up to 9 h, while microbeads formulated with uncomplexed diltiazem showed drug release up to 4 h. The tamarind gum-alginate IPN microbeads formulated by higher glutaraldehyde concentration were found to be released the drug slowly as compared to that formulated by lower glutaraldehyde concentration. The in vivo pharmacokinetic parameters were measured in the Wistar rats, and the results showed that a relatively higher *AUC* values of these tamarind gum-alginate IPN microbeads of diltiazem-Indion 254® complex signifying the better bioavailability of diltiazem.

5.5 Tamarind Gum-Pectinate Gelled Mucoadhesive Beads for Sustained Metformin HCl Release

A novel ionotropically gelled mucoadhesive beads for the use in controlled sustained release delivery of metformin HCl were formulated using low methoxy pectin–tamarind gum polymer blends (Nayak et al. 2014a). In this work, tamarind gum was employed as polymer blends with low methoxy pectin as mucoadhesive agent as well as sustained drug releasing agent (i.e., release retardant). $CaCl_2$ was used as ionotropic cross-linker to prepare these mucoadhesive beads of metformin HCl. For the formulation optimization of tamarind gum-pectinate gelled mucoadhesive beads of metformin HCl, a computer-aided optimization technique based on the 3^2 (2 factors and 3 levels) factorial design was employed, where the influences of low methoxy pectin and tamarind gum amounts on drug encapsulation efficiency and cumulative metformin HCl release after 10 h were investigated based on the response surface methodology. Through the analyses by the response surface methodology, it was clear that increase in the drug encapsulation efficiency and decline in the cumulative metformin HC release after 10 h were observed with the increasing amounts of both the polymers (tamarind gum and low methoxy pectin) in these beads. The drug encapsulation efficiency of all these tamarind gum-pectinate gelled beads of metformin HCl was measured within the range of 68.80 ± 3.27–95.12 ± 4.26%. The amounts of both polymers used as polymer blends in these beads formulation significantly influenced the drug encapsulation efficiency ($p < 0.05$). The average bead sizes of these formulated dried tamarind gum-pectinate gelled beads of metformin HCl ranged 1.52 ± 0.12–1.93 ± 0.26 mm. An increasing bead sizes was found as the amounts of both polymers used as polymer blends in these bead formulation (tamarind gum and low methoxy pectin) were increased in the preparation of these beads of metformin HCl. Also, it was noticed that the increment in average bead size was proportional to the increase in drug encapsulation efficiency. The optimized tamarind gum-pectinate gelled beads of metformin HCl showed drug encapsulation efficiency of 95.12 ± 4.26% and average bead size (average diameter) of 1.93 ± 0.26 mm.

The optimized tamarind gum-pectinate gelled beads of metformin HCl was characterized by SEM and FTIR analyses. In the SEM analyses, the bead morphology showed spherical shaped beads with rough surface; while cracks and wrinkles were shown in the surface topography (Fig. 23). The FTIR analyses suggested that the tamarind gum-pectinate beads containing metformin HCl possessed significant characters of metformin HCl with the absence of any kinds of interaction in between metformin HCl and tamarind gum-pectinate polymer blends used.

The in vitro drug releasing pattern of these newly developed tamarind gum-pectinate gelled beads of metformin HCl was studied in the acidic medium (0.1 N HCl, pH 1.2) for initial 2 h and then in intestinal medium (phosphate buffer, pH 7.4) for next hours of the study. The in vitro drug releasing results displayed a slower and sustained drug releasing behavior of these tamarind gum-pectinate

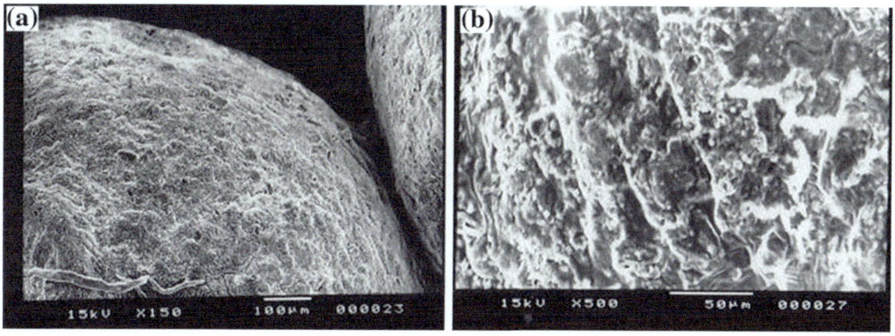

Fig. 23 SEM photographs of optimized tamarind gum-pectinate gelled beads of metformin HCl (Nayak et al. 2014a). Copyright © 2013 with permission from Elsevier Ltd.

gelled beads of metformin HCl over a prolonged period of 10 h (Fig. 24). During the initial period of in vitro drug releasing, very negligible amounts of metformin HCl (15% after 2 h) were observed. This occurrence might be because of the compactness of gel network through the shrinkage of pectinate gel at acidic pH. After that, comparatively faster drug releasing pattern was observed in the intestinal medium (phosphate buffer, pH 7.4). The cumulative percentage of in vitro drug releasing after 10 h from these tamarind gum-pectinate gelled beads of metformin HCl was measured within the range of 46.53 ± 3.28–90.45 ± 4.26%. These gelled beads of metformin HCl demonstrated controlled releasing (zero-order) pattern with super-case-II transport mechanism of drug release over a period of 10 h.

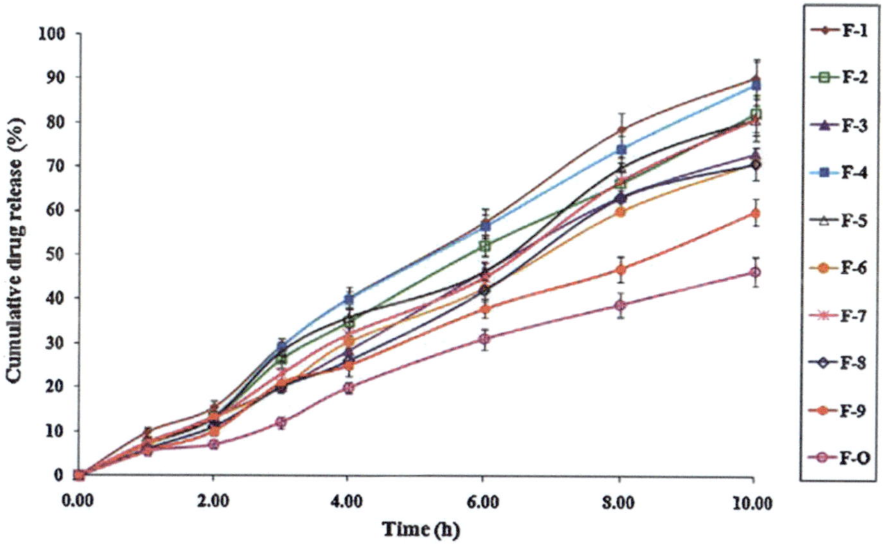

Fig. 24 In vitro drug releasing patterns of various tamarind gum-pectinate gelled beads of metformin HCl (Nayak et al. 2014a). Copyright © 2013 with permission from Elsevier Ltd.

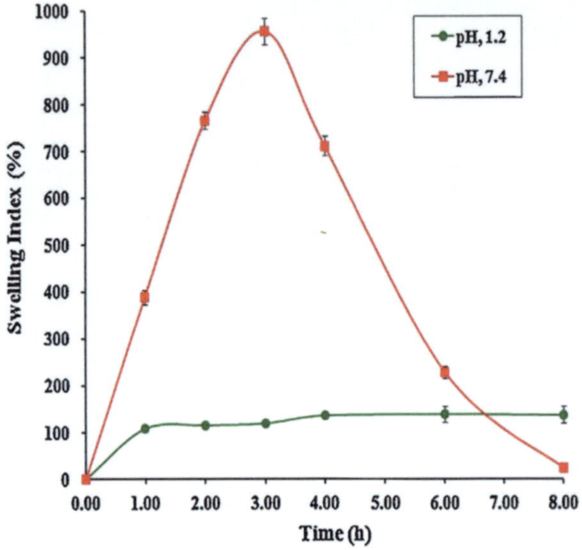

Fig. 25 Swelling behavior of optimized tamarind gum-pectinate gelled beads of metformin HCl in acidic medium (0.1 N HCl, pH 1.2) and alkaline medium (phosphate buffer, pH 7.4) (Nayak et al. 2014a). Copyright © 2013 with permission from Elsevier Ltd.

The optimized tamarind gum-pectinate gelled beads of metformin HCl also exhibited pH-dependent swelling (Fig. 25). The swelling index of these optimized beads of metformin HCl was found lower in the acidic medium (0.1 N HCl, pH 1.2) in comparison with that of in the alkaline medium (phosphate buffer, pH 7.4), initially. This occurred because of the compactness of gel network through the shrinkage of pectinate gel at the acidic pH medium. Maximum swelling index was seen in between 2 and 3 h in the alkaline medium (phosphate buffer, pH 7.4), and after which, erosion and dissolution of the swelled optimized beads occurred. The ex vivo wash off evaluation of these newly developed optimized tamarind gum-pectinate gelled beads of metformin HCl was performed using goat intestinal mucosal membrane in the gastric pH (1.2) and intestinal pH (7.4). The results of the ex vivo wash off showed that the percentage of optimized beads attached to the goat intestinal mucosa at the gastric pH was comparatively greater than that at the intestinal pH. The fewer percentage of optimized beads attached to the goat intestinal mucosa at the intestinal pH might be because of the ionization of $-COO^-$ groups and other functional groups of the polymers employed, which enhanced their solubility reducing the adhesive strength. The overall results of the ex vivo wash off evaluation suggested that the developed optimized tamarind gum-pectinate gelled beads of metformin HCl possessed good mucoadhesivity.

The in vivo pharmacodynamic study of these optimized tamarind gum-pectinate gelled beads of metformin HCl was carried out in the alloxan-induced diabetic male albino rats. The results of the in vivo pharmacodynamic study revealed significant hypoglycemic effect in the alloxan-induced diabetic rats over a prolonged period (10 h in this study) after oral administration of the optimized tamarind gum-pectinate gelled mucoadhesive beads of metformin HCl (Fig. 26). The results

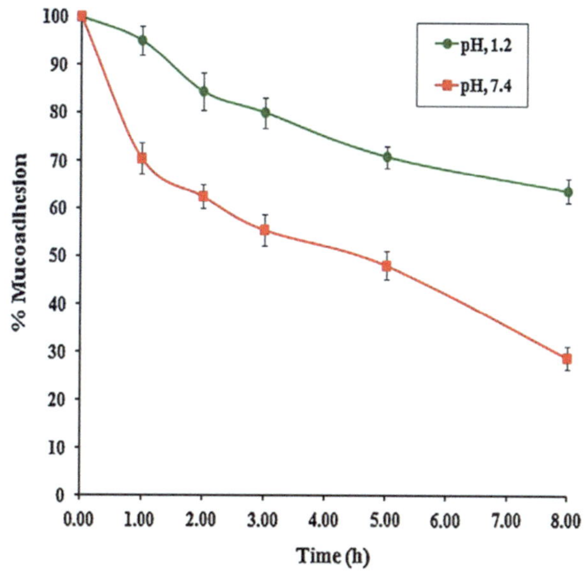

Fig. 26 Ex vivo wash off mucoadhesive behavior of optimized tamarind gum-pectinate gelled mucoadhesive beads of metformin HCl in acidic medium (0.1 N HCl, pH 1.2) and alkaline medium (phosphate buffer, pH 7.4) using goat intestinal mucosal membrane (Nayak et al. 2014a). Copyright © 2013 with permission from Elsevier Ltd.

of the study also clearly showed that these formulated gelled mucoadhesive beads containing metformin HCl made of tamarind gum-pectinate blends have the quality to maintain tight blood glucose level and improved patient compliance through improving, controlling, and prolonging systemic absorption of metformin HCl.

5.6 *Tamarind Gum–Gellan Gum Gelled Mucoadhesive Beads for Sustained Metformin HCl Release*

Gellan gum is an anionic biopolysaccharide produced by pure culture of *Psedomonas eloda* as a fermentation product, which is composed of tetrasaccharide, (1 → 4)-L-rhamnose-α (1 → 3)-D-glucose-β-(1 → 4)-D-glucuronic acid-β-(1 → 4)-D-glucose as a repeating unit in the molar ratio of 2:1:1 (Emeje et al. 2010; Vijan et al. 2012). In the current years, a great deal of research has been directed to the use of gelled gellan gum microparticles/beads for sustained drug releasing (Ahuja et al. 2010; Narker et al. 2010). The increasing use of gellan gum in designing various gelled gellan gum microparticles/beads is due to its biodegradability, biocompatibility, and ionotropically gel formation capacity by the influence of di-/trivalent metal cations (Nayak et al. 2014d, e). The mechanism of the ionotropic gelation characteristics of gellan gum by the influence of di-/trivalent metal cations entails the development of double-helical junctional zones followed by the aggregation of double-helical segments to structure three-dimensional network through the complexation by di-/trivalent metal cations as well as hydrogen bonding with the water molecules (Hamcerencu et al. 2008). These ionotropically gelled gellan gum microparticles/beads are found stable in the acidic pH than the alkaline pH because of comparatively speedy swelling in the alkaline pH than the acidic pH (Babu et al. 2010). Thus, these ionotropically gelled gellan gum microparticles/beads tend to drop their stability over the term because of the diffusion leading to the exchange of di-/trivalent metal cations for the monovalent ones. This results a premature release of encapsulated drugs from ionotropically gelled gellan gum microparticles/beads in the intestinal alkaline pH. To overcome the problem associated with the premature release of encapsulated drugs, numerous modifications of ionotropically gelled gellan gum microparticles/beads have been investigated by the researchers (Maiti et al. 2011; Jana et al. 2013b; Kulkarni et al. 2011, 2013). Recently, numerous researches based on the blendings of other natural polymers with the gellan gum in the development of ionotropically gelled gellan gum microparticles/beads for sustained drug releasing drug delivery have also been studied (Ahuja et al. 2010; Nayak et al. 2014c, d, e).

On the basis of the polymer-blendings approach, Nayak et al. (2014c) have developed ionotropically gelled tamarind gum–gellan gum mucoadhesive beads for sustained metformin HCl delivery over a longer period. In this research, Nayak et al. (2014c) have loaded metformin HCl within these tamarind gum–gellan gum beads through ionotropic gelation using Ca^{2+}-ion cross-linking. A computer-aided optimization technique based on the 3^2 (2 factors and 3 levels) factorial design was employed, where the influences of gellan gum-to-tamarind gum ratio and cross-linker ($CaCl_2$) concentration on the drug encapsulation efficiency and cumulative metformin HC release after 10 h of ionotropically gelled tamarind gum-gellan gum mucoadhesive beads were investigated based on the response surface methodology. Through the analyses by the response surface methodology, it was clear that an increase in the drug encapsulation efficiency was observed with the decrease in both gellan gum-to-tamarind gum ratio and $CaCl_2$ concentration as ionotropic cross-linker; a decline in the cumulative metformin HC release after 10 h was found to be decreased with the decrease in both gellan gum-to-tamarind gum ratio and $CaCl_2$ concentration. By applying numerical optimization technique, the optimal process variable setting was selected for the formulation of optimized ionotropically gelled tamarind gum–gellan gum beads of metformin HCl. The optimal process variable setting was selected as gellan gum-to-tamarind gum ratio of 1.14 and $CaCl_2$ concentration 1.91% w/v. The drug encapsulation efficiency and the mean diameter of these optimized beads of metformin HCl were found as 95.73 ± 4.02% and 1.70 ± 0.24 mm, respectively.

The optimized ionotropically gelled tamarind gum–gellan gum beads of metformin HCl were characterized for SEM and FTIR analyses. The spherically shaped discrete beads with an irregular and rough surface consisting of characteristic large wrinkles and cracks were seen in the SEM photograph of the optimized beads (Fig. 27). On the bead surface, few polymeric derbies and drug crystals were also seen. The FTIR analyses suggested the absence of any kinds of interaction in between metformin HCl and the polymers used as blends in the beads preparation (i.e., tamarind gum and gellan gum).

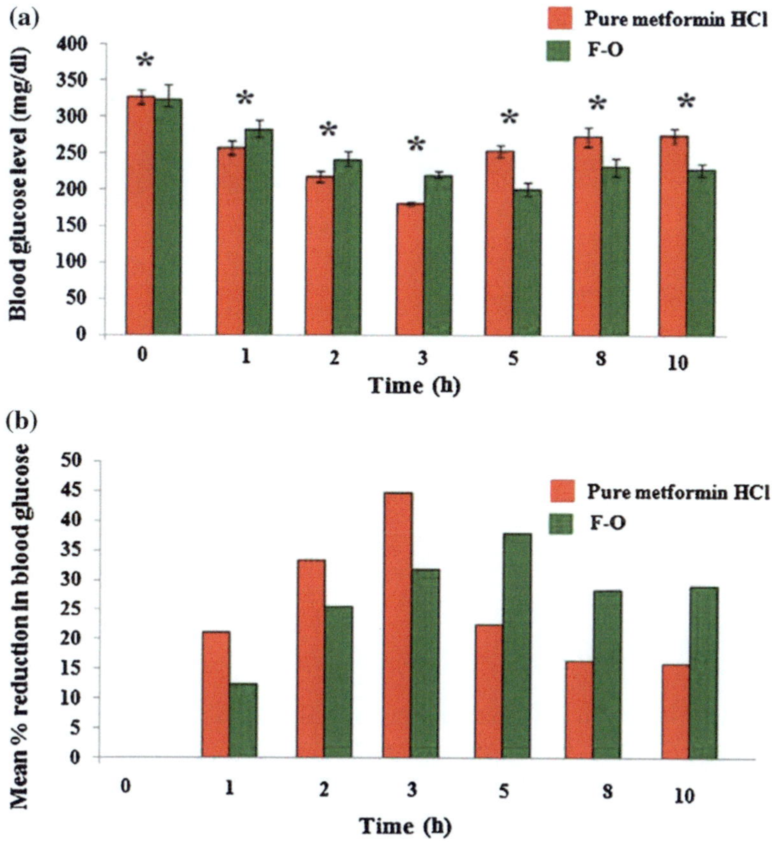

Fig. 27 **a** Comparative in vivo blood glucose level in alloxan-induced diabetic rats after oral administration of pure metformin HCl (standard) and optimized tamarind gum-pectinate gelled mucoadhesive beads of metformin HCl; **b** comparative in vivo mean percentage reduction in blood glucose level in alloxan-induced diabetic rats after oral administration of pure metformin HCl (standard) and optimized tamarind gum-pectinate gelled mucoadhesive beads of metformin HCl (Nayak et al. 2014a). Copyright © 2013 with permission from Elsevier Ltd.

The in vitro drug release study of these tamarind gum–gellan gum gelled beads of metformin HCl was carried out in the acidic medium (0.1 N HCl, pH 1.2) for initial 2 h and then in intestinal medium (phosphate buffer, pH 7.4) for next hours of the study. From the results of this in vitro drug release study, a prolonged sustained metformin HCl release from all these gelled beads was observed over a period of 10 h of study (Fig. 28). From these gelled beads, metformin HCl release was found as slower in the 0.1 N HCl (pH, 1.2), which was measured as less than 15.30% after 2 h of the in vitro drug release. After that, in the phosphate buffer (pH, 7.4), a comparatively faster metformin HCl release from these beads was noticed. This occurrence might be because of the fact that these beads swelled rapidly in the alkaline release medium than that in the acidic release medium, which could led to

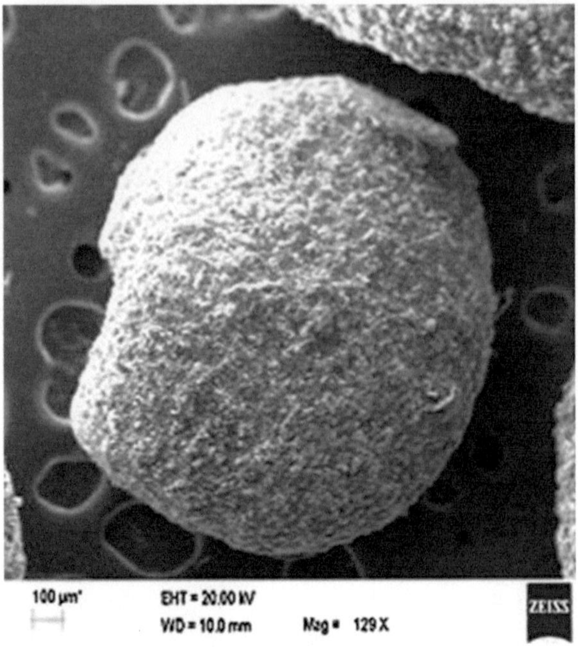

Fig. 28 SEM photographs of optimized tamarind gum–gellan gum gelled beads of metformin HCl (Nayak et al. 2014c). Copyright © 2013 with permission from Elsevier Ltd.

comparatively increased drug release in the alkaline release medium than that in the acidic release medium. In the alkaline release medium, perhaps a larger swelling force might be generated by the electrostatic repulsion between the ionized $–COO^{-}$ groups of the gellan gum structure. The major amount of metformin HCl release from these tamarind gum–gellan gum gelled beads at the initial stage of the in vitro drug release study could possibly be because of the presence of the surface adhered crystals of drug. These beads demonstrated a controlled release (zero-order) pattern with super-case-II transport mechanism controlled by the swelling and relaxation of polymeric-blends matrix.

The swelling pattern of optimized ionotropically gelled tamarind gum–gellan gum beads of metformin HCl was found lesser in the acidic gastric pH (1.2) as compared with that of in the alkaline intestinal pH (7.4) (Fig. 29). Within the 2–3 h of swelling in the alkaline pH, a maximum swelling of these beads was noticed. After that, erosion of the polymeric matrix and dissolution of the swelled beads occurred. The results of swelling pattern suggested that these tamarind gum-gellan gum beads of metformin HCl beads are able to swell in the stomach after oral administration. As they are subsequently transferred to upper part of intestine, the beads will begin to swell more and behave as matrices for the controlled releasing of incorporated drug. The optimized ionotropically gelled tamarind gum-gellan gum beads of metformin HCl also show good ex vivo mucoadhesivity with goat intestinal mucosa and found faster ex vivo wash off in the alkaline pH than that in the acidic pH (Fig. 30). This occurrence might be because of ionization of $–COO^{-}$ and other

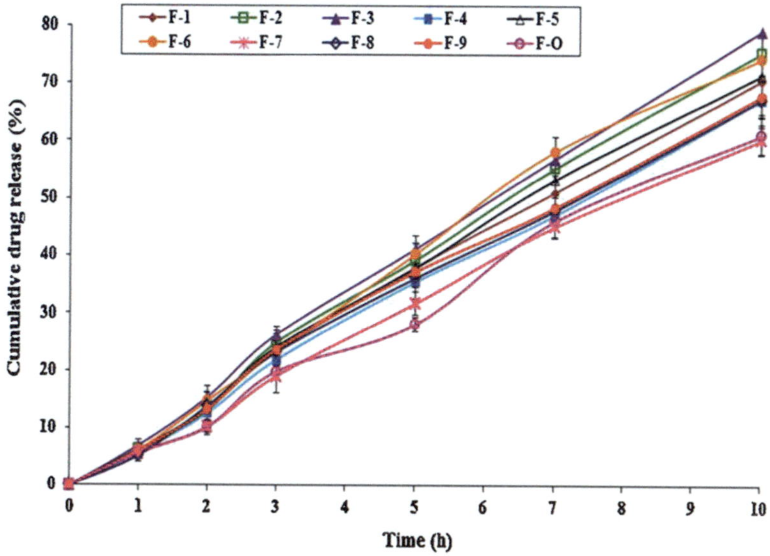

Fig. 29 In vitro drug releasing patterns of different tamarind gum–gellan gum gelled beads of metformin HCl (Nayak et al. 2014c). Copyright © 2013 with permission from Elsevier Ltd.

Fig. 30 Swelling behavior of optimized tamarind gum–gellan gum gelled beads of metformin HCl in acidic medium (0.1 N HCl, pH 1.2) and alkaline medium (phosphate buffer, pH 7.4) (Nayak et al. 2014c). Copyright © 2013 with permission from Elsevier Ltd.

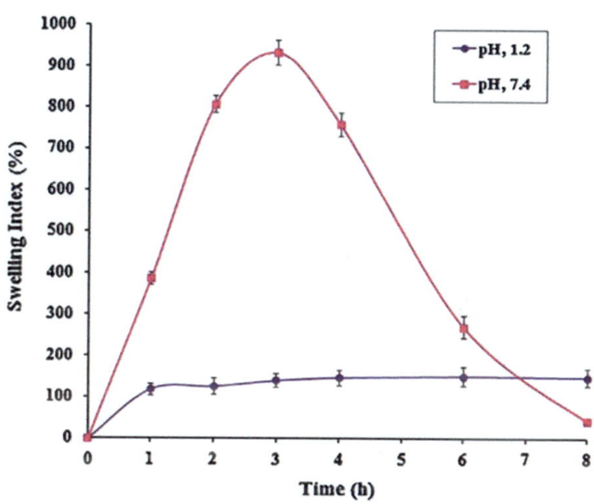

functional groups of the matrix structure. This also might enhance their solubility by dropping the adhesive strength.

The in vivo pharmacodynamic study of these optimized tamarind gum–gellan gum gelled mucoadhesive beads of metformin HCl was carried out in the alloxan-induced diabetic male albino rats using pure metformin HCl as a standard. Diabetic rats treated with pure metformin HCl exhibited a rapid reduction in the

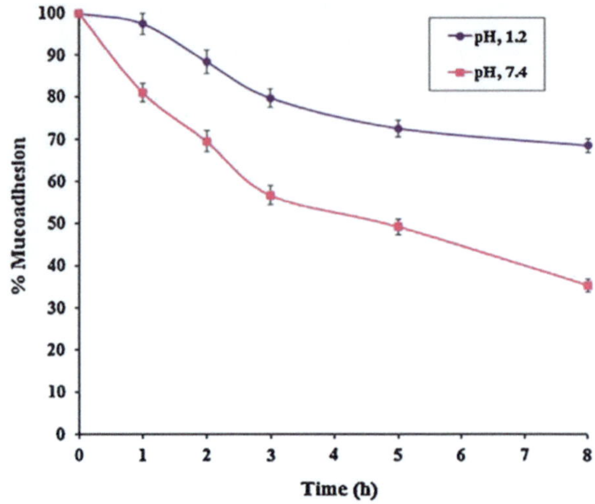

Fig. 31 Ex vivo wash off mucoadhesive behavior of optimized tamarind gum–gellan gum gelled mucoadhesive beads of metformin HCl in acidic medium (0.1 N HCl, pH 1.2) and alkaline medium (phosphate buffer, pH 7.4) using goat intestinal mucosal membrane (Nayak et al. 2014c). Copyright © 2013 with permission from Elsevier Ltd.

blood glucose level, which was identified within 3 h of oral administration. After that, the blood glucose level was found to be recovered rapidly toward the normal level. Diabetic rats treated with optimized mucoadhesive beads of metformin HCl exhibited a reduction in the blood glucose level, which was measured as slower up to 3 h of oral administration than that of the diabetic rats treated with pure metformin HCl (Fig. 31). After oral administration of pure metformin HCl and optimized tamarind gum–gellan gum gelled mucoadhesive beads of metformin HCl at each time point measured, significant differences ($p < 0.05$) were found between the blood glucose levels of the treated diabetic rats. However, the reduction in the blood glucose level was increased gradually with the increment of time in case of the diabetic rats treated with optimized mucoadhesive beads of metformin HCl and was observed to be sustained over a period of 10 h. Thus, a significant hypoglycemic effect in alloxan-induced diabetic rats by the effect of these newly developed optimized tamarind gum–gellan gum gelled mucoadhesive beads of metformin HCl was seen over prolonged period after oral administration.

5.7 Chitosan-Tamarind Gum IPN Microparticles for Sustained Release of Aceclofenac

In an investigation, Jana et al. (2013a) have developed chitosan-tamarind gum-based IPN microparticles for the sustained release of aceclofenac. These IPN microparticles of chitosan-tamarind gum were prepared through covalent cross-linking by glutaraldehyde at pH 5.5. The drug entrapment efficiency of these chitosan-tamarind gum IPN microparticles of aceclofenac was within the range of 85.84 ± 1.75 to 91.97 ± 1.30%. With the increasing concentration of glutaraldehyde as cross-linker, increasing drug entrapment efficiency in these IPN microparticles of aceclofenac was measured and this could be because of high degree of cross-linking. The highest drug entrapment efficiency was observed in case of the IPN microparticles of aceclofenac prepared using 400 mg of aceclofenac, 200 mg of chitosan, 400 mg of tamarind gum, and 2 ml of glutaraldehyde. The average sizes of these chitosan-tamarind gum IPN microparticles of aceclofenac were measured within the range of 490.55 ± 23.24 to 621.60 ± 53.57 μm. With the increasing concentration of glutaraldehyde as cross-linker, decrease in the microparticle sizes was observed. This occurrence could be because of the development of more rigid polymeric network at the higher cross-linking density when more concentration of glutaraldehyde was employed during preparation.

These chitosan-tamarind gum IPN microparticles of aceclofenac were characterized by SEM, FTIR, and DSC analyses. The SEM photographs of these IPN microparticles demonstrated almost spherically shaped morphology devoid of any agglomeration (Fig. 32). The surface topography of these IPN microparticles of aceclofenac was found very rough with the presence of some characteristic wrinkles. The characteristic wrinkles might be caused because of the partially collapsing of the polymeric gel network during drying. The development of IPN structure between tamarind gum and chitosan was suggested by the results of FTIR spectroscopic analyses. The probable reaction for the development of IPN structure between tamarind gum and chitosan was also projected by Jana et al. (2013a) (Fig. 33). DSC analyses indicated the stability status of aceclofenac in the chitosan-tamarind gum IPN matrix. The results of the DSC analyses demonstrated that aceclofenac present within the IPN microparticles was in stable condition, indicating no physical as well as chemical interactions between the aceclofenac and the chitosan-tamarind gum IPN polymeric matrix.

In vitro drug releasing behavior of these chitosan-tamarind gum IPN microparticles of aceclofenac was evaluated using dialysis bag diffusion technique in phosphate buffer (pH 6.8). The results of in vitro drug release from these IPN microparticles of aceclofenac demonstrated sustained release of aceclofenac over a period of 8 h (Fig. 34). With the increasing amount of glutaraldehyde employed for the preparation of chitosan-tamarind gum IPN microparticles of aceclofenac, increase in sustained aceclofenac release pattern from these IPN microparticles was observed. This occurrence might be because of higher degree of cross-linking by glutaraldehyde. These IPN microparticles of aceclofenac followed the Korsmeyer-Peppas model with anomalous (non-Fickian) diffusion mechanism indicating both diffusion and swelling controlled drug release.

In vivo anti-inflammatory effect of the chitosan-tamarind gum IPN microparticles of aceclofenac was evaluated by carrageenan-induced rat paw edema model using male Sprague Dawley rats. After oral administration of pure aceclofenac at a dose of 10 mg/kg body weight and chitosan-tamarind gum IPN microparticles of aceclofenac at a dose of equivalent to aceclofenac 10 mg/kg body weight, rats of all groups were challenged by a subcutaneous injection of 0.05 ml of 1% (w/v) solution of carrageenan in saline into the plantar site of left hind paw. The pure

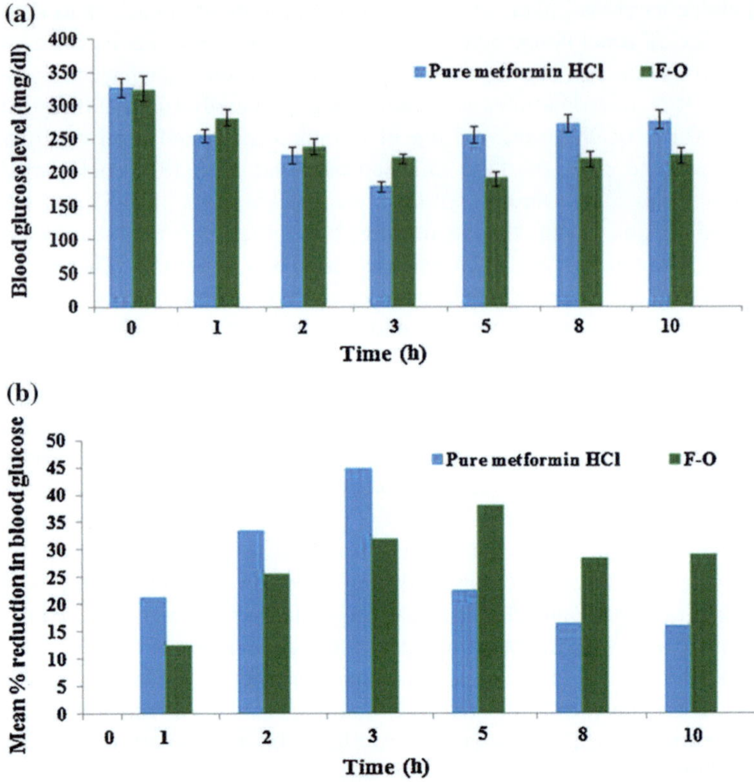

Fig. 32 **a** Comparative in vivo blood glucose level in alloxan-induced diabetic rats after oral administration of pure metformin HCl (standard) and optimized tamarind gum–gellan gum gelled mucoadhesive beads of metformin HCl; **b** comparative in vivo mean percentage reduction in blood glucose level in alloxan-induced diabetic rats after oral administration of pure metformin HCl (standard) and optimized calcium tamarind gum–gellan gum gelled mucoadhesive beads of metformin HCl (Nayak et al. 2014c). Copyright © 2013 with permission from Elsevier Ltd.

aceclofenac showed comparatively fast inhibition of swelling of rat paw edema than that of the chitosan-tamarind gum IPN microparticles of aceclofenac tested demonstrating a faster anti-inflammatory action up to the third hour and, after that, this anti-inflammatory action was diminished. The chitosan-tamarind gum IPN microparticles of aceclofenac demonstrated a slower inhibition of swelling of rat paw edema and maintained increasing inhibition of rat paw edema over 6 h after oral administration indicating sustained action of aceclofenac (Fig. 35).

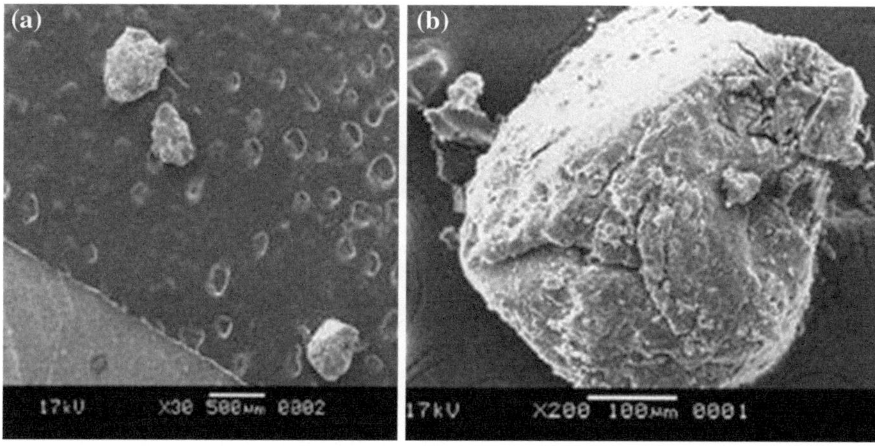

Fig. 33 SEM photographs of chitosan-tamarind gum IPN microparticles of aceclofenac (Jana et al. 2013a). Copyright © 2013 with permission from Elsevier Ltd.

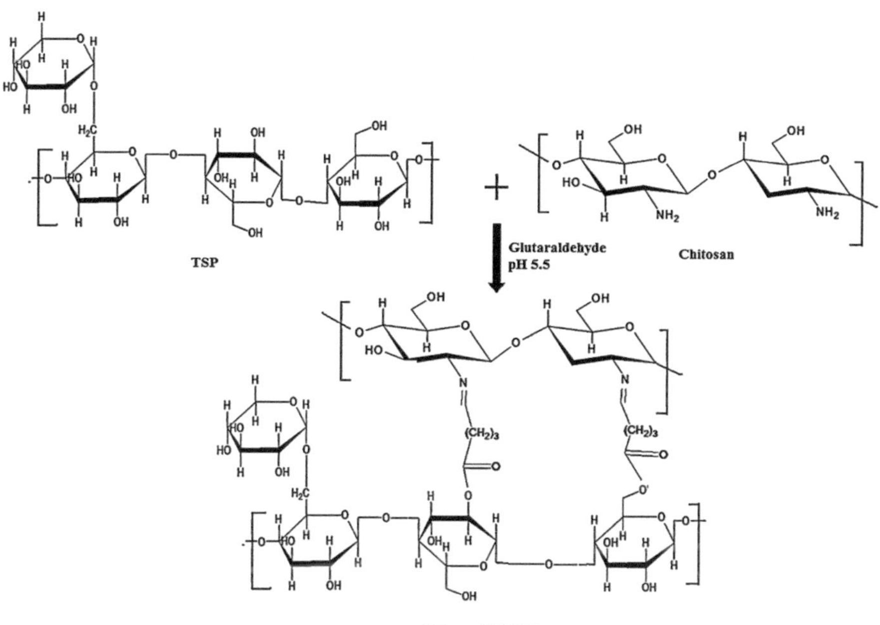

Fig. 34 Probable reaction for the development of IPN structure between tamarind gum and chitosan was also projected by Jana et al. (2013a). Copyright © 2013 with permission from Elsevier Ltd.

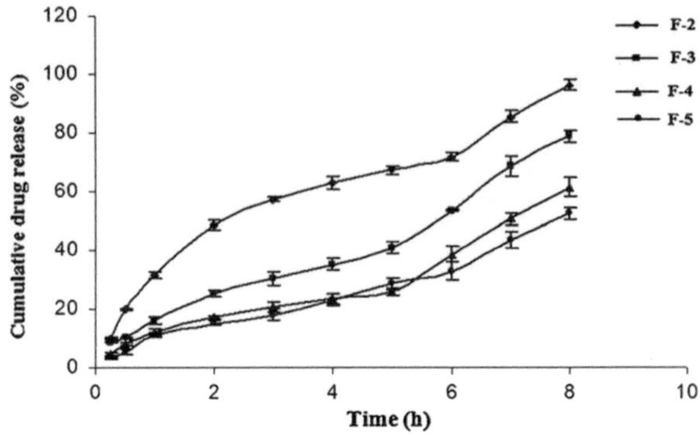

Fig. 35 In vitro drug release from various chitosan-tamarind gum IPN microparticles of aceclofenac in phosphate buffer (pH 6.8) (Jana et al. 2013a). Copyright © 2013 with permission from Elsevier Ltd.

6 Conclusion

In recent years, drug delivery researchers and formulation scientists are attracted to utilize various plant-derived polysaccharides as pharmaceutical excipients to design various pharmaceutical dosage forms. Among various plant polysaccharides used as pharmaceutical excipients, sterculia gum and tamarind gum are two commonly utilized biopolymers employed in designing various sustained drug releasing multiple unit dosage forms for oral use. The present chapter revealed a comprehensive discussion on the different approaches to the utility of sterculia gum and tamarind gum for the development of gelled microparticles/beads for the use in sustained drug releasing applications. These sterculia gum- and tamarind gum-based microparticle/bead systems for sustained drug delivery were formulated through different techniques such as ionotropic gelation, ionotropic emulsion gelation, covalent cross-linking, combined ionotropic gelation/covalent cross-linking. All these sterculia gum- and tamarind gum-based microparticles/beads were found capable of encapsulating higher quantity of drugs and releasing encapsulated drugs in sustained manner over a prolonged period. Some of these sterculia gum- and tamarind gum-based microparticles/beads exhibited pH-dependent swelling and good mucoadhesivity with the biological membrane. The in vivo performances of these microparticles/beads made of sterculia gum and tamarind gum encapsulated with different kinds of drugs exhibited significant results to support the controlled systemic absorption of drugs over a prolonged period. This chapter will be helpful in applying appropriate strategies or achieving desired controlled sustained drug release profiles from microparticles/beads made of sterculia gum and tamarind gum.

References

Ahuja M, Yadav M, Kumar S (2010) Application of response surface methodology to formulation of ionotropically gelled gum cordial/gellan beads. Carbohydr Polym 80:161–167

Al-Kassas R, Al-Gohary OMN, Al-Fadhel MM (2007) Controlling of systemic absorption of gliclazide through incorporation into alginate beads. Int J Pharm 341:230–237

Anderson DMW (1989) Evidence of safety of gum karaya (*Sterculia* spp.) as a food additive. Food Additive Contamin: Part A 6:189–199

Anderson DMW, Wang WP (1994) The tree exudate gums permitted in foodstuffs as emulsifiers, stabilisers and thickeners. Chem Ind Forest Products 14:73–83

Angadi SC, Manjeswar LS, Aminabhavi TM (2010) Interpenetrating polymer network blend microspheres of chitosan and hydroxethyl cellulose for controlled release of isoniazid. Int J Biol Macromol 47:171–179

Atyabi F, Majzoob S, Iman M, Salehi M, Dorkoosh F (2005) In vitro evaluation and modification of pectinate gel beads containing trimethyl chitosan, as a multi-particulate system for delivery of water-soluble macromolecules to colon. Carbohydr Polym 61:39–51

Avachat AM, Dash RR, Shrotriya SN (2011) Recent investigations of plant based natural gums, mucilages and resins in novel drug delivery systems. Indian J Pharm Educ Res 45:86–99

Avachat A, Gujar KN, Wagh KV (2013) Development and evaluation of tamarind seed glucan-based mucoadhesive buccal films of rizatriptan benzoate. Carbohydr Polym 91:537–542

Babu VR, Sairam M, Hosamani KM, Aminabhavi TM (2007) Preparation of sodium alginate-methyl cellulose blend microspheres for controlled release of nifedipine. Carbohydr Polym 69:241–250

Banerjee S, Chaurasia G, Pal D, Ghosh AK, Ghosh A, Kaity S (2010) Investigation on crosslinking density for development of novel interpenetrating polymer network (IPN) based formulation. J Sci Ind Res 69:777–784

Banerjee S, Siddiqui L, Bhattacharya SS, Kaity S, Ghosh A, Chattopadhyay P, Pandey A, Singh L (2012) Interpenetrating polymer network (IPN) hydrogel microspheres for oral controlled release application. Int J Biol Macromol 50:198–206

Banerjee S, Singh S, Bhattacharya SS, Chattopadhyay P (2013) Trivalent ion cross-linked pH sensitive alginate-methyl cellulose blend hydrogel beads from aqueous template. Int J Biol Macromol 57:297–307

Behall KM, Schofield DJ, Lee K, Powell AS, Mores PB (1987) Mineral balance in adult men: effect of four fibers. Am J Clin Nutri 46:307–314

Bera H, Boddupalli S, Nayak AK (2015a) Mucoadhesive-floating zinc-pectinate-sterculia gum interpenetrating polymer network beads encapsulating ziprasidone HCl. Carbohydr Polym 131:108–118

Bera H, Kandukuri SG, Nayak AK, Boddupalli S (2015b) Alginate-sterculia gum gel-coated oil-entrapped alginate beads for gastroretentive risperidone delivery. Carbohydr Polym 120:74–84

Capron JP, Zeitoun P, Julien DA (1981) A multicenter controlled trial of a combination of kaolin, sterculia gum, meprobamate and magnesium salts, in the irritable bowel syndrome (Authors Transl.). Gastroenterol Clin Biol 5:67–72

Cerf DL, Irinei F, Muller G (1990) Solution properties of gum exudates from *Sterculia urens* (karaya gum). Carbohydr Polym 13:375

Chakraborty S, Khandai M, Sharma A, Khanam N, Patra CN, Dinda SC, Sen KK (2010) Preparation, in vitro and in vivo evaluation of algino-pectinate bioadhesive microspheres: An investigation of the effects of polymers using multiple comparison analysis. Acta Pharma 60:255–266

Chanda R, Mahapatro SK, Mitra T, Roy A, Bahadur S (2008) Development of oral mucoadhesive tablets of tarbutalline sulphate using some natural materials extracted from *Albelmoschus esculeatus* and *Tamarindus indica*. Res J Pharm Res 1:46–51

Chandramouli Y, Firoz S, Vikram A, Padmaja C, Chakravarthi RN (2012) Design and evaluation of controlled release matrix tablets of acyclovir sodium using tamarind seed polysaccharide. J Pharm Biol 2:55–62

Das S, Ng K-Y, Ho PC (2010) Formulation and optimization of zinc-pectinate beads for the controlled delivery of resveratrol. AAPS PharmSciTech 11:729–742

Das B, Dutta S, Nayak AK, Nanda U (2014) Zinc alginate-carboxymethyl cashew gum microbeads for prolonged drug release: development and optimization. Int J Biol Macromol 70:505–515

Datta R, Bandyopadhyay AK (2006) A new nasal drug delivery system for diazepam using natural mucoadhesive polysaccharide obtained from tamarind seeds. Saudi Pharm J 14:115–119

Desai KGH (2005) Preparation and characteristics of high-amylose corn starch/pectin blend microparticles: a technical note. AAPS PharmSciTech 6:article 30

Deshmukh VN, Jadhav JK, Sakarkar DM (2009) Formulation and in vitro evaluation of theophylline anhydrous bioadhesive tablets. Asian J Pharm 3:54–58

Deveswaran R, Abraham S, Bharath S, Basavraj BV, Furtado S, Madhavan V (2009) Design and characterization of diclofenac sodium tablets containing tamarind seed polysaccharide as release retardant. Int J PharmTech Res 1:191–195

Emeje MO, Franklin-Ude PIF, Odoefule SI (2010) Evaluation of fluid uptake kinetics and drug release from gellan gum tablets containing metronidazole. Int J Biol Macromol 47:158–163

Gauthami S, Bhat VR (1992) A monograph on gum karaya. National Institution of Nutrition, Indian Council of Medical Research, Hyderabad, India

Gheraldi E, Tavanti A, Celandroni F, Lupetti A, Blandizzi C, Boldrini E, Campa M, Senesi S (2000) Effect of novel mucoadhesive polysaccharide obtained from tamarind seeds on the intraocular penetration of gentamicin and ofloxacin in rabbits. J Antimicrob Chemother 46:831–834

Goh CH, Heng PWS, Chan LW (2012) Alginates as a useful natural polymer for microencapsulation and therapeutic applications. Carbohydr Polym 88:1–12

Guerre J, Neuman M (1979) Treatment of cronic colonic diseases with a new topical digestive agent, mucilage (karaya gum) combined with polyvinyl polypyrrolidone (PVPP). Med Chirurgie Digest 8:679–682

Gupta V, Puri R, Gupta S, Jain S, Rao GK (2010) Tamarind kernel gum: an upcoming natural polysaccharide. Sys Rev Pharm 1:50–54

Guru PR, Nayak AK, Sahu RK (2013) Oil-entrapped sterculia gum–alginate buoyant systems of aceclofenac: development and in vitro evaluation. Colloids Surf B: Biointerf 104:268–275

Hamcerencu M, Desbrieres J, Khoukh A, Popa M, Riess G (2008) Synthesis and characterization of new unsaturated esters of gellan gum. Carbohydr Polym 71:92–100

Hasnain MS, Nayak AK, Singh R, Ahmad F (2010) Emerging trends of natural-based polymeric systems for drug delivery in tissue engineering applications. Sci J UBU 1:1–13

Huttel E (1983) Treatment of acute diarrhoea in general practice. Therapeutic experiences with karaya bismuth. Die Medizinishe Welt 34:1383–1384

Jana S, Saha A, Nayak AK, Sen KK, Basu SK (2013a) Aceclofenac-loaded chitosan-tamarind seed polysaccharide interpenetrating polymeric network microparticles. Colloids Surf B: Biointerf 105:303–309

Jana S, Das A, Nayak AK, Sen KK, Basu SK (2013b) Aceclofenac-loaded unsaturated esterified alginate-gellan gum microspheres: in vitro and in vivo assessment. Int J Biol Macromol 57:129–137

Jana S, Gangopadhaya A, Bhowmik BB, Nayak AK, Mukhrjee A (2015) Pharmacokinetic evaluation of testosterone-loaded nanocapsules in rats. Int J Biol Macromol 72:28–30

Jenkins AD, Kratochivil P, Stepto RFT, Suter UW (1996) Glossary of basic terms in polymer science (IUPAC recommendations 1996). Pure Appl Chem 68:2287–2311

Joseph J, Kanchalochana SN, Rajalakshmi G, Hari V, Durai RD (2012) Tamarind seed polysaccharide: a promising natural excipient for pharmaceuticals. Int J Green Pharm 6:270–278

Kaity S, Isaac J, Ghosh A (2013) Interpenetrating polymer network of locust bean gum-poly (vinyl alcohol) for controlled release drug delivery. Carbohydr Polym 94:456–467

Kaur H, Ahuja M, Kumar S, Dilbaghi N (2012a) Carboxymethyl tamarind kernel polysaccharide nanoparticles for ophthalmic drug delivery. Int J Biol Macromol 50:833–839

Kaur H, Yadav S, Ahuja M, Dilbaghi N (2012b) Synthesis, characterization and evaluation of thiolated tamarind seed polysaccharide as a mucoadhesive polymer. Carbohydr Polym 90:1543–1549

Khanna M, Nandi RC, Sarin JP (1987) Standardization of tamarind seed polyose for pharmaceutical use. Indian Drugs 24:268–269

Kulkarni GT, Gowthamarajan K, Dhobe RR, Yohanan F, Suresh B (2005) Development of controlled release spheroids using natural polysaccharide as release modifier. Drug Deliv 12:201–206

Kulkarni RV, Mangod BS, Mutalik S, Sa B (2011) Interpenetrating polymer network microcapsules of gellan gum and egg albumin entrapped with diltiazem-resin complex for controlled release application. Carbohydr Polym 83:1001–1007

Kulkarni RV, Mutalik S, Mangond BS, Nayak UY (2012) Novel interpenetrated polymer network microbeads of natural polysaccharides for modifi ed release of water soluble drug: in vitro and in vivo evaluation. J Pharm Pharmacol 64:530–540

Kulkarni RV, Nagathan VV, Biradar PR, Naikawadi AA (2013) Simvastatin loaded composite polyspheres of gellan gum and carrageenan: in vitro and in vivo evaluation. Int J Biol Macromol 57:238–244

Kulkarni RV, Patel FS, Nanjappaiah HM, Naikawadi AA (2014) In vitro and in vivo evaluation of novel interpenetrating polymer network microparticles containing repaglinide. Int J Biol Macromol 69:514–522

Kumar P, Singh I (2010) Formulation and characterization of tramadol-loaded IPN microgels of alginate and gelatin: optimization using response surface methodology. Acta Pharm 60:295–310

Lang P, Masci G, Dentini M, Crescenzi V, Cooke D, Gidley MJ, Fanutti C, Reid JSG (1992) Tamarind seed polysaccharide: preparation, characterization and solution properties of carboxylated, sulphated and alkalaminated derivatives. Carbohydr Polym 17:185–198

Lankalapalli S, Kolapalli RM (2012) Biopharmaceutical evaluation of diclofenac sodium controlled release tablets prepared from gum karaya-chitosan polyelectrolyte complexes. Drug Dev Ind Pharm 38:815–824

Lee SJ, Kim SS, Lee YM (2000) Interpenetrating polymer network hydrogels based on poly (ethylene glycol) macromer and chitosan. Carbohydr Polym 41:197–205

Leung AY (1980) Encyclopedia of common natural ingredients used in foods, drugs and cosmetics. Wiley, New York

Maiti S, Ranjit S, Mondol R, Ray S, Sa B (2011) Al^{3+} ion cross-linked and acetalated gellan hydrogel network beads for prolonged release of glipizide. Carbohydr Polym 85:164–172

Malakar J, Nayak AK (2012) Formulation and statistical optimization of multiple-unit ibuprofen-loaded buoyant system using 2^3-factorial design. Chem Eng Res Des 9:1834–1846

Malakar J, Nayak AK, Pal D (2012) Development of cloxacillin loaded multiple-unit alginate-based floating system by emulsion–gelation method. Int J Biol Macromol 50:138–147

Malakar J, Nayak AK, Jana P, Pal D (2013) Potato starch-blended alginate beads for prolonged release of tolbutamide: development by statistical optimization and in vitro characterization. Asian J Pharm 7:43–51

Manchanda R, Arora SC, Manchanda R (2014) Tamarind seed polysaccharide and its modification-versatile pharmaceutical excipients—a review. Int J PharmTech Res 6:412–420

Meier P, Seiler WO, Stahelin HB (1990) Bulk-forming agents as laxatives in geriatric patients. Schweizerische Medizinische Wochenschrift 120:314–317

Mishra MU, Khandare JN (2011) Evaluation of tamarind seed polysaccharide as a biodegradable carrier for colon specific drug delivery. Int J Pharm Pharmaceut Sci 3:139–142

Munarin F, Tanzi MC, Petrini P (2012) Advances in biomedical applications of pectin gels. Int J Biol Macromol 51:681–689

Munday DL, Philip JC (2000) Compressed xanthan gum and karaya gum matrices: Hydration, erosion and drug release mechanism. Int J Pharm 203:179–192

Nandi RC (1975) A process for preparation of polyose from the seeds of *Tamarindus indica*. Indian Patent 142092

Narker M, Sher P, Pawar A (2010) Stomach-specific controlled release gellan beads of acid-soluble drug prepared by ionotropic gelation method. AAPS PharmSciTech 11:267–277

Nayak AK (2016) Tamarind seed polysaccharide-based multiple-unit systems for sustained drug release. In: Kalia S, Averous L (eds) Biodegradable and bio-based polymers: environmental and biomedical applications. Wiley-Scrivener, USA, pp 469–492

Nayak AK, Pal D (2011) Development of pH-sensitive tamarind seed polysaccharide-alginate composite beads for controlled diclofenac sodium delivery using response surface methodology. Int J Biol Macromol 49:784–793
Nayak AK, Pal D (2012) Natural polysaccharides for drug delivery in tissue engineering. Everyman's Sci XLVI:347–352
Nayak AK, Pal D (2013) Ionotropically-gelled mucoadhesive beads for oral metformin HCl delivery: formulation, optimization and antidiabetic evaluation. J Sci Ind Res 72:15–22
Nayak AK, Pal D (2014) Blends of jackfruit seed starch-pectin in the development of mucoadhesive beads containing metformin HCl. Int J Biol Macromol 62:137–145
Nayak AK, Pal D (2015) Chitosan-based interpenetrating polymeric network systems for sustained drug release. In: Tiwari A, Choi J-W (eds) Advanced theranostics materials. Wiley-Scrivener, USA, pp 207–232
Nayak AK, Pal D (2016a) Sterculia gum-based hydrogels for drug delivery applications. In: Kalia S (ed) Polymeric hydrogels as smart biomaterials. Springer, Switzerland, pp 105–151
Nayak AK, Pal D (2016b) Sterculia gum-based hydrogels for drug delivery applications. In: Kalia S (ed) Polymeric hydrogels as smart biomaterials. Springer, Switzerland, pp 105–151
Nayak AK, Hasnain MS, Beg S, Alam MI (2010) Mucoadhesive beads of gliclazide: design, development and evaluation. Sci Asia 36:319–325
Nayak AK, Khatua S, Hasnain MS, Sen KK (2011) Development of alginate-PVP K 30 microbeads for controlled diclofenac sodium delivery using central composite design. DARU J Pharm Sci 19:356–366
Nayak AK, Das B, Maji R (2012) Calcium alginate/gum Arabic beads containing glibenclamide: development and in vitro characterization. Int J Biol Macromol 51:1070–1078
Nayak AK, Pal D, Das S (2013a) Calcium pectinate-fenugreek seed mucilage mucoadhesive beads for controlled delivery of metformin HCl. Carbohydr Polym 96:349–357
Nayak AK, Pal D, Hasnain MS (2013b) Development and optimization of jackfruit seed starch-alginate beads containing pioglitazone. Curr Drug Deliv 10:608–619
Nayak AK, Pal D, Pradhan J, Hasnain MS (2013c) Fenugreek seed mucilage-alginate mucoadhesive beads of metformin HCl: design, optimization and evaluation. Int J Biol Macromol 54:144–154
Nayak AK, Pal D, Malakar J (2013d) Development, optimization and evaluation of emulsion-gelled floating beads using natural polysaccharide-blend for controlled drug release. Polym Eng Sci 53:338–350
Nayak AK, Pal D, Santra K (2014a) Development of calcium pectinate-tamarind seed polysaccharide mucoadhesive beads containing metformin HCl. Carbohydr Polym 101:220–230
Nayak AK, Pal D, Santra K (2014b) Development of pectinate-ispagula mucilage mucoadhesive beads of metformin HCl by central composite design. Int J Biol Macromol 66:203–221
Nayak AK, Pal D, Santra K (2014c) Tamarind seed polysaccharide-gellan mucoadhesive beads for controlled release of metformin HCl. Carbohydr Polym 103:154–163
Nayak AK, Pal D, Santra K (2014d) Ispaghula mucilage-gellan mucoadhesive beads of metformin HCl: development by response surface methodology. Carbohydr Polym 107:40–41
Nayak AK, Pal D, Santra K (2014e) *Artocarpus heterophyllus* L. seed starch-blended gellan gum mucoadhesive beads of metformin HCl. Int J Biol Macromol 65:329–339
Nayak AK, Pal D, Santra K (2015) Screening of polysaccharides from tamarind, fenugreek and jackfruit seeds as pharmaceutical excipients. Int J Biol Macromol 79:756–760
Nayak AK, Pal D, Santra K (2016) Swelling and drug release behavior of metformin HCl-loaded tamarind seed polysaccharide-alginate beads. Int J Biol Macromol 82:1023–1027
Oliveira GF, Ferrari PC, Carvalho LQ, Evangelista RC (2010) Chitosan–pectin multiparticulate systems associated with enteric polymers for colonic drug delivery. Carbohydr Polym 82:1004–1009
Pal D, Mitra S (2010) A preliminary study on the in vitro antioxidant activity of the stems of *Opuntia vulgaris*. J Adv Pharm Technol Res 1:268–272
Pal D, Nayak AK (2011) Development, optimization and anti-diabetic activity of gliclazide-loaded alginate-methyl cellulose mucoadhesive microcapsules. AAPS PharmSciTech 12:1431–1441

Pal D, Nayak AK (2012) Novel tamarind seed polysaccharide-alginate mucoadhesive microspheres for oral gliclazide delivery: in vitro-in vivo evaluation. Drug Deliv 19:123–131
Pal D, Nayak AK (2016) Interpenetrating polymer networking systems of natural polymeric blends: drug delivery. In: Mishra M (ed) Encyclopedia of biomedical polymers and polymeric biomaterials. Taylor & Francis Group, New York
Pal D, Nayak AK, Kalia S (2010) Studies on *Basella alba* L. leaves mucilage: evaluation of suspending properties. Int J Drug Discov Technol 1:15–20
Pal D, Banerjee S, Ghosh A (2012) Dietary-induced cancer prevention: an expanding research arena of emerging diet related to healthcare system. J Adv Pharm Technol Res 3:16–24
Park CR, Munday PL (2004) Evaluation of selected polysaccharide excipients in buccoadhesive tablets for sustained release of nicotine. Drug Dev Ind Pharm 30:609–617
Pliszczak D, Bordes C, Bourgeois S, Marote P, Zahouani H, Tupin S et al (2012) Mucoadhesion evaluation of polysaccharide gels for vaginal application by using rheological and indentation measurements. Colloids Surf B: Biointerf 92:168–174
Prajapati VD, Jani GK, Moradiya NG, Randeria NP (2013) Pharmaceutical applications of various natural gums, mucilages and their modified forms. Carbohydr Polym 92:1685–1699
Prezotti FG, Cury BSF, Evangelista RC (2014) Mucoadhesive beads of gellan gum/pectin intended to controlled delivery of drugs. Carbohydr Polym 113:286–295
Rao PS, Srivastava HC (1973) Tamarind. In: Whistler RL (ed) Industrial gums, 2nd edn. Academic Press, New York, pp 369–411
Rao PS, Ghosh TP, Krishna S (1946) Extraction and purification of tamarind seed polysaccharide. J Sci Ind Res 4:705
Rees DA (1981) Polysaccharide shapes and their interactions-some recent advances. Pure Appl Chem 53:1–14
Rokhade AP, Patil SA, Aminabhavi TM (2007) Synthesis and characterization of semi-interpenetrating polymer network microspheres of acrylamide grafted dextran and chitosan for controlled release of acyclovir. Carbohydr Polym 67:605–613
Samal PK, Dangi JS (2014) Isolation, preliminary characterization and hepatoprotective activity of polysaccharides from *Tamarindus indica* L. Carbohydr Polym 102:1–7
Saravanan M, Rao KP (2010) Pectin–gelatin and alginate–gelatin complex coacervation for controlled drug delivery: Influence of anionic polysaccharides and drugs being encapsulated on physicochemical properties of microcapsules. Carbohy Polym 80:808–816
Saravanan M, Kishore GS, Ramachandran S, Rao GS, Sridhar SK (2002) Preparation and characterization of nimesulide microcapsules by gelatin–pectin complex coacervation. Indian Drugs 39:368–372
Sharma R, Ahuja M (2011) Thiolated pectin: synthesis, characterization and evaluation as a mucoadhesive polymer. Carbohydr Polym 85:658–663
Singh B, Chauhan D (2011) Barium ions crosslinked alginate and sterculia gum-based gastrortentive floating drug delivery system for use in peptic ulcers. Int J Polymer Mater 60:684–705
Singh B, Pal L (2012) Sterculia crosslinked PVA and PVA-poly(AAm)hydrogel wound dressings for slow drug delivery: mechanical, mucoadhesive, biocompatible and permeability properties. J Mech Behav Biomed Mater 9:9–21
Singh B, Sharma N (2008) Modification of sterculia gum with methacrylic acid to prepare a novel drug delivery system. Int J Biol Macromol 43:142–150
Singh B, Sharma N (2011) Design of sterculia gum based double potential antidiarrheal drug delivery system. Colloids Surf B: Biointerf 82:325–332
Singh B, Vashishtha M (2008) Development of novel hydrogels by modification of sterculia gum through radiation cross-linking polymerization for use in drug delivery. Nucl Instr Methods Phys Res Sec B: Beam Interact Mater Atoms 266:2009–2020
Singh B, Sharma V, Chouhan D (2010) Gastroretentive floating sterculia-alginate beads for use in antiulcer drug delivery. Chem Eng Res Des 88:997–1012
Singh S, Sharma V, Pal L (2011) Formation of sterculia polysaccharide networks by gamma rays induced graft copolymerization for biomedical applications. Carbohydr Polym 86:1371–1380

Sinha P, Ubaidulla U, Hasnain MS, Nayak AK, Rama B (2015a) Alginate-okra gum blend beads of diclofenac sodium from aqueous template using $ZnSO_4$ as a cross-linker. Int J Biol Macromo 79:555–563

Sinha P, Ubaidulla U, Nayak AK (2015b) Okra (*Hibiscus esculentus*) gum-alginate blend mucoadhesive beads for controlled glibenclamide release. Int J Biol Macromol 72:1069–1075

Siva DA, Brito ACF, Paula RCMD, Feitosa JPA, Paula HCB (2003) Effect of mono and divalent salts on gelation of native, Na and deacetylated *Sterculia striate* and *Sterculia urens* polysaccharide gels. Carbohydr Polym 54:229–236

Soares GA, de Castro AD, Cury BSF, Evangelista RC (2013) Blends of cross-linked high amylase starch/pectin loaded with diclofenac. Carbohydr Polym 91:135–142

Sreenivasa B, Prasanna RY, Mary S (2000) Design and studies of gum karaya matrix tablet. Int J Pharm Excip 239–242

Sriamornsak P, Sungthongjeen S (2007) Modification of theophylline release with alginate gel formed in hard capsules. AAPS PharmSciTech 8:E1–E8

Sriamornsak P, Sungthongjeen S, Puttipipatkhachorn S (2007) Use of pectin as a carrier for intragastric floating drug delivery: carbonate salt contained beads. Carbohydr Polym 67:436–445

Sriamornsak P, Nunthanid J, Cheewatanakornkool K, Manchun S (2010) Effect of drug loading method on drug content and drug release from calcium pectinate gel beads. AAPS PharmSciTech 11:1315–1319

Vijan V, Kaity S, Biswas S, Isaac J, Ghosh A (2012) Microwave-assisted synthesis and characterization of acrylamide grafted gellan, application in drug delivery. Carbohydr Polym 90:496–506

Weiping W (2000) Tragacanth and karaya. In: Williams PA, Philips GO (eds) Handbook of hydrocolloids. Cambridge, Woodhea, pp 155–168

Wong TW, Nurjaya S (2008) Drug release property of chitosan–pectinate beads and its changes under the influence of microwave. Eur J Pharm Biophar 69:176–188

Yoo S-H, Song Y-B, Chang P-S, Lee HG (2006) Microencapsulation of α-tocopherol using sodium alginate and its controlled release properties. Int J Biol Macromol 38:25–30

Zide BM, Bevin AG (1980) Treatment of shallow soft tissue ulcers with an infrequent dressing technique. Anal Plastic Surg 4:79–83

MIX
Papier aus verantwortungsvollen Quellen
Paper from responsible sources
FSC® C105338

If you have any concerns about our products,
you can contact us on
ProductSafety@springernature.com

In case Publisher is established outside the EU,
the EU authorized representative is:
Springer Nature Customer Service Center GmbH
Europaplatz 3, 69115 Heidelberg, Germany

Printed by Libri Plureos GmbH
in Hamburg, Germany